An Introduction to Disorder

Disorder is everywhere—in alloys, in thermal fluctuations, and in the amorphous structures of glasses. Understanding it is essential to understanding real materials. This book offers an introduction to the physics of disordered systems, from their foundations in entropy and statistical mechanics to modern applications in information and quantum information theory.

It begins by discussing the generation and use of random numbers, Monte Carlo and quantum Monte Carlo methods, and impurity effects ranging from single-atom substitutions to large-scale defects such as dislocations and grain boundaries. Entropy is explored as a unifying concept linking statistical mechanics, information theory, and quantum information, including Shannon and von Neumann entropies and quantum entanglement.

Further chapters examine amorphous materials, glasses and spin glasses, Anderson localization, and metal–insulator transitions, as well as fractals and chaotic dynamics. It concludes with the impact of disorder on critical phenomena.

Designed for advanced undergraduates and first-year graduate students, this book serves as a comprehensive entry point into the study of disorder. It provides a solid conceptual foundation while pointing readers to key research literature for deeper exploration, making it both an accessible textbook and a valuable reference for future study. Throughout, the mathematics is kept as simple as possible, with more advanced methods introduced only when needed.

Key Features:

- Presents a basic introduction for advanced undergraduate and first-year graduate students in physics and engineering, both theoretical and experimental.
- Covers topics of the most interest to researchers in nanotechnology, material science, statistical physics, information theory, and quantum information theory.
- Explores the variety of different types of disorder which are found in physical systems.

Arthur R. McGurn, CPhys, FInstP, is a Fellow of the Institute of Physics, a Fellow of the American Physical Society, a Fellow of Optica (formerly Optical Society of America), a Fellow of the Electromagnetics Academy, and an Outstanding Referee for the journals of the American Physical Society. He earned his PhD in Physics in 1975 from the University of California, Santa Barbara, followed by postdoctoral studies at Temple University, Michigan State University, and George Washington University (NASA Langley Research Center). Prof. McGurn's research interests include the theory of magnetism in disordered materials, electron conductivity, the properties of phonons, ferroelectrics and their nonlinear dynamics, Anderson localization, amorphous materials, the scattering of light from disordered media and rough surfaces, the properties of speckle correlations of light, quantum optics, nonlinear optics, the dynamical properties of nonlinear systems, photonic crystals, and meta-materials. He has over 150 publications spread amongst these various topics. He has taught physics for 38 years at Western Michigan University, where he is currently a Professor Emeritus of Physics and a WMU Distinguished Faculty Scholar. A number of PhD students have graduated from Western Michigan University under his supervision.

An Introduction to Disorder

Entropy, Information, and Morphologies

Arthur R. McGurn

CRC Press
Taylor & Francis Group
Boca Raton London New York

CRC Press is an imprint of the
Taylor & Francis Group, an **informa** business

Designed cover image: Shutterstock

First edition published 2027
by CRC Press
2385 NW Executive Center Drive, Suite 320, Boca Raton FL 33431

and by CRC Press
4 Park Square, Milton Park, Abingdon, Oxon, OX14 4RN

CRC Press is an imprint of Taylor & Francis Group, LLC

ISBN: 978-1-032-66981-6 (hbk)
ISBN: 978-1-032-67490-2 (pbk)
ISBN: 978-1-032-67491-9 (ebk)

DOI:10.1201/9781032674919

Typeset in Times
by Deanta Global Publishing Services, Chennai, India

To my wife, Maria

Contents

Preface

This book is intended to be an introduction for undergraduate and graduate students to some of the basic ideas of the theories of disorder and information. These topics are illustrated by the consideration of a variety of the major methodologies recently developed in these areas of research. In this regard, an attempt is made to sample the most important concepts of the subjects with a particular focus on analytical treatments. Due to the current importance of computer simulation methods in engineering applications, however, we also present ideas involved with simulation techniques. In this regard, the formulations of computer methods are given as a general introduction, but they are also themselves important studies in the nature of disorder.

The book begins with an introduction to the basic ideas of random number generation and how random numbers are used in basic Monte Carlo simulations of classical and quantum mechanical systems as well as in numerical integration routines. The techniques involved in the statistical testing of sequences of random numbers are introduced as well as the application of sequences of random numbers. For example, the development of the Metropolis sampling algorithm is explained as it is applied in both classical and quantum Monte Carlo studies. As we shall also see, statistical testing of sequences of random numbers is also an important topic since the results obtained from simulation studies are only as accurate as the sequences of random numbers applied in them are truly random. In addition, the development of designer sequences of random numbers, for example, simulating Poisson and Gaussian distributions is also discussed as a useful technique in simulation studies.

A discussion of information theory is also presented. Applications of the ideas of information for the study of information content, measurement, and transmission are developed along with the concepts of error analysis and for treatments involved with system noise. The concept of information entropy is outlined and introduced in the development of the methods of statistical mechanics. In addition, maximum entropy methods are discussed as well as their applications to techniques of reverse Monte Carlo and the development of genetic algorithms applied in the design of systems with desired operational characteristics.

A treatment is given of lattice imperfections in otherwise crystalline materials. This includes a discussion of Green's function applications to the single site impurity problem and the related coherent potential approximation in the study of electrical conductivity. An application is also made for the discussion of the Kondo problem arising from the presence of magnetic impurities in a metal. Line imperfections in the crystal lattice, known as dislocations, are introduced as well as considered for their role in the determination of the plastic properties of materials. In this regard, it is shown that the strength of a perfect crystalline material under stress is much greater than that of a crystalline material containing dislocations. Consequently, dislocations and their dynamics are considered as a basis for understanding the strengths of materials found in the real world. The origins of the presence of dislocations in materials are also explained as arising from a variety of crystalline physical processes.

In a more extreme consideration of disorder, some of the properties of amorphous materials are treated. Due to the lack of translational symmetry of the atomic order in amorphous materials, the excitations in amorphous systems are not as restricted in their forms as those found in crystalline materials. For example, the Bloch forms of the wavefunctions no longer apply in amorphous materials. Instead, the properties of the materials are determined to a greater degree by the local properties of the materials, i.e., the approximate symmetries of the local atomic environment. In this regard, we shall present an introduction to recursive methods applied for calculating the density of states of the excitations in amorphous media. Related systems known as quasicrystals are also considered. These represent materials of long-range spatial order which are not defined on lattices but rather exist on tiling patterns. Such tiling patterns have fewer restrictions on the local point group symmetries they support. The atomic order of the tiling patterns is found to contribute to a variety of interesting non-crystalline properties.

In line with the discussion of amorphous materials, the ideas of Anderson localization are also introduced. Anderson localization involves a type of phase transition between extended and localized modes of excitations in disordered systems. Extended modes range over an infinite region of a material, whereas localized modes are confined to a bounded region of space. This has great significance for the low-temperature transport properties of electron and vibrational systems. In electron systems, for example, the localization effects can be observed as a metal–insulator transition known as the Anderson transition. In this regard, we shall see that the transport properties of systems are characterized by the properties of the two-particle Green's functions of the system excitations. This is distinctly different from the Mott metal–insulator transition, which is present in ordered systems and arises from the nature of the electron bonding in the systems. In optics, localization also enters into a number of interesting scattering effects. In this regard, localization can be shown to be possible in all systems that support wave excitations, whether or not they are characterized by classical or quantum mechanical dynamics.

Another interesting type of material is glass. Glass is essentially a metastable state of materials which form as extremely slowly varying liquids so that, on time scales of interest, they appear to be fixed random arrays of atoms. Nevertheless, on long-time scales, glasses behave like liquids. Glasses solidify from the liquid to the glass phase by a transition known as the glass transition, involving a transition to a metastable state which occurs over a range of temperatures. In this regard, the glass transition differs from a typical phase transition. Specifying these ideas to magnetic materials, we shall also look at a particular magnetic system known as a spin glass. In these materials, the atomic spins of the material order themselves spatially in a random way so that there is a phase transition from a paramagnetic phase to an ordered spin glass phase in which the spins are ordered in time. The various effective field theories used to understand this ordering process are presented.

Another type of important process of disorder is that of chaotic dynamics. This is a type of dynamics which arises in nonlinear systems. In such systems, it is found that a small perturbation in the initial conditions of a dynamical process can quickly lead to a radical change in the system dynamics. The dynamics of such systems appear random in nature, and consequently, it is often referred to as deterministic chaos. The divergence of chaotic trajectories as functions of time is characterized by a Lyapunov exponent, which provides an important measure of the system. Chaos has been studied both in nonlinear difference equations and in systems of nonlinear differential equations.

Related to chaotic dynamics is fractal geometry. This is a geometry involving self-similarity so that upon changing the length scale on which a system is observed, it looks the same. This is then a form of scale invariance. Statistical correlations of the disorder in such systems are generally of the form of power laws. Examples of fractal geometry are found at the second order phase transition of systems undergoing an order–disorder transition. At the second order transition, the system loses any concept of a natural length in the system. Fractals may be spatially disordered or may have geometries generated algorithmically by a well-defined process involving no disorder. In time sequences, fractals are observed in, for example, the study of $\frac{1}{f}$–noise.

As a final topic on disorder, we shall introduce some of the basic ideas of scaling in the study of disorder. This is done in the context of percolation theory, which treats systems with a purely geometric disorder. These systems are found to undergo a geometric transition which is similar to a second order phase transition in thermodynamic systems. Specifically, we consider a periodic lattice of points in which the points are randomly linked by bonds (i.e., bond percolation) or in which the sites of the lattice are randomly occupied (i.e., site percolation). The question of interest for these systems is when an infinite cluster of connected points exists on the lattice. It turns out that the probability of an occupied site being in an infinite cluster exhibits a phase transition similar to those found in the ferromagnetic transitions of a magnetic material. In this regard, scaling and renormalization group methods applied in statistical mechanics can be directly taken over to the study of the percolation theory transition. The particular example of bond percolation and its relation to the study of the q-state Potts model will be discussed.

All the systems of disorder we have chosen to present here have contributed significantly to the understanding of physical processes in the last 70 years. It is hoped that these studies are helpful in leading to a better understanding of future encounters with disorder.

Arthur R. McGurn, Rancho Mirage, 2026

Thank you for the puzzles.
They were very entertaining.
And now...

1 Introduction

Disorder is an essential aspect of all physical processes and enters to a certain degree into any theoretical considerations of physical systems [1–3]. Its effects may be large or small, but disorder in a system always affects the quality of the system response generated to a stimulus input. This makes it a fundamental feature in the study of quantum physics, which itself is a study of uncertain, probabilistic processes found in nature. Add to this the disorder due to the random composition of the material of the system or stochastic time processes the system is subject to. In this book, the focus will be on classifying quantitatively the idea of disorder, developing its measures along with their relationship to the ideas of information and information theory, and presenting its study in a variety of interesting physical systems that have been of current interest. In the course of this study, the idea will be to understand the different sources of disorder and to explore their nature as found in a variety of contemporary physical theories.

Quantum mechanics, as based on the uncertainty principle, is a probabilistic theory even in the consideration of idealized well-defined theoretical models of physical processes and thus entails a degree of inherent disorder. However, added to this probabilistic undertaking, additional uncertainty arises in the formulation of the model of a physical system. These added uncertainties come from certain statistical features of a theoretical model originating in our inability to fully describe the system. That is, the system may interact with a disordered or poorly characterized medium or have time-dependent stochastic interactions with an external heat reservoir. These statistically characterizable uncertainties allow us only to develop a limited representation of physical systems and their properties. Consequently, disorder always plays a part in measurement considerations and our theoretical conceptions regarding them.

Associated with the nature of probability and disorder are additional ideas of information and information theory [1] and how to use these to quantify and study disorder. More specifically, information theory is the study of the quantification, measurement, and transmission properties of information or signal communications. It provides a means for the characterization and quantification of random processes and what is learned upon making a measurement on a system. Consequently, a specific measure of information contained in a physical system can be developed in terms of a well-defined function of the probability distribution describing the possible outcomes of measurements made upon the system. This function, as we shall see, is known as the information entropy or just the entropy. We shall show that it provides a numerical quantification of the information content of the system. It acts as a measure of how unpredictable the result obtained from a measurement made on the system is. In this sense, it can also act as a quantification of disorder in physical processes.

In this context, the more improbable a measured outcome, the more we learn about the system from it, i.e., the more information it gives us regarding the system [1]. Consequently, more information is contained in the unexpected outcomes of a measurement than in those we expect. In accord with this, outcomes which are highly improbable are found to exhibit larger information entropies or information content, while the situation in which we know the exact state of the system before a measurement is even made are states of zero information entropy. In this regard, if we know the answer before the measurement is made, we learn nothing from the measurement, and it has no information content. These ideas of information and entropy as a measure of information are, hence, useful in quantitatively determining the information contained in a system, in studying information as it is transmitted from one system to another, and in providing the requirements for error correction arising during information processing and transmission. They can also be used to measure the disorder of a system and our lack of ability to predict its subsequent development.

DOI: 10.1201/9781032674919-1

As an example of information content and disorder, in both statistical mechanics and information theory disorder arises from our inability to know completely every detail of the system under consideration. Consequently, it is necessary to view the system as represented by a probability distribution characterizing the probable outcomes of measurements made upon it. In this regard, in statistical mechanics the probability distribution of a thermodynamic measurement made on a system is directly related to its Boltzmann probability distribution of possible states [4, 5]. Consequently, the entropy of the thermodynamic system can be defined directly in terms of its Boltzmann distribution. As we shall see, the entropy, in this sense, is a measure of our knowledge of the properties of the system.

This conversely applies so that from our general ideas about information entropy the distribution of thermodynamic states can also be arrived at. As we shall ultimately see, from ideas regarding the nature of the entropy of a thermodynamic system the Boltzmann distribution of the system is obtained as a problem of maximizing a general form of the entropy function subject to the constraints of a specific system [4, 5]. This makes statistical physics a particular application of the general ideas of information theory [1]. In both the studies of statistical mechanics and information theory we therefore find that an entropy function enters prominently in understanding the nature of the system and its processes. From these ideas the maximum entropy principle will also be generalized to the study of the distribution of the properties of random systems generated from measured data.

Disorder is also an essential feature in the transport properties exhibited by materials and sets the nature of these as being ballistic, diffusive, or of a localized nature [6–8]. These aspects of transport are highly dependent on the quantity of impurities introduced into a material, the phase coherence of the scattering in the system, and the disorder arising from thermal considerations. Similarly, we shall find that the effects of disorder can also be heavily determined by the dimensionality of the system. These ideas will be explained in our treatment of each of the types of transport processes and the variety of media in which they occur.

Specifically, dimensional considerations as to whether the transport involves volume states, surface states, or states in nanoscale wires are fundamentally important determinators of the nature of the transport. For example, ballistic transport can become dominant in applications in nanoscience [6–8]. Here, due to the small scales over which transport characteristics occur, device applications can be made in the absence of impurities or for a very small number of impurities. On the other hand, the ideas of localization may dominate the study of metal–insulator transitions displayed in bulk three-dimensional disordered media. As a result, the nature of the transport process is seen to be classified as ballistic, diffusive, or localized, dependent on the properties of the disorder and the system dimensionality [6–8].

Disorders can be classified into a variety of types [6–9] and one of our focuses will be an introduction to these different types. For example, disorder can be represented by isolated imperfections in a crystal lattice, a concentration of impurities in a crystalline material, an amorphous atomic network, or a system composed in a fractal geometry.

In crystalline-based materials, disorder is often naturally developed during the course of solidification or in processes involving a change of state of the solid material. In this regard, topological disorder develops in a material due to a variety of factors affecting the rate of growth of an ordered or disordered structure and in the fluctuations to which such processes are subject [10]. This is seen in the development of dislocations and grain boundaries during the process of solidification of crystalline materials. The nature of rapid condensation into the solid state is also an important factor in the formation of so-called quasicrystals which are composed as metastable atomic systems containing axes of rotational symmetries which are inconsistent with the translational symmetries found in crystalline materials. This leads to the development of the ideas of patterning in contrast to the arrangement of atoms and molecules on a spatial lattice. Such disorders are encountered in nature in addition to the more mundane disorder of single-site or multiple-site impurities found in otherwise crystalline materials.

Amorphous materials, on the other hand, present another type of atomic ordering which opens a new set of theoretical challenges [8, 9]. An example of such systems is glasses. These types of materials are not described as regular arrangements of atoms periodically distributed on a lattice but are atomic orderings which seemingly are statistically distributed in a random spatial network. This greatly complicates the study of the theory of the physical properties of amorphous materials [6–9]. Unlike crystalline materials which have translational symmetries that restrict the nature of the dynamical solutions of the excitations, amorphous materials are not restricted by such symmetries. Nevertheless, as we shall see, a variety of approaches have been developed to study amorphous types of materials. These techniques reside in theories which focus initially on the origins of the physical properties in the local atomic geometries of the network [6–9].

Fractal geometry is an additional class of geometry that is of great importance in characterizing disordered physical systems [11, 12]. Systems with fractal geometry display what is termed self-similarity. This implies that the system geometrically looks the same or exhibits the same type of behavior on all length scales.

An example of such a self-similar system is found in the properties of power law functions of the form r^{α} where r is a length and α is a constant exponent. These are geometric functions which give a value solely related to the length under consideration, i.e., r. Consequently, electrostatic and gravitational forces originating from point sources are developed from force fields exhibiting a fractal r^{-2} geometry. These force laws are seen not to be characterized by a law involving the introduction of specific length λ into the formulas characterizing them.

For the power law function [11, 12] we see that changing the length scale so that $r \to \frac{r}{a}$ causes the corresponding power law to change $r^{\alpha} \to \left(\frac{1}{a}\right)^{\alpha} r^{\alpha}$. Here the essential power law form of the spatial dependence is found to be preserved in the r^{α} length dependence.

On the other hand, a function such as $e^{-r/\lambda}$ has a natural length scale λ built into it. This picks the particular length λ as being an important characteristic of the function. Consequently, changing the length scale so that $r \to \frac{r}{a}$ causes $e^{-r/\lambda} \to e^{-r/(a\lambda)}$ which changes the functional dependence on r from a spatial decay characterized by the length λ to a spatial decay characterized by the length $a\lambda$. In this regard, a change of the length scale changes the functional form of the length dependence.

It is interesting to note that an example of scale invariance and fractals is found in the study of phase transitions. Specifically, an earlier example of the scale invariance of the power law forms is observed in the properties of a thermodynamic system at the critical point of a second order phase transition. At the point of the phase transition, the system develops a fractal geometry which is typified by its properties exhibiting the forms of power law functions. Away from the critical point, however, the functions characterizing the system develop an additional dependence on a correlation length ξ which acts as a characteristic length scale in the system. As a result, the system loses its properties of self-similarity away from the critical point [11, 12]. This will be further explained in the course of the text.

In terms of time disordered processes, a recent set of important processes involving nonlinear dynamics have been of great interest in physics and engineering [11–13]. These involve the study of chaos and chaotic systems arising from the study of nonlinear dynamics. In nonlinear dynamics, an apparently disordered dynamic can arise in a system due to the presence of nonlinearities. In this regard, the dynamics can be developed from a completely nonrandom, deterministic algorithm but is found to nevertheless display an apparently erratic time behavior which can only be described statistically [11–13]. This dynamical behavior is referred to as chaos or chaotic dynamics. It represents a high degree of motional instability.

Specifically, in chaotic systems it is found that two trajectories with slightly different initial conditions exhibit an exponential divergence from one another during the long-time development of their motion. This dynamics presents a range of problems associated with the characterization of motion which, while appearing to be random, is generated by a deterministic, nonrandom, algorithm [11–13].

We will begin our discussions of disorder with a consideration of one of the simplest and most basic random systems that can be devised for the study of random disorder and which illustrates many of the common features found in all disordered systems. This involves the generation of sequences of random numbers and the characterization of the statistical nature of these disordered sequences. It is a problem of great interest and fundamental to mathematics, computer science, and as well as to physics. It is at the root of many important methods of computer simulation and an essential factor in setting the accuracy of numerical integration, Monte Carlo [14], and quantum Monte Carlo methods of computer simulation [15, 16]. The problem in the generation and study of random numerical sequences is focused on developing simple, rapid methods for the generation of random numbers and then developing a variety of tests to determine the quality of the randomness of such sequences [2].

To summarize, in this book we shall consider a variety of disordered systems and processes illustrating the presence and importance of disorder in physical systems. In addition, a focus is also made on the information content of a system, processes involved in information transmission, the generation and characterization of randomly disordered numerical sequences, and the application of these numerical sequences in the development of computer simulation methodologies. As we shall see, the study of disorder is, consequently, a vast interdisciplinary effort involving applied mathematics, physics, and engineering. Here we shall only highlight and review some of its basic features which have recently been considered as the more interesting topics in the field.

1.1 SEQUENCES OF RANDOM NUMBERS: MONTE CARLO METHODS

We begin our discussions with an introduction to the study of randomly disordered numerical sequences [2, 14–18]. This is an important study as most numerical simulation considerations for physics and engineering applications are based on introducing randomness into simulation programs designed to model the physical properties of the system being studied. In this regard, the elements of randomness used in such considerations are ultimately based on the randomness of sequential lists of random numbers [2, 18]. Consequently, a basis of numerical methods involves the generation of such randomly disordered sequences.

Specifically, an initial focus is on the generation and characterization of sequences of random numbers [2, 18]. This involves an effort to generate by an algorithmic process disordered sequences of numbers exhibiting no statistical correlations between numbers separated by fixed intervals in the sequences. Such random lists of numbers are often referred to as pseudo-random numbers because they arise from a well-defined algorithmic process which is not based on randomness. The resulting random sequences of numbers are found to be essential to the development of certain types of computer simulation studies of the properties of systems studied in statistical physics and information theory [1, 2, 18]. In addition, they have important numerical applications in mathematical physics, entering into a variety of numerical integration schemes [2, 18]. Here they provide improvements over standard methods of numerical integration (e.g., the trapezoidal rule or Gaussian quadrature method) based on nonrandom techniques. This becomes particularly true in the study of integration performed in higher dimensional systems.

In an important application in the numerical simulation of many-body systems, random numbers are essential to the method of Monte Carlo and quantum Monte Carlo simulation of thermodynamic systems [2, 14–18]. In such computer simulation treatments of many-body systems, a large but finite model of a system is studied based on a technique involving the sampling of randomly generated configurations of the system. In this regard, due to the limitations of the computer memory, the necessity of considering a finite model only offers an approximation to the infinite system. However, this limitation can be remedied through methods available to correct for errors due to so-called finite size effects of the model. These methods for the treatment of finite size effects will be explained in our later considerations. In this basic methodology then resides the general basis of most simulation work in many-body physics [2, 14–18].

In the Monte Carlo process, random configurations of the dynamical variables of the Hamiltonian are generated based on methods using sequences of random numbers [2, 18]. From these random configurations, a criterion based on the values of the Hamiltonian they generate is used to select the states entering into the calculation of the properties of the model. This involves accepting or rejecting a generated configuration based on a sampling algorithm. In statistical physics applications, the sampling algorithm and its criterion are known as Metropolis sampling [18] and allows for the introduction of statistical properties into the accepted configurations. This makes the accepted configurations representative of, for example, the Maxwell-Boltzmann distribution of thermodynamic states of the system. The set of accepted configurations generated in this way then forms the basis for approximating the properties of the statistical mechanical model. Specifically, averaging the properties of the system over the set of accepted configurations offers an approximation of the thermodynamic properties of the system as they are characterized by the model Hamiltonian.

The process just outlined forms the basis for the numerical study of the thermodynamics of many statistical mechanical models. In this format, the Monte Carlo method [14] has been commonly applied to approximate the properties of numerous systems in classical mechanics. The method, however, can also be extended to treat quantum mechanical systems. In this regard, the statistical mechanics of quantum systems can be formulated in terms of the evaluations of path integrals. In this process, the treatment of a quantum mechanical model involving non-commuting operations is rewritten in terms of path integrals involving classical variables [15–18]. This, however, is made at the expense of an increase in the dimensionality of the system. Ultimately, the properties of the system are obtained from the path integral studied by means of standard Monte Carlo methods [14, 18]. The resulting reformulation of the quantum mechanical problem then is evaluated in the context of a Metropolis sampling technique of the new classical variables of the path integral and is referred to as quantum Monte Carlo [15–17].

Other methods involving Monte Carlo sampling are the so-called maximum entropy and reverse Monte Carlo methods [1, 19, 20]. These methods are not based on statistical mechanics but are often used as routines meant to determine the best fit of a curve to a set of data. They are based on the idea that the data is characterized by a probability distribution [19]. The fitting is then done by choosing a probability distribution, as a representation of the known properties of the system, which has a maximum probable information entropy subject to the constraints on the experimental data. The method has found some applications in developing models of atomic systems and random disordered media which reproduce, for example, the X-ray or electromagnetic scattering cross sections observed on these types of experimental systems as they are measured in the laboratory [19]. While the methods generate models of the system, these models may differ from the actual atomic configurations of the experimental system.

As an example, the reverse Monte Carlo is often applied to find the best atomic configuration to reproduce the scattering data obtained in the experimental study of the structure of crystalline and amorphous materials or surfaces formed at the interface between media. In the reverse Monte Carlo, the properties of the experimental scattering data are compared with the scattering computed for randomly generated configurations of the system [19, 20]. The comparison is made using a sampling algorithm based on random numerical sequences and a selection criterion applied to guide the process into determining the most likely configuration reproducing the scattering data.

In a related but different statistical method known as genetic algorithms, an evolutionary approach to the design of systems exhibiting certain favored characteristics is made. In this approach, a process of natural selection is used, which is based on random mutations made on the configurations of the system and on the application of a criterion governing the selection of the mutated states. The evolutionary procedure of modifying the system is then accomplished by mimicking the process of natural selection as found in biological populations. In this way, the evolutionary approach has found successful applications in a variety of engineering design problems.

We shall begin the study of disorder by considering its simplest examples, followed by a progressive study of more complex disorders in physical systems. Consequently, following the treatment of random numerical sequences [2, 18], the simplest disorder found in condensed matter physics is that of a single site disorder in an otherwise perfectly ordered material, i.e., a site imperfection in a crystal medium [3]. This is important for many spectroscopic studies and in beginning the treatment of transport properties. In addition, it forms a starting point for the development of the analytic techniques applied to the study of disordered media.

1.2 SINGLE SITE IMPURITIES IN CRYSTALS

A topic of considerable interest is that of a single atomic impurity placed within an otherwise perfect crystal [3, 7, 21]. This is one of the basic problems in solid state physics which can in many cases be treated exactly in an analytic format. In our presentation we shall develop the problem within the context of a tight-binding model, involving a Hamiltonian which is composed of an atomic site occupancy energy and an energy for hopping between nearest neighbor atomic sites [3]. This is a simple versatile model, containing the basic elements of an energy associated with position and an energy associated with motion. It provides a relevant format for the study of the modal excitations characterizing the conduction electrons, atomic lattice vibrations known as phonons, and magnetic spin wave excitations commonly known as magnons [3]. These constitute the modal excitations of their respective dynamical systems. In this context the tight-binding model allows for the development of a straightforward treatment of the single site impurity problem.

The treatment for the excitations in the electron, phonon, or magnon systems in versions of the tight-binding model is very similar to one another [3]. In this regard, the study of the modes of the tight-binding model is made within the context of a basic Green's function theory which allows for the study of the modal solutions to be made in terms of closed-form analytic expressions. In the course of the discussion, a brief introduction or review of Green's function methodology is given so that the treatment does not require a background in Green's functions. Ultimately, Green's function applications are made for the problem of multiple impurities introduced into the tight-binding Hamiltonians. A focus of these Green's function applications is on the consideration of the density of energy states and the energy and lifetimes of the impurity mode excitations in dynamical systems and their development in terms of Green's function methods [2, 7, 21].

A generalization of the single site impurity problem can also be made to offer an approximation for systems containing a concentration or density of impurity atoms. An interesting and commonly used analytic approach is based on an effective field formulation developed on the ideas of the mean field theory treatment used in the study of the statistical mechanics of, for example, ferromagnetic materials. This generalization of the ideas of mean field theory to dynamical systems is used to obtain an effective field theory for the dynamics of the many impurity problem and is known as the coherent potential approximation [3, 21]. It is a natural mean field treatment generalizing the single site impurity problem.

In the coherent potential approximation, an effective Green's function is formulated at each frequency in the dynamics of the disordered system, generating its average or effective frequency response [3, 21]. This average frequency dependent response of the system in turn allows us to compute the frequency dependent properties of the disordered materials in terms of standardized Green's function methods. This yields a coherent potential approximation of the density of states in energy, a frequency renormalization of the modes of the system, and a determination of the modal lifetimes of the excitations.

Following these discussions, as an example of a next level of problems involving impurities, we will study an important class of problems considering the effects of low concentrations of localized magnetic impurities on the conductivity of metals. This is a famous problem known as the Kondo problem [22, 23] and directly involves many of our discussions of the single impurity problem and the mean field treatment of systems with low concentrations of magnetic impurities.

1.3 THE KONDO EFFECT

An interesting application of the study of site impurities is in the development of the Anderson model and the closely related s-d model of magnetism [22]. These models are applied in the treatment of the theory of the scattering of conduction electrons of a metal from a single magnetic site impurity. In these considerations, the conduction electrons are treated in the context of an electron gas model, and the impurity is represented as a single atomic energy level which can be occupied by at most two electrons, each of opposite spin from the other. The single magnetic impurity portrayed in these models could be realized, for example, by aspects of an impurity atom with a partially filled $d-$ or $f-$ atomic orbital introduced into and interacting with the conduction electrons of a metal host characterized by a standard free electron gas model [22].

As a result of the magnetic impurity scattering [22, 23], the electrons can be exchanged between the gas and the impurity so as to create an enhanced magnetic moment localized about the site of the atomic impurity. It is found that as the temperature is reduced, the scattering cross section of the electrons with the impurity is enhanced, accentuating the effects of the impurities on the conduction electrons. Ultimately, at low concentrations of magnetic impurities in the metal host, the enhanced scattering is observed as an increase in the resistivity with decreasing temperature at temperatures lower than the Debye temperature [22, 23]. In this context, for a low concentration of magnetic impurities the increase in resistivity arises as a generalization on the incoherent single site scattering from a collection of magnetic impurities. This resistivity increase is known as the Kondo effect, and it is an example of a phenomenon in which the solution of the single impurity problem can be generalized to explain effects coming at low impurity concentrations.

The Kondo effect is regarded as an anomalous behavior of the resistivity exhibited in systems with paramagnetic magnetic impurities. By comparison, in typical systems containing nonmagnetic impurities, Matthiessen's rule tells us that the low-temperature resistivity below the Debye temperature decreases with decreasing temperature [22]. The resistivity, in this regard, is the sum of a temperature-independent resistivity from impurity scattering and a temperature-dependent resistivity arising from phonon and/or other possible modes of excitation in the system. Specifically, for metallic systems it is often found in the absence of magnetic impurities that the temperature-dependent resistivity is characterized by the general form [22]

$$\rho(T) = \rho_0 + aT^5 \tag{1.1}$$

where ρ_0 is a temperature independent term coming from the presence of the nonmagnetic impurities, and the $aT^5 > 0$ term arises from phonon scattering of the conduction electrons. This form of the resistivity is generally true for metals with conduction electrons modeled by the standard electron gas model.

While systems with nonmagnetic impurities have resistivities which monotonically decrease with decreasing temperature, contrary to this, the Kondo systems exhibit regions of increasing resistivity with decreasing temperature. In this regard, the Kondo problem [22, 23] is an interesting illustration of some of the unusual properties of the electron gas model observed in the context of interactions with magnetic impurities.

While the single site and Kondo problems illustrate the properties of systems with small concentrations of impurities, more elaborate forms of impurities are important to understanding the general physical properties of disordered systems. We shall next look at some types of topological impurities which enter into many of the important treatments of the strengths of materials as well as into their transport properties. These extend the range of our study of disorder to new types of imperfections occurring within the crystal lattice itself.

1.4 DISLOCATIONS

Another type of imperfection that is commonly found in crystalline materials is a lattice line imperfection, also known as a dislocation [10, 24, 25]. It is an imperfection introducing a fundamental geometric change into the structure of the crystal lattice itself. As we shall see, it represents a topological imperfection which introduces new nearest neighbor connectivities between the system of lattice points. As an example, if we consider a crystal lattice as formed by stacking identical planes of atoms on one another in a type of layering, a dislocation known as an edge dislocation can be made by replacing one of the layered planes with a half plane [10, 24, 25]. In this example, the periodic translational symmetry of the crystal lattice is destroyed at the edge of the half plane but is gradually restored with distance away from the edge. (See Figure 1.1a for a schematic representation of an edge dislocation.) Similarly, if we cut a crystal lattice along a plane halfway through its bulk, then shift the edges of the cut relative to each other by one unit of lattice periodicity along the axis of the cut and rejoin the cut edges; another type of line imperfection is made. This is known as a screw dislocation [10, 24, 25], and it is again found to distort the lattice geometry in an essential way by introducing a type of spiral staircase in the nearest neighbor lattice points. (See Figure 1.1b for a schematic representation of a screw dislocation.) This again destroys the periodicity of the lattice near the edge of the lattice cut. In both of these examples it is found that the nearest neighbor lattice points are changed from those of the lattice in the absence of the dislocation. The most radical changes are found along a line of highly distorted connections between the nearest neighbor lattice sites of the system, so that the imperfections are known as line imperfections.

Dislocations are very important as they determine the irreversible plastic deformation properties of materials [10, 24, 25]. In a very early consideration of plasticity, it was found that if the plastic deformation of crystals is considered to occur by dragging or sliding complete sets of atomic planes over one another, the stress required to plastically deform a crystal would generally be ten thousand times that observed experimentally. This presented a problem to the theory of the strengths of materials as though the sliding of planes would account for plastic deformation, materials generally are much easier to plastically deform than found in the simplistic consideration of sliding atomic planes. Consequently, early on it was realized that a more subtle mechanism must be involved in the irreversible deformation of crystalline materials. These would involve more complex, spatially localized, changes in crystal structure such as those observed in dislocations.

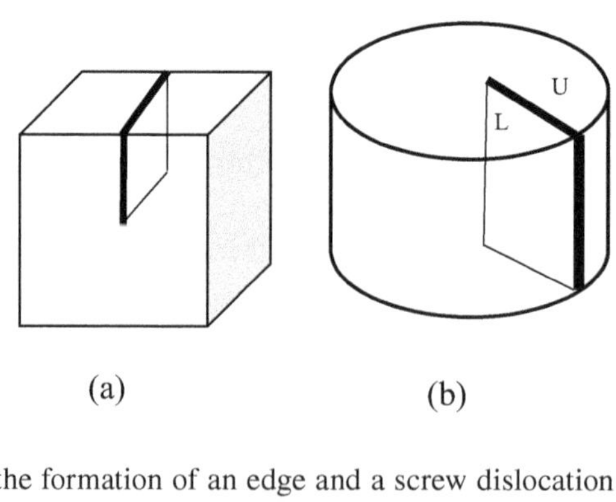

FIGURE 1.1 Schematic of the formation of an edge and a screw dislocation in a bulk solid. In (a) an edge dislocation is introduced into a bulk cube of solid by adding an extra half-plane of atoms (represented by a half plane with an edge at the center of the solid). In (b) a screw dislocation is introduced into a bulk cylinder of solid by cutting along a half-plane and moving the opposite sides of the cut one unit of the lattice periodicity relative to each other along the axis of the cut passing through the center of the cylinder. (Here the cut is represented by a half plane with an edge at the center of the solid cylinder; one side of the cut is denoted L, and the other is denoted U, and the cut axis is the line along the axis of the cylinder.) In the screw dislocation, the final state of the crystal lattice forms a type of spiral staircase centered on the axis of the cut.

Eventually it was found that dislocations in the crystal structure could account for the plastic behavior of crystals. In this picture, the reason that plastic deformation occurs at much smaller stress is that the mechanism of plastic deformation arises from the motion of dislocations in the crystal and not through the relative motion of complete sets of atomic lattice planes over one another. Dislocation motion is found to require less stress than sliding atomic planes and can account for all forms of shape change found in crystalline materials. Consequently, a consideration of the dynamics of dislocations and systems of dislocations is fundamental to treating topics in the strengths of materials. As a crystal deforms, single dislocations move through the crystal structure. In addition, as the motion progresses the dislocations collide and entangle with each other as well as entangling on site and multiple site imperfections. We shall consider in our study some of the basic properties of the so-called edge and screw dislocations and how their motions give rise to the phenomenon known as work hardening.

Another important point to note in regard to the existence of dislocations in crystals is how and why they come to be present in a crystal lattice [10, 24, 25]. The answer to this is that they facilitate the formation of crystals as the crystals condense out of their gas or liquid phases. In the process of crystal condensation, the mechanisms involved in condensation are greatly facilitated energetically and logistically by the natural introduction of dislocations into the lattice geometry of the forming solid. This is an example of how disorder is favored in the formation process or self-assembly of a material. Furthermore, in the dynamics of plastic deformation of a material, additional dislocations are created within a material through a variety of mechanisms involving the generation and entanglement of dislocations in the system [10, 24, 25]. This process of material hardening through the creation of dislocations is part of the process known as work hardening [10, 24, 25].

As an example of the creation of dislocations during condensation, consider a perfectly flat surface, formed between a solid and the liquid it is condensing from. (See Figure 1.2 for a schematic

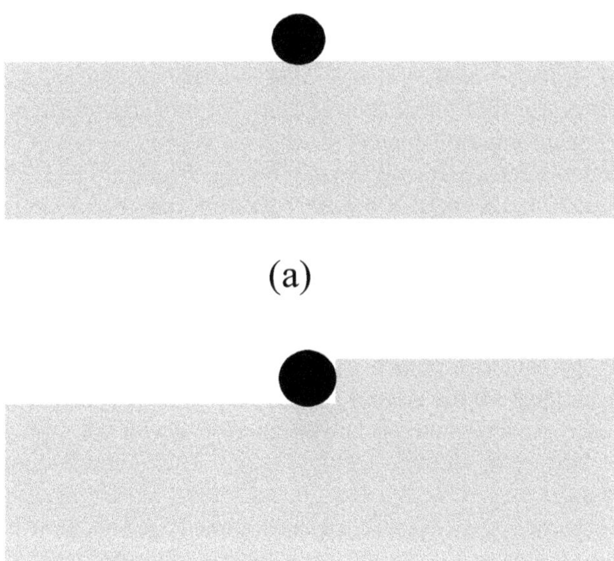

(a)

(b)

FIGURE 1.2 Schematic for an atom (represented by a circle) bonding from out of the liquid state to a solid surface. In (a) a single atom bonds to a flat planar surface, and in (b) a single atom bonds to the two surfaces formed at a ridge or ledge at the solid–liquid interface. Notice that in (a) a single bond is made while in (b) two bonds are formed.

representation of a condensation process.) The rate of bonding of a single atom in the liquid to the solid surface is the same everywhere along the perfectly flat surface. On the other hand, if the surface contains a step or ridge offering an atom two interfaces to bond to, the rate of attachment to the two interfaces forming the step or ridge will be greater than on the perfectly flat surface [10]. This causes the condensation process to seek out disorder or imperfections in the surface, which increases the rate of condensation. Such effects ultimately favor the introduction of dislocations into the solid.

In our continued studies of increasingly more complex forms of disorder in materials, it becomes helpful at this point to introduce a measure which quantifies the disorder in a system. This will allow us to make quantitative studies of disorder, its related information content, and the study of information transmission. Consequently, we next introduce an important general measurement of disorder which also serves as a measure of the information. The measure is known as entropy and arises in the context of mathematics as a characterization of probability distributions. In this regard, it is a tool for studying the information content of a probability distribution and, as we shall see, also acts as a basis for the development of the ideas of statistical mechanics and statistical physics [4, 5, 19, 20, 26].

1.5 ENTROPY

An important general type of quantification of disorder is the entropy function. Entropy is a measure of the information content of a system, taken in the sense of how much we learn about it when we make a measurement on the system [4, 5, 19, 20, 26]. In this respect, the possible results of a measurement are characterized by a probability distribution defined over the possible outcomes of the measurement. This distribution then indicates the relative occurrence of outcomes within the initial premeasurement system. The nature of the distribution of the probabilities over the possible states of the system determines our knowledge of the system and what we can learn about the system by a single measurement [19, 20].

For example [19, 20]: If the probability distribution over the possible measured states is very narrow (i.e., involving only a few states) and skewed heavily to a single particular outcome, it should exhibit a small entropy or information content. This is because the result of a measurement on this type of system is skewed to a particular result so that not much is really learned from a measurement of the expected outcome. Consequently, in the extreme limit that the system exists only in a single state (i.e., a single outcome with probability one), the entropy should be zero as nothing is learned from the measurement. Specifically, we know the result without even making the measurement.

On the other hand [19, 20], for the other extreme in which the probability distribution is very broad (i.e., defined over a large number of states) with a near equal distribution over all of the many possible states, the entropy should approach a maximum value for the system studied. In this case, the outcome of a measurement on the system supplies us with a considerable amount of information. In the limit that each possible state of the system occurs with the same equal probability, the maximum value of the entropy is attained for the system. This is the maximal uncertainty in the realization of the system.

From a general consideration of the type of functional form that best quantifies the state of the entropy of a system, a general functional form for the entropy is arrived upon [4, 5, 19, 20, 26]. These considerations include the discussions of entropy just made as well as some additional properties one intuitively regards as necessary in defining a measure of information and what information should encompass. Specifically, from such a process the following formulation is made: if a system is composed of N different possible outcomes represented by the set of outcomes $\{x_i\}$ for $i = 1,2,\ldots,N$ and each outcome x_i occurs in the system with probability p_i forming the set of outcome probabilities $\{p_i\}$ for $i = 1,2,\ldots,N$ where $\sum_{i=1}^{N} p_i = 1$, then the entropy E of the system is taken to be of the form [4, 5, 19, 20, 26]

$$E = -K\sum_{i=1}^{N} p_i \ln p_i \tag{1.2}$$

where $K > 0$ is a constant chosen to set the scale of the entropy. Notice, in this regard that the case involving the singleton sets $\{x_1\}$ with $\{p_1 = 1\}$ gives

$$E = 0, \tag{1.3a}$$

and the case of $\{x_i\}$ for $i = 1,2,\ldots,N$ with $\left\{p_i = \frac{1}{N}\right\}$ for $i = 1,2,\ldots,N$ gives the maximum entropy value

$$E = K \ln N \tag{1.3b}$$

for the system of N equally probable states [19, 20].

Regarding entropy from another point of view, if a system has N possible states $\{x_i\}$ for $i = 1,2,\ldots,N$ we can determine the probabilities $\{p_i\}$ for $i = 1,2,\ldots,N$ of these states using Eq. (1.2) and applying some general considerations of the system properties [19, 20]. This allows us to determine, by a constrained maximization of the entropy, the probability distribution which most likely represents the system in regard to all the limitations which we know apply to the system. Such considerations work on a variety of systems, whether they be physical, economic, etc., for the determination of the probability distribution. For example, by maximizing the entropy in Eq. (1.2) of a system subject to a few general constraints on the distribution (e.g., its average value and/or a few of the moments of the distribution), we arrive at general forms for the Gaussian and Poisson distributions.

In this regard, of particular interest to physicists are thermodynamic constraints which lead to the derivation of the canonical and grand canonical distributions used in statistical mechanics [26]. This essentially introduces the idea of the Boltzmann distribution [26] into the discussion of thermodynamics. In addition, such ideas of maximizing the entropy subject to constraints lead to the formulation of the maximum entropy method, the Monte Carlo and quantum Monte Carlo methods, and the reverse Monte Carlo methods [19]. These generate the most likely resolutions of statistical distributions subject to constraints. In addition, they are often helpful in curve fitting to experimental data [19].

Entropy is also an important factor in studying the encoding of information and the transmission of information along transmission lines. Here the efficiency of coding a message and the rate of information transmission become factors as well as the study of errors introduced into the information during its transmission and the correction of these errors in the received signal [19, 20]. All of the processes considered above depend heavily on the entropy formulation as outlined, acting as a measure of information [19, 20].

The next more complex form of disorder that enters into solid state systems is that of amorphous disorder [3, 8, 9]. This characterizes a class of non-crystalline materials that have important technological applications. In these systems one loses the convenience of translational symmetry which sets the functional form of the modes of excitation in crystalline systems [6, 7]. This makes theoretical treatments of amorphous materials [3, 8, 9] much more difficult to handle so that a new approach to their study is consequently required. Closely related to such amorphous materials are polymer materials [27] and more recently discovered states of matter known as quasicrystals [28].

These also differ fundamentally from crystalline solids and will additionally be a focus in our study of non-crystalline forms of matter.

1.6 NON-CRYSTALLINE MATERIALS: AMORPHOUS MATERIALS, POLYMERS, AND QUASICRYSTALS

Another topic of disordered media involves the study of amorphous materials [3, 8, 9, 27, 28]. These are important materials typified, for example, by glasses, amorphous semiconductors, and amorphous molecular solids. They are not characterized by a crystal lattice which, in its lowest equilibrium energy state, uniquely defines the underlying long-range structure of a crystalline solid as a state of translational symmetry. Unlike crystalline materials, in amorphous materials the atoms exist in one of a number of possible metastable networks of bonded atoms in which the atoms appear to be randomly connected to one another through chemical interactions. As a result of the absence of long-range translational order, in amorphous materials there are fewer restrictions set on the structure of the local geometric properties. In this regard, the nature of the local point group symmetries found throughout the solid are helpful in understanding the bulk properties of the amorphous material. Consequently, unlike crystalline materials, many of the properties displayed by amorphous materials are to a greater extent influenced by local factors such as the number of bonds that an atom forms with its neighbors in the amorphous network or the statistical distribution of the bond coordinations within the material [3, 8, 9, 27, 28]. In general, because of these differences with crystalline materials, new types of theoretical methods are required to understand the nature of the excitations in amorphous media [21, 29]. These methods, slanted towards a focus on local properties, are developed during the course of the discussions.

Amorphous materials are often formed from the liquid state through a process of condensation into the solid state system known as a glass [30]. Whereas the condensation of a liquid to a solid crystalline state is characterized as a first-order phase transition, the nature of the condensation from the liquid to the glass state exhibits a different, more complex set of characteristics. In this regard, the condensation into the glass solid state is termed a glass transition and constitutes an important study in the field of condensed matter physics.

Glasses are formed when a liquid is quickly cooled to a metastable amorphous state [30]. In a sense a glass can be thought of as a liquid which flows extremely slowly over the time scales that are of interest to us during the study of the glass. The glass transition, considered as a function of temperature, then, is not a phase transition but rather occurs over a region of temperatures in which some of the thermodynamic properties of the material undergo an inflection from liquid to solid-like behaviors. Associated with this inflectional change is a glass transition temperature T_g. This is a temperature which is assigned within the transition region and which is used to roughly characterize the process of going from the liquid to the glassy solid [30].

Some important theoretical methods used in the study of amorphous materials are the recursive method [29] and methods of computer simulation. Here a focus will be made on recursive methods as these involve solutions of quantum mechanical systems in analytic terms. In this way, they illustrate methodologies focused on obtaining exact treatments of systems from considerations of the local properties of their disordered models. We shall also make an introduction to another set of models known as the so-called two-states models [31]. These are developed for the discussion of the low-temperature thermodynamic properties of amorphous networks. They represent important models which have been applied to study the tunneling of amorphous networks of atoms between their many metastable configurations. In these models slight changes occur in the amorphous atomic network as the system hops between metastable changes in the network geometry. The two-states models illustrate the importance of the metastability of the amorphous network in understanding the low-temperature anomalous behavior found in these systems.

Other types of disorder related to those in amorphous media are found in polymer materials [27]. These are molecular systems formed of polymer chains in which the system can exhibit a variety of complex, intertwining configurations. The sets of different configurations of the molecules are called molecular conformations, and their properties and statistics are an interesting topic in thermodynamics. Polymer solids can exist as arrays of long chain molecules or as networks of interconnecting chain molecules forming a sort of fishnet or web [27]. The molecular conformations of these structures are also a basis for the explanation of many of the important elastic behaviors displayed by some polymer materials such as rubber. Our focus on polymers will then be on the basics of polymer conformations and how these enter into determining the elastic properties of polymer materials [27].

Specifically, the elastic relaxation of rubber comes from the change in entropy of the polymers composing it [27]. In the stretching process, the number of possible conformations of the rubber molecules in the material is changed, and as the number of molecular conformations is directly related to the entropy of the system, the entropy is also changed. The change in the conformational entropy is shown from basic thermodynamic considerations to then be directly related to the stretching force of the rubber. This elastic force of rubber is unlike the elastic properties of crystalline materials in which the elastic response of the material results primarily from the change in energy of the interacting atoms as they change their relative separation from one another. Consequently, in crystals, the entropy change upon stretching is a much smaller effect than that found in rubber.

Also related to the study of amorphous materials is the relatively new topic of quasicrystals [28]. These are new types of materials which exhibit long-range order in the absence of translational symmetry. In this regard, they are not defined on a crystal lattice but are associated with some of the properties of mathematical features known as tilings [28]. They exhibit long-range order as the atomic positions in a quasicrystal can be generated algorithmically, but the structures generated are not lattice structures as they do not display translational symmetry. An interesting property of quasicrystals is that they are not limited in their point group symmetries in the same way crystalline materials are. A large class of alloys have been shown to exhibit quasicrystal behavior.

A particular example of a set of two-dimensional points exhibiting long-range order in the absence of translational symmetry can be generated as follows [28]: Start by considering a cubic lattice of points with unit distance between the nearest neighbor points of the lattice. Pick a lattice point as the origin of coordinates, i.e., the point labeled $(0,0,0)$. Choose a plane passing through $(0,0,0,)$ and containing a line having an irrational slope along the x−axis and an irrational slope along the y−axis. The intersection of the plane with lines parallel to the z−axis containing lattice points of the cubic lattice is an example of such a plane of long-range order. In this system, we find an absence of two-dimensional periodicity but nevertheless the points of the lattice are deterministically related to one another.

Another important focus of our discussions on disordered materials will be the topic of spin glass [32]. This is a type of amorphous material in which the feature of interest is the unusual magnetic ordering of the material. In these materials, the magnetic ordering in the system encounters a degree of frustration due to the presence of both ferromagnetic and antiferromagnetic interactions between the atoms composing the material.

1.7 SPIN GLASS

An important related problem to that of amorphous materials is the problem of magnetic systems known as spin glass [32]. These are materials composed of atoms with magnetic moments which at low temperatures freeze into an apparently randomly ordered configuration of spins, i.e., on average all the spins in the material have no preferred direction but each spin individually exhibits a preferred average spatial orientation. In this random structure the frozen spins maintain their spatial orientations so that they do not change their individual average spin directions over time.

Consequently, the lack of long-range translational order in these systems is a property of the random spin ordering of the material. In fact, as in amorphous materials, spin glass is not a state of absolute minimum free energy of the system but is one of numerous metastable states of the free energy [32]. In the absence of an externally applied magnetic field, the spin glass does not display a net magnetization, but in an applied magnetic field, it can exhibit interesting hysteresis effects associated with its existence in metastable equilibrium configurations. This is a further indication that there are many possible configurations in which the system may settle as it condenses into a spin glass phase.

The spin orientations throughout the spin glass freeze about different spatial directions, but the freezing of the spins occurs in what appears as random directions in space. The direction in which a given spin freezes, in this regard, is set by its interaction with the local neighborhood in which it finds itself. This arises from the interactions with its neighboring spins through random spin exchange interactions which vary in nature between ferromagnetic and antiferromagnetic types of spin couplings [32]. In this regard, the discussions of these so-called spin glass systems are similar in some points to the theoretical framework used in the study of polymer elasticity. However, a variety of new theoretical ideas will also be introduced to handle the detailed nature of the transitions to the spin glass phase.

A particularly interesting theory of spin glass that we shall treat is the Edward-Anderson theory [33]. This involves an application of the ideas of mean field theory and a trick known as the replica method [27, 33] to obtain an analytic study of the properties of the ordering in the spin glass phase. In this theory, the interactions between the spins of the system are characterized by a statistical distribution of the exchange couplings between the spins of the material. The distribution ranges over both ferromagnetic and antiferromagnetic interactions, and the nature of the ordering can arise from these conflicting orientational tendencies.

In addition, in another class of systems, the development of spin glass behavior is found to occur in materials for specific lattice types with solely antiferromagnetic couplings [32]. Here the antiparallel spin ordering arising from the antiferromagnetic interactions is found to be in conflict with the presence of antiferromagnetic ordering in the geometry of the lattice. This type of material then introduces a new method of frustration to the system in which the development of magnetic order is inconsistent with the lattice type.

A dynamical process that is next treated is found as an explanation of the dynamics observed in a variety of disordered materials. This is the type of dynamics known as diffusion [3, 6, 21]. We shall treat the general development of diffusion, after which we will consider some applications. A particularly interesting application includes an application in the study of the formation of Turing patterns [34–36].

1.8 DIFFUSION AND TURING PATTERNS

An introduction to diffusive processes is presented in terms of the Boltzmann equation formulation for transport in a randomly disordered medium [4, 5]. For this approach, a brief introduction to Boltzmann transport theory is given in which an equation of motion of the probability density for a set of particles characterized in their space, momentum, and time coordinates is developed [4, 5]. The treatment is made for both classical and quantum mechanics approaches and is based on following the space and momentum coordinates of each particle as they develop in time. From this time-evolving distribution, the time dependence of the various thermodynamic averages of the system of particles is ultimately obtained.

Discussions are given of the Fermi-Dirac and Maxwell-Boltzmann distributions treated by the Boltzmann transport approach [4, 5]. These provide insight into the dynamics of conduction electrons and holes in both semiconductors and metals. A specific focus is on the limits of the Boltzmann transport theory leading to the diffusion equation for particle transport. The differences between the diffusion equation in the Fermi-Dirac and Maxwell-Boltzmann distributions are discussed, and the Einstein equation for the diffusion coefficient is obtained [4, 5].

In another application of diffusion, a discussion is given of the origins of Turing patterns [34–36]. These involve morphological considerations in which diffusion in systems involving the equilibrium of several reacting constituents and their relative concentration are found to manifest in a spatial pattering of the concentration of constituents in the system. In this regard, while populations can exist in an equilibrium state based on their reaction rates, the presence of diffusive motion in the constituents can cause them to exhibit patterns of spatial variation of the relative concentration of constituents. Such types of patterns are found in physical, chemical, and biological populations. The study of these morphologies is made in the context of an application of stability theory to a coupled set of nonlinear dynamical equations characterizing the reactions and their rates of diffusion [34–36].

A next important application of theoretical physics to disordered materials involves the study of Anderson localization [3, 8, 9, 21]. Anderson localization involves the study of a phase transition associated with the nature of the wavefunctions found in disordered media. Specifically, considered as a function of the system disorder, the wavefunctions of the medium can change from propagating states (extended over the volume of the material) to localized states (confined to a bound region of the material). At zero temperature the extended states are found to engage in transport processes, while the localized states do not. The ideas of Anderson localization then form the basis for one form of metal–insulator transitions observed in solids [3, 8, 9, 21].

1.9 ANDERSON LOCALIZATION

Another topic of importance in disordered media is that of Anderson localization [3, 8, 9, 21]. In the classical mechanics of particles, the localization or confinement of a particle to a bounded region of space is directly related to its kinetic and potential energies. On the other hand, in the case of waves and quantum mechanical particles, the localization of an excitation to a bounded region of space is no longer as straightforward. Due to the phase-coherent scattering properties of the phase information carried by wave-like excitations, a different set of criteria act in determining whether or not an excitation becomes localized. This is due to the fact that a particle carries a phase which causes it to interact with itself during scattering processes and is a direct consequence of the wave properties of the excitation. In this regard, such phase-coherent phenomena arise in the treatment of so-called double passage effects involving time-reversed trajectories present in the random medium and have been observed in a variety of optical, astronomical, meteorological, and electronic phenomena. As we shall see, double passage processes preserve the phase of the excitation and cause a phase-coherent enhancement of backscattering from the random medium [3, 8, 9, 21].

In general, two types of localization effects [3, 8, 9, 21] are found in disordered media. These are known as weak and strong localization effects. If the enhanced backscattering is small, the system exhibits what is known as weak localization. This is an enhancement in the scattering of the wave back into its incident wave and can be observed in, for example, the scattering of light from rough surfaces or in some classes of atmospheric optical effects. On the other hand, if the enhanced backscattering is large, the system exhibits strong localization. This is a confinement of the excitation within a bounded region of space. It arises because the backscattering becomes strong enough to stop the propagation of the wave excitation forward into the medium. In electron systems, strong localization shows up as the Anderson metal–insulator transition. It occurs experimentally in some disordered systems as a pressure-dependent metal–insulator transition.

For the study of Anderson localization, we will develop the Kubo linear response theory for the treatment of electric conductivity in disordered conducting media [3, 8, 9, 21]. This is a Green's function theory in which the conductivity is obtained from a consideration of two-particle Green's functions characterizing the dynamics of two different modal excitations through the medium. Likewise, in addition to the conductivity, the two-particle Green's functions are also shown to be related to the probabilities of two particles propagating through a random disordered medium, moving from a set of initial points to a set of final coordinates. These two aspects and interpretations of two-particle Green's functions are useful to facilitate their application in developing the studies of electrical conductivity. In this regard, they also naturally enter into the treatment of many of the other transport properties of disordered media involving a variety of types of excitations in the system, e.g., these may include the transport of photons and phonons similarly considered by these means.

Earlier in our study of the single impurity problem we based the studies on the consideration of single particle Green's functions [3]. There it was shown that these determined the density of states of the system and the dispersion relations and lifetimes of the single particle excitations. In our present treatment of Anderson localization, however, we extend the discussions to the treatment of the properties of two-particle Green's functions [3, 8, 9]. In this extension, it is now shown that the two-particle Green's functions determine the transport properties of the system and for electron modes are directly related to the electrical conductivity of the Anderson localized system.

Specifically, from the considerations of two-particle Green's functions, it is found that Anderson localization is observed at $T = 0$ as an absence of the system conductivity which for $T \neq 0$ develops into a weak conductivity due to electron hopping between localized states [3, 8, 9]. These considerations ultimately allow us to use localization properties of the excitations to explain the classification of disorder media into metals, semiconductors, and insulators [8, 9, 21]. This enables our understanding of the transitions between these states with changes in the system disorder and temperature.

In our studies, it is also shown that the spectroscopy of a system can be obtained by considering a few simple operator combinations averaged over the entire scattering system [8, 37]. These averages are again directly related to the two-particle optical Green's functions, i.e., the photonic Green's functions for light propagation in the system. In this way, it is found that a treatment of weak localization effects observed in the scattering of optical radiation from rough surfaces and/or disordered optical media is related to two-particle photon Green's functions for propagation in the disordered system [8, 37]. Specifically, in the optics problems, the differential reflection coefficient characterizing the scattering is found to be given by expressions involving two-particle Green's functions for the light propagation in the system.

In our next studies we will consider new types of geometry that have been of recent interest in the study of physics [11–13]. Specifically, we will treat geometries which exhibit the properties of self-similarity, i.e., an absence of a natural length scale. These types of geometries have, for example, entered into the study of systems at second-order phase transitions and in other systems exhibiting power-law dependent spatial correlations.

1.10 FRACTAL GEOMETRY

Many of the properties of physical systems are characterized by sets of length scales resulting from the nature of the geometry the systems are defined over, the dimensionality of the systems themselves, and the nature of the interactions acting within the systems [11–13, 21]. For example, in statistical physics the X-ray, neutron, or electron scattering from a gas, liquid, or solid displays characteristic features which are related to lengths associated with the statistical correlations within them. Arising from this and depending on the nature of the experiment, features in the scattering data are associated with, for example, the spin-spin correlation length, the density-density correlation length, the superconductor coherence length, the periodicity of the lattice, etc. Such lengths all enter to a greater or lesser extent in setting the scattering properties as well as the other basic thermodynamic properties of the materials. They characterize the response of materials to external probes, set their transport properties, and enter into the interactions between systems.

As an example, in the case of the study of the neutron scattering from magnetic materials, one often encounters the spin-spin correlation function [3, 21]. This is a measure of how much a spin at one point in the system is affected by the presence of a spin at another point in the system. It is directly related to the diffuse elastic scattering from the magnetic features of the material so that the space-time Fourier transform of the correlation function generates the scattering cross section from the magnetic material. This allows us to study the correlations between the spins by means of scattering experiments.

For instance, in terms of the z−components of spin in a magnetic material, the $z-z$ matrix element of the spin-spin correlation function has an entry of the form [3, 21, 38, 39]

$$\langle S_z(0)S_z(\vec{r})\rangle - \langle S_z(0)\rangle\langle S_z(\vec{r})\rangle \propto \frac{e^{-\frac{r}{\xi}}}{r^{\eta}} \tag{1.4a}$$

Here $S_z(0)$ is the spin component at the origin of space, $S_z(\vec{r})$ is the spin component at the position $\vec{r}$, $r = |\vec{r}|$, $\langle\ \rangle$ denotes a thermodynamic average, η is a critical exponent describing the component of power law decay of the correlation function, and ξ is the spin-spin correlation length of the exponential part of the space dependence. In this regard, notice in Eq. (1.4a) that the exponential factor [38, 39]

$$e^{-\frac{r}{\xi}}, \tag{1.4b}$$

sets a rate of spatial decay in the correlation function in terms of the length ξ. This represents a dependence on the particular length ξ in an absolute sense, so that the correlation has a preferred length ξ arising in its characterization of the system.

On the other hand, the power law factor [38, 39]

$$\frac{1}{r^{\eta}} \tag{1.4c}$$

in Eq. (1.4a) is not characterized by any preset length scale. The only spatial information in the factor in Eq. (1.4c) is that if we double $r \to 2r$ we find that $\frac{1}{r^{\eta}} \to \frac{1}{2^{\eta}}\frac{1}{r^{\eta}}$. Specifically, it is seen that changing the length in Eq. (1.4c) generates a relative change in the power law factor and not an absolute change governed by a preset length scale such as ξ. The form of the power law dependence, consequently, does not change.

Notice that the geometry of the system characterized by the factor in Eq. (1.4b) is quite different from the geometry of the system characterized by the factor in Eq. (1.4c). In this regard, it is found that for the $\xi \to \infty$ limit, we have from Eq. (1.4a) that [38, 39]

$$\langle S_z(0)S_z(\vec{r})\rangle - \langle S_z(0)\rangle\langle S_z(\vec{r})\rangle \propto \frac{1}{r^{\eta}} \tag{1.5}$$

and the length scale ξ drops out of the problem. For this limit, then, the correlation function is solely a power law. In physical systems at a second-order phase transition, it is generally observed that $\xi \to \infty$, so that the correlation functions of the system exhibit a scale-independent, power law behavior. The second-order transition in this sense is one between two different characteristic geometries of the material. One has a natural length scale ξ and the other is scale independent.

A second order transition occurs at a specific temperature of the system known as the critical temperature and is a continuous transition between an ordered and a disordered state of the system. For example, in ferromagnetic materials, it $\xi \to \infty$ occurs at the Curie point transition and represents a continuous change between an ordered ferromagnetic phase (i.e., a state of permanent magnetization) and a disordered paramagnetic phase [38, 39]. (Note, specifically, that at the point of this disordering transition $\langle S_z(0)\rangle = \langle S_z(\vec{r})\rangle = 0$ and $\langle S_z(0)S_z(\vec{r})\rangle - \langle S_z(0)\rangle\langle S_z(\vec{r})\rangle \propto \frac{1}{r^{\eta}}$). Consequently, at the point of a second order phase transition the material generally loses its dependence on a fixed length scale and only exhibits power law behavior. This represents an anomalous change in the geometry of the system at the critical point [38].

In this regard, the power law dependence in the correlation functions indicates that the geometry of the ferromagnetically ordered system at the critical point of its transition to the paramagnetic state is made on an infinitely connected set of spins with a new type of geometry. This new scale-independent geometry hosting the remnants of the ferromagnetic ordering is known as a fractal geometry [38].

Fractal geometries occur in geometric objects that appear to be the same when viewed on any of a set of length scales [11–13]. They are so-called self-similar structures because if we view them at one length scale or magnification and then change to view them on a different length scale or magnification, they look the same in both viewings. In this sense, they have the same appearance to an observer when viewed on all length scales. The ordering of spins at the Curie point, consequently, occurs on infinitely connected clusters which appear to have the same geometric properties when viewed at many different scales.

Fractal systems have a number of interesting properties which are developed in the course of their study [11–13]. They can exist as randomly disordered systems as in the case of the ferromagnetic-paramagnetic transition or as regular structures generated by an algorithmic process. In fractals built as regular, nonrandom structures, an iterative procedure is used to orderly construct a geometric structure which is self-similar on a variety of length scales [38]. In general, the geometric nature of fractals is characterized by certain exponents developed from their geometric self-similarity. This is indicative of the power law scaling of their geometry. In this regard, one can introduce a generalization of the ideas of spatial dimension so as to represent fractals as structures which exist in fractional spatial dimensions. These fractional dimensions are often located between our traditional notations of the $0-, 1-, 2-$, or $3-$dimensions of points, lines, planes, and bulk space [11–13].

Fractal objects can be formulated in the $2-$dimensional plane, for example, which are more filling of the $2-$dimensional space than is a $1-$dimensional line segment but which do not entirely fill the $2-$dimensional plane. These geometric objects, which partially fill the plane, will have a fractal dimension between 1 and 2 depending on how completely they are able to occupy the plane. As an example of an algorithmically generated curve which is more than a $1-$dimensional structure but less than a $2-$dimensional plane, consider the Koch curve. To form this structure, one begins with the one-dimensional line segment in Figure 1.3a, modifies the segment as in Figure 1.3b, and then modifies each of the line segments in Figure 1.3b in a similar way to obtain Figure 1.3c. This is repeated over and over an infinite number of times to the line segments found after each iteration until after an infinite set of iterations the Koch curve is generated. As we shall see later, the fractal dimension of the Koch curve is 1.26. This is between $1-$ and $2-$ dimensions and indicates that geometrically the Koch curve is more complex than a simple line segment but considerably less filling of the plane than a truly $2-$dimensional structure. In addition, we shall also find that the Koch curve is a line which is nowhere differentiable [11–13].

As shall be seen, fractals occur in many important physical systems. One which shall be a focus to us will be in the chaotic dynamics of nonlinear systems. In these systems the dynamics can be represented in terms of trajectories evolving on spatial structures involving fractal geometry. We shall next turn to a consideration of nonlinear dynamics, its manifestation in the form of chaos, and the relationship of chaos to fractal geometry. In the course of our discussion, we shall see that the ideas of chaos and fractal geometry are of fundamental importance in understanding the nature of nonlinear dynamics.

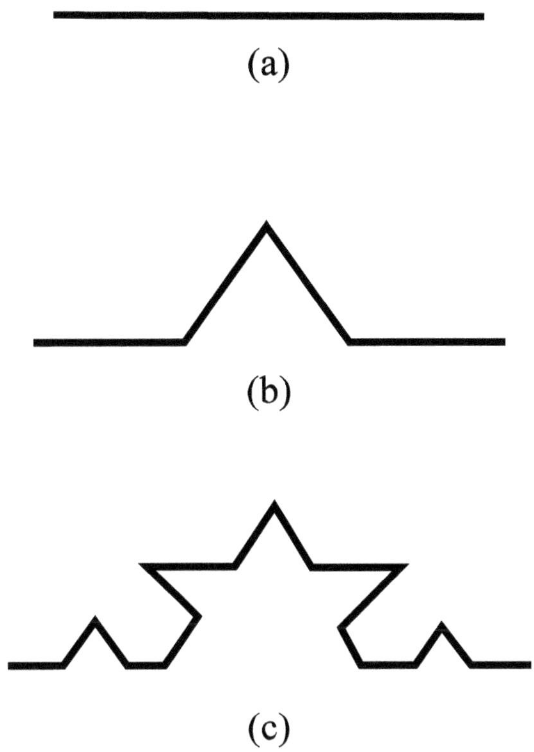

FIGURE 1.3 A schematic process for generating a Koch curve through an infinite iterative process. The iterative process is as follows: (a) we start with a line segment, (b) the segment is broken into thirds and the middle segment is replaced by two sides of an equilateral triangle, (c) the operation in (a) and (b) is performed on each of the segments of (b). This replacement of (a) with (b) is repeated again on the segments forming (c) to form a new curve. The process of replacing (a) with (b) is continued for an infinite number of times to generate the Koch curve.

1.11 CHAOS

A new type of disorder is that associated with chaotic dynamics [11–13]. Disorder in chaotic systems arises solely due to nonlinearity, and the apparent disorder of the system is generated in the course of executing a well-defined algorithm process for determining the motion. While the algorithm gives the same dynamics for a given set of initial conditions, the dynamics appear to be somewhat uncorrelated and haphazard. In this regard, von Neumann once considered nonlinear algorithms as possible codes for generating sequences of random numbers [2, 11–13]. Chaotic systems represent surprising processes because the origin of the disorder arises from the nonlinearity of the dynamics internal to the algorithm and not from randomness introduced externally to the algorithmic process determining the motion. In this regard, the disorder generated by chaotic processes is sometimes referred to as deterministic chaos.

A famous example of the simplicity of the nonlinear algorithms exhibiting chaotic behavior is the logistic mapping, which involves a single particle moving in $1-$dimension. For a particle in $1-$dimensional motion along the $x-$axis, the logistic mapping has the form of the difference equation given by the nonlinear recursive form [11–13]

$$x_{n+1} = \mu x_n \left(1 - x_n\right). \tag{1.6}$$

Here the particle positions $\{x_i\}$ (for $i \geq 0$ an integer) are an ordered sequence of equal time steps generated on the $x-$axis, $0 \leq x_i \leq 1$ for each time step i, and $0 \leq \mu \leq 4$. In this mapping, regions of chaotic disorder in the particle trajectories $\{x_i\}$ of the system can be found in the region where $\mu \geq 3.56995$ and for initial conditions within the range $0 \leq x_0 \leq 1$. Under these conditions the system wanders through a nonrepeating sequence of values $0 \leq x_n \leq 1$ in a seemingly chaotic process. In addition, the system is also found to be extremely sensitive to small changes in the initial condition x_0 used to generate a chaotic sequence $\{x_i\}$. This is an important aspect of chaotic motion.

The sensitivity of the system to initial conditions is characterized in chaotic dynamics by the so-called Lyapunov exponents of the system [11–13]. In this representation, small changes in the value of the initial condition x_0 are found to lead to subsequent divergent behaviors between the sequences generated by the two slightly different choices of initial conditions. This last property is a definition of chaotic motion. For example, for the logistic mapping in Eq. (1.6) if for small $\delta(0)$ we make the change of initial conditions $x_0 \rightarrow x_0 + \delta(0)$, we find after the nth iteration that $x_n \rightarrow x_n + \delta(n)$. The distance between the initial conditions is then given by [11–13]

$$d_0 = |\delta(0)| \tag{1.7a}$$

and the distance between the nth iterations is

$$d_n = |\delta(n)|. \tag{1.7b}$$

In a chaotic motion of the logistic mapping, the separations in Eq. (1.7) are found to be divergent for long-time iterations so that for $n \gg 1$

$$d_n = d_0 e^{\lambda n} \tag{1.8}$$

where $\lambda > 0$ is the Lyapunov exponent characterizing the chaotic motion of the system.

We shall see in our studies that, with increasing μ, regions of chaos found in $0 \leq \mu \leq 4$ can be broken up between regions of periodic motion. Consequently, the logistic equation exhibits both periodic motions and chaotic motions depending on the values of μ under consideration. This type of complexity in the dynamical process is quite surprising because the recursive form as Eq. (1.6) is quite simple. However, the complexity exists due to nonlinearity and may be found in a variety of nonlinear systems of difference equations and nonlinear systems of differential equations. In this regard, in our discussion here we have treated a $1-$dimensional system, but chaotic systems are also found in higher dimensional systems of equations.

Consequently, it is found that higher dimensional chaotic systems have increasingly more difficult dynamics to describe than do linear systems and $1-$dimensional nonlinear mappings. A type of representation available to characterize such higher dimensional systems is the Poincaré mappings, which help in providing a pictorial summary of some of the dynamics of chaotic systems. In these mappings, certain geometric features occur which are characterized as having fractal geometries. In these fractal structures, the Poincare mappings [11–13] will allow us to link the study of chaos with the study of fractal geometry.

Continuing the discussions of disordered systems, a recent set of important studies involve the computer simulation treatment of quantum mechanical systems [16, 39, 40]. These studies are complicated due to the operator nature of quantum systems. We shall next look at modifying the Monte Carlo simulation study of classical mechanical systems to treat quantum mechanical systems [32]. Following these considerations, the applications of the quantum Monte Carlo methods to understanding the low-temperature properties of the boson system ^4He and the fermion system ^3He will be made [41, 42].

1.12 QUANTUM MONTE CARLO SIMULATION

An interesting modification of Monte Carlo methods can be made for applications in the determination of the thermodynamics of quantum mechanical systems [16, 39, 40]. This is important since unlike classical mechanical systems [32], where all the variables of the theory involve real or complex numbers, quantum mechanical systems are written in terms of operators obeying non-commutative algebras. Consequently, quantum systems are much more complicated to evaluate. Fortunately, however, a variety of mathematical tricks have been developed to overcome the difficulties arising from non-commutative algebras. Following from these, the study of the thermodynamics of quantum systems can be made based on methods involving the conversion of the quantum mechanical partition function into the form of a path integral expressed solely in real and complex variables. The resultant path integral and the thermodynamic properties computed from it can then be evaluated using standard Monte Carlo methods. This transformation of the problem is known as the path integral formulation for the quantum statistical mechanics of the system [16, 39, 40]. In the conversion of the problem, however, it follows that a partition function of a quantum mechanical model in d −dimensions is expressed as a path integral in $d+1$ −dimensions. For the reformulation, the complication of non-commuting algebras is exchanged for the complication of having to treat a new problem in an increased number of dimensions.

As an application of quantum Monte Carlo, we shall consider an interesting many-body problem involving the study of the low-temperature properties of the superfluid Helium II phase of liquid ^4He. The ^4He isotope is composed of two neutrons and two protons with a total nuclear spin 0 so that liquid ^4He is a Bose-Einstein fluid consisting of interacting hard-core helium atoms with weak residual attractive interactions between atoms [41, 42]. At low temperatures (below $4.2^0 K$ at atmospheric pressure) ^4He forms a Helium I liquid phase which again below the so-called lambda transition ($2.17^0 K$ at atmospheric pressure) transforms into a superfluid phase known as Helium II which is mixed and coexistent with a Helium I component. The new Helium II phase of the system is known as a superfluid, and it accounts for the component of liquid helium exhibiting zero viscosity [41, 42]. The Helium II phase arises as a condensate forming a macroscopic quantum state of the fluid. The condensate is shown to display a variety of unusual properties arising from its nature as a macroscopic quantum state.

Early in the study of Helium II, a number of phenomenological models were proposed to understand the thermodynamic properties of superfluid ^4He. In our treatment, we will focus on three such important phenomenological theories [41, 42]. These include: the two-fluid model, which considers helium at temperatures below the lambda transition $2.17^0 K$ (at atmospheric pressure) as a mix of viscous Helium I and zero viscosity Helium II; the Ginzburg model characterizing the form of the free energy in the vicinity of the second order lambda phase transition between liquid Helium I and the mixed phase of Helium I and Helium II; and a proposal by Landau, made to account for the zero viscosity below the lambda transition, for the single particle excitation spectrum of the mixed Helium I and Helium II superfluid state. This will be followed by discussions of recent quantum Monte Carlo (computer simulation) studies which have provided a first principles justification of these earlier phenomenological treatments of liquid ^4He.

Another superfluid system of interest is that of liquid ^3He. The superfluid phase in this system occurs at much lower temperatures than that of the Helium II phase in liquid ^4He and arises from a completely different set of mechanisms from those in the ^4He system. The ^3He isotope is composed of two protons and one neutron and has total spin $\frac{1}{2}$. Consequently, ^3He is a Fermi-Dirac system with its superfluidity arising from a pairing interaction similar to the pairing interaction in electronic superconducting systems. However, whereas the superconductivity in low-temperature electronic superconductors comes from s-wave pairing, the superfluidity in low-temperature ^3He comes from p-wave pairing. In this regard, notice that in low-temperature electronic superconductors the superconductivity is manifest in the flow of electrical current; however, in liquid ^3He superfluidity is concerned with the flow of charge neutral atoms, i.e., with the fluid viscosity. A discussion of the theory of both s-wave and p-wave pairing will be given with a focus on the application to understanding superfluidity in ^3He as contrasted with electronic superconductivity in low-temperature electronic materials [41, 42].

REFERENCES

1. Alencar, M. S. 2014. *Information Theory* (Momentum Press, New York).
2. Kneusel, R. T. 2018. *Random Numbers and Computers* (Springer Nature, Cham).
3. Economou, E. N. 2006. *Green's Functions in Quantum Physics* (Springer-Verlag, Berlin).
4. Vasilyev, A. 1983. *An Introduction to Statistical Physics* (Mir Publishers, Moscow).
5. Landau, L. D. and Lifshitz, E. M. 1958. *Statistical Physics* (Pergamon Press LTD, London).
6. Kittle, C. 1986. *Introduction to Solid State Physics,* 6th Edition (John Willey & Sons, Inc., New York).
7. Ashcroft, N. W. and Mermin, N. D. 1976. *Solid State Physics* (Harcourt Brace College Publishers, Fort Worth).
8. Sheng, P. 1990. *Scattering and Localization of Classical Waves in Random Media* (World Scientific, Singapore).
9. Mott, N. F. and Davis, E. A. 2012. *Electronic Processes in Non-Crystaline Materials* (Oxford University Press, Oxford).
10. Reed-Hill, R. E. 1973. *Physical Metallurgy Principles,* 2nd Edition (D. Van Nostrand Company, New York).
11. Strogatz, S. H. 1994. *Nonlinear Dynamics and Chaos* (Perseus Books, Reading).
12. Feldman, D. P. 2012. *Chaos and Fractals: An Elementary Introduction* (Oxford University Press, Oxford).
13. Rasband, S. N. 1990. *Chaotic Dynamics of Nonlinear Systems* (John Wiley & Sons, Inc, New York).
14. Landau, D. P. and Binder, K. 2015. *A Guide to Monte Carlo Simulations in Statistical Physics, Fourth Edition* (Cambridge University Press, Cambridge).
15. Feynman, R. P. and Hibbs, A. R. 1965. *Quantum Mechanics and Path Integrals* (McGraw-Hill Book Company, New York).
16. Suzuki, M., ed. 1993. *Quantum Monte Carlo Methods in Condensed Matter Physics* (World Scientific, Singapore).
17. Pang, T. 2016. *An Introduction to Quantum Monte Carlo Methods* (Morgan & Claypool Publishers, San Rafael).
18. Press, W. H., Taukolsky, S. A., Vetterling, W. T., and Flannery, B. P. 1997. *Numerical Recipes in Fortran 77: The Art of Scientific Computing*, Second Edition (Cambridge University Press, Cambridge), Chapter 7.
19. Barnett, S. M. 2009. *Quantum Information* (Oxford University Press, New York).
20. Shannon, C. E. 1948. A Mathematical Theory of Communication. *The Bell System Technical Journal* 27, 379–423.
21. Marder, M. P. 2000. *Condensed Matter Physics* (John Willey & Sons, Inc., New York).
22. Hewson, A. C. 1997. *The Kondo Problem to Heavy Fermions* (Cambridge University Press, Cambridge).
23. Kuramoto, Y. 2020. *Quantum Many-Body Physics: A Perspective on Strong Correlations* (Springe Japan, KK).
24. Hull, D. and Bacon, D. J. 2001. *Introduction to Dislocations* (Elsevier Ltd., Amsterdam).
25. Nabarro, F. R. N. and Hirth, J. P. 2007. *Dislocations in Solids* (Elsevier, Amsterdam).
26. Jaynes, E. T. 1957. Information Theory and Statistical Mechanics. *Physical Review* 106, 620–630.
27. Perepechko, I. I. 1981. *An Introduction to Polymer Physics* (Mir Publishers, Moscow).
28. Macia-Barber, E. 2021. *Quasicrystals: Fundamentals and Applications* (CRC Press, Boca Raton).
29. Haydock, R. 1980. The Recursion Solution of the Schrodinger Equation. *Solid State Physics* 35, 216–294 (Academic Press Inc. New York).
30. Klinger, M. 2013. *Glassy Disordered Systems: Glass Formation and Universal Anomalous Low-Energy Properties* (World Scientific Publisher, Singapore).
31. Phillips, W. A. 1987. Two-Level States in Glass. *Reports on Progress in Physics* 50, 1657–1708.
32. Binder, K. and Young, A. P. 1986. Spin Glasses: Experimental Facts, Theoretical Concepts, and Open Questions. *Reviews of Modern Physics* 59, 801–963.

33. Edwards, S. F. and Anderson, P. W. 1975. Theory of Spin Glasses. *Journal of Physics F: Metal Physics* 5, 965–974.
34. Kinashita, S. 2013. *Pattern Formation and Oscillatory Phenomena* (Elsevier Inc., Amsterdam).
35. Cross, M. C. and Hohenberg, P. C. 1993. Pattern Formation Outside of Equilibrium. *Reviews of Modern Physics* 65(3), 854–1086.
36. Turing, A. M. 1952. The Chemical Basis of Morphgenesis. Philosophical Transactions of the Royal Society 237, 37–72.
37. McGurn, A. R. 1990. Enhances Retroreflection Effects in the Reflection of Light from Randomly Rough Surfaces. *Surface Science Reports* 10(8), 357–410.
38. Plischke, M. and Bergersen, B. 1989. *Equilibrium Statistical Physics* (Prentise Hall, New Jersey).
39. Ceperley, D. M. 1995. Path Integrals in the Theory of Condensed Helium. *Reviews of Modern Physics* 67, 279–353.
40. McGurn, A. R., Ryan, P., Maradudin, A. A., and Wallis, R. F. 1989. Quantum Monte Carlo Calculation of the Thermodynamic Functions of a Lennard-Jones Chain of Atoms. *Physical Review B* 40, 2407–2413.
41. Annett, J. F. 2004. *Superconductivity, Superfluids and Condensates* (Oxford University Press, New York).
42. Huang, K. 2010. *Introduction to Statistical Physics*, 2nd edition (CRC Press, Boca Raton).

2 Sequences of Random Numbers

In this chapter we consider a basic type of randomness or disorder that has important applications in understanding physical processes and in numerical simulation approaches to the study of physical and engineering systems. This involves the generation and study of the characterization of ordered sequences of so-called pseudo-random numbers [1]. The sequences generated are known as pseudo-random because they arise from computer routines, composed of an ordered array of numerical steps, but which are formulated to exhibit statistical properties associated with randomness and disorder. In this regard, for their applications the numbers are sequentially ordered, i.e., the numbers occur positioned in their set in the order in which they are generated by the computer routine. However, in spite of this sequential ordering the presence of a number in the random set seems to bear little or no dependence on its fellow members in the sequence. The quality of randomness of a given sequence for applications is, consequently, tested with a number of well-defined statistical measures [1].

We begin by discussing the use of sequences of random numbers in the numerical evaluation of integrals. Here the application of random numbers to numerical integration is known as Monte Carlo integration [1–3]. A comparison of the Monte Carlo method of integration with standard numerical methods based on the definition of the Riemann integral (i.e., the Simpson and the trapezoidal rule methods) is made [2, 3]. In practical applications, it is found that for the evaluation of lower dimensional systems, the accuracy and efficiency of numerical integration favor the use of the trapezoidal rule method. On the other hand, for the evaluation of multidimensional integrals, Monte Carlo methods become more favorable in treating integrations involving four or more dimensions. This has particular interest in thermodynamic systems, which are generally focused on multidimensional integrals of very high dimensionality. Later, in Chapter 4, the discussion of Monte Carlo methods and the associated topics of Metropolis-Hastings [2] sampling for the study of thermodynamic properties will be considered in detail. For now, however, other applications of the Monte Carlo will be considered in this chapter.

Following the treatment of Monte Carlo integration [1–3], considerations will turn to the actual generation of sequences of random numbers. In what seems to be a contradiction, it is found that sequentially random sequences of numbers are generated by computer algorithms, and there are a number of different algorithms that have been written for their generation.

The discussions of computer algorithms begin with a treatment of the linear congruent generator which has a long history in the development of sets of random sequences of numbers. Specifically, the linear congruent generator [1–3] is based on a linear recurrence relation which generates random sequences of positive integers. The ability of the linear recurrence relation to generate random sequences is found to be highly dependent on the values of the fixed coefficients defining the linear recurrence relation. From the generated sequence of random integers, by a linear transformation, it is in turn possible to generate a random sequence of rational numbers on a segment of the real axis [1].

In addition to the linear congruent generator, several other generators exist which are based on different principles of operation [1]. These are different generators, and their methodologies are briefly described, with references to their respective literatures given. For all of the random number generators, statistical tests on the sequences they generate are the ultimate arbiters of their usefulness as generators of random numbers [1].

 DOI: 10.1201/9781032674919-2

Random sequences generated by computer algorithms must be statistically checked to assure that the sequences are in fact random [1]. Our initial focus is on generators of sequences of uniform random numbers [1, 3] which have values that occur with equal probability over their set range of values, i.e., random integers $\{m_j\}$ developed uniformly on the interval $0 \le m_j \le N$ or random rational numbers $\{x_j\}$ developed uniformly on the interval $0 \le x_j \le 1$. The first test of a sequence of random numbers will, consequently, determine their uniformity over an appropriate range of values. In addition, the numbers of the sequence should occur in the sequence independent of the presence of the other numbers in the sequence. This is determined by measures of the correlations within the sets of random numbers.

Following the discussions of uniformly distributed sequences of random numbers, a treatment is given of random sequences in which the numbers occur with some nonuniform statistical distribution. These discussions are focused on exponential and Gaussian distributions of sequences of random variables [1, 3]. As an application, these distributed sequences of numbers will then be used in an example from surface physics [4]. In this discussion, the generation of a random disordered surface which exhibits Gaussian random statistics will be made, and the statistical properties of the surface disorder investigated.

2.1 THE APPLICATION OF RANDOM NUMBERS IN NUMERICAL INTEGRATIONS

One application of random numbers is in numerical integration routines, commonly known as Monte Carlo integration methods [2, 3]. In this approach, an integral over an interval or volume is expressed in terms of a sum over a set of randomly generated points. As we shall see, applications of random numbers in the approximation of integrals can be more effective than, for example, standard numerical methods based on evaluations involving regular integration grids of points. We shall begin our discussions with the application of random numbers in the evaluation of a one-dimensional integral [2, 3]. This is followed by discussions of the application of Monte Carlo integration methods to higher dimensional integrals and a comparison of the accuracy of the numerical results from the Monte Carlo approach with those obtained using the trapezoidal rule method [2, 3].

As an example of the application of random numbers in the Monte Carlo evaluation of an integral defined in one dimension, we focus on the numerical integration of a function over the interval $0 < x < 1$ on the x-axis. This restriction can be made because an integral over the interval $0 < x < 1$ is related to an integral over the general interval $a < x' < b$ over an x'-axis through a change of variables. Specifically, the two intervals map into one another under a transformation of the form [2, 3]

$$x' = (b-a)x + a \text{ or } x = \frac{1}{b-a}(x'-a) \tag{2.1a}$$

so that it follows that for $f(x')$ considered on the interval $a < x' < b$

$$I = \int_a^b f(x')dx' = \int_0^1 f(x'(x))\frac{dx'(x)}{dx}dx \tag{2.1b}$$

$$= (b-a)\int_0^1 f(x'(x))dx.$$

As a result, every problem of the form $\int_a^b f(x')dx'$ is ultimately related to a problem of the form $\int_0^1 f(x'(x))dx$.

To begin with, consider then that we are asked to obtain a numerical approximation for the integral [2, 3]

$$I = \int_0^1 f(x)\,dx \tag{2.2}$$

using a set composed as a sequence of uniformly distributed random numbers $\{x_0, x_1, x_2, \ldots\ldots\ldots, x_N\}$ where $0 < x_i < 1$ for $i = 0,1,2,\ldots\ldots\ldots, N$ and for which $N \gg 1$. The random numbers are generated uniformly distributed over the interval $0 < x < 1$ with a distribution characterized by the probability density [2, 3]

$$p_u(x) = 1, \text{ for } 0 < x < 1 \tag{2.3a}$$

$$p_u(x) = 0, \quad \text{otherwise,} \tag{2.3b}$$

where

$$\int_0^1 p_u(x)\,dx = 1. \tag{2.4a}$$

(See Figure 2.1 for a plot of $p_u(x)$ versus x.) From Eqs. (2.2) and (2.3) it then follows that in terms of the probability density our integral is written as

$$I = \int_0^1 f(x)\,dx = \int_0^1 p_u(x) f(x)\,dx. \tag{2.5}$$

Consequently, in the discussions to be given in the following, the better that the set of randomly distributed numbers is characterized by the distribution in Eq. (2.3), the more accurately the value of the integral will be approximated. Let us next consider Eq. (2.5) written in terms of the set $\{x_0, x_1, x_2, \ldots\ldots\ldots, x_N\}$.

The integral in Eqs. (2.2) and (2.5) can also be represented in terms of the set $\{x_0, x_1, x_2, \ldots\ldots\ldots, x_N\}$ of discrete points distribution by Eq. (2.3). In terms of the set of $N+1$ randomly distributed points, it then follows that [2, 3]

$$I = \int_0^1 p_u(x) f(x)\,dx \cong \frac{1}{N+1}\sum_{i=0}^{N} f(x_i) = \langle f \rangle = I_{N+1}. \tag{2.6}$$

Here in the approximation using $N+1$ points the approximate value of the integral $I_{N+1} = \langle f \rangle \cong I$ is seen to be an average of the set $\{f(x_0), f(x_1) f(x_2) \ldots\ldots\ldots, f(x_N)\}$ of function values evaluated over the set $\{x_0, x_1, x_2, \ldots\ldots\ldots, x_N\}$. The sum, I_{N+1}, in Eq. (2.6) consequently becomes an increasingly better approximation of I as $N \to \infty$. This will now be shown to be the case.

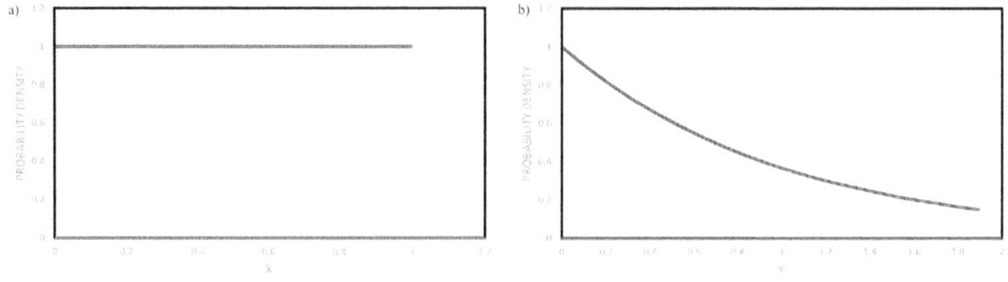

FIGURE 2.1 Plot of Probability Densities for: a) a plot of $p_u(x)$ versus x and b) a plot of $p_e(y)$ versus y.

Since the sequence $\{x_0, x_1, x_2, \ldots\ldots\ldots, x_N\}$ offers only an approximation to the uniform distribution $p_u(x)$ on $0 < x < 1$, it is interesting to estimate the tolerance or uncertainty of the approximation in Eq. (2.6). This is given by studying the variance about the average $\langle f \rangle$ under different realizations of the random sequence $\{x_0, x_1, x_2, \ldots\ldots\ldots, x_N\}$. In this way, the variance σ_{N+1} for the Monte Carlo result can be shown to be given by [3]

$$\sigma_{N+1}{}^2 = \frac{\langle (f - \langle f \rangle)(f - \langle f \rangle) \rangle}{N+1} = \frac{1}{(N+1)^2} \sum_{i=0}^{N} \left(f(x_i) - \langle f \rangle\right)^2 \tag{2.7}$$

$$= \frac{\langle f^2 \rangle - \langle f \rangle^2}{N+1}$$

where

$$\langle f^2 \rangle = \frac{1}{N+1} \sum_{i=0}^{N} f(x_i)^2 . \tag{2.8}$$

From Eq. (2.7) the variance is found to represent the uncertainty per Monte Carlo point of the value of the integral in Eq. (2.6). In this context as $N \to \infty$ the measure of uncertainty in the value of I_{N+1} is indicated by [3]

$$\sigma_{N+1} \propto \frac{1}{\sqrt{N}} \to 0 \tag{2.9}$$

and the approximation becomes increasingly better as the number of Monte Carlo points N becomes large.

For a one-dimensional integral, the error in the Monte Carlo result given in Eq. (2.9) does not represent an improvement over the approximate value of the integral obtained by using, for example, the standard trapezoidal rule method. In the application of the trapezoidal rule, the difference between the true value I of the integral and the numerical value obtained using a regular integration grid of $N+1$ integration points, I_{N+1}, is shown to decrease with increasing N as [3]

$$\left| I - I_{N+1} \right| \propto \frac{1}{N^2} \to 0. \tag{2.10}$$

The Monte Carlo uncertainty in Eq. (2.9) is found to compare unfavorably with this result. This advantage of the trapezoidal rule, however, changes upon considering applications to higher dimensional integrals [3].

The integration in Eq. (2.2) can be extended to higher dimensions by treating a multidimensional integral defined over a volume V. In this regard, consider the approximation of an integral in $d-$dimensional space based on N random uniformly distributed points in the $d-$dimensional volume V. Our problem is then to approximate [3]

$$I = \int_V f(\vec{x})\, dV \tag{2.11a}$$

where in terms of the uniform random distribution of points as in Eq. (2.6) it similarly follows that an approximation of the integral based on the N points, I_N, is obtained as [3]

$$I \cong I_N = V \langle f \rangle \tag{2.11b}$$

with

$$\langle f \rangle \cong \frac{1}{N}\sum_{i=1}^{N} f\left(\vec{x}_i\right) \tag{2.11c}$$

for the set of N uniformly distributed points $\left\{\vec{x}_i\right\}$ over V. As in the case of the one-dimensional problem, the value of the integral I is given by the average of the set $\left\{f\left(\vec{x}_i\right)\right\}$ generated from the set of random points $\left\{\vec{x}_i\right\}$. Similar to the one-dimensional case, the error in approximating the multidimensional integral in Eq. (2.11) with N random points is found to be given by [3]

$$\sigma_N = V\sqrt{\frac{\langle f^2 \rangle - \langle f \rangle^2}{N}} \tag{2.12}$$

where here again

$$\langle f^2 \rangle \cong \frac{1}{N}\sum_{i=1}^{N}\left[f\left(\vec{x}_i\right)\right]^2. \tag{2.13}$$

An example of the application of Monte Carlo integration to a multidimensional integral that commonly occurs in the literature [1] involves an integration over the range shown in Figure 2.2. In Figure 2.2 is a unit square in the two-dimensional space (x, y) and within the unit square is a quarter of a circle of unit radius. For this geometry, consider a function $f(x, y)$ defined over the region in Figure 2.2 so that $f(x, y) = 1$ within the quarter of circle but $f(x, y) = 0$, otherwise outside the quarter circle region. The integral of $f(x, y)$ over the square is then given by [1, 2]

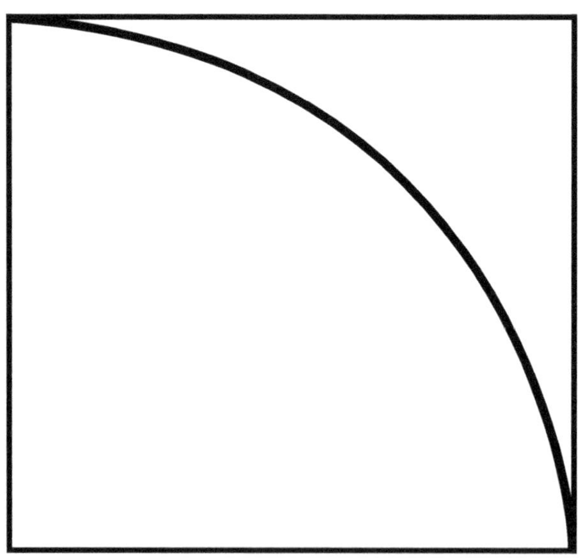

FIGURE 2.2 Plot in the $x - y$ plane of a square with unit sides along the $x-$ and $y-$axes containing a quarter circle of unit radius. The circle is centered on the origin at $(0,0)$ which is located in the lower left corner of the square so that a quarter of the circle is contained within the square.

$$I = \int_{unit\,square} f(x,y)\mathrm{d}x\mathrm{d}y = \frac{\pi}{4} = \langle f \rangle \cong \frac{1}{N}\sum_{i=1}^{N} \langle f \rangle (\vec{x_i}) = \frac{\#\,in\,quartter\,circle}{\#\,in\,the\,square} \tag{2.14}$$

where $\vec{x_i} = x_i\hat{i} + y_i\hat{j}$ for x_i and y_i randomly distributed over $0 < x < 1$ and $0 < y < 1$, respectively, in the unit square. The approximation in Eq. (2.14) is directly obtained by dividing the number of points in the quarter circle by the total number of points in the unit square. In one study [1] using 256 random points, Eq. (2.14) was found to give an approximate value for $\pi \approx 3.1578$.

Now let us consider the comparison of the accuracy of the higher dimensional N –point Monte Carlo integration with that obtained using the N –point trapezoidal method [3]. For integrals in $d > 1$ the expression for the variance of the N – point Monte Carlo approximation remain unchanged from that of the one-dimensional case and continue to be given by

$$\sigma_{N+1} \propto \frac{1}{\sqrt{N}} \to 0. \tag{2.15}$$

Upon consideration of the N – point trapezoidal approximation, however, it can be shown that [3]

$$\left|I - I_{N+1}\right| \propto \frac{1}{N^{\frac{2}{d}}} \to 0 \tag{2.16}$$

where I is the true value of the integral and I_N is obtained from an application of the trapezoidal rule with an N integration grid. The trapezoidal rule result is found to depend on the dimension of the system being considered and becomes increasingly poor with increasing d. In this regard, notice that for $d > 4$ the exponent in Eq. (2.16) depends on d with $\frac{2}{d} < \frac{1}{2}$ so that we expect the Monte Carlo simulation to begin to prevail over the trapezoidal rule method [3].

The Monte Carlo evaluation of higher dimensional integrals is, consequently, to be preferred as the dimension increases. This brings us again to the topic of randomly distributed numbers and, specifically, the considerations of how to generate them.

2.2 GENERATION OF RANDOM NUMBERS

Random numbers are usually generated by computer algorithms. In the following, we shall discuss some of the typical algorithms developed for the generation of sets of sequentially random numbers and the methods used to statistically quantify and assure their degree of randomness. While it may seem contradictory that sequentially random numbers are obtained from algorithms, it is the statistical measures used to determine the degree of randomness of a generated sequence that are important in finding good sequences, not necessarily the processes of their generation. In this sense, the measures of randomness ultimately set the considerations of the suitability of a generated sequence of numbers as being appropriate for any given computational application requiring randomness. Consequently, in recognition of the fact that orderly processes are involved in the development of sets of sequentially random numbers, such generated sets are often referred to as sets of pseudo-random numbers [1, 2]. A focus of our initial discussions of sequentially generated random numbers will be on numbers uniformly distributed over the interval $0 < x < 1$ on the x –axis with a probability density given in Eq. (2.3). This is followed by an introduction to the statistical properties required of random numbers and a discussion of the standard tests applied to such sequences.

A common type of sequentially random number generator is the linear congruential generator (LCG) [1–3]. This generator is developed to produce a set of sequentially random positive integers. From this sequence, in turn, a set of sequentially random rational numbers, uniformly distributed over the interval $0 < x < 1$ on the x –axis, is obtained by a linear mapping. As a consequence of this mapping relationship, the randomness properties of the integer and rational number sequences are intimately connected to one another.

The LCG algorithm for the sequential generation of integers is based on the recurrence relation between integers given by [1–3]

$$X_{n+1} = (aX_n + b) \, mod \, m, \text{for } n = 0,1,2\ldots \tag{2.17}$$

where the integer constant m is known as the modulus, the constant coefficients $0 < a < m$ and $0 \le b < m$ are both integers, and the initial integer starting value of the sequence $0 \le X_o < m$ is known as the seed. The algorithm in Eq. (2.17) generates a sequence of integers

$$\{X_0, X_1, X_2, \ldots\ldots\ldots, X_n, \ldots\} \tag{2.18}$$

which are periodic with a period which is less than or equal to m. For this process, in the case that $b = 0$ the generator in Eq. (2.17) is known as the multiplicative congruential generator and when $b \ne 0$ it is known as the mixed congruential generator.

The integers in the set in Eq. (2.18) can in turn be mapped onto a set of rational numbers $\{x_0, x_1, x_2, \ldots\ldots\ldots, x_n, \ldots\}$ over the interval $0 < x < 1$ so that

$$\{x_0, x_1, x_2, \ldots\ldots\ldots, x_n, \ldots\} = \left\{\frac{X_0}{m}, \frac{X_1}{m}, \frac{X_2}{m}, \ldots\ldots\ldots, \frac{X_n}{m}, \ldots\right\}. \tag{2.19}$$

Both Eqs. (2.18) and (2.19) are found to have sequential properties which strongly depend on a, b, and X_0. To develop credible random number sequences, these parameters must be carefully set.

The application of the sets in Eqs. (2.18) and (2.19) for the generation of sets of random integers and their related random rational numbers is a bit of an experimental process that we shall now begin to consider. Specifically, one needs to choose the constants in Eq. (2.17) so that a long sequence of random integers and, consequently, their related random rational sequence are generated by the recurrence process. This involves trials of different sets of the constant parameters in Eq. (2.17). It also involves the application of a variety of statistical tests to the generated sequences in order to determine the parameters which minimize the correlation between the generated numbers while maximizing the repeat period of the generated sequence.

For many choices of a, b, m, and X_0 the generator in Eq. (2.17) yields sequences of numbers which cannot be used as models of sequential random numbers. However, in a few cases, there exist good sets of a, b, m, and X_0 which allow for the generation of long useful sets of random integers $\{X_0, X_1, X_2, \ldots\ldots\ldots, X_n, \ldots\}$. Some general guidelines in the choice of a, b, m, and X_0 in order to generate good sequences of random integers are [1]:

1. The integers b and m should be relatively prime.
2. The integer $a - 1$ should be divisible by the prime factors of m.
3. If m is divisible by 4, the integer $a - 1$ should also be divisible by 4.
4. In general, aside from the conditions on (a, b, m), some conditions specific to the routine may be placed on the range of seeds X_0 needed for the generation of good sequences.

Through the years a number of sequences have been found and tested in their ability to generate random sequences based on the LCG. These include [1] LCG found in Numerical Recipes [3] which gives a generator based on the set with $(a, b, m) = (1664525, 1013904223, 2^{32})$ and a generator in Microsoft Visual C++ bits 30…16 with parameters $(a, b, m) = (214013, 2531011, 2^{32})$. All of these generators and their seeds must pass certain statistical tests which validate that their generated sequences are approximations of truly random sequences.

We now look at the application of statistical measures to indicate the suitability of generated sequences as representations of random sequences.

2.2.1 Uniformity of the Distribution

One of the requirements of generated random sequences is that their numbers are uniformly distributed over their interval of generation. For the sequences of rational numbers $\{x_0, x_1, x_2, \ldots\ldots\ldots, x_{N-1}\}$ over the interval $0 < x_j < 1$ with $j = 0, 1, 2, \ldots\ldots, N-1$ this requires that their distribution is characterized by the probability density [1]

$$p_u(x) = 1, \text{ for } 0 < x < 1 \tag{2.20a}$$

$$p_u(x) = 0, \text{ otherwise.} \tag{2.20b}$$

Consequently, if the interval $0 < x < 1$ contains N sequentially random rational numbers and is divided into $n \gg 1$ equal segments of length $\Delta x = \frac{1}{n} \ll 1$, then the number of points in the ith segment, ΔN_i, must satisfy

$$O_i = \Delta N_i \cong E_i = N\Delta x. \tag{2.21}$$

Here O_i is the number of random numbers in the ith interval and E_i is the expected number of random numbers in the ith interval under a perfectly uniform distribution. Two general methods have been developed to test sequences for the property of uniformity. These are now discussed.

The first method is the Chi-Square Test [1–3]. In this method, the series

$$\chi^2 = \sum_{i=1}^{n} \frac{(O_i - E_i)^2}{E_i} \tag{2.22}$$

is computed, where i runs over the n segments (each of length Δx) of the interval $0 < x < 1$. It is seen from Eq. (2.22) that for $O_i = E_i$ the value of $\chi^2 = 0$ and the points become uniformly distributed. For $\chi^2 \neq 0$, however, the generated points are not uniformly distributed over the interval and a tolerance must be set on χ^2 as a determination of what is acceptable for computational purposes.

Another method that is commonly used as a test of uniformity is the Kolmogorov-Smirnov Test. In this test, the uniform probability distribution in Eq. (2.20) is integrated to give for $0 < x < 1$

$$F(x) = \int_0^x p_u(x)\,dx = x. \tag{2.23}$$

The function $F(x)$ is then the probability of finding a uniformly distributed point in the region between 0 and x. Next, the actual distribution $\{x_0, x_1, x_2, \ldots\ldots\ldots, x_{N-1}\}$ of N uniformly distributed random numbers is considered, and the number of them, $L_N(x)$, less that x is determined. From this, the ratio [1]

$$S_N(x) = \frac{L_N(x)}{N} \tag{2.24}$$

is computed for a comparison with the ideal number $F(x)$.

In a perfect uniform distribution Eqs. (2.23) and (2.24) should agree. Consequently,

$$D = max\left|F(x) - S_N(x)\right| \tag{2.25}$$

provides a measure of the disagreement in the uniformity of the distribution of the generated sequence with the ideal distribution. This is then the number to minimize in developing a usable sequence of random numbers.

2.2.2 Sequential Correlations in the Distribution

The next set of measures determines the predictability of a number occurring at a particular position in the random sequence of numbers. Specifically, the focus is now on how correlated the numbers are in their sequence positions. In this regard, in a truly random sequence of N uniformly distributed random numbers of the form $\{x_0, x_1, x_2, \ldots\ldots\ldots, x_{N-1}\}$ defined over the interval $0 < x < 1$ there should be no correlation or predictability in sequencing or positioning of the numbers from the numbers found earlier in the sequence. In the following, two standard tests for correlations in the number sequence are considered. They are the Gap Test and the Autocorrelation Test [1].

The first test considered is known as the Gap Test [1]. In this test, the numbers in the interval $0 < x < 1$ are divided into n bins of equal lengths $\Delta x = \frac{1}{n}$ and the occupancies of the numbers in the random sequence $\{x_0, x_1, x_2, \ldots\ldots\ldots, x_{N-1}\}$ in each of the bins are studied. Specifically, in this test, it is asked: if x_i is in a particular bin and the next number of the sequence found in that same bin is x_{i+k} where $k > 0$, how many random numbers of the sequence are there between x_i and x_{i+k}? It is seen that the integer $k-1$ is the number of sequence numbers occurring between the two numbers x_i and x_{i+k} which occupy the same bin. For a truly random sequence over the n bins of the interval $0 < x < 1$, the sequence intervals (i.e., $k-1$) must be distributed [1] with the binomial probabilities $p_{k-1} = \frac{1}{n}\left(1 - \frac{1}{n}\right)^{k-1}$ where $\frac{1}{n}$ is the probability that a number of the sequence falls within any particular bin of the interval $0 < x < 1$.

A second criterion for independence is known as the Autocorrelation Test [1]. In this test, the product of pairs of numbers in the random sequence, which are separated by a fixed number of intervening members of the sequence, is considered. A sum known as the autocorrelation function of these pairs over the sequence is then constructed. The sum is designed to be zero in a totally random set in which the randomness of each member of the sequence is independent of the other members of the sequence. Specifically, if $\{x_0, x_1, x_2, \ldots\ldots\ldots, x_{N-1}\}$ is a sequence of random numbers over the interval $0 < x < 1$, then the autocorrelation function is defined as [1]

$$AC_{i,m} = \frac{1}{M+1}\sum_{k=0}^{M} x_{i+km}\, x_{i+(k+1)m} - \left(\frac{1}{2}\right)^2. \tag{2.26}$$

Here the interval between the numbers in a chosen pair of random numbers is $m-1$, the integer $m \ll N$, $M = \frac{N}{m}$, and periodic boundary conditions are applied in computing the sum over k. In regard to the periodic boundary conditions, the periodicity of the sequence is defined so that $x_N = x_0$, $x_{N+1} = x_1$, etc.

Notice that each element of the sequence x_j is distributed randomly and uniformly over $0 < x_j < 1$ with the average value $\langle x_j \rangle = \frac{1}{2}$. Consequently, in a perfect random system, it is expected that $AC_{i,m} = 0$, and to the extent that $AC_{i,m} \neq 0$ the sequence $\{x_0, x_1, x_2, \ldots\ldots\ldots, x_{N-1}\}$ is a less accurate representation of a randomly disordered sequence. A criterion of how nonzero an autocorrelation function can be in the generation of an acceptable sequence of random numbers is then required.

It is interesting to note that many of the statistical ideas related to the characterization of random numbers that have just been introduced in our discussions are generalizable to the study of disorder in materials. These include both the treatment of averages of disordered properties as well as correlations in the disorder found in physical systems and their properties. As we shall see later, such properties have great influence on the transport and thermodynamic characteristics of disordered materials. The disorder in materials includes both uniform disorder of our earlier discussions but also other types of correlated disorder which will now be considered. These characterizations of random disorder will occur again in other contexts in the remaining chapter of the text.

To conclude the discussion of tests for random sequences, we note that there are additional tests of both uniformity and sequential correlations that have been developed. For these, the interested reader is referred to the literature [1–3].

We next turn to a brief discussion of some other methods of generating uniform distributions of random numbers, pointing out to the reader more detailed discussions in the literature. This will be followed by treatments of additional types of sequentially random distributions of numbers. In particular, we look at random sequences that are distributed with exponential or Gaussian probability distributions [1, 3]. These are related to the uniform random distribution by transformations. The chapter will end with some examples of applications of randomness to physical and engineering systems.

2.3 OTHER METHODS OF GENERATING UNIFORM RANDOM NUMBERS

A number of other methods have been developed for the generation of random sequences of numbers. In developing generator algorithms, a requirement is that they run fast and efficiently while at the same time producing long sequences of uniform and uncorrelated random numbers. This places a fundamental restriction on their implementation and complexity. In this respect, the LCG algorithms are certainly simple and quick to operate on the computer. However, the LCG is not the only method that is appropriate, and through the years a number of other methodologies have been developed which give improvements over the sequences developed by the LCG. In the following, some generators which have been developed are briefly described. The considerations give a brief summary of some of the highlights in recent developments, and the reader is referred to the literature for a more detailed treatment [1–3].

A modification of the LCG can be made to give what are known as the Multiple-with-Carry (MWC) and Complimentary Multiple-with-Carry (CMWC) generators. In the MWC generator, the recurrence relation in Eqs. (2.17) and (2.18) is modified to the form [1]

$$X_{n+1} = (aX_n + b_n)\, mod\, m, \text{ for } n = 0,1,2,\ldots \tag{2.27a}$$

$$b_{n+1} = \frac{(aX_n + b_n)}{m}, \text{ for } n = 0,1,2,\ldots \tag{2.27b}$$

which generates the two sequences

$$\{X_0, X_1, X_2, \ldots\ldots\ldots, X_n, \ldots\} \tag{2.28a}$$

$$\{b_0, b_1, b_2, \ldots\ldots\ldots, b_n, \ldots\} \tag{2.28b}$$

from the seeds X_0 and b_0 where a and m are integers. In Eq. (2.27) $ab-1$ is chosen to be a prime number, and the sequence in Eq. (2.28a) is the generated sequence of random positive integers. It is seen that an additional bit of uncertainty is introduced into the system by the set in Eq. (2.28b); however, the generated sequence in Eq. (2.28a) must still be examined by the series of statistical tests to be discussed later.

The Complimentary Multiple-with-Carry (CMWC) generators are a further generalization of the recurrence relation in Eq. (2.17). They are generalizations of the form [1]

$$X_{n+1} = b - 1 - (aX_{n-r} + b_n)\, mod\, m, \text{for } n = 0,1,2,\ldots \tag{2.29a}$$

$$b_{n+1} = \frac{(aX_{n-r} + b_n)}{m}, \text{ for } n = 0,1,2,\ldots \text{ and integer } r \leq n. \tag{2.29b}$$

and produce the sequences

$$\{X_0, X_1, X_2, \ldots\ldots\ldots, X_n, \ldots\} \tag{2.30a}$$

$$\{b_0, b_1, b_2, \ldots\ldots\ldots, b_n, \ldots\} \tag{2.30b}$$

from the seeds $X_0, X_1, \ldots, X_{r-1}$ and b_o where a and m are integers. In Eq. (2.29) $ab^r - 1$ is chosen to be a prime number. Again, the sequence in Eq. (2.30a) is that of the generated random positive integers. Here the ideas are essentially the same as in the MWC but an improvement in the statistical properties is arrived at by the changes in Eq. (2.29a). The next classes of generators we consider are no longer centered around the ideas leading to the LCG.

Some additional random number generators operate on the binary forms of numbers (i.e., numbers expressed in base 2) in which the numbers are expressed in terms of 0's and 1's. Here the object is to sequentially generate random sequences of 0's and 1's, a sort of heads or tails process [1, 3]. In the operation of these algorithms, various shift operations which move the positions of the digits in a binary number and XOR binary arithmetic operations are used in the process of random sequence generation [1, 3]. One such variant based on these techniques is the so-called Xorshift generator, which is commonly used in numerical work. Based on the particular method of the Xorshift generator, a random sequence has been generated which is periodic with a period $2^{128} - 1$. This is very high compared to the LCG.

Another type of routine that has been developed and recently commonly applied in the study of random processes is the Mersenne Twister [1, 3]. This generates random sequences of integers which are periodic with a period $2^{19937} - 1$. Here $2^{19937} - 1$ is one of the Mersenne primes discussed in number theory. The routine was specifically designed to have a Mersenne prime as its period where Mersenne primes M_n are defined as prime number which can be written in the form $M_n = 2^n - 1$ for n is an integer.

2.4 EXPONENTIAL DISTRIBUTION OF RANDOM NUMBERS

Consider a set of random numbers with values between $x = 0$ and $x = 1$ that are uniformly distributed on the x-axis over the interval $0 < x < 1$. The probability density of the set of values x is given by the normalized functions [3]

$$p_u(x) = 1, \text{for } 0 < x < 1 \tag{2.31a}$$

$$p_u(x) = 0, \text{otherwise}, \tag{2.31b}$$

so that $p_u(x)dx$ is the probability of finding the random number x in the interval dx centered about x. (See Figure 2.1 for a plot of the density function $p_u(x)$). It then follows from Eq. (2.31) that

$$\int_0^1 p_u(x)\,dx = \int_0^1 dx = 1. \tag{2.32}$$

Another type of probability density of random numbers is for an exponential distribution defined over the y-axis and which is nonzero on the interval $0 \leq y < \infty$. In this probability density for the set of exponentially distributed random numbers, the distribution is characterized by the normalized probability density [3]

$$p_e(y) = e^{-y}, \text{ for } 0 \le y < \infty \tag{2.33a}$$

$$p_e(y) = 0, \text{ otherwise,} \tag{2.33b}$$

so that $p_e(y)dy$ is the probability of finding the random number y in the interval dy centered about y. (See Figure 2.1 for a plot of $p_e(y)$ versus y.) It then follows that the probability density is normalized so that

$$\int_0^\infty p_e(y)dy = \int_0^\infty e^{-y}dy = 1. \tag{2.34}$$

The two probability densities (i.e., the one for the uniform distribution defined in Eqs. (2.31) and (2.32) and the one for the exponential distributions defined in Eqs. (2.33) and (2.34)) are related to one another by a spatial transformation. In this regard, consider the invertible transformation between the x and y variables defined by [1, 3]

$$x = 1 - e^{-y} \text{ or } y = -\ln(1-x). \tag{2.35}$$

From Eq. (2.35) it is seen that the interval $0 < y < \infty$ is mapped onto the interval $0 < x < 1$ or the interval $0 < x < 1$ is mapped onto the interval $0 < y < \infty$ so that $y = 0$ is associated with $x = 0$ and $y = \infty$ is associated with $x = 1$. Consequently, under Eq. (2.35) it follows from Eqs. (2.31) through (2.34) that [3]

$$\int_0^1 p_u(x)dx = \int_0^1 dx = 1 = \int_0^\infty \frac{dx}{dy}dy = \int_0^\infty e^{-y}dy = \int_0^\infty p_e(y)dy = 1 \tag{2.36}$$

where from Eq. (2.35) we have used

$$\frac{dx}{dy} = e^{-y} = p_e(y). \tag{2.37}$$

In addition, we see from Eq. (2.37) that a subinterval of length Δx on the x –axis is mapped onto a subinterval Δy on the y –axis of length [3]

$$\Delta y = \frac{\Delta x}{e^{-y}} = \frac{\Delta x}{1-x}. \tag{2.38}$$

Consequently, both the Δx and Δy intervals in Eq. (2.38) are related to one another by

$$p_u(x)\Delta x = \Delta x = p_e(y)\Delta y = e^{-y}\Delta y \tag{2.39}$$

so that both intervals represent the same probability.

Consequently, by means of the transformations in Eq. (2.35) a random array of points $\{x_0, x_1, x_2, \ldots\ldots\ldots, x_N\}$ representing the uniform density $p_u(x)$ defined over $0 < x < 1$ can be converted into a random array of points $\{y_0, y_1, y_2, \ldots\ldots\ldots, y_N\}$ representing the exponential density $p_e(y)$ defined over $0 < y < \infty$. Specifically, between the two distributions of random points it follows that [3]

$$y_i = -\ln(1 - x_i) \quad \text{for } i = 0, 1, 2, \ldots\ldots\ldots, N \tag{2.40}$$

and from each point x_i of the uniform distribution a corresponding point y_i is generated by the mappings.

2.5 GAUSSIAN RANDOM NUMBERS

The ideas used in developing the one-dimensional set of exponentially distributed numbers are straightforwardly extended to other distributions and to the development of higher dimensional sets of random numbers with nonuniform spatial distributions. In this way, we now consider the generation of a nonuniform distribution of random numbers treated in higher dimensional systems. To see how this is done, consider mapping a uniform distribution of random numbers defined over the square region $0 < x_1, x_2 < 1$ into a Gaussian distribution of random numbers defined over the square region $0 < y_1, y_2 < \infty$. For our considerations, a Gaussian distribution with a maximum centered on $(y_1, y_2) = (0,0)$ will be generated by devising an appropriate mapping from the uniform distribution [3].

Specifically, we start by considering the uniform distribution of random points defined by the probability density [3]

$$p_u(x_1, x_2) = 1 \quad \text{for} \quad 0 < x_1, x_2 < 1 \tag{2.41}$$

$$p_u(x_1, x_2) = 0 \text{ , otherwise.}$$

The probability density is normalized so that

$$\iint_0^1 p_u(x_1, x_2) dx_1 dx_2 = 1. \tag{2.42}$$

Similarly, the Gaussian distribution of random points is defined by the probability density [3]

$$p_g(y_1, y_2) = \frac{1}{\sqrt{2\pi}} e^{-\frac{y_1^2}{2}} \frac{1}{\sqrt{2\pi}} e^{-\frac{y_2^2}{2}} \quad \text{for} \quad 0 < y_1, y_2 < \infty \tag{2.43}$$

$$p_g(y_1, y_2) = 0 \text{ , otherwise}$$

so that

$$\iint_0^\infty p_g(y_1, y_2) dy_1 dy_2 = 1. \tag{2.44}$$

Here the maximum of the density in Eq. (2.43) is at the origin of the (y_1, y_2) coordinate system, and the density is normalized over the square region $0 < y_1, y_2 < \infty$.

These two distributions of random points are related to one another through a mapping which sends the square $0 < x_1, x_2 < 1$ in Eq. (2.41) into the square $0 < y_1, y_2 < \infty$ in Eq. (2.43). Specifically, consider the mapping [3]

$$y_1 = \sqrt{-2 \ln x_1} \cos 2\pi x_2 \tag{2.45a}$$

$$y_2 = \sqrt{-2 \ln x_1} \sin 2\pi x_2 \tag{2.45b}$$

which is inverted to give [3]

$$x_1 = \exp\left[-\frac{1}{2}\left(y_1^2 + y_2^2\right)\right] \tag{2.46a}$$

$$x_2 = \frac{1}{2\pi}\arctan\frac{y_2}{y_1}. \tag{2.46b}$$

These two mappings project the probability densities into one another as well as their related random distributions of points.

To see this relationship, consider the following. Under the change of coordinates defined by Eq. (2.46) it follows that the Gaussian and uniform probability densities are related to one another by the standard Jacobian change of variables given by [3]

$$p_g(y_1,y_2)dy_1dy_2 = p_u(x_1,x_2)\left|\frac{\partial(x_1,x_2)}{\partial(y_1,y_2)}\right|dy_1dy_2 \tag{2.47}$$

$$= -p_u(x_1,x_2)Det\begin{vmatrix}\frac{\partial x_1}{\partial y_1} & \frac{\partial x_1}{\partial y_2}\\ \frac{\partial x_2}{\partial y_1} & \frac{\partial x_2}{\partial y_2}\end{vmatrix}dy_1dy_2$$

where $\frac{\partial(x_1,x_2)}{\partial(y_1,y_2)}$ is the Jacobian determinant. From Eq. (2.46) it then follows that

$$\frac{\partial(x_1,x_2)}{\partial(y_1,y_2)} = -\frac{1}{\sqrt{2\pi}}e^{-\frac{y_1^2}{2}}\frac{1}{\sqrt{2\pi}}e^{-\frac{y_2^2}{2}} \tag{2.48}$$

so that, taking $p_u(x_1,x_2)=1$ in the region $0<x_1,x_2<1$, from Eqs. (2.47) and (2.48) we find that over square $0<y_1,y_2<\infty$

$$p_g(y_1,y_2)dy_1dy_2 = \frac{1}{\sqrt{2\pi}}e^{-\frac{y_1^2}{2}}\frac{1}{\sqrt{2\pi}}e^{-\frac{y_2^2}{2}}dy_1dy_2 \tag{2.49}$$

$$= \frac{1}{\sqrt{2\pi}}e^{-\frac{y_1^2}{2}}dy_1\frac{1}{\sqrt{2\pi}}e^{-\frac{y_2^2}{2}}dy_2.$$

Consequently, if we are given a set of random points $\{(x_{1,i},x_{2,j}),\ \ for\, i,j=1,2,\ldots\ldots N\}$ which are uniformly distributed in the square $0<x_1,x_2<1$, then the transformation [3]

$$y_{1,i} = \sqrt{-2\ln x_{1,i}}\cos 2\pi x_{2,j} \tag{2.50a}$$

$$y_{2,j} = \sqrt{-2\ln x_{1,i}}\sin 2\pi x_{2,j} \tag{2.50b}$$

maps the point $(x_{1,i},x_{2,j})$ uniformly distributed in the $0<x_1,x_2<1$ square to the point $(y_{1,i},y_{2,j})$ that is Gaussian distributed in the $0<y_1,y_2<\infty$ square. Under this mapping, the set of random points $\{(x_{1,i},x_{2,j}),\ \ for\, i,j=1,2,\ldots\ldots N\}$ is uniformly distributed over the square $0<x_1,x_2<1$ while the generated set $\{(y_{1,i},y_{2,j}), for\, i,j=1,2,\ldots\ldots N\}$ is Gaussian distributed over the $0<y_1,y_2<\infty$ square and the maximum of the Gaussian density of points is at $(y_1,y_2)=(0,0)$.

2.6 MODELS OF SURFACE DISORDER

As an example of random number applications, we discuss the use of random numbers to generate simply disordered surfaces which are designed to have certain required statistical properties [4]. These can be considered as models for applications in simulations related to surface physics properties. In this regard, some applications include studies related to surface scattering, the properties associated with impurities binding at surfaces, and the study of surface waves which are bound excitations propagating on the interface between two separate media.

Specifically, we treat surfaces in which the topography is described by a single-valued surface profile function of the form [4]

$$z = \xi(x, y). \tag{2.51}$$

Here $z = \xi(x, y)$ gives the position of the surface defined about the infinite $x, y-$plane (i.e., the $z = 0$ plane), and it is understood that there are no tunnels or overhanging features on the surface. The statistical properties of the surface are described by the average of its profile function and the correlations between its points averaged over the surface. Consequently, averaging the profile over the entire surface gives

$$\langle \xi(x, y) \rangle = 0 \tag{2.52a}$$

so that on mean the surface is the $(x, y)-$plane. Products of the surface profile functions can, similarly, be averaged over the surface to characterize its correlations. In this way it is found that the pair correlation function has the form [4]

$$\langle \xi(x_1, y_1)\xi(x_2, y_2) \rangle = \sigma^2 W\left(\sqrt{(x_1 - x_2)^2 + (y_1 - y_2)^2}\right) \tag{2.52b}$$

where $W(0) = 1$, $d_{1,2} = \sqrt{(x_1 - x_2)^2 + (y_1 - y_2)^2}$ is the distance between the points (x_1, y_1) and (x_2, y_2), and σ^2 characterizes the deviation of the surface from the $(x, y)-$plane. At higher levels of correlation, the $k-$point correlation function defined between k different points on the $(x, y)-$plane takes the form [4]

$$\langle \xi(x_1, y_1)\xi(x_2, y_2)\ldots\ldots\ldots\xi(x_k, y_k) \rangle. \tag{2.52c}$$

In these expressions $\langle\ \rangle$ denotes an average over the infinite disordered surface. In computing the surface averages of the correlation functions in Eq. (2.52) the points (x_1, y_1), (x_2, y_2),........, (x_k, y_k) of the correlator are fixed relative to one another in their spatial separations and the correlator averages over the surface are determined as the fixed configurations of points move over the entire infinite $(x, y)-$plane. In this way the disorder of the entire rough surface is sampled in the average. As we shall see later, the pair correlation function and the higher order correlators computed in this way enter the treatment of many important dynamical processes on random surfaces.

One particular class of randomly rough surfaces is known as Gaussian random surfaces. These enter into many analytic treatments of surface disorder because their mathematical properties facilitate the application of field theory approaches in the determination of their physical properties. Specifically, these types of disorder allow for the application of Wick's Theorem. We shall now turn to a study of the generation of Gaussian random surfaces and a review of some of their important properties.

For Gaussian random rough surfaces, the correlators in Eq. (2.52c) of products of an even number of surface profile functions can be expressed in terms of sums involving the products of pair correlation functions. In this regard, Eq. (2.52c) becomes [4] for $k = 2n > 2$

$$\langle \xi(x_1, y_1)\xi(x_2, y_2)\ldots\ldots\ldots\xi(x_{2n}, y_{2n})\rangle = \tag{2.53a}$$

$$= \sum_P \langle \xi(x_{P1}, y_{P1})\xi(x_{P2}, y_{P2})\rangle\ldots\ldots\langle \xi(x_{P(2n-1)}, y_{P(2n-1)})\xi(x_{p2n}, y_{P2n})\rangle$$

where the sum over P represents a sum over all the different permutations of the indices $1, 2, 3, \ldots\ldots, 2n$. On the other hand, for general products of an odd number of surface profile functions, the correlators are zero so that [4]

$$\langle \xi(x_1, y_1)\xi(x_2, y_2)\ldots\ldots\ldots\xi(x_{2n+1}, y_{2n+1})\rangle = 0. \tag{2.53b}$$

In this way, later it is seen that all the properties of the random surface are ultimately determined by the pair correlation function in Eq. (2.52b). As a particular example of the factorization in Eq. (2.53a), consider the specific case of the four-point correlation function and its factorization, which is written as

$$\begin{aligned}\langle \xi(x_1, y_1)\xi(x_2, y_2)\xi(x_3, y_3)\xi(x_4, y_4)\rangle &= \langle \xi(x_1, y_1)\xi(x_2, y_2)\rangle\langle \xi(x_3, y_3)\xi(x_4, y_4)\rangle \\ &+ \langle \xi(x_1, y_1)\xi(x_3, y_3)\rangle\langle \xi(x_2, y_2)\xi(x_4, y_4)\rangle \\ &+ \langle \xi(x_1, y_1)\xi(x_4, y_4)\rangle\langle \xi(x_2, y_2)\xi(x_3, y_3)\rangle.\end{aligned} \tag{2.53c}$$

This example shows how the correlator separates into the sum of products involving all the possible productions of two pair correlator functions. We now turn to a general discussion of the numerical generation of Gaussian random surfaces as they are expressed in terms of any given pair correlation function. This process produces random surfaces which can be used in computer simulation studies based on Gaussian random surface profiles.

To generate a model of a Gaussian randomly rough surface, start by considering the $(x, y)-$plane. We begin our study by discretizing the $(x, y)-$plane by introducing an integration mesh. The points in the two-dimensional $(x, y)-$plane are discretized into a lattice of mesh points with lattice translation vectors of the form [4]

$$\vec{r}(m, n) = (m\hat{i} + n\hat{j})\Delta x \tag{2.54}$$

where $\Delta x \ll 1$ is an infinitesimal length, and $m, n = 0, \pm 1, \pm 2, \ldots$ range over the integers. Starting from the origin of the mesh at $(m, n) = (0, 0)$ all of the points, denoted by the integer pairs (m, n), in the lattice of the two-dimensional $(x, y)-$plane are then defined from the set of translation vectors in Eq. (2.54). Next, associate with each point (m, n) of the lattice mesh defined by Eq. (2.54) an independent, zero-mean, Gaussian distributed random number $X_{m,n}$ with a standard deviation of unity, defined by the properties that on average [3, 4]

$$\langle X_{m,n}\rangle = 0 \tag{2.55a}$$

and

$$\langle X_{m,n}X_{m',n'}\rangle = \delta_{m,m'}\delta_{n,n'} \tag{2.55b}$$

where $\delta_{n,m} = 1 \; if \; n = m$, $\delta_{n,m} = 0 \; if \; n \neq m$, and here $\langle \; \rangle$ represents an average over the set of random variables $X_{m,n}$. The set of random variables $X_{m,n}$ defined with these properties on the mesh of the (x, y)–plane can now be used in the limit $\Delta x \to 0$ to generate a rough surface profile function with Gaussian random statistics.

In terms of $X_{m,n}$ the surface profile function $\xi(\vec{r}(m,n))$ at $\vec{r}(m, n) = (m\hat{i} + n\hat{j})\Delta x$ can be written in the form [4]

$$\xi(\vec{r}(m,n)) = \sigma \sum_{k,l} W_{k,l} X_{k+m,l+n}. \tag{2.56}$$

In this linear form, the profile function in Eq. (2.56) can be shown—for a correctly defined set of coefficients $\{W_{k,l}\}$—to exhibit Gaussian random statistics in its average properties over the random surface with correlations characterized by Eqs. (2.52) and (2.53). In this regard, the Gaussian randomness of the surface follows from the statistical properties of the set of $\{X_{m,n}\}$. The detailed nature of the correlations, on the other hand, follows from relating $\{W_{k,l}\}$ to the pair correlator of interest.

Finally, in Eq. (2.56) $W_{m,n}$ must be related to the pair correlation function characterizing the correlation in the disorder of the surface. This relationship is set by the Fourier integral [4]

$$W_{m,n} = \Delta x \int \frac{d^2k}{(2\pi)^2} g^{1/2}(|\vec{k}|) e^{i\vec{k}\cdot\vec{r}(m,n)} \tag{2.57}$$

where from Eq. (2.52b) the pair correlation function enters the surface statistics through the relationship

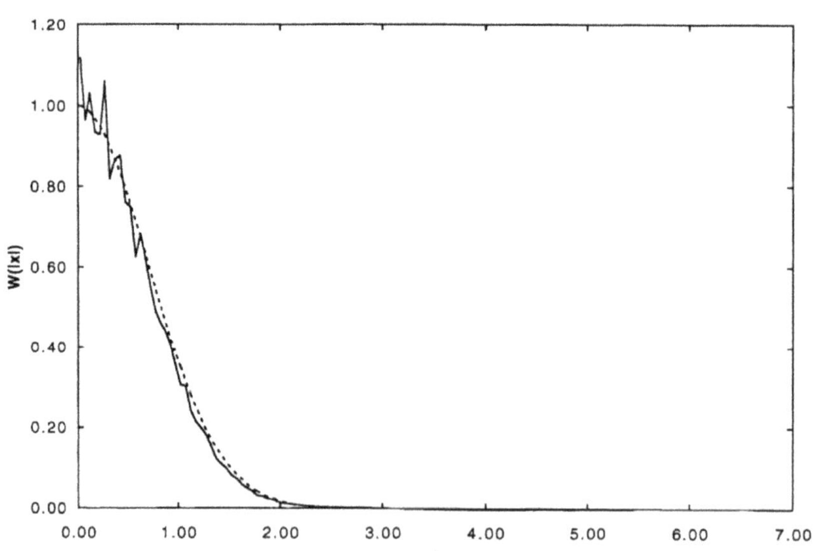

FIGURE 2.3 Plot of $W(|\vec{r}|) = W(|x|)$ versus $\frac{r}{a} = \frac{x}{a}$ measured along the x–axis in the notation of Ref. 4. The dashed line is the pair correlation function in Eq. (2.59) where we associate $x \leftrightarrow r$ between the notation of Ref. 4 and our notation in Eq. (2.59). The solid line is the result of the simulation generated surface curve where again we make the association $x \leftrightarrow r$ between the notation of Ref. 4 and our notation. The agreement between the correlations of the generated surfaces and Eq. (2.59) is good. Reproduced from [4]. Copyright 1997 SPIE.

$$g\left(|\vec{q}|\right) = \int d^2 r W\left(|\vec{r}|\right) e^{-i\vec{q}\cdot\vec{r}}. \tag{2.58}$$

Consequently, from Eqs. (2.56) through (2.58), a Gaussian random surface based on the pair correlation function in Eq. (2.52b) is obtained.

In Ref. 4 the theory in Eqs. (2.54) through (2.58) was used to study a variety of statistical properties of Gaussian random surfaces characterized by different pair correlation functions. Here we present a single illustrative result from the study presented there [4]. In Figure 2.3 results are shown for a pair correlation function in Eq. (2.52b) of the form

$$W\left(\vec{r}\right) = e^{-\frac{r^2}{a^2}}. \tag{2.59}$$

In the figure results are shown for the pair correlation function (dashed line) in Eq. (2.59) and for the same correlation function calculated from the simulated surface profile (solid line) in Eq. (2.56). The small differences in the two profiles are from the numerical approximations. For the details of the numerical parameters used in the simulation, the reader is referred to the original reference [4].

REFERENCES

1. Kneusel, R. T. 2018. *Random Numbers and Computers* (Springer Nature, Cham).
2. Pang, T. 2016. *An Introduction to Quantum Monte Carlo Methods* (Morgan & Claypool Publishers, San Rafael).
3. Press, W. H., Taukolsky, S. A., Vetterling, W. T., and Flannery, B. P. 1997. *Numerical Recipes in Fortran 77: The Art of Scientific Computing*, 2nd Edition (Cambridge University Press, Cambridge).
4. Simeonov, S., McGurn, A., and Maradudin, A. 1997. The Transverse Correlation Length for Randomly Rough Surfaces: Two-Dimensional Roughness. *Proceedings of SPIE* 3141, 152–163.

3 Disorder in Crystals

In this chapter, a discussion is given of impurities introduced into crystalline materials [1–11]. These include single-site and multiple-site replacement impurities [1–6] occupying crystal lattice sites as well as a variety of so-called line imperfections or dislocations [7–12], stacking faults [7], and grain boundaries [7] introduced into the structure of the crystal lattice itself. While the first class of replacement impurities does not change the nature of the crystal lattice, the second type involves modification of the basic geometry of the lattice structure. Each of these types of imperfections is commonly encountered in the study of crystals in the real world, and each can contribute to a variety of interesting physical processes and properties exhibited by crystals. In fact, many of these imperfections occur as necessary participants in the condensation processes involved in the formation of a crystal from the liquid or gas state [7]. Once the crystal has been formed, however, impurities also continue to act to determine the basic mechanical and plasticity properties [3–7] of the fully condensed crystalline solids as well as entering into setting the transport and spectroscopic processes of the material.

A focus in our treatment is on an introduction to basic theoretical formulations that have been developed to understand the physical properties of crystals originating in the presence of each of the two basic types of imperfections. In this regard, for replacement impurities a number of basic models of electronic, vibrational, mechanical, and magnetic materials are introduced and considered in a variety of different spatial dimensions [1–12]. Here the focus is on the effects of site disorder on the dispersive properties of the excitations. The study of these effects involves a basic introduction to Green's function methods and their applications in the treatment of replacement imperfections. On the other hand, the effects of imperfections in the lattice geometry are handled in terms of the ideas of the theory of continuum elasticity which allows us to determine some of the general features of the stresses and strains introduced into the crystal lattice by a distortion of the crystal lattice geometry. This leads, eventually, to a consideration of the theory of plasticity which develops the mechanisms involved in the irreversible deformation of crystals.

Site and multiple site impurities are created in the crystal by atomic replacement in the crystalline lattice [1–7], and the discussions of these site imperfections focus on their influence on the propagation of excitations in the crystal as the excitations move through the disordered media. These are responsible for many of the spectroscopic properties exhibited by impurity-containing materials. Illustrations are made for excitations propagating in electronic, vibrational, and magnetic (spin) media [1–5, 7] which are shown to be closely related to the excitations of the general form of a so-called tight-binding Hamiltonian. The discussions presented involve an introduction to basic Green's function theory [4] and its applications in the determination of the dispersion relations of the modes of a tight-binding model system, with a focus on analytic results. This involves a study of the energies of the modal excitations as well as the lifetimes of the excitations arising from their interaction with the impurities in the crystal lattice. Treatments involving both classical and quantum field theory approaches are included in our treatments [1, 4], and the methods considered are ultimately generalized, for systems of uniformly distributed impurities, to a mean field theory based treatment of Green's functions known as the coherent potential approximation.

Discussions are given for crystalline materials in 1−, 2−, and 3−dimensions with a focus on the density of states of these systems and the localized bound state modes associated with their imperfections [1–7]. The localized impurity modes are shown under specific conditions to develop outside the continuum bands of extended state excitations and do not extend infinitely over the space of the crystal. Consequently, localized modes are non-propagating [1–7], characterized in terms of both the properties of the atomic impurities and the materials in which they are embedded, and at

DOI: 10.1201/9781032674919-3

nonzero temperatures generally do not participate in transport processes. In the present discussion, impurities are studied in terms of simple models treated in the context of solvable analytic theories. The development of computer simulation techniques for applications in more complex impurity models is not discussed here but is the topic of another chapter [1–7].

An introduction is given to the Anderson model for the discussion of localized magnetic impurities in metallic systems [2–6]. A focus is on understanding the development of magnetic moments on the impurities arising from their interaction with conduction electrons of a host metal. This is a phenomenon that has been of particular interest in the study of impurities with partially filled d- and f-shell localized electron states introduced into metallic materials characterized by the electron gas model. The conditions for magnetic moments to exist on the impurities lead to the rewriting of the Anderson model in terms of the so-called s-d model [1–4]. This model describes the electron gas as interacting with the magnetic moments developed on the impurity via spin dependent interactions. The study of the s-d model in turn leads to the treatment of the Kondo effect or Kondo problem which is observed in low temperature (e.g., typically temperatures less than $60^{\circ} K$) resistivity anomalies of so-called Kondo materials [1–6].

A brief introduction to the Kondo problem is given [2, 3, 5, 6]. This is observed in metals containing magnetic impurities as a low-temperature anomalous increase in the resistivity with decreasing temperature. In general, in the absence of magnetic impurities, metallic systems exhibit a decreasing resistivity with decreasing temperatures, ultimately ending at zero temperature in a residual concentration dependent impurity resistivity. This general behavior is known as Matthiessen's rule [1–3]. The temperature decrease in resistivity arises due to the freezing out of electron-phonon scattering in the systems as the temperature decreases and the phonon population decreases, while the concentration dependent impurity scattering is generally temperature independent. On the other hand, in the Kondo effect [5] it is found that the scattering from localized magnetic moments developed on the impurities embedded in the metal accounts for the increase in the resistivity with decreasing temperature. In this regard, as the temperature decreases, unlike the decrease in phonon scattering, the magnetic scattering increases.

In addition, an introduction is given to the basic ideas of line imperfections [7–12] (also known as dislocations) in the crystal structure as well as to stacking faults, grain boundaries, and the operation of slip planes. These features of disorder in crystals are fundamental in determining their mechanical and plasticity properties. A detailed discussion of the elastic properties of screw and edge dislocations is presented in terms of classical continuum elasticity theory [7–12]. Discussions of the breakdown of the elasticity theory description of dislocations are given with a brief presentation provided of the atomic theories needed for a complete treatment of the dislocations' motion. A focus is on an account of the role of dislocations in the plastic deformation of crystals [7]. This plastic deformation is explained as arising from the motion of dislocations along their slip planes. The goal of the discussions is the development of the basic models and properties of the line imperfections in the crystal lattice.

In addition to setting the plastic properties of crystals, the condensation of atoms into a crystalline form often depends to a great extent on the presence of such types of line imperfections. The role of the dislocations in facilitating the condensation of solids at their interfaces with liquid and gas phases is explained [7]. This role is shown to make the formation of perfect crystals without dislocations a most difficult task.

3.1 SINGLE SITE IMPURITIES

We begin with the treatment of single site impurities in crystalline materials focusing on their electronic, vibrational, and magnetic properties and the elementary excitations associated with them [1–6]. These features are considered in the context of the study of the linear dynamics of excitations as it is described by simple models, and for a consideration of the effects of single site impurities introduced into such systems. In this respect, the study of nonlinear dynamics will be the topic of later chapters. A goal is to establish a general mathematical formulation [4] which can be easily extended to the study of other more complex dynamical models. Consequently, many of the ideas we treat here can be directly generalized to handle specific models of interest in practical applications. The focus is on the determination of new impurity modes associated with bound states located about the impurity site [4]. These consist of excitations bound and localized about the impurities and are found at energies outside the bands of continuum excitation of the system in the absence of the site impurities.

Our initial considerations are of a tight-binding model of electrons propagating through a crystal [4] In this model, the electrons move through the material by hopping between nearest neighbor atomic sites in the crystalline lattice. This is done in the context of the independent electron model, which is often applied in the electron gas description of electron conductivity. The model involves linear dynamics of the electrons in which electron-electron interactions are ignored [1, 4], and its form and ideas can be directly extended to a consideration of the phonon modes in vibrational materials and magnons in magnetic materials [1–3].

In the electron gas treatment [1–4], the electrons do not interact with one another, and each electron moves through the crystal with a single particle energy having a dispersion relation of the form

$$\epsilon_{\vec{p}} = \frac{p^2}{2m} \tag{3.1}$$

where $\vec{p}$ is the momentum of the electron and m is its effective mass. The electrons in the gas then occupy the energy orbits of the system according to the Pauli principle. On the other hand, in the tight-binding model treatment the electron energy levels in Eq. (3.1) are replaced by those of the single electron dispersion relation obtained from the tight-binding Hamiltonian. Aside from the differences in the dispersive properties of the electrons in the two models, the treatment of the two is handled in much the same way.

Closely related to the tight-binding model of electrons are the models developed to handle lattice vibrations (i.e., phonon modes) and spin waves (i.e., magnon modes) [1–4]. These models in their simplest, basic forms look much like the tight-binding model except now they describe boson excitations rather than fermion excitations [4]. Nevertheless, in spite of their difference in statistics, the excitation spectra of electrons, phonons, and magnons look similar to one another, and the mathematics of their site impurity problems are much the same. Consequently, we will first focus on the electron systems and after this deal with its modifications to handle phonon and magnon systems.

In our considerations, we will first treat one-dimensional models of electrons, phonons, and magnons [1–4]. The one-dimensional models are the simplest systems to treat and illustrate the essential aspects of the theory of impurity states. Once these models are mastered, the higher dimensional systems are easier to understand.

3.2 TIGHT-BINDING MODEL OF ELECTRONS

We begin by considering the one-dimensional tight-binding model of electrons [1–4]. In the treatment of this system, an atomic chain composed of identical atoms is studied in which the conduction electrons of the chain are modeled as hopping between the atomic sites. This is initially done in the absence of impurities and is then followed by a consideration of atomic chains containing impurity atoms introduced into the system by atomic substitution. For these studies, the model is treated at zero temperature, and the focus is on the dynamics of the conduction electrons and the electrons associated with the impurity atoms.

Consider a chain of N identical atoms with atomic site labels $1,2,3,\ldots,N$ taken in the limit that $N \to \infty$ and in which the system is subject to periodic boundary conditions. In the tight-binding model for this chain, an electron moves between the atoms by hopping between nearest-neighbor sites, and the sites 1 and N are treated as nearest neighbors. (See Figure 3.1 for a schematic of the atomic chain.) Each electron then moves independently of the other electrons in the system, i.e., there is no electron-electron scattering. The tight-binding Hamiltonian describing the electron motion is then of the form [1–6]

$$H_0 = \sum_{n=1}^{N} U \,|\, n\rangle\langle n \,| + \sum_{n=1}^{N} t\left(|\, n\rangle\langle n+1 \,| + |\, n+1\rangle\langle n \,|\right) \tag{3.2}$$

where $|\,n\rangle$ and $\langle n\,|$ represent the electron site basis states of the nth atom on the chain, $\langle n|m\rangle = \delta_{n,m}$, and the periodic boundary conditions require that $|\,N+1\rangle = |\,1\rangle$ and $\langle N+1\,| = \langle 1\,|$. In this formulation, the first term on the right represents an energy the electron receives upon occupying a site of the lattice whereas the second term represents an energy the electron receives as it travels along the chain.

The solutions for the single electron wavefunctions or orbitals of the Hamiltonian in Eq. (3.2) are of the form [1–4]

$$|\,\epsilon_k\rangle = \frac{1}{\sqrt{N}} \sum_{n=1}^{N} e^{ikn} \,|\, n\rangle \tag{3.3a}$$

and

$$\langle \epsilon_k \,| = \frac{1}{\sqrt{N}} \sum_{n=1}^{N} e^{-ikn} \langle n \,| \tag{3.3b}$$

where we have taken a unit lattice constant (i.e., nearest neighbor separation on the lattice) and for periodic boundary conditions on the chain it is required that

$$e^{-ik(N+1)} = e^{-ik}. \tag{3.3c}$$

Consequently, the values of k for the solutions in Eq. (3.3) are given by $-\pi \le k = \frac{2\pi}{N} m \le \pi$ where m is an integer. Substituting, for example, Eq. (3.3a) into Eq. (3.2), it then follows that

$$H_0 \,|\, \epsilon_k\rangle = \epsilon_k \,|\, \epsilon_k\rangle \tag{3.4}$$

where the eigenvalue for the solution of the equation is [4]

$$\epsilon_k = U + 2t\cos k. \tag{3.5}$$

In addition, we find an orthonormality relationship between the eigenstates given by

$$\langle \epsilon_k | \epsilon_l \rangle = \delta_{k,l} \tag{3.6}$$

so that the states of the chain in Eq. (3.3) form a complete basis of the system.

In addition to the relations in Eqs. (3.2) through (3.6), we can also define a Green's function for Eq. (3.2). This is given by the expression [1–6]

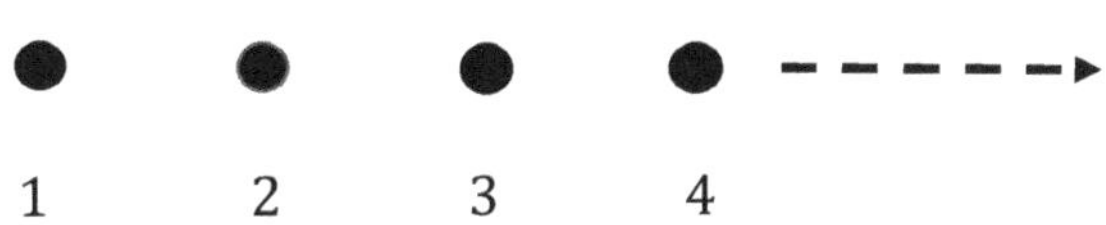

FIGURE 3.1 Chain of sites labeled $n = 1, 2, 3, \ldots.., N$ where $N \to \infty$.

$$G_0(t) = e^{-i\frac{H_0}{\hbar}t}, \tag{3.7}$$

and represents the time translation operator of the system, translating the system through the time interval starting at $t = 0$ and ending at $t = t$. Specifically, in this representation the probability amplitude for an electron to propagate from the state $|n\rangle$ to the state $|m\rangle$ during the time interval t is given by

$$\left\langle m\left|e^{-i\frac{H_0}{\hbar}t}\right|n\right\rangle = \langle m|G_0(t)|n\rangle. \tag{3.8}$$

We shall soon see that the basic form of the Green's functions in Eq. (3.7) and its associated transition matrices in Eq. (3.8) are most helpful in studying the properties of the eigenmodes discussed in Eqs. (3.3) through (3.6) as well as the nature of the impurity states developed in our model upon the introduction of impurities. In this respect, the Green's function representation allows for an understanding of the detailed dynamical properties of the Hamiltonian characterizing the system.

Of interest regarding the Green's function defined in Eq. (3.8) is its Fourier transformation in time. This is given by the expression [1–6]

$$G_0(E) = \frac{1}{i\hbar}\int_0^\infty dt e^{i\frac{E}{\hbar}t} G_0(t) = \frac{1}{i\hbar}\int_0^\infty dt e^{i\frac{E}{\hbar}t} e^{-i\frac{H_0}{\hbar}t} \tag{3.9}$$

$$= \frac{1}{E - H_0} = \sum_k \frac{|\epsilon_k\rangle\langle\epsilon_k|}{E - \epsilon_k}$$

where the sum over k runs over the eigenmodes of the Hamiltonian obtained from Eq. (3.4), $E = \hbar\omega$ is the energy of the Green's function, ω is the frequency of the Fourier transform, and 1 in the second to last equality on the right is the unit matrix. This form of the Green's function, written in terms of the eigenmodes and eigen-energies of the chain in the absence of impurities, becomes important in understanding the dispersion relations of the eigenmodes of our Hamiltonian. It is also valuable in determining how these change upon the introduction of impurities into the system. These properties of the systems follow directly from the nature of the energy singularities of the Green's function considered as a function of E.

In this regard, notice from Eq. (3.9) that $G_0(E)$ is singular at the eigenvalues of Eq. (3.4). Because of this, it is customary to analytically continue the energy E into the complex energy plane, i.e., taking $E \to E + i\epsilon$ where $\epsilon \to 0^+$. This reduces the singular nature of Eq. (3.9) on the real energy axis and from Eq. (3.7) is seen to introduce an infinitesimal time rate of decay into the time evolution of the system [4–6]. Applying this in Eq. (3.9) it then follows that

$$G_0(E + i\epsilon) = \sum_k \frac{|\epsilon_k\rangle\langle\epsilon_k|}{E - \epsilon_k + i\epsilon} = \sum_k |\epsilon_k\rangle\langle\epsilon_k| \left[\frac{E - \epsilon_k}{(E - \epsilon_k)^2 + \epsilon^2} + \frac{-i\epsilon}{(E - \epsilon_k)^2 + \epsilon^2}\right] \tag{3.10}$$

$$= \sum_k |\epsilon_k\rangle\langle\epsilon_k| \left[\frac{E - \epsilon_k}{(E - \epsilon_k)^2 + \epsilon^2}\right] - i\pi\sum_k |\epsilon_k\rangle\langle\epsilon_k|\delta(E - \epsilon_k),$$

and, as a result, the expression in Eq. (3.10) is obtained as the principal value of the Green's function in the limit that $\epsilon \to 0^+$. Consequently, the imaginary part of the Green's function is found to be intimately related to the singularities at the eigenvalues generated in Eq. (3.4). These develop as poles in the complex energy plane.

Taking the imaginary parts of both sides of Eq. (3.10) it follows that [1–6]

$$\sum_k |\epsilon_k\rangle\langle\epsilon_k|\delta(E-\epsilon_k) = -\frac{1}{\pi} ImG_0(E+i\epsilon) \tag{3.11}$$

where the sum on the left of Eq. (3.11) is now shown to be related to the density of electronic states of the chain. This demonstration involves a study of Eq. (3.11) expressed in the $\{|n\rangle\}$ basis states.

To relate Eq. (3.11) to the density of states, consider its diagonal matrix elements in the $\{|n\rangle\}$ basis states. These elements are of the general form [1–6]

$$\left\langle n\left|-\frac{1}{\pi} ImG_0(E+i\epsilon)\right|n\right\rangle = -\frac{1}{\pi} Im\langle n|G_0(E+i\epsilon)|n\rangle \tag{3.12}$$

$$= \sum_k \langle n|\epsilon_k\rangle\langle\epsilon_k|n\rangle\delta(E-\epsilon_k) = \sum_k \langle\epsilon_k|n\rangle\langle n|\epsilon_k\rangle\delta(E-\epsilon_k).$$

Next by tracing over the matrix elements $-\frac{1}{\pi} Im\langle n|G_0(E+i\epsilon)|n\rangle$ it follows that

$$Tr\left\{-\frac{1}{\pi}\langle n|G_0(E+i\epsilon)|n\rangle\right\} = -\frac{1}{\pi} Tr\, Im\langle n|G_0(E+i\epsilon)|n\rangle \tag{3.13}$$

$$= \sum_{k,n} \langle\epsilon_k|n\rangle\langle n|\epsilon_k\rangle\delta(E-\epsilon_k) = \sum_k \delta(E-\epsilon_k)$$

where Tr represents the trace of the Green's function matrix over the $\{|n\rangle\}$ basis states. The sum over delta functions on the far right of Eq. (3.13) is now shown to be a realization of the density of energy states of the system.

The density of electronic states $D(E)$ is defined as the number of electronic energy levels per energy so that the number of energy levels $N_{[\epsilon_a,\epsilon_b]}$ in the energy interval $[\epsilon_a,\epsilon_b]$ is obtained from

$$N_{[\epsilon_a,\epsilon_b]} = \int_{\epsilon_a}^{\epsilon_b} dE D(E). \tag{3.14}$$

From Eq. (3.13) it then follows that [4]

$$N_{[\epsilon_a,\epsilon_b]} = \int_{\epsilon_a}^{\epsilon_b} dE \sum_i \delta(E-\epsilon_i) = -\frac{1}{\pi} Tr \int_{\epsilon_a}^{\epsilon_b} dE\; Im\langle n|G_0(E+i\epsilon)|n\rangle, \tag{3.15}$$

and we can identify

$$D(E) = \lim_{\epsilon\to 0^+}\left\{-\frac{1}{\pi} Tr\, Im\langle n|G_0(E+i\epsilon)|n\rangle\right\} \tag{3.16a}$$

in the limit that $\epsilon \to 0^+$ and for Tr representing the trace over the $\{|n\rangle\}$ basis states. The resulting expression in Eq. (3.16a) expresses the density of states of the Hamiltonian in terms of its eigenvectors and eigenvalues or alternatively in terms of its Green's function.

Notice that the density of states in Eq. (3.16a) is a trace sum of terms of the form [4]

$$\lim_{\epsilon\to 0^+}\left\{-\frac{1}{\pi} Im\langle n|G_0(E+i\epsilon)|n\rangle\right\} \tag{3.16b}$$

over the electron site states $\{|n\rangle\}$. It is therefore natural to define the density of states at each site as

$$D_n(E) = \lim_{\epsilon\to 0^+}\left\{-\frac{1}{\pi} Im\langle n|G_0(E+i\epsilon)|n\rangle\right\} \tag{3.17}$$

so that Eq. (3.16a) takes the form

$$D(E) = \sum_{n=1}^{N} D_n(E) \tag{3.18}$$

where n sums over the lattice sites. For the case of the uniform chain with periodic boundary conditions, we note that the site density of states $D_n(E)$ of each of the sites on the chain is identical.

For the tight-binding model of Eq. (3.2) in the absence of impurities, exact analytic forms for Eqs. (3.16) through (3.18) can be obtained. Specifically, from Eq. (3.10) we consider [1–6]

$$\langle n|G_0|m\rangle = \sum_k \frac{\langle n|\epsilon_k\rangle\langle\epsilon_k|m\rangle}{E-\epsilon_k+i\epsilon} = \frac{1}{N}\sum_l \frac{e^{i\frac{2\pi l}{N}(n-m)}}{E-\epsilon_{\frac{2\pi l}{N}}+i\epsilon} \tag{3.19}$$

where $-\pi \le k = \frac{2\pi}{N}l \le \pi$ for l an integer and we have used $\langle n|\epsilon_k\rangle = \frac{1}{\sqrt{N}}e^{i\frac{2\pi l}{N}n}$ and $\langle\epsilon_k|m\rangle = \frac{1}{\sqrt{N}}e^{-i\frac{2\pi l}{N}m}$. In the continuum limit that $N\to\infty$ the expression on the far right of Eq. (3.19) becomes an integral given by

$$\frac{1}{N}\sum_l \frac{e^{i\frac{2\pi l}{N}(n-m)}}{E-\epsilon_{\frac{2\pi l}{N}}+i\epsilon} \to \int_{-\pi}^{\pi}\frac{dk}{2\pi}\frac{e^{ik(n-m)}}{E-U-2t\cos k} = \int_0^{2\pi}\frac{dk}{2\pi}\frac{e^{ik(n-m)}}{E-U-2t\cos k} \tag{3.20}$$

so that

$$\langle n|G_0|m\rangle = \int_0^{2\pi}\frac{dk}{2\pi}\frac{e^{ik(n-m)}}{E-U-2t\cos k}. \tag{3.21}$$

The integral in Eq. (3.21) is evaluated in closed analytic form using complex variables.

To make the evaluation of Eq. (3.21), consider a change of variables in Eq. (3.21) made by defining the new complex variable [1–6]

$$z = e^{ik} \text{ so that } dk = \frac{dz}{iz}. \tag{3.22}$$

Applying this in Eq. (3.21) we then find that

$$\langle n|G_0|m\rangle = \int_0^{2\pi}\frac{dk}{2\pi}\frac{e^{ik(n-m)}}{E-U-2t\cos k} = \oint\frac{dz}{2\pi iz}\frac{z^{|n-m|}}{E-U-t\left(z+\frac{1}{z}\right)}, \tag{3.23}$$

where the integration on the far left is performed around a unit circle centered about the origin in the complex plane.

To evaluate the complex integral in Eq. (2.23) we need to consider two separate cases. In the first case, which is of most interest to us, we treat the energy E as being located outside the band of continuum energy states given in Eqs. (3.2) through (3.5). For this case $(E-U)^2 > 4t^2$ and the integral is evaluated using residues. In this way, it follows that [1–6]

$$\langle n|G_0|m\rangle = \oint \frac{dz}{2\pi i} \frac{z^{|n-m|}}{(E-U)z - tz^2 - t} = -\oint \frac{dz}{2\pi i} \frac{z^{|n-m|}}{t(z-z_+)(z-z_-)} \tag{3.24a}$$

where

$$z_+ = \frac{E-U+\sqrt{(E-U)^2-4t^2}}{2t} \tag{3.24b}$$

and

$$z_- = \frac{E-U-\sqrt{(E-U)^2-4t^2}}{2t}. \tag{3.24c}$$

From the residue theorem we find that for $E-U>0$

$$\langle n|G_0|m\rangle = -2\pi i\left[\frac{1}{2\pi i}\frac{z_-^{n-m}}{t(z_- - z_+)}\right] = \left[\frac{E-U}{2t} - \frac{t}{|t|}\sqrt{\left(\frac{E-U}{2t}\right)^2 - 1}\right]^{|n-m|} \frac{1}{\sqrt{(E-U)^2-4t^2}}, \tag{3.25a}$$

and for $E-U<0$ we find that

$$\langle n|G_0|m\rangle = -2\pi i\left[\frac{1}{2\pi i}\frac{z_+^{n-m}}{t(z_+ - z_-)}\right] = -\left[\frac{E-U}{2t} + \frac{t}{|t|}\sqrt{\left(\frac{E-U}{2t}\right)^2 - 1}\right]^{|n-m|} \frac{1}{\sqrt{(E-U)^2-4t^2}}. \tag{3.25b}$$

For the particular case in which $n=m$ the Green's function is seen to be antisymmetric in $E-U$.

As we shall presently see, the energies of the bound state impurity modes are located in the region $(E-U)^2 > 4t^2$ which is outside the continuum band of energy states. To calculate these new modes of the impurity system, we need to use the Green's function solutions in Eq. (3.25). This, however, is not the only region of our interest, and we shall briefly mention Green's function solutions for the case in which $(E-U)^2 < 4t^2$. These are related to the density of states of the continuum modes and to so-called resonant modes in the continuum. This is followed by considerations of bound state impurity modes, which will be treated in greater detail.

The second set of Green's function solutions for $(E-U)^2 < 4t^2$ do not enter into the impurity mode calculations and will only briefly be discussed here. These second solutions, however, are important as they enter into the renormalization of the solutions within the band of continuum states. Specifically, they are used to calculate the so-called resonant states which are almost bound states within the continuum band. These resonant states exist over long but finite times and eventually decay into the continuum. The reader interested in resonant states is referred to the literature [2–6] for a further discussion of these states. Here we only list the forms of the second set of Green's functions and discuss their applications in evaluating the density of states of the system in the absence of impurities.

In the second case of Eq. (3.23) the energy E is in the continuum of states. Consequently, for this limit $(E-U)^2 < 4t^2$ and the solutions of the Green's functions are found to involve branch cut considerations. From the literature [2–6] the case $U,t > 0$ is known to have the general form [1–6]

$$\langle n|G_0|m\rangle = -i\left[-\frac{E-U}{2t} + i\sqrt{1-\left(\frac{E-U}{2t}\right)^2}\right]^{|n-m|} \frac{1}{\sqrt{4t^2-(E-U)^2}}. \tag{3.26}$$

Notice the solution in Eq. (3.26) is a complex function, and this allows us to determine the density of continuum energy states from Eqs. (3.16) through (3.18).

As an important result from Eq. (3.26), now knowing the form of the continuum band Green's function, the density of states of the continuum energy spectrum of H_0 can be determined from Eqs. (3.16) and (3.17). In this regard, it follows from Eq. (3.26) that the diagonal site Green's function matrix elements are given by the form [1–6]

$$\langle n|G_0|n\rangle = -i\frac{1}{\sqrt{4t^2-(E-U)^2}}. \tag{3.27}$$

Consequently, from Eq. (3.16a) we next find within the band of continuum states that the density of states of our system is given by

$$D(E) = \lim_{\epsilon \to 0^+}\left\{-\frac{1}{\pi} Tr\, Im\langle n|G_0(E+i\epsilon)|n\rangle\right\}. \tag{3.28a}$$

$$= N\frac{1}{\pi}\frac{1}{\sqrt{4t^2-(E-U)^2}} = N\frac{1}{2\pi|t|}\frac{1}{\sqrt{1-\left(\frac{E-U}{2t}\right)^2}}.$$

Consistent with this result it follows from Eq. (3.17) that the site density of states at each site of the chain is [1–6]

$$D_n(E) = \frac{1}{\pi}\frac{1}{\sqrt{4t^2-(E-U)^2}} = \frac{1}{2\pi|t|}\frac{1}{\sqrt{1-\left(\frac{E-U}{2t}\right)^2}}. \tag{3.28b}$$

Notice that the independence of $D_n(E)$ on the site index n is a result of the translational symmetry of the chain.

In Figure 3.2 a plot of the density of states per site of the tight-binding chain of atoms is given as a function of $\frac{E-U}{2t}$. It is found to be a symmetric function which is largest at the top and bottom of the continuum band of states, being a minimum at the center of the band.

For a further treatment of the details of the tight-binding solutions in the continuum band, the reader is referred to Ref. 4. We shall now extend our consideration to a detailed treatment of the tight-binding model with impurities.

3.2.1 Bound Impurity Modes in the Tight-Biding Model

The addition of an impurity to the tight-binding Hamiltonian by replacing an atom at one of its sites with an impurity atom is made by considering a new Hamiltonian of the form [1–6]

$$H = H_0 + H_I, \tag{3.29a}$$

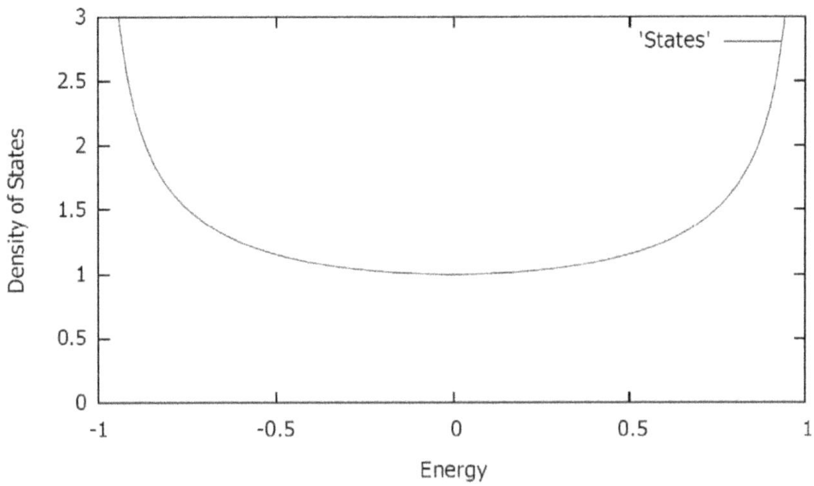

FIGURE 3.2 Plot of the Density of States per site, $\frac{2\pi|t|}{N}D(E)=\frac{1}{\sqrt{1-\left(\frac{E-U}{2t}\right)^2}}$, versus the Energy, E, for the electron tight-binding model.

and its associated eigenvalue problem is defined by

$$[H_0+H_I]|\epsilon_i\rangle=\epsilon_i|\epsilon_i\rangle. \tag{3.29b}$$

Here H_0 is the tight-binding Hamiltonian in Eq. (3.2), H_I is the change in H_0 due to the impurity replacement, $|\epsilon_i\rangle$ is an eigenvector of the impurity problem, and ϵ_i is the corresponding eigenvalue. The focus in discussing this problem is to determine the energies of the bound state modes and the form of the density of states outside the continuum band. This involves discussing the Green's functions of the new system, which are related to the density of energy states in the energy regions outside the continuum energy band of H_0. We begin by looking at the Green's functions of the impurity Hamiltonian in Eq. (3.29).

As in Eq. (3.7) the Green's function of our impurity Hamiltonian is defined from its time translation operator. This has the form [1–6]

$$G_0(t)=e^{-i\frac{H}{\hbar}t}, \tag{3.30}$$

and its time Fourier transformation is given, from similar considerations to those in Eq. (3.9), by

$$\begin{aligned}G(E)&=\frac{1}{i\hbar}\int_0^\infty dte^{i\frac{E}{\hbar}t}G(t)=\frac{1}{i\hbar}\int_0^\infty dte^{i\frac{E}{\hbar}t}e^{-i\frac{H}{\hbar}t}\\&=\frac{1}{E-H}=\sum_i\frac{|\epsilon_i\rangle\langle\epsilon_i|}{E-\epsilon_i}.\end{aligned} \tag{3.31}$$

Notice here that to obtain Eq. (3.31) we have just replaced H_0 in Eq. (3.9) with the new Hamiltonian H. Consequently, from these forms

$$G(E)=\frac{1}{E-H}=\sum_i\frac{|\epsilon_i\rangle\langle\epsilon_i|}{E-\epsilon_i}, \tag{3.32}$$

and it follows that the Green's function is singular at the new eigenvalues of the modified system so that we again analytically continue the energy E into the complex plane by taking $E \to E + i\epsilon$ where $\epsilon \to 0^+$.

Let us begin the treatment of Eq. (3.32) by rewriting it in the form [4]

$$[E - H]G(E) = 1 \tag{3.33a}$$

where 1 on the right is the unit matrix. Next, using $H = H_0 + H_I$ Eq. (3.33a), it can be written as

$$[E - H_0]G(E) = 1 + H_I G(E) \tag{3.33b}$$

Dividing Eq. (3.33b) by $[E - H_0]$ and using the definition $G_0(E) = \dfrac{1}{E - H_0}$ in Eq. (3.9) then gives [4]

$$\begin{aligned} G(E) &= G_0(E) + G_0(E)H_I G(E) \qquad (3.34)\\ &= G_0(E) + G_0(E)H_I G_0(E) + G_0(E)H_I G_0(E)H_I G_0(E) + \ldots\\ &= G_0(E) + G_0(E)H_I G_0(E)\sum_{i=0}^{\infty}\left[H_I G_0(E)\right]^i\\ &= G_0(E) + G_0(E)H_I G_0(E)\frac{1}{1 - H_I G_0(E)}. \end{aligned}$$

The final form of $G(E)$ on the far-right side of Eq. (3.34) is important in our considerations. It will be found to be key in determining the bound state energies introduced into the system by the impurity.

3.2.2 Example of a Bound Impurity Mode in the Tight-Biding Model

As a simple example of a site impurity problem, consider an impurity characterized by the form [4]

$$H_I = |j\rangle U_j \langle j|. \tag{3.35}$$

This introduces an impurity at the jth site of the chain, changing the diagonal interactions of the jth site in the Hamiltonian from $| j\rangle U \langle j |$ of the pure system to $| j\rangle [U + U_j]\langle j |$ of the system in the presence of the added impurity.

Next consider the Green's function element $\langle j | G(E) | j\rangle$ with regards to its new singularities. We find from Eq. (3.34) that [4]

$$\langle j|G(E)|j\rangle = \langle j\left|\left[G_0(E) + G_0(E)H_I G_0(E)\frac{1}{1 - H_I G_0(E)}\right]\right|j\rangle. \tag{3.36}$$

Focusing on the second term in Eq. (3.36), it follows for this term that

$$\begin{aligned} \langle j|\left[G_0(E)H_I G_0(E)\frac{1}{1 - H_I G_0(E)}\right]|j\rangle &= \langle j|\left[G_0(E)H_I G_0(E)\right]|j\rangle \qquad (3.37)\\ &+ \langle j|\left[G_0(E)H_I G_0(E)H_I G_0(E)\right]|j\rangle + \langle j|\left[G_0(E)H_I G_0(E)H_I G_0(E)H_I G_0(E)\right]|j\rangle + \ldots \end{aligned}$$

Applying the diagonal potential for H_I in Eq. (3.35) we then find that [4]

$$\langle j|\left[G_0(E)H_IG_0(E)\frac{1}{1-H_IG_0(E)}\right]|j\rangle=\langle j|\left[G_0(E)|j\rangle U_j\langle j|G_0(E)\right]|j\rangle \tag{3.38}$$

$$+\langle j|\left[G_0(E)|j\rangle U_j\langle j|G_0(E)|j\rangle U_j\langle j|G_0(E)\right]|j\rangle+\ldots$$

$$=\langle j|G_0(E)|j\rangle U_j\langle j|G_0(E)|j\rangle\frac{1}{1-U_j\langle j|G_0(E)|j\rangle}.$$

Considering Eqs. (3.36) and (3.38) it is found for the last term on the right side of Eq. (3.38) that the zeros of the denominator $1-U_j\langle j|G_0(E)|j\rangle$ give rise to new singularities in the system. When these occur outside the band of continuum solutions of H_0, they represent new energy modes associated with bound states of the impurity atom. Consequently, to obtain these new modes we need to study the solutions of [4]

$$1-U_j\langle j|G_0(E)|j\rangle=0 \tag{3.39}$$

occurring outside the band of continuum modes.

From Eq. (3.25) we find that [4]

$$\langle j|G_0|j\rangle=\pm\frac{1}{\sqrt{(E-U)^2-4t^2}} \tag{3.40}$$

where the positive sign is for energies above the upper band edge of the continuum states and the negative sign is for energies below the lower band edge of the continuum states. Substituting Eq. (3.40) into Eq. (3.39), it then follows that

$$1\mp U_j\frac{1}{\sqrt{(E-U)^2-4t^2}}=0 \tag{3.41}$$

so that

$$E-U=\pm\sqrt{U_j^2+4t^2} \tag{3.42}$$

where the positive sign is for energies at the upper band edge for $U_j>0$ and the negative sign is for energies at the lower band edge for $U_j<0$. Since the edges of the continuum band of states are located at

$$E-U=\pm\sqrt{4t^2}=\pm 2|t| \tag{3.43}$$

it follows that, depending on its sign, a nonzero U_j can introduce localized bound states at either the top or bottom edges of the continuum of states.

3.2.3 More Complex Example of Bound Impurity Mode in the Tight-Binding Model

In the next level of complexity, consider an impurity problem involving a single impurity bond in the system. This is characterized by an impurity matrix of the form

$$H_I = t_j \left[| j-1\rangle\langle j | + | j\rangle\langle j-1 | \right] \tag{3.44}$$

where j and $j-1$ are two nearest neighbor sites on the chain which are connected by the impurity bond t_j. For this case the H_I term of the Hamiltonian in Eq. (3.29a) introduces a change in the hopping interaction between the jth and $(j-1)^{th}$ sites of the chain. Specifically, the term

$$t\left[|j-1\rangle\langle j| + |j\rangle\langle j-1| \right] \tag{3.45a}$$

in H_0 is now replaced in our new Hamiltonian H by

$$(t+t_j)\left[| j-1\rangle\langle j | + | j\rangle\langle j-1 | \right]. \tag{3.45b}$$

The resulting impurity problem represented in H, consequently, becomes a matrix problem, ultimately involving 2×2 matrices.

To approach the impurity problem arising from Eq. (3.44), consider the 2×2 Green's function matrix $G_0(E)$ of the Hamiltonian H_0 in the absence of the impurity. This has the form [1–6]

$$\begin{vmatrix} \langle j|G_0(E)|j\rangle & \langle j|G_0(E)|j-1\rangle \\ \langle j-1|G_0(E)|j\rangle & \langle j-1|G_0(E)|j-1\rangle \end{vmatrix} = \begin{vmatrix} \tilde{G}_{0,0}(E) & \tilde{G}_{0,1}(E) \\ \tilde{G}_{0,1}(E) & \tilde{G}_{0,0}(E) \end{vmatrix} \tag{3.46}$$

where $\tilde{G}_{0,0} = \langle j|G_0(E)|j\rangle = \langle j-1|G_0(E)|j-1\rangle$ are the site diagonal Green's functions, $\tilde{G}_{0,1}(E) = \langle j|G_0(E)|j-1\rangle = \langle j-1|G_0(E)|j\rangle$ are the Green's functions between nearest-neighbor sites along the chain, and j and $j-1$ are the sites involving the bond impurity. Note here that the symmetry properties of the lattice are used to simplify the form of the Green's function on the far right of the equation. In this regard, the translation and reflection symmetries of the chain require that all the site diagonal Green's functions $\langle n|G_0(E)|n\rangle$ along the chain are the same and all the nearest-neighbor Green's function matrix elements $\langle n|G_0(E)|n-1\rangle = \langle n-1|G_0(E)|n\rangle$ are the same.

Continuing to write our problem in this way, in the 2×2 representation the matrix H_I can be written as [4]

$$\begin{vmatrix} \langle j|H_I|j\rangle & \langle j|H_I|j-1\rangle \\ \langle j-1|H_I|j\rangle & \langle j-1|H_I|j-1\rangle \end{vmatrix} = \begin{vmatrix} 0 & t_j \\ t_j & 0 \end{vmatrix} \tag{3.47}$$

and also the Green's function matrix $G(E)$ has the 2×2 form

$$\begin{vmatrix} \langle j|G(E)|j\rangle & \langle j|G(E)|j-1\rangle \\ \langle j-1|G(E)|j\rangle & \langle j-1|G(E)|j-1\rangle \end{vmatrix} \tag{3.48}$$

in the $\{|j\rangle, |j-1\rangle\}$ basis. The impurity problem can now be reduced to a treatment in terms of 2×2 matrices.

In this view of the impurity Green's function $G(E)$, Eq. (3.34) generalizes to a 2×2 matrix equation given by

$$\langle n|G(E)|m\rangle = \langle n| \left[G_0(E) + G_0(E)H_I G_0(E)\frac{1}{1-H_I G_0(E)} \right] |m\rangle \tag{3.49}$$

where $n,m = j-1, j$, and H_I is the 2×2 matrix in Eq. (3.47), and $G_0(E)$ is from Eq. (3.46). As in the case of our earlier impurity example in Subsection 3.1.2, the solutions of Eq. (3.49) allow for the determination of the bound state impurity modes introduced by H_I.

Considering Eq. (3.49), it is seen that the new singularities of the bound states in the system must be located at the singularities of the second term in the sum on the far right of the equation, i.e., the term arising in the form $G_0(E)H_I G_0(E)\dfrac{1}{1-H_I G_0(E)}$. These are the contributions to the system from the impurity scattering and set the nature of the impurity interactions in the system. Consequently, let us focus on the second term on the right side of Eq. (3.49), which written explicitly in the notation of 2×2 matrices has the form [4]

$$\left\{\begin{vmatrix}\tilde{G}_{,00}(E) & \tilde{G}_{1,0}(E)\\ \tilde{G}_{1,0}(E) & \tilde{G}_{0,0}(E)\end{vmatrix}\begin{vmatrix}0 & t_j\\ t_j & 0\end{vmatrix}\begin{vmatrix}\tilde{G}_{0,0}(E) & \tilde{G}_{0,1}(E)\\ \tilde{G}_{0,1}(E) & \tilde{G}_{0,0}(E)\end{vmatrix}\right\} * \tag{3.50}$$

$$\left[\begin{vmatrix}1 & 0\\ 0 & 1\end{vmatrix}-\begin{vmatrix}0 & t_j\\ t_j & 0\end{vmatrix}\begin{vmatrix}\tilde{G}_{0,0}(E) & \tilde{G}_{0,1}(E)\\ \tilde{G}_{0,1}(E) & \tilde{G}_{0,0}(E)\end{vmatrix}\right]^{-1},$$

and investigate its energy singularities. If these occur outside the band of continuum solutions of H_0, they represent new energy modes associated with bound states of the impurity bond atoms.

Consequently, to determine the bound states of the impurity we need to consider the inverse of the matrix [4]

$$\begin{vmatrix}1 & 0\\ 0 & 1\end{vmatrix}-\begin{vmatrix}0 & t_j\\ t_j & 0\end{vmatrix}\begin{vmatrix}\tilde{G}_{0,0}(E) & \tilde{G}_{0,1}(E)\\ \tilde{G}_{0,1}(E) & \tilde{G}_{0,0}(E)\end{vmatrix}, \tag{3.51a}$$

with a focus on where its singularities occur. From the general formula for the inverse of a 2×2 matrix, it is known that the singularities of the inverse are located at the zeros of the determinant of the matrix. A consideration then of the solutions of

$$det\left\{\begin{vmatrix}1 & 0\\ 0 & 1\end{vmatrix}-\begin{vmatrix}0 & t_j\\ t_j & 0\end{vmatrix}\begin{vmatrix}\tilde{G}_{0,0}(E) & \tilde{G}_{0,1}(E)\\ \tilde{G}_{0,1}(E) & \tilde{G}_{0,0}(E)\end{vmatrix}\right\}=0 \tag{3.51b}$$

for solutions occurring outside the band of continuum modes will yield the bound states of the impurity problem, provided they exist.

To simplify Eq. (3.51b) we note that the unitary matrix

$$\mathcal{U}=\frac{1}{\sqrt{2}}\begin{vmatrix}1 & 1\\ 1 & -1\end{vmatrix} \tag{3.52a}$$

diagonalizes H_I in Eq. (3.47) so that

$$\mathcal{U}^{+}\begin{vmatrix}0 & t_j\\ t_j & 0\end{vmatrix}\mathcal{U}=\frac{1}{\sqrt{2}}\begin{vmatrix}1 & 1\\ 1 & -1\end{vmatrix}\begin{vmatrix}0 & t_j\\ t_j & 0\end{vmatrix}\frac{1}{\sqrt{2}}\begin{vmatrix}1 & 1\\ 1 & -1\end{vmatrix}=\begin{vmatrix}t_j & 0\\ 0 & -t_j\end{vmatrix}. \tag{3.52b}$$

This is now used to simplify the problem in Eq. (3.51b). In this regard, consider the unitary transformation applied to Eq. (3.51a) given by [4]

$$\mathcal{U}^{+}\left\{\begin{vmatrix}1 & 0\\ 0 & 1\end{vmatrix}-\begin{vmatrix}0 & t_j\\ t_j & 0\end{vmatrix}\begin{vmatrix}\tilde{G}_{0,0}(E) & \tilde{G}_{0,1}(E)\\ \tilde{G}_{0,1}(E) & \tilde{G}_{0,0}(E)\end{vmatrix}\right\}\mathcal{U} \tag{3.53}$$

$$= \mathcal{U}^{+}\begin{vmatrix}1 & 0\\0 & 1\end{vmatrix}\mathcal{U} - \mathcal{U}^{+}\begin{vmatrix}0 & t_j\\t_j & 0\end{vmatrix}\mathcal{U}\mathcal{U}^{+}\begin{vmatrix}\tilde{G}_{0,0}(E) & \tilde{G}_{0,1}(E)\\\tilde{G}_{0,1}(E) & \tilde{G}_{0,0}(E)\end{vmatrix}\mathcal{U}$$

$$= \begin{vmatrix}1 & 0\\0 & 1\end{vmatrix} - \begin{vmatrix}t_j & 0\\0 & -t_j\end{vmatrix}\begin{vmatrix}\tilde{G}_s(E) & 0\\0 & \tilde{G}_a(E)\end{vmatrix}$$

$$= \begin{vmatrix}1 - t_j\tilde{G}_s(E) & 0\\0 & 1 + t_j\tilde{G}_a(E)\end{vmatrix}$$

where

$$\tilde{G}_s(E) = \tilde{G}_{0,0}(E) + \tilde{G}_{0,1}(E) \tag{3.54a}$$

and

$$\tilde{G}_a(E) = \tilde{G}_{0,0}(E) - \tilde{G}_{0,1}(E). \tag{3.54b}$$

It follows in the treatment of Eq. (3.51b) that the determinant of the matrix is invariant under a unitary transformation and as a result we find that there are two possible bound state solutions to our problem. These are obtained as the solutions of

$$1 - t_j\tilde{G}_s(E) = 0 \tag{3.55a}$$

and

$$1 + t_j\tilde{G}_a(E) = 0 \tag{3.55b}$$

for E outside the continuum band of states of H_0.

3.2.4 Other Tight-Binding Like Models: Magnons and Phonons

In this subsection, the generalization of the form of the tight-binding model to treat phonon and magnon excitations in crystalline materials is considered. Phonon and magnon models are very similar to electron tight-binding models even though the quantum statistics of boson excitations are quite different from the fermion statistics of electrons. In all of these systems, however, the excitation spectrums and their zero temperature properties can be developed independently of their statistical properties.

In the case of vibrational systems [1–6], we consider a chain of N atoms with atomic sites labeled $1,2,3,\ldots,N$ in the limit that $N \to \infty$, and for considerations in which the chain is subject to periodic boundary conditions. (See Figure 3.1 for a schematic of the atomic chain.) The atoms along the chain are coupled together by nearest neighbor spring couplings, and for simplicity, the atoms are constrained to move only along the axis of the chain. In its basic form, a Lagrangian for the vibrational motion of the atoms about their equilibrium lattice positions is then given by the form

$$L = \sum_{n=1}^{N}\frac{m}{2}\dot{x}_n^2 - \sum_{n=1}^{N}\frac{K}{2}(x_n - x_{n+1})^2. \tag{3.56}$$

Here x_n is the displacement from equilibrium of the nth atom of the chain along the direction of the chain axis taken to be the x–axis, m is the atomic mass of each atom, and $K > 0$ is the spring constant between nearest atomic neighbors. For periodic boundary conditions, it is required that $x_{N+1} = x_1$.

From the Lagrange equations of motion, it follows from Eq. (3.56) that [1–6]

$$M\ddot{x}_n = K\left[-2x_n + x_{n+1} + x_{n-1}\right] \tag{3.57}$$

where $n = 1,2,3,\ldots,N$ and for the boundary conditions $x_{N+1} = x_1$ and $x_0 = x_N$. Assuming a harmonic time dependence of the form $x_n \propto e^{-i\omega t}$ Eq. (3.57) becomes

$$M\omega^2 x_n = K\left[2x_n - x_{n+1} - x_{n-1}\right]. \tag{3.58}$$

This difference equation can be rewritten into a matrix eigenvalue problem for the frequencies ω^2 and corresponding sets of orthonormal eigenmodes $\{x_n\}$ of the problem.

Rewriting Eq. (3.58) in the format of a matrix eigenvalue problem for the eigenvalue ω^2, we find

$$\sum_{n=1}^{N}\omega^2 \,|\, n\rangle\langle n \,| = \sum_{n=1}^{N} 2\frac{K}{M} \,|\, n\rangle\langle n \,| - \sum_{n=1}^{N}\frac{K}{M}\left(|\, n\rangle\langle n+1 \,| + |\, n+1\rangle\langle n \,|\right) \tag{3.59}$$

$$= D_M\left(\frac{K}{M}\right)$$

where $\{|\, n\rangle\}$ and $\{\langle n \,|\}$ are the site basis states of the atoms such that $\langle n | l\rangle = \delta_{n,l}$, the periodic boundary conditions require that $|\, N+1\rangle = |\, 1\rangle$ and $\langle N+1 \,| = \langle 1 \,|$, and $D_M\left(\frac{K}{M}\right)$ is the matrix representing the dynamics of the system. Note that the form of Eq. (3.59) is essentially the same as that of the eigenvalue problem for the electron energy E of the electron tight-binding Hamiltonian in Eq. (3.2). In this regard, aside from the quantum statistics that are eventually applied to the modes of the systems, the determination of the excitation spectrums of the systems is essentially isomorphic to one another. In this isomorphism between the tight-binding electron and vibrational problems, we make the associations

$$E \leftrightarrow \omega^2 \tag{3.60a}$$

$$U \leftrightarrow 2\frac{K}{M} \tag{3.60b}$$

and

$$t \leftrightarrow -\frac{K}{M} \tag{3.60c}$$

which transform the modal eigenvalue problem of one system into that of the other.

As in the case of the electron system, the solutions for the single phonon modal wavefunctions of the eigenvalue problem in Eq. (3.59) are of the form [1–6]

$$|\, \omega_k^2\rangle = \frac{1}{\sqrt{N}}\sum_{n=1}^{N} e^{ikn} \,|\, n\rangle \tag{3.61a}$$

and

$$\langle \omega_k^2 \,| = \frac{1}{\sqrt{N}}\sum_{n=1}^{N} e^{-ikn} \langle n \,| \tag{3.61b}$$

where for periodic boundary conditions on the chain it is required that

$$e^{-ik(N+1)} = e^{-ik}. \tag{3.61c}$$

Consequently, the values of k of the solutions in Eq. (3.61) are given by $-\pi \le k = \frac{2\pi}{N} r \le \pi$ for r an integer. Substituting, for example, Eq. (3.61a) into Eq. (3.59), it then is found that

$$D_M\left(\frac{K}{M}\right) | \omega_k^2 \rangle = \omega_k^2 | \omega_k^2 \rangle \tag{3.62}$$

where the eigenvalue for the solution of the equation is

$$\omega_k^2 = 2\frac{K}{M}[1-\cos k] = 4\frac{K}{M} sin^2\left(\frac{k}{2}\right). \tag{3.63}$$

In addition, the orthonormality relationship between the eigenstates are given by

$$\langle \omega_k^2 | \omega_l^2 \rangle = \delta_{k,l} \tag{3.64}$$

so that the wavevector eigenstates of the chain form a complete basis of the system.

Notice that, as in the case with the electron system, we can define a Green's function for the eigenvalue problem of the vibrational model in Eq. (3.59). This is given through analogy with Eq. (3.9) by the form

$$G_0\left(\omega^2\right) = \frac{1}{\omega^2 - D_M\left(\frac{K}{M}\right)} = \sum_k \frac{| \omega_k^2 \rangle \langle \omega_k^2 |}{\omega^2 - \omega_k^2} \tag{3.65}$$

where the relations in Eq. (3.60) have been used to make the extension. As in the electron problem, it follows that the vibrational Green's function in Eq. (3.65) is singular at the eigenvalues of the vibrational system in Eq. (3.59). Consequently, we analytically continue the frequency squared ω^2 into the complex plane by taking $\omega^2 \to \omega^2 + i\epsilon$ where $\epsilon \to 0^+$.

The density of states $D\left(\omega^2\right)$ of the continuum frequency spectrum of the matrix $D_M\left(\frac{K}{M}\right)$ can be determined from Eqs. (3.27), (3.28), and (3.60), and is defined such that the number of modes $N_{\left[\omega_a^2,\omega_b^2\right]}$ in the frequency interval $\left[\omega_a^2,\omega_b^2\right]$ is obtained from the integral [1–6]

$$N_{\left[\omega_a^2,\omega_b^2\right]} = \int_{\omega_a^2}^{\omega_b^2} d\omega^2 D\left(\omega^2\right). \tag{3.66}$$

In this regard, we see from Eq. (3.26) that the site diagonal Green's function matrix elements are given (i.e., making the identifications in Eq. (3.60)) by

$$\langle n | G_0 | n \rangle = -i\frac{1}{\sqrt{4\left(-\frac{K}{M}\right)^2 - \left(\omega^2 - 2\frac{K}{M}\right)^2}}. \tag{3.67}$$

Similarly, from Eq. (3.28a) it follows that within the band of continuum states the density of states is given by

$$D\left(\omega^2\right) = \lim_{\epsilon\to 0^+}\left\{-\frac{1}{\pi} Tr\, Im\left\langle n \mid G_0\left(\omega^2 + i\epsilon\right) \mid n\right\rangle\right\} \tag{3.68a}$$

$$= N\frac{1}{\pi}\frac{1}{\sqrt{4\left(\frac{K}{M}\right)^2 - \left(\omega^2 - 2\frac{K}{M}\right)^2}} = N\frac{M}{2\pi K}\frac{1}{\sqrt{1-\left[\left(\sqrt{\frac{M}{2K}}\omega\right)^2 - 1\right]^2}}.$$

In addition, from Eq. (3.28b) the site density of states at each site of the chain is [4]

$$D_n\left(\omega^2\right) = \frac{1}{\pi}\frac{1}{\sqrt{4\left(\frac{K}{M}\right)^2 - \left(\omega^2 - 2\frac{K}{M}\right)^2}} = \frac{M}{2\pi K}\frac{1}{\sqrt{1-\left[\left(\sqrt{\frac{M}{2K}}\omega\right)^2 - 1\right]^2}}. \tag{3.68b}$$

Again, due to the symmetry of the chain, $D_n\left(\omega^2\right)$ is found to be independent of the site index n. Here we have also used the associations made in Eq. (3.60).

In Figure 3.3a a plot of $D\left(\omega^2\right) = ND_n\left(\omega^2\right)$ versus ω^2 is given for the $1-$dimensional vibrational atomic chain. The density of states is found to be divergent at the upper and lower band edges of the continuum spectrum. The minimum of the density of states is located at the center of the frequency spectrum of the continuum band, and $D_n\left(\omega^2\right)$ is symmetric about this minimum.

Notice that the number of states dN found in the interval $d\omega^2$ is related through Eq. (3.68) by

$$dN = D\left(\omega^2\right)d\omega^2 = N\frac{1}{\pi}\frac{1}{\sqrt{4\left(\frac{K}{M}\right)^2 - \left(\omega^2 - 2\frac{K}{M}\right)^2}}d\omega^2. \tag{3.69}$$

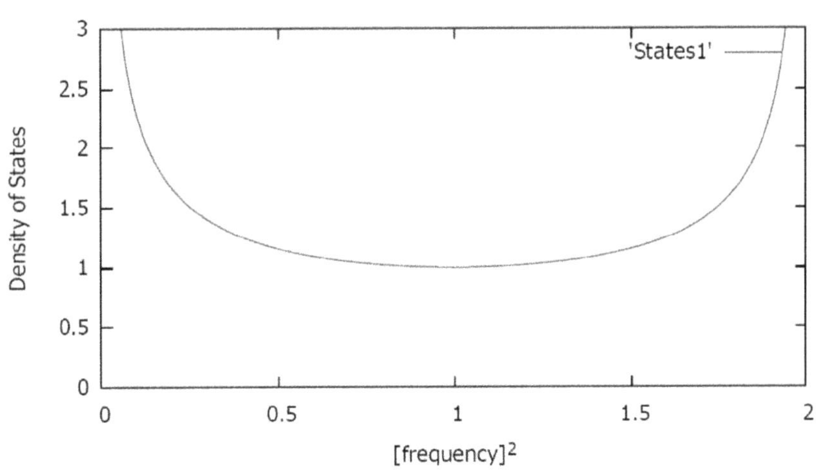

FIGURE 3.3A Plot of the Density of States per site, $\frac{2\pi K}{M}\frac{D\left(\omega^2\right)}{N} = \frac{1}{\sqrt{1-\left[\left(\sqrt{\frac{M}{2K}}\omega\right)^2 - 1\right]^2}}$, versus $\left(\sqrt{\frac{M}{2K}}\omega\right)^2$.

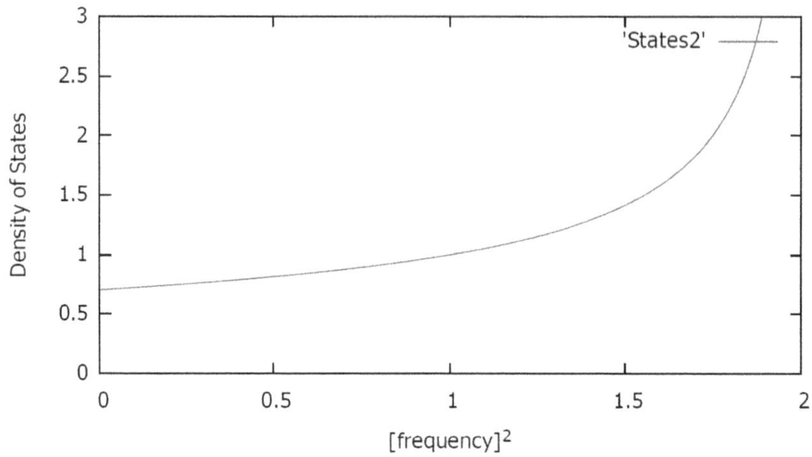

FIGURE 3.3B Plot of the Density of States per site, $\pi\sqrt{\frac{K}{2M}}\frac{\tilde{D}(\omega)}{N}=\frac{\sqrt{\frac{M}{2K}}\omega}{\sqrt{1-\left[\left(\sqrt{\frac{M}{2K}}\omega\right)^2-1\right]^2}}$, versus $\left(\sqrt{\frac{M}{2K}}\omega\right)^2$.

Similarly, we can also obtain a relationship between the number of states dN found in the interval $d\omega$. In this regard, it follows from Eq. (3.69) that [4]

$$dN=\tilde{D}(\omega)d\omega=N\frac{1}{\pi}\frac{2\omega}{\sqrt{4\left(\frac{K}{M}\right)^2-\left(\omega^2-2\frac{K}{M}\right)^2}}d\omega \tag{3.70}$$

$$=N\frac{1}{\pi}\sqrt{\frac{2M}{K}}\frac{\sqrt{\frac{M}{2K}}\omega}{\sqrt{1-\left[\left(\sqrt{\frac{M}{2K}}\omega\right)^2-1\right]^2}}d\omega,$$

where $\tilde{D}(\omega)$ is the density of states for frequency ω. This is also a commonly used form to discuss the density of states of vibrational modes, and for an illustration, in Figure 3.3b a plot is presented of $\tilde{D}(\omega)$ versus ω^2. Notice the difference in the basic forms between the plot of $D(E)$ versus E in Figure 3.2 and $\tilde{D}(\omega)$ versus ω^2. This arises from the factor 2ω that is found in the numerator of Eq. (3.70).

As another important generalization of the tight-binding model, a magnetic system composed as an array of magnetic atoms is discussed [1–4]. In particular, we consider an insulating magnetic material in which each of the magnetic atoms of the system carries a net angular momentum, and associated with these atomic angular momenta are atomic magnetic moments. The magnetic properties displayed by the material then arise from the sum of the magnetic responses of each of its atoms. Included in this response, the magnetic atoms of the materials are found to be linked together by chemical bonding interactions which ultimately affect the relative orientation of their atomic dipole moments [1–3]. Consequently, in a combination of external magnetic fields, crystal fields, and the internal bonding properties of the individual magnetic atoms of the material, a net total magnetic response of the material is developed. This response is the basis of the topic of the magnetic properties exhibited by insulating materials. It is commonly handled in the context of the Heisenberg model of magnetism developed in the context of spin wave theory.

In the Heisenberg model [1–4] a system of atomic angular momenta (often referred to in magnetism as atomic spins) are treated as being coupled together by means of chemical bonds and also individually interacting with crystal fields and with an external magnetic field. In this sense, the spins are considered as vectors arising from the atomic angular momenta of the individual magnetic atoms of the material. We shall begin the discussion by focusing on a chain of identical magnetic atoms, each having an atomic spin, and the spins on neighboring atoms interact with one another through simple dot product interactions. This is a highly idealized model, but it offers a simplification illustrating many of the theoretical techniques commonly used in the study of magnetic materials. In this simplest study of the excitations of the atomic spins, we will find that the spin dynamics is characterized by a tight-binding like model.

For the atomic spin chain, we consider a system of N identical magnetic atoms with atomic lattice sites labeled $1,2,3,\ldots,N$ for the limit that $N \to \infty$, and for the application of periodic boundary conditions to the chain. (See Figure 3.1 for a schematic of the lattice of the atomic chain displayed along the x-axis.) The atomic angular momenta of the chain are coupled together by dot product interactions involving nearest-neighbor spins. The sum of such interactions then forms the so-called Heisenberg Hamiltonian for insulating magnetic materials.

In our system the Heisenberg Hamiltonian is, consequently, written in the form [1–4]

$$H = -2J\sum_{n=1}^{N}\vec{S}_i \cdot \vec{S}_{i+1} \tag{3.71}$$

where $J > 0$ is known as the exchange constant and is taken positive for a ferromagnetic chain, $\{\vec{S}_i\}$ are the set of atomic spins or angular momenta, and for periodic boundary conditions it follows that $\vec{S}_{N+1} = \vec{S}_1$. Note that the ground state energy of the ferromagnetic system in Eq. (3.71) is given by the spin configuration in which all the atomic spins are aligned in the same direction. We shall take the spin ground state to be aligned along the z-axis.

As a simplification in computing the equations of motion of the spins of the chain characterized by Eq. (3.71), we will treat them as classical vectors of length S. This is a reasonable approximation for atomic spins which are large compared to those of a spin $-\frac{1}{2}$ chain, i.e., for $S \gg \frac{1}{2}$. For this limit, the classical mechanics approximation allows us to find the dispersion relations of the spin wave excitations (known as magnons) in a manner similar to that in our just finished study of the dispersion relations of the classical mechanics of the vibrational problem. However, before directly approaching the problem in Eq. (3.71), it is helpful to make a brief diversion and consider an even simpler problem involving a spin interacting with an externally applied magnetic induction. This is done in the context of the classical dynamics limit.

For an initial consideration of classical spin systems, let us begin by treating the classical mechanics equations of motion for a single spin in a uniform external magnetic induction. The dynamics of this system is characterized by the classical equation of motion [1–4]

$$\hbar\frac{d\vec{S}}{dt} = -g\mu_B\vec{S} \times \vec{B} \tag{3.72}$$

where $\hbar\vec{S}$ is the atomic angular momentum (note that $\hbar$ is included so that $\hbar\vec{S}$ has the units of angular momentum), g is the g-factor for the interaction of the spin with the magnetic induction, and $\vec{B}$ is an externally applied magnetic induction. We shall now show that the dynamics of a spin $\vec{S}_i$ in the chain Hamiltonian in Eq. (3.71) can be obtained from a consideration of the single spin problem in Eq. (3.72). This is done by regarding the interaction of $\vec{S}_i$ in Eq. (3.71) with its nearest-neighbor spins as reduced to an effective magnetic induction field interaction arising from its neighboring spins. This identification is now discussed.

Consider the interaction term associated with the spin $\vec{S}_i$ in the Heisenberg Hamiltonian in Eq. (3.71). It is given by the form [1–4]

$$-2J\vec{S}_i \cdot \left[\vec{S}_{i-1} + \vec{S}_{i+1}\right] = -\left(-g\mu_B\vec{S}_i\right) \cdot \left[-\left(\frac{2J}{g\mu_B}\right)\left(\vec{S}_{i-1} + \vec{S}_{i+1}\right)\right] \tag{3.73}$$

$$= -\vec{\mu}_i \cdot \vec{B}_{eff,i}$$

where

$$\vec{\mu}_i = -g\mu_B\vec{S}_i \tag{3.74a}$$

is the magnetic dipole moment of the atom at the ith site, and

$$\vec{B}_{eff,i} = -\left(\frac{2J}{g\mu_B}\right)\left(\vec{S}_{i-1} + \vec{S}_{i+1}\right) \tag{3.74b}$$

is an effective magnetic induction at the ith site arising from the interaction of the ith spin with its nearest-neighbor spins. Note from classical electrodynamics that the potential energy of the spin $\vec{S}_i$ in the system is then from Eqs. (3.73) and (3.74) given by

$$U_i = -\vec{\mu}_i \cdot \vec{B}_{eff,i}. \tag{3.75a}$$

For this form of potential energy, it is also shown in classical electrodynamics that the torque on the dipole moment of the ith spin is [1–4]

$$\vec{\tau}_i = \vec{\mu}_i \times \vec{B}_{eff,i} \tag{3.75b}$$

where the torque $\vec{\tau}_i$ on the moment $\vec{\mu}_i$ arises from $\vec{B}_{eff,i}$ and determines the motion of the angular momentum $\hbar\vec{S}_i$.

It then follows from Eqs. (3.72) and (3.75) that the equation of motion of $\vec{S}_i$ in terms of its nearest-neighbor spins is given by

$$\hbar\frac{d\vec{S}_i}{dt} = -g\mu_B\vec{S}_i \times \vec{B}_{eff,i} \tag{3.76}$$

$$= g\mu_B\vec{S}_i \times \left(\frac{2J}{g\mu_B}\right)\left(\vec{S}_{i-1} + \vec{S}_{i+1}\right)$$

$$= 2J\vec{S}_i \times \left(\vec{S}_{i-1} + \vec{S}_{i+1}\right).$$

Evaluating the classical cross product in Eq. (3.76), the set of equations of motion are found to be given in component form by

$$\hbar\frac{dS_i^x}{dt} = 2J\left[S_i^y\left(S_{i-1}^z + S_{i+1}^z\right) - S_i^z\left(S_{i-1}^y + S_{i+1}^y\right)\right], \tag{3.77a}$$

$$\hbar \frac{dS_i^y}{dt} = 2J\left[S_i^z \left(S_{i-1}^x + S_{i+1}^x \right) - S_i^x \left(S_{i-1}^z + S_{i+1}^z \right) \right], \tag{3.77b}$$

$$\hbar \frac{dS_i^z}{dt} = 2J\left[S_i^x \left(S_{i-1}^y + S_{i+1}^y \right) - S_i^y \left(S_{i-1}^x + S_{i+1}^x \right) \right] \tag{3.77c}$$

This represents the complete classical dynamics of the spin chain in terms of sets of difference equations.

Notice that for small deviations of the spins from the perfectly ordered ground state in which all the spins point along the z−axis, the x− and y− components of the spin are very small. Consequently, for a dynamics involving small deviations from the ordered state a perturbation treatment can be made in the small variables S_i^x and S_i^y. In this expansion, it follows that Eq. (3.77c) gives an expression of second order in the small variables of the system. Compared to this the two expressions in Eqs. (3.77a) and (3.77b) yield formulae which are first order equations in the small variables. Consequently, retaining only terms that are first order in the perturbation, we approximate the second order expression in Eq. (3.77c) as

$$\hbar \frac{dS_i^z}{dt} = 0. \tag{3.78}$$

We see that in this approximation S_i^z is a constant of the motion. Taking $S_i^z = S$, where S is the full length of the spin vector, we also find from Eqs. (3.77a) and (3.77b) that the first order terms in these expressions give

$$\hbar \frac{dS_i^x}{dt} = 2J\left[S_i^y 2S - S\left(S_{i-1}^y + S_{i+1}^y \right) \right] = 2JS\left[2S_i^y - \left(S_{i-1}^y + S_{i+1}^y \right) \right] \tag{3.79a}$$

and

$$\hbar \frac{dS_i^y}{dt} = 2JS\left[\left(S_{i-1}^x + S_{i+1}^x \right) - 2S_i^x \right]. \tag{3.79b}$$

Eqs. (3.79) then are the linearized equations of motion of the spin chain. These yield the so-called spin wave approximation for the system dynamics.

To simplify Eq. (3.79) we define new spin variables given by the linear combinations [1–4]

$$S_i^+ = S_i^x + iS_i^y \tag{3.80a}$$

and

$$S_i^- = S_i^x - iS_i^y. \tag{3.80a}$$

In terms of these new variables, the two coupled equations in Eq. (3.79) decouple into the two separate equations given by

$$-i\hbar \frac{dS_i^+}{dt} = 2JS\left[-2S_i^+ + \left(S_{i-1}^+ + S_{i+1}^+ \right) \right] \tag{3.81a}$$

and

$$-i\hbar \frac{dS_i^-}{dt} = 2JS\left[2S_i^- - \left(S_{i-1}^- + S_{i+1}^- \right) \right]. \tag{3.81b}$$

By assuming harmonic time dependences such that $S_i^+ \propto e^{-i\omega t}$ and $S_i^- \propto e^{i\omega t}$, it then follows that Eq. (3.81) reduce to the difference equations [1–4]

$$\hbar\omega S_i^+ = 2JS\left[2S_i^+ - \left(S_{i-1}^+ + S_{i+1}^+\right)\right] \tag{3.82a}$$

and

$$\hbar\omega S_i^- = 2JS\left[2S_i^- - \left(S_{i-1}^- + S_{i+1}^-\right)\right]. \tag{3.82b}$$

(Notice that Eq. (3.82b) is just an adjoint duplication of Eq. (3.82a) so that only Eq. (3.82a) need be treated.) These difference equations are now transformed into a matrix eigenvalue problem.

The difference equation in Eq. (3.82a) can be rewritten into a matrix eigenvalue problem for the frequencies ω and corresponding sets of orthonormal eigenmodes $\{S_i^+\}$ of the problem. Rewriting Eq. (3.82a) in the format of a matrix eigenvalue problem for the eigenvalue ω, we find

$$\sum_{n=1}^{N} \hbar\omega \mid n\rangle\langle n \mid = \sum_{n=1}^{N} 4JS \mid n\rangle\langle n \mid - \sum_{n=1}^{N} 2JS\left(\mid n\rangle\langle n+1 \mid + \mid n+1\rangle\langle n \mid\right) \tag{3.83}$$

$$= DS_M\left(JS\right)$$

where $\{\mid n\rangle\}$ and $\{\langle n \mid\}$ are the site basis states of the spins such that $\langle n \mid l\rangle = \delta_{n,l}$, the periodic boundary conditions require that $\mid N+1\rangle = \mid 1\rangle$ and $\langle N+1 \mid = \langle 1 \mid$, and $DS_M\left(JS\right)$ is the dynamical spin matrix.

The spin eigenvalue problem in Eq. (3.83) is isomorphic to that of our electron tight-binding problem in Eqs. (3.2) and (3.4). By comparing the two different formulations it is seen that an isomorphism between the tight-binding electron and spin problems is obtained by making the associations

$$E \leftrightarrow \hbar\omega \tag{3.84a}$$

$$U \leftrightarrow 4JS \tag{3.84b}$$

and

$$t \leftrightarrow -2JS. \tag{3.84c}$$

These associations directly transform the eigenvalue problem of one system into that of the other.

The solutions for the single spin wavefunctions in Eq. (3.83) are planewaves forms given by

$$\mid \omega_k\rangle = \frac{1}{\sqrt{N}} \sum_{n=1}^{N} e^{ikn} \mid n\rangle \tag{3.85a}$$

and

$$\langle \omega_k \mid = \frac{1}{\sqrt{N}} \sum_{n=1}^{N} e^{-ikn} \langle n \mid \tag{3.85b}$$

where $\{\mid n\rangle\}$ are the site basis states of the spin chain. For periodic boundary conditions it is also required that

$$e^{-ik(N+1)} = e^{-ik}. \tag{3.85c}$$

so that the values of k are given by $-\pi \leq k = \frac{2\pi}{N} m \leq \pi$ for integers m. Substituting, Eq. (3.85a) into Eq. (3.81) it then follows that

$$DS_M(JS)\,|\,\omega_k\rangle = \omega_k\,|\,\omega_k\rangle \tag{3.86}$$

where the eigenvalues are

$$\omega_k = 4JS(1-\cos k). \tag{3.87}$$

An orthonormality relationship also exists given by

$$\langle \omega_k\,|\,\omega_l\rangle = \delta_{k,l}, \tag{3.88}$$

so that the planewave basis of the system is complete.

Notice that, as in the case with the electron system, a Green's function can be defined for the eigenvalue problem of the Heisenberg model in Eq. (3.83). This is given through analogy with Eq. (3.9) by the form

$$G_0(\omega) = \frac{1}{\omega - DS_M(JS)} = \sum_k \frac{|\,\omega_k\rangle\langle\omega_k\,|}{\omega-\omega_k} \tag{3.89}$$

where the isomorphism relations in Eq. (3.84) have been used to make the extension. As in the electron problem, it follows that the spin Green's function in Eq. (3.89) is singular at the eigenvalues of the spin system in Eq. (3.83). Consequently, we analytically continue the frequency ω into the complex plane by taking $\omega \to \omega + i\epsilon$ where $\epsilon \to 0^+$.

The density of states $D(\omega)$ of the continuum frequency spectrum of the matrix $DS_M(\omega)$ can be determined from Eqs. (3.27), (3.28), and (3.84), and is defined such that the number of modes $N_{[\omega_a,\omega_b]}$ in the frequency interval $[\omega_a,\omega_b]$ is obtained from the integral

$$N_{[\omega_a,\omega_b]} = \int_{\omega_a}^{\omega_b} d\omega D(\omega). \tag{3.90}$$

In this regard, we see from Eq. (3.26) that the site diagonal Green's function matrix elements are given (making the identifications in Eq. (3.84)) by [4]

$$\langle n|G_0|n\rangle = -i\frac{1}{\sqrt{16(JS)^2-(\hbar\omega-4JS)^2}}. \tag{3.91}$$

Similarly, from Eq. (3.28a) it follows that within the band of continuum states the density of states is

$$D(\omega) = \lim_{\epsilon\to 0^+}\left\{-\frac{1}{\pi}Tr\,Im\langle n\,|\,G_0(\omega+i\epsilon)\,|\,n\rangle\right\} \tag{3.92a}$$

$$= N\frac{1}{\pi}\frac{1}{\sqrt{16(JS)^2-(\hbar\omega-4JS)^2}} = N\frac{1}{4JS\pi}\frac{1}{\sqrt{1-\left(\frac{\hbar\omega}{4JS}-1\right)^2}},$$

and associated with this from Eq. (3.28b) the site density of states at each site of the chain is, from the isomorphism with the electron system, given by the form

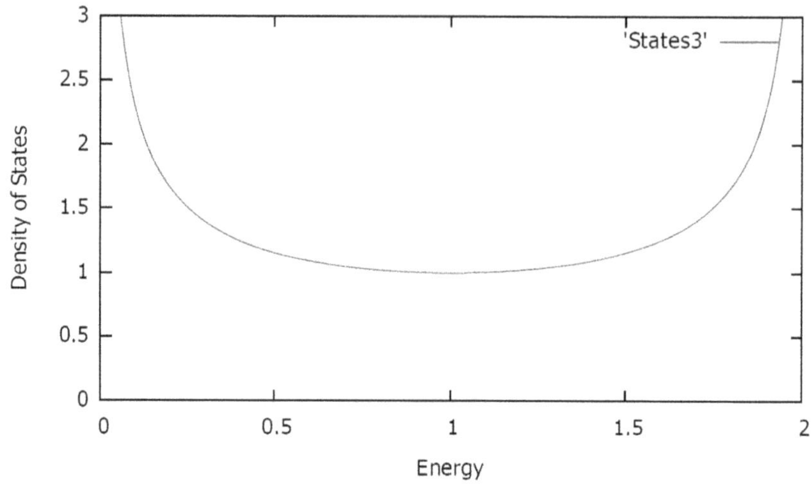

FIGURE 3.3C Plot of the Density of States per site, $\frac{4\pi JS}{N}D(\hbar\omega)=\frac{1}{\sqrt{1-\left(\frac{\hbar\omega}{4JS}-1\right)^2}}$, versus Energy $\hbar\omega$.

$$D_n(\omega)=\frac{1}{\pi}\frac{1}{\sqrt{16(JS)^2-(\hbar\omega-4JS)^2}}=\frac{1}{4JS\pi}\frac{1}{\sqrt{1-\left(\frac{\hbar\omega}{4JS}-1\right)^2}}. \tag{3.92b}$$

Due to the symmetry of the chain, $D_n(\omega)$ is again found to be independent of the site index n. In Figure 3.3c a plot is presented of $\frac{D(\omega)}{N}$ versus $\hbar\omega$ for an infinite spin chain.

We will not continue our treatment in this subsection to the consideration of impurities in vibrational and spin chains. These problems in the context of chain models are not as interesting as the treatment of impurities of higher dimensional systems and the impurities occurring in them. The generalization to higher dimensions will presently be discussed, with the study of some higher dimensional impurity problems. First, however, a brief mention is given regarding the quantization of our classical mechanical results. This is meant to help in understanding the quantum properties of our models.

3.2.5 Quantization of These Results and Temperature Dependent Systems

An element of quantum mechanics can be introduced into the models we have discussed. This is done initially by considering the Planck quantized excitations generated by them as being occupied according to the rules of quantum statistics. In this view, the modes of the system are treated as being present in the system in the context of Fermi-Dirac or Bose-Einstein statistics. Going a bit further, some quantum corrections to the dispersion and wavefunctions of the excitations would also be included. A good first approximation, however, can be made to the quantum statistics by including a simple Planck quantization of the energy levels of the system along with the introduction of quantum statistics in the occupancy of these levels. In the electron system, this is done by occupying the energy orbits $|\epsilon_k\rangle$ on the basis of Pauli statistics. On the other hand, the vibrational and the spin systems are bosons having excitation energies of the form $\hbar\omega_k$ and the energy states in these models are occupied on the basis of boson statistics of identical particles.

In addition, the total energy of the boson systems includes a zero-point energy, so the contribution of each excitation to the total energy of the system is obtained from the standard formula [1–4]

$$E_k = \left(n_k + \frac{1}{2}\right)\hbar\omega_k \tag{9.93}$$

where n_k is an integer indicating the occupancy of the frequency ω_k boson in the system. In this regard, associated with each boson is a zero-point energy which would of itself destroy the spatial order of the ground state of the 1−dimensional chain model, i.e., because of entropy considerations 1−dimensional models do not exhibit ordered states. This makes the chain models just an interesting illustration of the basic theory which will now be extended to more realistic higher dimensional models [1–3]. In higher dimensions, however, the ability of zero-point fluctuations to disorder the system can also lead to the important topics of quantum phase transitions. This is particularly of interest in 2−dimensional systems for the Berezinskii-Kosterlitz-Thouless transition which involves novel ground state orderings and in 2− and 3−dimensional systems exhibiting quantum phase transitions between different quantum phases of a material [1–4].

As a final point, an imposition of the above quantum statistical considerations goes a long way in providing a semi-quantitative approach to the basic properties of the chain as well as in higher dimensional dynamical systems. A full quantum field theoretical treatment of these models, however, is required to generate the exact quantum description of the quantum version of our models. In the full quantum field theoretical approach, further corrections to our considerations also show up as a renormalization of some of the parameters introduced in the generalizations we present based on the classical models.

3.3 TIGHT-BINDING MODEL OF ELECTRONS: HIGHER DIMENSIONAL SYSTEMS

In this section we discuss some considerations of tight-binding models in higher dimensions and their related single impurity problems [4]. First the generalization to 2− and 3−dimensions of the tight-binding Green's functions in the absence of impurities is presented [4]. In terms of these, results for the pure system density of states are then introduced and summarized. These discussions are followed by some treatments of single site impurities in the 2− and 3−dimensions electron systems. Ultimately, our focus on single impurities leads in the next section to the extension of the topic of impurities to the consideration of multiple impurities. This involves the study of multiple random impurities uniformly distributed throughout the crystal lattice and their study in the context of the coherent potential approximation.

As an initial consideration we reconsider the tight-binding electron Green's functions in terms of 2− and 3−dimensional models. These Green's functions are readily obtained from discussions based on the 1−dimensional Green's functions solutions we have previously obtained. A focus is on developing exact expressions for the Green's functions outside the bands of continuum states of the pure system tight-binding models. These are essential in the consideration of the properties of the bound state modes of the single site impurities of the tight-binding models.

We begin by defining a tight-binding model on a square lattice where the lattice sites or vertices are occupied by identical atoms, and the edges of the array are subject to periodic boundary conditions [4]. (See Figure 3.4 for a schematic of this system.) The lattice sites of the square lattice are composed for an $N \times N$ array of atoms with the atomic sites labeled by (i, j) where $i, j = 1,2,3,\ldots,N$ and where N is considered in the limit that $N \to \infty$. For applications of the boundary conditions on the sides of the square, the opposite sides of the square are subject to periodic boundary conditions. In this regard, the site $(N+1, j) = (1, j)$ and the site $(i, N+1) = (i,1)$ for $i, j = 1,2,3,\ldots,N$.

An electron moves in the square lattice by hopping between nearest-neighbor atomic sites, and, again, each electron moves independently of the other electrons in the system. There is, consequently,

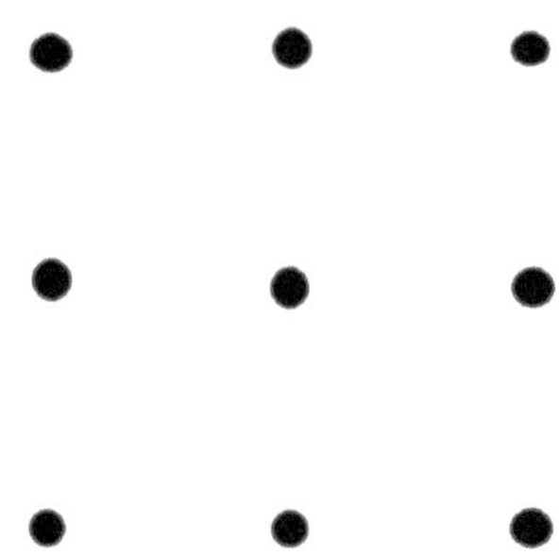

FIGURE 3.4 Schematic of a portion of a square lattice given as an infinite set of points labeled by (i,j) for $i,j = 1,2,3,\ldots,N$ with $N \to \infty$.

no electron-electron scattering. The tight-binding Hamiltonian describing the electron motion is then of the form [4]

$$H_0 = \sum_{n,m=1}^{N} U \,|\, n,m\rangle\langle n,m \,|+ \sum_{n,m=1}^{N} t\left(|\, n,m\rangle\langle n+1,m\,| + |\, n+1,m\rangle\langle n,m\,|\right) \tag{3.94}$$

$$+\sum_{n,m=1}^{N} t\left(|\, n,m\rangle\langle n,m+1\,| + |\, n,m+1\rangle\langle n,m\,|\right)$$

where $|\, n,m\rangle$ and $\langle n,m\,|$ represent the electron site basis states for the atom on the (n,m) square lattice site, and the electron states are normalized such that $\langle n,m|n',m'\rangle = \delta_{n,n'}\delta_{m,m'}$. In addition, the periodic boundary conditions require that $|\, n,N+1\rangle = |\, n,1\rangle$, $\langle n,N+1\,| = \langle n,1\,|$, $|\, N+1,m\rangle = |\, 1,m\rangle$, and $\langle N+1,m\,| = \langle 1,m\,|$. As in our earlier treatments the first term on the right represents an energy the electron receives upon occupying a site of the lattice whereas the second and third terms represent energies the electron receives as it travels between sites within the square lattice.

The solutions for the single electron wavefunctions or orbitals of the Hamiltonian in Eq. (3.94) are of the planewave forms [4]

$$|\,\epsilon_{\vec{k}}\rangle = \frac{1}{N}\sum_{n,m=1}^{N} e^{i(k_x n + k_y m)}\,|\, n,m\rangle \tag{3.95a}$$

and

$$\langle \epsilon_{\vec{k}}\,| = \frac{1}{N}\sum_{n,m=1}^{N} e^{-i(k_x n + k_y m)}\langle n,m\,| \tag{3.95b}$$

where $\vec{k} = (k_x, k_y)$, we have taken a unit lattice constant between nearest neighbor sites on the square lattice, and for periodic boundary conditions on the square it is required that

$$e^{-ik_x(N+1)} = e^{-ik_x} \text{ and } e^{-ik_y(N+1)} = e^{-ik_y} \tag{3.95c}$$

Consequently, the values of $\vec{k}$ in Eq. (3.95) are given by $-\pi \le k_x = \frac{2\pi}{N}m_x, k_y = \frac{2\pi}{N}m_y \le \pi$ for m_x and m_y integers.

Substituting, Eq. (3.95a) into Eq. (3.94) it then follows that the eigenvalues representing the orbital energies of the electrons in the system are determined from the eigenvalue equation

$$H_0\,|\,\epsilon_{\vec{k}}\rangle = \epsilon_{\vec{k}}\,|\,\epsilon_{\vec{k}}\rangle \tag{3.96}$$

where the eigenvalue for the solution of the equation is

$$\epsilon_{\vec{k}} = U + 2t\left(\cos k_x + \cos k_y\right). \tag{3.97}$$

In addition, an orthonormality relationship between the eigenstates is given by

$$\left\langle \epsilon_{\vec{k}} | \epsilon_{\vec{l}} \right\rangle = \delta_{\vec{k},\vec{l}} \tag{3.98}$$

so that the planewave states of the square form a complete basis of the system.

As in our earlier discussions of the 1−dimensional tight-binding model, we define a Green's function for Eq. (3.94) in terms of the expression for the time-translation operator of H_0. This is given by

$$G_0(t) = e^{-i\frac{H_0}{\hbar}t}. \tag{3.99}$$

In terms of Eq. (3.99) the probability amplitude for an electron to propagate from the state $|n,m\rangle$ to the state $|n',m'\rangle$ during the time interval t is given by the Green's function matrix element in the site representation [4]

$$\left\langle n,m \middle| e^{-i\frac{H_0}{\hbar}t} \middle| n',m' \right\rangle = \langle n,m|G_0(t)|n',m'\rangle. \tag{3.100}$$

Considering the system in energy-space, we then find from Eq. (3.99) that

$$\begin{aligned} G_0(E) &= \frac{1}{i\hbar}\int_0^\infty dt e^{i\frac{E}{\hbar}t} G_0(t) = \frac{1}{i\hbar}\int_0^\infty dt e^{i\frac{E}{\hbar}t} e^{-i\frac{H_0}{\hbar}t} \\ &= \frac{1}{E - H_0} = \sum_{\vec{k}} \frac{\left|\epsilon_{\vec{k}}\right\rangle\left\langle\epsilon_{\vec{k}}\right|}{E - \epsilon_{\vec{k}}} \end{aligned} \tag{3.101}$$

where the sum on $\vec{k}$ runs over the eigenmodes from Eq. (3.96) of the Hamiltonian. In the discussion of the Green's function, it is again customary to analytically continue the energy E into the complex plane so that $E \to E + i\epsilon$ where $\epsilon \to 0^+$.

From Eqs. (3.100) and (3.101) we now consider the transition matrix element [4]

$$\langle n,m|G_0|n',m'\rangle = \sum_{\vec{k}} \frac{\left\langle n,m|\epsilon_{\vec{k}}\right\rangle\left\langle\epsilon_{\vec{k}}|n',m'\right\rangle}{E - \epsilon_{\vec{k}} + i\epsilon} = \frac{1}{N^2}\sum_{\vec{k}} \frac{e^{i\left[k_x(n-n') + k_y(m-m')\right]}}{E - \epsilon_{\vec{k}} + i\epsilon} \tag{3.102}$$

where $-\pi \le k_x = \frac{2\pi}{N}m_x, k_y = \frac{2\pi}{N}m_y \le \pi$ for m_x and m_y integers, and we have used the relations $\left\langle n,m|\epsilon_{\vec{k}}\right\rangle = \frac{1}{N}e^{i\left[k_x n + k_y m\right]}$ and $\left\langle \epsilon_{\vec{k}}|n,m\right\rangle = \frac{1}{N}e^{-i\left[k_x n + k_y m\right]}$. In the continuum limit that $N \to \infty$ the expression on the far right of Eq. (3.102) becomes an integral given by [4]

$$\frac{1}{N^2}\sum_{\vec{k}} \frac{e^{i\left[k_x(n-n') + k_y(m-m')\right]}}{E - \epsilon_{\vec{k}} + i\epsilon} \to \int_{-\pi}^{\pi}\frac{dk_x}{2\pi}\int_{-\pi}^{\pi}\frac{dk_y}{2\pi}\frac{e^{i\left[k_x(n-n') + k_y(m-m')\right]}}{E - U - 2t\left(\cos k_x + \cos k_y\right)} \tag{3.103}$$

$$= \int_0^{2\pi} \frac{dk_x}{2\pi} \int_0^{2\pi} \frac{dk_y}{2\pi} \frac{e^{i[k_x(n-n')+k_y(m-m')]}}{E-U-2t(\cos k_x + \cos k_y)}$$

so that

$$\langle n,m|G_0|n',m'\rangle = \int_0^{2\pi} \frac{dk_x}{2\pi} \int_0^{2\pi} \frac{dk_y}{2\pi} \frac{e^{i[k_x(n-n')+k_y(m-m')]}}{E-U-2t(\cos k_x + \cos k_y)}. \tag{3.104a}$$

The integral in Eq. (3.104a) is now evaluated using our results for the 1−dimensional Green's function.

Focusing from Eq. (3.104a) on the diagonal Green's function matrix element [4]

$$\langle 0,0|G_0|0,0\rangle = \int_0^{2\pi} \frac{dk_x}{2\pi} \int_0^{2\pi} \frac{dk_y}{2\pi} \frac{1}{E-U-2t(\cos k_x + \cos k_y)}, \tag{3.104b}$$

we note the cosine identity

$$\cos k_x + \cos k_y = 2\cos\frac{1}{2}(k_x+k_y)\cos\frac{1}{2}(k_x-k_y). \tag{3.105}$$

This identity allows us to rewrite Eq. (3.104b) into the form

$$\langle 0,0|G_0|0,0\rangle = \int_0^{2\pi} \frac{dk_x}{2\pi} \int_0^{2\pi} \frac{dk_y}{2\pi} \frac{1}{E-U-4t\cos\frac{1}{2}(k_x+k_y)\cos\frac{1}{2}(k_x-k_y)}, \tag{3.106}$$

and changing variables to

$$q_1 = \frac{1}{2}(k_x+k_y) \text{ and } q_2 = \frac{1}{2}(k_x-k_y) \tag{3.107}$$

gives

$$\langle 0,0|G_0|0,0\rangle = \int_0^{\pi} \frac{dq_1}{2\pi} \int_0^{2\pi} \frac{dq_2}{2\pi} \frac{2}{E-U-4t\cos q_1 \cos q_2} \tag{3.108}$$

where we have applied some geometric and symmetry considerations in ultimately arriving at the form of Eq. (3.108).

Notice in Eq. (3.108) that

$$\int_0^{2\pi} \frac{dq_2}{2\pi} \frac{1}{E-U-4t\cos q_1 \cos q_2} = \frac{1}{\sqrt{(E-U)^2 - 16t^2 cos^2 q_1}} \frac{E-U}{|E-U|} \quad \text{for } |E-U| > 4|t| \tag{3.109a}$$

and

$$\int_0^{2\pi} \frac{dq_2}{2\pi} \frac{1}{E-U-4t\cos q_1 \cos q_2} = -i\frac{1}{\sqrt{16t^2 cos^2 q_1 - (E-U)^2}} \quad \text{for } |E-U| < 4|t|. \tag{3.109b}$$

have already been evaluated in our consideration of the 1−dimensional Green's functions for the $n = m = 0$ case of Eq. (3.21) through Eq. (3.26). Consequently, from Eqs. (3.108) and (3.109a) it follows for $|E-U| > 4|t|$ (i.e., outside the band of continuum states) that [4]

$$\langle 0,0|G_0|0,0\rangle = 2\frac{E-U}{|E-U|}\int_0^{\pi}\frac{dq_1}{2\pi}\frac{1}{\sqrt{(E-U)^2-16t^2\cos^2 q_1}} \tag{3.110}$$

$$= \frac{1}{\pi}\frac{2}{(E-U)}K\left(\frac{4|t|}{E-U}\right)$$

where $K(x)$ is the Jacobian elliptic integral of the first kind defined by

$$K(x) = \int_0^{\frac{\pi}{2}}\frac{d\varphi}{\sqrt{1-x^2\cos^2\varphi}}. \tag{3.111}$$

Notice also that, as in the case of the $1-$dimensional Green's function, the diagonal matrix element in Eq. (3.110) is an odd function of $E-U$. Similarly, from a consideration of Eqs. (3.108) and (3.109b) it also follows that [4]

$$\langle 0,0|G_0|0,0\rangle = -i\int_0^{\pi}\frac{dq_1}{2\pi}\frac{1}{\sqrt{16t^2\cos^2 q_1-(E-U)^2}} \tag{3.112}$$

$$= -i\frac{1}{2\pi|t|}K\left(\sqrt{1-\left(\frac{E-U}{4t}\right)^2}\right)$$

for $|E-U|<4|t|$, i.e., for a consideration of the states within the band of continuum states.

Within the continuum band of eigenstates (i.e., for $|E-U|<4|t|$) the density of states is obtained from the diagonal elements of the Green's function matrix. Following a similar discussion to that in Eqs. (3.27) and (3.28) for the $1-$dimensional model, we next find within the band of continuum states that the density of states of our $2-$dimensional system (i.e., for $|E-U|<4|t|$) is given by [4]

$$D(E) = \lim_{\epsilon\to 0^+}\left\{-\frac{1}{\pi}Tr\, Im\langle n,m\,|\,G_0(E+i\epsilon)\,|\,n,m\rangle\right\} \tag{3.113a}$$

$$= N^2\frac{1}{2\pi^2|t|}K\left(\sqrt{1-\frac{(E-U)^2}{16t^2}}\right).$$

Consistent with this result it follows from Eq. (3.17) that the on-site density of states at each lattice site (n,m) of the square lattice is [4]

$$D_{n,m}(E) = \frac{1}{2\pi^2|t|}K\left(\sqrt{1-\frac{(E-U)^2}{16t^2}}\right). \tag{3.113b}$$

Notice that the independence of $D_{n,m}(E)$ of the site index (n,m) is a result of the translational symmetry of the square lattice and the periodic boundary conditions. For further details of obtaining this density of states, the reader is referred to the literature [4].

The $3-$dimensional Green's function can be generated by applying a similar reasoning to that involving the determination of our results in Eqs. (3.108) through (3.113) for the $2-$dimensional tight-binding Green's function [4]. In this regard, consider generalizing Eq. (3.104) for the square lattice Green's function to the site diagonal elements of the cubic lattice Green's function. This is done in the limit as $N \to \infty$ for a $N \times N \times N$ cubic lattice occupied by identical atoms and subject to periodic boundary conditions between opposite faces of the cube.

The states at the cubic lattice sites are now labeled by three integers so that $|n,m,p\rangle$ and $\langle n,m,p|$ are the site states labeled by the indices (n,m,p) of the cubic lattice. Focusing on the diagonal Green's function matrix element at the origin of the cubic lattice, we generalize Eq. (3.104) to become [4]

$$\langle 0,0,0|G_0|0,0,0\rangle = \int_0^{2\pi} \frac{dk_x}{2\pi} \int_0^{2\pi} \frac{dk_y}{2\pi} \int_0^{2\pi} \frac{dk_x}{2\pi} \frac{1}{E-U-2t(\cos k_x + \cos k_y + \cos k_z)} \tag{3.114a}$$

$$= \int_0^{2\pi} \frac{dk_z}{2\pi} \frac{1}{E-U-2t\cos k_z} \int_0^{2\pi} \frac{dk_x}{2\pi} \int_0^{2\pi} \frac{dk_y}{2\pi} \frac{1}{1-2t'(k_z)(\cos k_x + \cos k_y)}$$

where

$$t'(k_z) = \frac{t}{E-U-2t\cos k_z}. \tag{3.114b}$$

From Eqs. (3.104) through (3.111) it then follows for $|E-U| > 4|t|$ that

$$\int_0^{2\pi} \frac{dk_x}{2\pi} \int_0^{2\pi} \frac{dk_y}{2\pi} \frac{1}{E-U-2t(\cos k_x + \cos k_y)} = \frac{1}{\pi} \frac{2}{(E-U)} K\left(\frac{4|t|}{E-U}\right) \tag{3.114c}$$

so that Eq. (3.114a) becomes

$$\langle 0,0,0|G_0|0,0,0\rangle = \int_0^{2\pi} \frac{dk_z}{t\pi^2} t'(k_z) K\left(4|t'(k_z)|\right). \tag{3.115}$$

The resulting expression in Eq. (3.115) can then be evaluated numerically as a simple integral.

The result in Eq. (3.115) is for the site diagonal Green's functions with $|E-U| > 6|t|$. This will be of application in studying the impurity bound state modes of our model which are found outside the band of continuum states and will be a focus of our considerations in the next subsection. However, before this consideration of the impurity problem, let us discuss the results for the density of continuum energy states of the pure system.

To obtain an expression of the continuum density of states of the 3–dimensional system, our Green's function considerations must be extended to the continuum states. These are at energies $|E-U| < 6|t|$ and involve a treatment of the full continuum spectrum of the Green's function. For such a further treatment, the reader is referred to the literature [4] for a discussion of the exact density of states of the 3–dimensional tight-binding model. In addition, it is a common study to generate a numerical evaluation of the integral form in Eq. (3.114). Here, however, only an approximation of the 3–dimensional tight-binding density of states is given. This involves the use of an approximate form of the Green's functions in this energy region.

A rough approximation to the 3–dimensional site diagonal tight-binding Green's function has been developed and is of some useful application [4]. In this regard, the approximate form of the site diagonal Green's functions for the cases in which $t > 0$ can be written in the closed analytic form

$$\langle n,m,p|G_0(E)|n,m,p\rangle = \frac{2}{E-U+\sqrt{(E-U)^2-(6t)^2}} \quad \text{for } E-U > 6t \tag{3.116}$$

$$= \frac{2}{E-U-\sqrt{(E-U)^2-(6t)^2}} \quad \text{for } E-U<-6t$$

$$= \frac{2}{E-U+i\sqrt{(6t)^2-(E-U)^2}} \quad \text{for} -6t<(E-U)<6t.$$

Using the approximation in Eq. (3.116), an approximation can now be made for the density of states of the 3−dimensional system. Specifically, it follows from the definition of the density of energy states in Eq. (3.113) extended to the 3−dimensional tight-binding model that [4]

$$D(E) = \lim_{\epsilon\to 0^+}\left\{-\frac{1}{\pi}Tr\,Im\langle n,m,p\,|\,G_0(E+i\epsilon)\,|\,n,m,p\rangle\right\}. \tag{3.117}$$

$$\approx N^3\frac{2}{\pi(6t)^2}\sqrt{(6t)^2-(E-U)^2}.$$

The single site density of states is then also obtained as

$$D_{n,m,p}(E) = \frac{2}{\pi(6t)^2}\sqrt{(6t)^2-(E-U)^2}. \tag{3.118}$$

which, expectedly, is independent of the site indices [4].

3.3.1 Bound Impurity Modes in the Tight-Binding Model

The impurity modes of the 2− and 3−dimensional tight-binding models are obtained from a similar consideration to that given for the 1−dimensional tight-binding model in Eqs. (3.29) through (3.34). In this respect, the tight-binding impurity Hamiltonian again is of the form [4]

$$H = H_0 + H_I. \tag{3.119a}$$

Here H_0 is defined in Eq. (3.94) for the 2−dimensional model or, in the case of a 3−dimensional treatment, is the generalization considered in Eqs. (3.114) through (3.118) of the cubic tight-binding model, and H_I is the change in H_0 due to the addition of a site impurity to the system. The eigenvalue problem for the impurity Hamiltonian in Eq. (3.119a) then becomes

$$[H_0 + H_I]\,|\,\epsilon_i\rangle = \epsilon_i\,|\,\epsilon_i\rangle. \tag{3.119b}$$

where $|\,\epsilon_i$ is an eigenvector of the impurity problem, and ϵ_i is its corresponding eigenvalue.

The Green's function of the impurity Hamiltonian is again defined from its time translation operator by the form [4]

$$G_0(t) = e^{-i\frac{H}{\hbar}t}, \tag{3.120}$$

with a time Fourier transformation given by

$$G(E) = \frac{1}{i\hbar}\int_0^\infty dt e^{i\frac{E}{\hbar}t}G(t) = \frac{1}{i\hbar}\int_0^\infty dt e^{i\frac{E}{\hbar}t}e^{-i\frac{H}{\hbar}t} \tag{3.121}$$

$$= \frac{1}{E-H} = \sum_i \frac{|\epsilon_i\rangle\langle\epsilon_i|}{E-\epsilon_i}.$$

Consequently, in terms of the energy eigenstates, the Green's function becomes

$$G(E) = \frac{1}{E-H} = \sum_i \frac{|\epsilon_i\rangle\langle\epsilon_i|}{E-\epsilon_i}, \tag{3.122}$$

where again we consider its limiting form $E \to E + i\epsilon$ where $\epsilon \to 0^+$.

The Green function in Eq. (3.122) then satisfies the relationship of the form

$$[E-H]G(E) = [E-H_0-H_I]G(E) = 1. \tag{3.123}$$

From this, by some algebra, it then follows that

$$G(E) = G_0(E) + G_0(E)H_IG(E) = G_0(E) + G_0(E)H_IG_0(E)\frac{1}{1-H_IG_0(E)}. \tag{3.124}$$

The final form of $G(E)$ on the far-right side of Eq. (3.124), through its singularities, is again found to determine the bound state energies introduced into the system by the impurity [4].

Specifically, in our 2−dimensional model with an impurity located at the origin of the lattice and characterized by the impurity contribution [4]

$$H_I = |0,0\rangle U_0 \langle 0,0|, \tag{3.125}$$

the bound state modes of the impurity are the solutions of

$$\langle 0.0|1 - H_IG_0(E)|0,0\rangle = 0 = 1 - U_0\langle 0,0|G_0(E)|0,0\rangle. \tag{3.126}$$

The solutions of Eq. (3.126) for energies outside the continuum set of modes represent the energy levels of modes bound to and localized about the impurities.

Consider the solutions of Eqs. (3.125) and (3.126) for the 2−dimensional impurity problem. An approximation to the problem can be made using an asymptotic form for the divergence of $\langle 0,0|G_0(E)|0,0\rangle$ as a function of E at the upper and lower band edges of the continuum band of modes. Specifically, at the upper and lower band edges, it can be shown from Eq. (3.112) for $t > 0$ that at the upper and lower edges of the continuum band of states [4]

$$\langle 0,0|G_0(E)|0,0\rangle \approx \mp\frac{1}{4\pi t}\ln\left[\pm\frac{\delta E_\mp}{t}\right] \tag{3.127}$$

where $\delta E_\pm = E - U \mp 4t$. Applying this in the bound state condition in Eq. (3.126) we find that

$$E - U = \pm 4t \pm te^{-\frac{4\pi t}{|U_0|}} \tag{3.128}$$

where the upper signs are for the upper band edge for $U_0 > 0$ and the lower signs are for the lower band edge for $U_0 < 0$.

Similarly, in our 3−dimensional model with an impurity located at the origin of the lattice and defined by $H_I = |0,0,0\rangle\, U_0 \langle 0,0,0|$ the bound state modes of the impurity are the solutions of

$$\langle 0,0,0|1 - H_IG_0(E)|0,0,0\rangle = 0 = 1 - U_0\langle 0,0,0|G_0(E)|0,0,0\rangle \tag{3.129}$$

localized outside the continuum band of states [4].

Our next considerations will be the generalization of the theory of single impurities to handle multiple impurities. This will involve the development of the ideas of mean field theory and, ultimately, the related theory of the coherent potential approximation. We begin with a review of the mean field theory applied to study the thermodynamics of the Ising model. The generalization of the ideas developed in the mean field theory to treat dynamical properties is then the focus of the coherent potential approximation.

3.4 MEAN FIELD THEORY AND THE IDEAS OF THE COHERENT POTENTIAL APPROXIMATION

The ideas of the mean field theory were first applied to the study of the ordering process in the thermodynamics of ferromagnetic materials [1–4]. The goal of these studies was to determine how, specifically, the system transitions from a disordered paramagnetic state to a fully ordered ferromagnet. The answer is found in terms of the development in the system of an average nonzero effective field at each magnetic atom of the system. This effective field characterizes the chemical interaction of an atom with its neighbors. In this sense, however, the effective field is not a magnetic field, but an ordering field generated from the chemical bonding of an atom with its neighboring magnetic atoms in the material.

As an illustration of a simple magnetic system with a paramagnet-ferromagnet transition consider the Ising model defined on a square lattice [1–4]. (See Figure 3.5a for a schematic of the model.) It describes a system of magnetic dipoles arranged on a square lattice in the $x-y$-plane that individually orient either along the positive or negative z-axis. An energy of interaction exists between neighboring atoms depending on their relative orientations, and the atoms in the system also can interact with an external applied magnetic field applied to the system along the z-axis. In the presence of thermal disorder, the object is to study the transition between the paramagnetic and ferromagnetic states of the system.

The Hamiltonian of the Ising model is given by [1–4]

$$H = -J\sum_{\langle i,j \rangle}\sigma_i\sigma_j - h\sum_i \sigma_i \tag{3.130}$$

where $\langle\ \rangle$ represents a sum over nearest-neighbor pairs of magnetic atoms on the square lattice, $\sigma_i = \pm 1$ for atoms with magnetic moments pointing in the negative (positive) z-direction, $J > 0$ is the exchange constant arising from the chemical bond between nearest-neighbor atoms, and h is the interaction from an external magnetic field. In a statistical mechanical study, the system is found to transition between paramagnetic and ferromagnetic states at a critical temperature denoted as T_c. An approximate treatment of the thermodynamics of this phase transition and the value of T_c can be made in the context of a so-called mean field theory approximation [1–4].

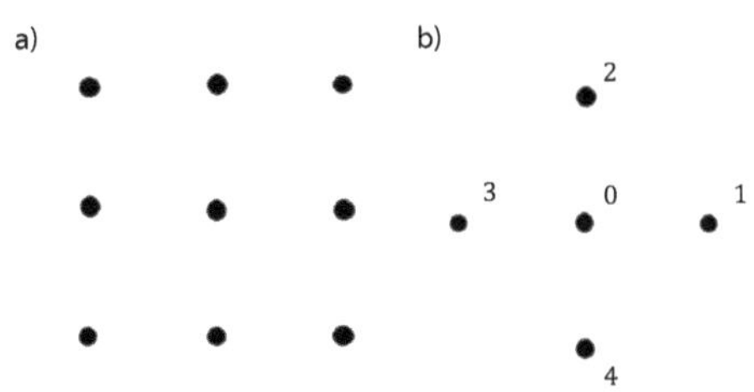

FIGURE 3.5 Schematic of: a) a portion of a square lattice given as an infinite set of points labeled by (i, j) for $i, j = 1,2,3,\ldots,N$ with $N \to \infty$. The lattice constant is taken as unity. b) a site of the square lattice labeled 0 and its four nearest neighbor points labeled $1,2,3,4$.

Let us begin the mean field treatment by focusing on the atom labeled σ_0. (See Figure 3.5b.) Its energy of interaction with its four nearest-neighbor atoms (labeled σ_1, σ_2, σ_3, σ_4) is characterized by the form

$$E = -\sigma_0\left(J\sum_{i=1}^{4}\sigma_i + \mathrm{h}\right). \tag{3.131}$$

We see that the energy of interaction of σ_0 with its neighbors fluctuates considerably depending on the configurations of the set of $\{\sigma_1,\sigma_2,\sigma_3,\sigma_4\}$. Over the entire system, however, there is developed an average value of E for each spin of the system. We focus on studying this average value.

Next, consider introducing an effective field to characterize the average interaction of the atoms $\{\sigma_1,\sigma_2,\sigma_3,\sigma_4\}$ with σ_0. Specifically, we define the effective field as [1–4]

$$m = \langle\sigma_j\rangle \tag{3.132}$$

where $\langle\ \rangle$ represents a statistical mechanical average over all the sites in the system. The effective field then arises in the system as an average value of the σ_j's of the material. In this regard, in the completely disordered state of the system the effective field is zero, but in the ordered system the effective field is nonzero. Notice in these considerations that replacing the neighboring atoms with an average field should become a better approximation in systems with greater numbers of nearest neighbors.

In terms of the effective field, E in Eq. (3.131) can be rewritten as

$$E = -\sigma_0\left(J\sum_{i=1}^{4}\sigma_i + \mathrm{h}\right) = -\sigma_0\left(4Jm + h\right) - \sigma_0 J\sum_{i=1}^{4}\left(\sigma_i - m\right) \tag{3.133}$$

where the effective field has been added into the system in the first term on the far right and subtracted out of the system in the second term on the far right. The first term on the far right represents the interaction energy of σ_0 with the effective field, and the second term represents the fluctuation of the real spin configurations (i.e., σ_1, σ_2, σ_3, σ_4) from that of the effective field. The mean field approximation is obtained by ignoring the second term in Eq. (3.133) and treating the energy of interaction of σ_0 in the system as given by

$$E = -\sigma_0\left(4Jm + h\right). \tag{3.134}$$

Consequently, in this approximation the fluctuations in the system about the mean field are dropped.

The system is now characterized from Eq. (3.134) by the effective Hamiltonian [1–4]

$$H_{eff}\left(\sigma_0\right) = -\sigma_0\left(4Jm + h\right) \tag{3.135}$$

which is essentially that of a paramagnetic spin σ_0 interacting in the field $4Jm + h$. This effective Hamiltonian is next used as an approximation to compute $m = \langle\sigma_j\rangle$ defined in Eq. (3.132). We shall do this, considering the $h = 0$ case of our model.

Taking $h = 0$ it follows that

$$H_{eff}\left(\sigma_0\right) = -\sigma_0 4Jm. \tag{3.136}$$

The effective field is next computed from Eq. (3.136) by the application of standard statistical mechanics. In this way we find [1–4]

$$m = \frac{Tr\left\{\sigma_0 e^{-\beta H_{eff}(\sigma_0)}\right\}}{Tr\left\{e^{-\beta H_{eff}(\sigma_0)}\right\}} = \tanh[4\beta Jm] \tag{3.137}$$

where $\beta = \frac{1}{k_B T}$, k_B is the Boltzmann constant, and T is the temperature. This provides a self-consistent equation to be solved for the effective field m. We shall consider the solution of Eq. (3.137) in the limit of small m.

In the limit of small m, the far right side of Eq. (3.137) is Taylor series expanded to give

$$m = 4\beta Jm - \frac{1}{3}(4\beta Jm)^3. \tag{3.138}$$

This algebraic form then has the solutions

$$m = 0 \tag{3.139a}$$

and

$$m = \pm\sqrt{3}\left(\frac{k_B T}{4J}\right)^{\frac{3}{2}}\left(\frac{4J}{k_B T} - 1\right)^{\frac{1}{2}} = \pm\sqrt{3}\left(\frac{T}{T_c}\right)^{\frac{3}{2}}\left(\frac{T_c}{T} - 1\right)^{\frac{1}{2}} \tag{3.139b}$$

where $T_c = \frac{4J}{k_B}$ is an expression for the critical temperature at which the paramagnetic-ferromagnetic transition occurs, i.e., the temperature at which m in Eq. (3.139b) changes to zero. The solution in Eq. (3.139a) is shown to be the lowest energy of the system when $T > T_c$, and the solutions in Eq. (3.139b) are the lowest energy solutions of the system when $T < T_c$. Notice that in the region where $T < T_c$ the symmetry of the Hamiltonian is broken so that the system has a choice of two ground state solutions to drop into. The two solutions are reflections of each other in the $x-y$-plane.

The ideas of the effective field study discussed here are readily modified to treat an effective field approximation for the Green's function solutions of the tight-binding electron model in a random disorder system. This involves a consideration of multiple scattering impurities uniformly distributed throughout an electronic material. The development is step by step the same as that in the ferromagnetic-paramagnetic transition considered earlier.

3.4.1 Coherent Potential Approximation

In this subsection we introduce an effective medium description of a disordered electronic material and its frequency responses [1–4]. This is done in a manner somewhat like the mean field theory treatment of the Ising model which was developed to explain the behavior of the magnetization with temperature in ferromagnetic materials. Specifically, an effective medium characterization of the tight-binding electron system is generated to represent an average response from the system to the disorder within it, ultimately allowing for the study of the dispersion relations and lifetimes of the electron excitations as they propagate through the disordered medium. The effective medium in the electron system is formulated for a consideration of a statistical set of uniformly distributed random impurities present in the tight-binding electron Hamiltonian [1–4].

For simplicity, the electron Hamiltonian of interest will be for a 2-dimensional system defined on a square lattice with atoms located at square lattice vertices of the form (n,m) where $n,m = 1,2,3,\ldots,N$ for $N \to \infty$. Notice, again, we consider systems limited to nearest-neighbor electron hopping through the lattices, and the lattice constant characterizing the nearest-neighbor separation is taken as unity. Per our earlier treatment, periodic boundary conditions are applied to the opposite sites of the square lattice.

Under these conditions, the Hamiltonian of the tight-binding model of the disordered medium is of the form [1–4]

$$H = H_0 + H_I \tag{3.140a}$$

where from Eq. (3.94)

$$H_0 = \sum_{n,m=1}^{N} U \,|\, n,m\rangle\langle n,m \,| + \sum_{n,m=1}^{N} t\left(|\, n,m\rangle\langle n+1,m \,| + |\, n+1,m\rangle\langle n,m \,|\right) \tag{3.140b}$$

$$+\sum_{n,m=1}^{N} t\left(|\, n,m\rangle\langle n,m+1 \,| + |\, n,m+1\rangle\langle n,m \,|\right)$$

is the Hamiltonian in the absence of impurities, and

$$H_I = \sum_{n,m=1}^{N} c_{n,m}\left(U_{n,m} - U\right) |\, n,m\rangle\langle n,m \,| = \sum_{n,m=1}^{N} c_{n,m}\delta U_{n,m} \,|\, n,m\rangle\langle n,m \,| \tag{3.140c}$$

describes the change in the model with the introduction of impurities. Here

$$c_{n,m} = 1 \text{ if an impurity is present at } (n,m) \tag{3.141}$$

$$= 0 \text{ otherwise,}$$

is a site occupation index indicating the presence or absence of an impurity at the site (n,m), and $U_{n,m} \,|\, n,m\rangle\langle n,m \,|$ is the site diagonal potential of the impurity atom at the site (n,m). Notice that for simplicity we consider only the case of site diagonal disorder in our Hamiltonian, and the nearest-neighbor hopping between sites is considered to not be disordered.

We start by rewriting the Hamiltonian in Eq. (3.140a) in the form [2, 4]

$$H = H_0 + H_I = \left[H_0 + \tilde{\Sigma}\right] + \left[H_I - \tilde{\Sigma}\right] \tag{3.142}$$

where

$$\tilde{\Sigma} = \sum_{n,m=1}^{N} \Sigma \,|\, n,m\rangle\langle n,m \,|, \tag{3.143}$$

and $\tilde{\Sigma}$ is an effective field which is currently determined. Defining

$$H_0^{\Sigma} = H_0 + \tilde{\Sigma} \tag{3.144a}$$

and

$$H_I^{\Sigma} = H_I - \tilde{\Sigma}, \tag{3.144b}$$

it then follows that Eq. (3.142) takes the form [2, 4]

$$H = H_0^{\Sigma} + H_I^{\Sigma}. \tag{3.145}$$

Notice that there is no disorder in H_0^Σ and all the disorder in the system is contained in H_I^Σ. The object next is to determine $\tilde{\Sigma}$ so it best represents through H_0^Σ the average response of the system to disorder while removing the individual random scattering in H_I^Σ from the considerations.

Specifically, to do this consider the Green's function of our model in Eq. (3.145). This is defined by [2, 4]

$$G(E) = \frac{1}{E - H} \tag{3.146}$$

so that from Eqs. (3.142) and (3.145) we can write Eq. (3.146) as

$$\left[E - H\right]G(E) = \left[E - H_0^\Sigma - H_I^\Sigma\right]G(E) = 1. \tag{3.147}$$

Next, Eq. (3.147) is rewritten into the form of the relationship

$$\left[E - H_0^\Sigma\right]G(E) = 1 + H_I^\Sigma G(E) \tag{3.148}$$

which, upon dividing by $E - H_0^\Sigma$ and applying some algebra becomes [2, 4]

$$\begin{aligned} G(E) &= \frac{1}{E - H_0^\Sigma}\left[1 + H_I^\Sigma G(E)\right] = G_0(E) + G_0(E) H_I^\Sigma G(E) \\ &= G_0(E) + G_0(E) H_I^\Sigma G_0(E) + G_0(E) H_I^\Sigma G_0(E) H_I^\Sigma G_0(E) + \ldots \\ &= G_0(E) + G_0(E) T(E) G_0(E). \end{aligned} \tag{3.149}$$

Here we have defined the Green's function characterizing the time translation operator $e^{-i\frac{H_0^\Sigma}{\hbar}t}$ generated from the effective field Hamiltonian H_0^Σ by

$$G_0(E) = \frac{1}{E - H_0^\Sigma} = \frac{1}{i\hbar}\int_0^\infty dt e^{i\frac{E}{\hbar}t} e^{-i\frac{H_0^\Sigma}{\hbar}t} \tag{3.150a}$$

and the T-matrix

$$T(E) = H_I^\Sigma + H_I^\Sigma G_0(E) H_I^\Sigma + \ldots = H_I^\Sigma \frac{1}{1 - G_0(E) H_I^\Sigma}. \tag{3.150b}$$

Notice that all the impurity scattering from the explicit disorder in H_I is contained within the T-matrix while $G_0(E)$ contains all the dynamics associated with the effective field. The focus is next on the behavior of the T-matrix in Eq. (3.150) and its use in defining the effective field in H_0^Σ.

The object in the coherent potential approximation is summarized in the application of two steps to Eqs. (3.149) and (3.150). These allow for the determination of the effective medium $\tilde{\Sigma}$. The first step is to average Eq. (3.149) over the disorder in the random system. This gives [2, 4]

$$\langle G(E)\rangle = G_0(E) + G_0(E)\langle T(E)\rangle G_0(E) \tag{3.151}$$

where $\langle\ \rangle$ represents an average over all random configurations consistent with our statistical knowledge of the disorder of the system. This usually includes an understanding of the density of impurities within the volume of the medium and whether or not there are some statistical correlations in their spatial distribution. The next step involves choosing the effective field $\tilde{\Sigma}$ such that the average of the T-matrix over the configurations of the random system consistent with its statistical properties is zero, i.e., by setting [2, 4]

$$\langle T(E)\rangle = 0. \tag{3.152}$$

The solution of Eq. (3.152) is then made for $\tilde{\Sigma}$ as a function of E and the concentration and correlations of the impurities in the material.

With the generation from Eq. (3.152) of this choice of $\tilde{\Sigma}$, the configuration average of the Green's function is directly related to the effective field Green's function, i.e.,

$$\langle G(E)\rangle = G_0(E). \tag{3.153}$$

The average properties of the random disordered model are now characterized by $G(E)$ and ultimately by $G_0(E)$ so that from Eq. (3.150a) we find [2, 4]

$$G_0(E) = \frac{1}{E - H_0^{\Sigma}} = \frac{1}{E - Re\left(H_0^{\Sigma}\right) - iIm\left(H_0^{\Sigma}\right)}, \tag{3.154}$$

written in terms of the real and imaginary parts of the effective field Hamiltonian. In this interpretation, the electron energy levels and the lifetimes of the electron energy levels in the random medium are given by the poles of $G_0(E)$ in the complex energy plane. Specifically, $Re\left(H_0^{\Sigma}\right)$ represents the energy dispersion relation of the electrons in the system and $Im\left(H_0^{\Sigma}\right)$ is related to the lifetime of the electron states, i.e., the typical length of time an electron spends in a particular energy state before it is scattered by the impurities into another electronic energy state. This description of the system is seen directly from the relationship of $G_0(E)$ to the time translation operator $e^{-i\frac{H_0^{\Sigma}}{\hbar}t}$ of the effective medium Hamiltonian given on the far right side of Eq. (3.150a).

The problem in Eq. (3.152) is still very complicated. Due to the hopping terms in the H_0 and H_0^{Σ}, the T-matrix always tends to link many different lattice sites together and these sites have different disorders. In this regard, as a further simplification of the mathematics in Eq. (3.152), let us make an approximation by writing [2, 4]

$$\langle T(E)\rangle \approx \left\langle \sum_{n,m=1}^{N} T_{n,m}(E)\right\rangle \approx \sum_{n,m=1}^{N} \langle T_{n,m}(E)\rangle = 0, \tag{3.155a}$$

where $T_{n,m}(E)$ are the T-matrix terms in $T(E)$ only involving multiple scatterings from the (n,m) lattice sites, i.e., they are the T-matrix elements $T_{n,m}(E) = T_{n,m} \mid n,m\rangle\langle n,m \mid$ where $T_{n,m}$ only contains scattering involving the site (n,m). These terms arise directly from a consideration of Eqs. (3.140c) and (3.144b). The form in Eq. (3.155a) then treats the multiple scattering from each site exactly while ignoring processes involving intermediate encounters with other sites in the system.

Consequently, we approximate our effective medium solution from the condition that [2, 4]

$$\langle T_{n,m}(E)\rangle = 0. \tag{3.155b}$$

Employing this approximation, from Eq. (3.150b) it then follows that

$$T_{n,m}(E) = \langle n,m| H_I^{\Sigma} |n,m\rangle \frac{1}{1 - \langle n,m| G_0(E) |n,m\rangle \langle n,m| H_I^{\Sigma} |n,m\rangle} \tag{3.155c}$$

$$= \left[c_{n,m}\left(U_{n,m}-U\right)-\Sigma\right]\frac{1}{1-\langle n,m|G_0(E)|n,m\rangle\left[c_{n,m}\left(U_{n,m}-U\right)-\Sigma\right]}$$

where Eqs. (3.140c) and (3.144b) are applied for the determination of $\langle n,m|H_I^{\Sigma}|n,m\rangle$.

If $\langle c_{n,m}\rangle = x$ is the fractional concentration of lattice impurity sites present in the material, and the value of the impurity potential $U_{n,m}$ at each of the impurity sites is distributed by the distribution $P\left(U_{n,m}\right)dU_{n,m}$ such that $\int_{-\infty}^{\infty}P\left(U_{n,m}\right)dU_{n,m}=1$, then from Eq. (3.155c)

$$\langle T_{n,m}(E)\rangle \approx x\int_{-\infty}^{\infty}P\left(U_{n,m}\right)dU_{n,m}\left(U_{n,m}-U-\Sigma\right)\frac{1}{1-\langle n,m|G_0(E)|n,m\rangle\left(U_{n,m}-U-\Sigma\right)} \tag{3.155d}$$

$$-(1-x)\Sigma\frac{1}{1+\langle n,m|G_0(E)|n,m\rangle\Sigma}=0.$$

(Note: We emphasize here that $P\left(U_{i,j}\right)$ is the same at each impurity site in the material.) The last equality on the right in Eq. (3.155d) is then the required condition used for the determination of the effective medium parameter Σ.

Our restriction of the considerations in Eqs. (3.155a) through (3.155d) to the single random site (n,m) on the lattice essentially reduces the problem of the random system by keeping the random (n,m) site and replacing the disorder on all the other sites by the effective medium $\tilde{\Sigma}$. The coherent potential approximation is then obtained from this reduced model by setting the average T-matrix on the random (n,m) lattice site to zero. This is exactly the same idea as that used in the mean field approach to the Ising model.

We have considered here only a simple system with site diagonal impurities. More complex models can be developed with the introduction of hopping disorder into the system. These are more difficult to treat and require effective media with more parameters characterizing the effective dynamics between the sites. For further treatments of these more complex models the reader is referred to the literature [2, 4].

We next turn to the consideration of a problem in the theory of magnetic impurities in metallic systems known as the Kondo effect [1–6]. This is a problem of considerable importance in low-temperature physics which leads to new properties in electron conductivity. It is essentially a single site impurity problem but is made more complex as the impurity interacts with an electron gas in a way which requires treatment in a full many-body approach to the dynamics of the electrons in the system.

3.5 KONDO PROBLEM

In this section we study a class of phenomena known as the Kondo effect [1–6]. The Kondo effect is generally found as an anomalous set of behaviors in the low-temperature resistivity of certain types of metallic systems containing magnetic impurities. It involves an important class of features in low-temperature physics where certain impure metallic materials display an increase in resistivity with decreasing temperature. This is an anomalous behavior as in non-superconducting metals at low temperatures the metallic resistivity generally decreases with decreasing temperature T rather than exhibiting an increase. As we shall see, the increase is intimately connected to the magnetic properties associated with the impurities and their interactions with the conduction electrons of the metallic matrix in which they are embedded.

Specifically, the low-temperature resistivity of a typical (non-anomalous) metal is generally expected to be of the form [1–6]

$$\rho(T)=\rho_0+aT^5 \tag{3.156}$$

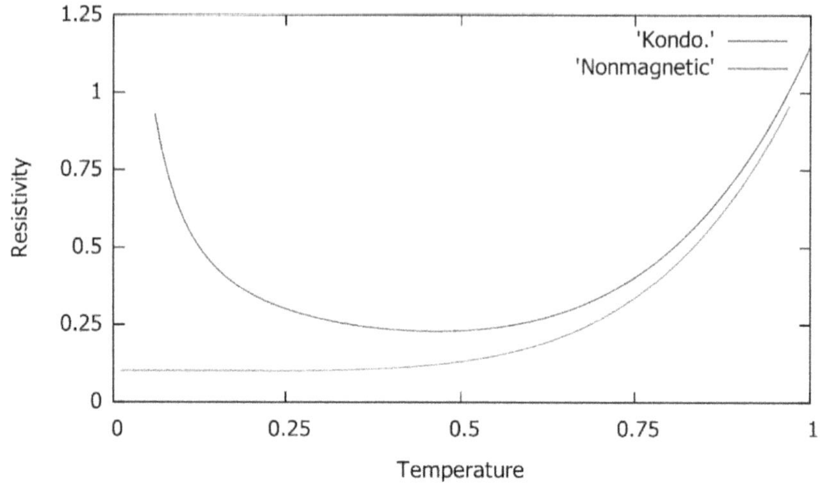

FIGURE 3.6 Schematic plot of the Resistivity versus Temperature at temperatures well below the Debye temperature. Results are shown for a Kondo material (upper curve) and a typical nonmagnetic metal (lower curve).

where a is a constant and ρ_0 is a constant zero-temperature residual resistivity. (Note: This separation between a constant residual resistivity and a temperature dependent term is known as Matthiessen's rule and is generally found in impure metals at temperatures less than the Debye temperature.) In Eq. (3.156) the residual resistivity ρ_0 arises from the scattering of electrons by static impurities in the lattice, and the T^5 term comes from the scattering of electrons by thermally excited phonons in the material [2]. Additional T^2 contributions to the resistivity due to electron-electron scattering can also be found to enter the resistivity at much lower temperatures as one approaches absolute zero [2]. Consequently, the temperature dependent phonon and electron scattering is seen to rapidly become decreasingly effective at decreasing temperature while the impurity scattering is relatively independent of the temperature. (See Figure 3.6 for a schematic representing the difference in the general features of the resistivity in Eq. (3.156) and that observed in typical Kondo systems.)

As we shall presently see, in the Kondo effect the low-temperature anomalous increase in $\rho(T)$ with decreasing temperature arises from the increase of the temperature dependent magnetic moment of the impurities and their consequent increasing conduction electron scattering cross sections at low temperatures [4–6]. This is an opposite type of temperature scattering effect than that arising from the phonons and electrons. These types of increases in the magnetic systems are examples of the general Kondo effect arising in a variety of different models for magnetic scattering in metals.

We begin by looking at the problem of an electron gas interacting with a single localized magnetic impurity [4–6]. This is known as the Anderson model and is one of a class of problems ultimately related to increases with decreasing temperature observed in the resistivity of metals containing magnetic impurities. The object in the study of the Anderson model is to demonstrate conditions under which the interaction of an impurity atom with an electron gas can result in a net magnetic moment becoming localized about the impurity atom.

We start these treatments with a brief review of some of the properties of the electron gas model of conduction electrons that are useful in understanding the Anderson model.

3.5.1 Electron Gas and Its Anti-Commutators

In this subsection, the focus is on the electron gas model of conduction electrons and the development of a formalism helpful in understanding the interaction of the gas with a single electronic impurity. To begin, we introduce some useful notation that allows us to account for the occupancies of the electronic energy levels in a many-body system consistent with the electron statistical behavior as fermions. This is followed in the later subsections by applications of the formulation to a study of the Anderson model, the reformulation of the Anderson model in terms of the s-d model of magnetism, and some elementary discussions of the Kondo problem [4–6]. The focus here, however, is on developing a formalism which will assist us in an understanding of the basic physical mechanisms at play in arriving at an understanding of the low-temperature properties of these models of conduction electrons.

In the electron gas model, the electrons move independently of one another with a free particle energy $\epsilon_{\vec{k},\sigma}$ given by the spin independent form [1–6]

$$\epsilon_{\vec{k},\sigma} = \frac{1}{2m}\left(\hbar\vec{k}\right)^2, \tag{3.157}$$

and the corresponding wavefunctions are planewave states characterized by wavevectors $\vec{k}$ and spin σ. Consequently, $\hbar\vec{k}$ is the electron momentum, $\sigma = \uparrow$ or $\downarrow$ indicates a spin up or spin down electron, and m is the electron effective mass. For electrons in a 3-dimensional cubic box of side $L \to \infty$ subject to periodic boundary conditions, the wavevectors of the planewave states are of the form $k_x = \frac{2\pi n_x}{L}, k_y = \frac{2\pi n_y}{L}, k_z = \frac{2\pi n_z}{L}$, where n_x, n_y, n_z are integers. Each energy level $\epsilon_{\vec{k},\sigma}$ is then occupied by a spin up $\uparrow$ or a spin down $\downarrow$ electron, subject to the Pauli exclusion principle.

The Pauli exclusion principle can be represented in the system by introducing sets of non-commuting operators to indicate the presence or absence of a particular electron mode in the gas. These operators are denoted $c^+_{\vec{k},\sigma}$ for a creation operator which creates an electron of wavevector $\vec{k}$ and spin $\sigma = \uparrow$ or $\downarrow$ in the system and denoted $c_{\vec{k},\sigma}$ for a destruction operator which removes an electron of wavevector $\vec{k}$ and spin $\sigma = \uparrow$ or $\downarrow$ from the system. The creation operator $c^+_{\vec{k},\sigma}$ creates a plane wave eigenstate $\left|\epsilon_{\vec{k},\sigma}\right\rangle$ from a vacuum state $|0\rangle$ which is a state with no $\vec{k},\sigma$ electron present, i.e.,

$$\left|\epsilon_{\vec{k},\sigma}\right\rangle = c^+_{\vec{k},\sigma}|0\rangle, \tag{3.158a}$$

and the destruction operator $c_{\vec{k},\sigma}$ turns an occupied eigenstate $\left|\epsilon_{\vec{k},\sigma}\right\rangle$ into a vacuum state $|0\rangle$, i.e.,

$$|0\rangle = c_{\vec{k},\sigma}\left|\epsilon_{\vec{k},\sigma}\right\rangle. \tag{3.158b}$$

A general state of the system, for example, is then represented by products of single powers of the creation operators acting on a vacuum state as in

$$\left|\epsilon_{\vec{q},\sigma}, \epsilon_{\vec{k},\sigma'}, \epsilon_{\vec{p},\sigma''}\right\rangle = c^+_{\vec{q},\sigma} c^+_{\vec{k},\sigma'} c^+_{\vec{p},\sigma''}|0\rangle \tag{3.158c}$$

The resulting state of the system in Eq. (3.158c) is a state composed of three different electrons of different wavevectors and spin states.

The Pauli exclusion principle is then imposed on the system by requiring that the set of operators $\left\{c_{\vec{k},\sigma}\right\}$ and $\left\{c^+_{\vec{k},\sigma}\right\}$ obey a non-commutative operator algebra defined by the so-called anti-commutators. These are expressed for our electron creation and destruction operators by

$$\left\{c_{\vec{k},\sigma}, c^+_{\vec{q},\sigma'}\right\} = \left[c_{\vec{k},\sigma} c^+_{\vec{q},\sigma'} + c^+_{\vec{q},\sigma'} c_{\vec{k},\sigma}\right] = \delta_{\vec{k}\sigma,\vec{q},\sigma'} \tag{3.159a}$$

$$\left\{c_{\vec{k},\sigma},c_{\vec{q},\sigma'}\right\}=\left\{c^{+}_{\vec{k},\sigma},c^{+}_{\vec{q},\sigma'}\right\}=0 \tag{3.159b}$$

where $\{a,b\}=ab+ba$ defines the anticommutator of the operators a and b. Notice from Eq. (3.159b) it follows that a double application of either creation or destruction operators gives zero so that

$$\left(c^{+}_{\vec{k},\sigma}\right)^{2}=\left(c_{\vec{k},\sigma}\right)^{2}=0. \tag{3.160}$$

This requires that two electrons cannot occupy the same eigenstate in the gas and is a characteristic property of fermions.

In operator notation, the Hamiltonian of the free electron gas is given by [4–6]

$$H_0=\sum_{\vec{k}}\sum_{\sigma=\downarrow}^{\uparrow}\epsilon_{\vec{k},\sigma}c^{+}_{\vec{k},\sigma}c_{\vec{k},\sigma}=\sum_{\vec{k}}\sum_{\sigma=\downarrow}^{\uparrow}\epsilon_{\vec{k},\sigma}n_{\vec{k},\sigma} \tag{3.161}$$

where

$$n_{\vec{k},\sigma}=c^{+}_{\vec{k},\sigma}c_{\vec{k},\sigma} \tag{3.162}$$

is the number operator of the states of $\vec{k},\sigma$ in the system. The number operator in Eq. (3.162) then counts the number of $\vec{k},\sigma$ electrons in the system, so that

$$n_{\vec{k},\sigma}\left|\epsilon_{\vec{k},\sigma}\right\rangle=\left|\epsilon_{\vec{k},\sigma}\right\rangle \tag{3.163a}$$

and otherwise

$$n_{\vec{k},\sigma}\left|0\right\rangle=0. \tag{3.163b}$$

For example, considering the state $\left|\epsilon_{\vec{q},\uparrow},\epsilon_{\vec{k},\uparrow},\epsilon_{\vec{p},\downarrow}\right\rangle=c^{+}_{\vec{q},\uparrow}c^{+}_{\vec{k},\uparrow}c^{+}_{\vec{p},\downarrow}\left|0\right\rangle$ given in Eq. (3.158c), it follows that [4–6]

$$\sum_{\vec{r}}\sum_{\sigma}n_{\vec{r},\sigma}\left|\epsilon_{\vec{q},\uparrow},\epsilon_{\vec{k},\uparrow},\epsilon_{\vec{p},\downarrow}\right\rangle=3\left|\epsilon_{\vec{q},\uparrow},\epsilon_{\vec{k},\uparrow},\epsilon_{\vec{p},\downarrow}\right\rangle \tag{3.164a}$$

and

$$\begin{aligned}H_0\left|\epsilon_{\vec{q},\uparrow},\epsilon_{\vec{k},\uparrow},\epsilon_{\vec{p},\downarrow}\right\rangle&=\sum_{\vec{r}}\sum_{\sigma=\downarrow}^{\uparrow}\epsilon_{\vec{r},\sigma}n_{\vec{r},\sigma}\left|\epsilon_{\vec{q},\uparrow},\epsilon_{\vec{k},\uparrow},\epsilon_{\vec{p},\downarrow}\right\rangle\\&=\left(\epsilon_{\vec{q},\uparrow}+\epsilon_{\vec{k},\uparrow}+\epsilon_{\vec{p},\downarrow}\right)\left|\epsilon_{\vec{q},\uparrow},\epsilon_{\vec{k},\uparrow},\epsilon_{\vec{p},\downarrow}\right\rangle.\end{aligned} \tag{3.164b}$$

Here of course the sums over $\vec{r}$ and σ are over all the possible states in the gas.

Some particularly useful number operator relationships arising from Eq. (3.159) are

$$\left\{n_{\vec{k},\sigma},c_{\vec{q},\sigma'}\right\}=c_{\vec{k},\sigma'}\delta_{\vec{k},\vec{q}}\delta_{\sigma,\sigma'} \tag{3.165a}$$

and

$$\left\{n_{\vec{k},\sigma},c^{+}_{\vec{q},\sigma'}\right\}=c^{+}_{\vec{q},\sigma'}\delta_{\vec{k},\vec{q}}\delta_{\sigma,\sigma'}. \tag{3.165b}$$

These are presently found of interest in the study of the scattering states of the electron gas as it interacts with a magnetic impurity. This brings us to the treatment of the Anderson model.

3.5.2 Anderson Hamiltonian and the s-d Hamiltonian

The Hamiltonian of the Anderson model generalizes the electron gas model H_0 in Eq. (3.161) to treat the electrons of the gas as weakly interacting with an atomic impurity having unfilled electron energy levels [2, 4–6]. A common application, in this regard, is in the study of the conduction electrons scattering from an impurity with incompletely filled d – or f – localized electron orbitals. The specific form of the Anderson Hamiltonian is then given by [2, 4–6]

$$H = \epsilon_d \left[n_{d,\uparrow} + n_{d,\downarrow} \right] + U n_{d,\uparrow} n_{d,\downarrow} + \sum_{\vec{k},\sigma} \epsilon_{\vec{k},\sigma} n_{\vec{k},\sigma} + \sum_{\vec{k},\sigma} \left[v_{\vec{k}} c^+_{d,\sigma} c_{\vec{k},\sigma} + v^*_{\vec{k}} c^+_{\vec{k},\sigma} c_{d,\sigma} \right] \tag{3.166}$$

where $n_{d,\sigma} = c^+_{d,\sigma} c_{d,\sigma}$ is the number of electrons present on the impurity atom (characterized by the energy terms $\epsilon_d \left[n_{d,\uparrow} + n_{d,\downarrow} \right] + U n_{d,\uparrow} n_{d,\downarrow}$), $n_{\vec{k},\sigma} = c^+_{\vec{k},\sigma} c_{\vec{k},\sigma}$ is the number of $\vec{k},\sigma$ electrons present in the electron gas (characterized by the energy $\sum_{\vec{k},\sigma} \epsilon_{\vec{k},\sigma} n_{\vec{k},\sigma}$ where we assume $\epsilon_{\vec{k},\sigma} = \epsilon_{\vec{k}}$ is independent of σ), and $v_{\vec{k}}$ is a weak scattering interaction which transfers electrons between the impurity atom and the electron gas. A focus in the study of the Anderson model will be on systems which exhibit net magnetic moments developed on the impurity.

The first two terms on the right of Eq. (3.166) are the energy of the single impurity atom, which is dependent on the impurity electron occupation numbers $n_{d\sigma}$. It is seen that the possible energies associated with the isolated impurity are [2, 4–6]

$$E_0 = 0 \text{ when no electrons are on the impurity,} \tag{3.167a}$$

$$E_1 = \epsilon_d \text{ when one electron is on the impurity,} \tag{3.167b}$$

and

$$E_2 = 2\epsilon_d + U \text{ when two electrons are on the impurity.} \tag{3.167c}$$

In this regard, the impurity binds a spin up and/or a spin down electron or no electrons. The third term on the right is the energy from Eq. (3.161) of the electron states in the electron gas. Near zero temperature and for weak interaction with the impurity, the lowest energy orbitals of $\sum_{\vec{k},\sigma} \epsilon_{\vec{k},\sigma} n_{\vec{k}\sigma}$ are primarily occupied up to some maximum electron energy orbital denoted as the fermi energy ϵ_f. The fourth term on the right is a scattering interaction which moves electrons between the impurity and the electron gas. Notice that the scattering in the fourth term is the simplest form of a Hermitian operator conserving the number of particles in the system as they are shuttled back and forth between the impurity and the gas. It represents the new physics in the system arising from the impurity interaction with the electron gas.

We are interested in studying how a net magnetic moment can develop on the impurity atom in the system with the interaction between the electron gas and the impurity. Consequently, let us focus on the system [2, 4–6] parameterized so that (See the energy schematic in Figure 3.7).

$$\epsilon_d < \epsilon_f \le \mu < 2\epsilon_d + U \tag{3.168}$$

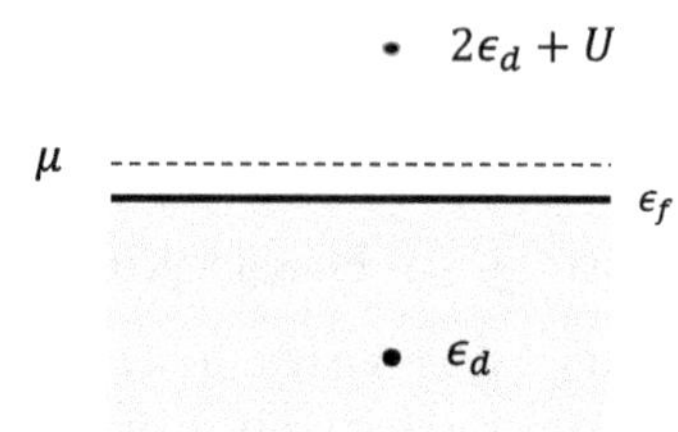

FIGURE 3.7 Schematic of the energy levels of the Anderson model where the energy of the levels increases vertically upward. The occupied energies of the electron gas are shaded and located below the fermi energy denoted ϵ_f. The energy level of the impurity with single electron occupancy is ϵ_d and with double electron occupancy is $2\epsilon_d + U$. The chemical potential of the system is denoted by μ.

where ϵ_f is the fermi energy of the electron gas and μ is the chemical potential of the system. For this configuration of the impurity (and in the case of a chemical potential μ which is much less than $2\epsilon_d + U$ but greater than or equal to ϵ_f) it is preferred for a single electron to occupy the impurity atom site. This is because it is the only electron occupancy of the impurity which has an energy below the fermi energy and the chemical potential. In our low-temperature calculations, the single electron impurity configuration is of interest as it leads to a net magnetic moment on the impurity atom.

The computation next determines the effects in Eq. (3.166) arising from the electron transfers between the electron gas and the impurity. This comes from the weak scattering of the fourth term in Eq. (3.166). Our considerations are also made at very low temperatures where the effects of scattering and temperature fluctuations introduce only small changes from the noninteracting system.

To begin the treatment of the eigenvalue problem of Eq. (3.166) subject to the conditions of Eq. (3.168), assume that we know the eigenvector solution to the problem is given by the wavefunction $|\psi\rangle$ and its eigenenergy is E. Consequently, with this notation the eigenvalue problem is [2, 4–6]

$$H|\psi\rangle = E|\psi\rangle. \tag{3.169}$$

From our conditions in Eq. (3.168), we expect in this eigenstate that the impurity will overwhelmingly have a single electron occupancy, with only a small mixture of the other possible impurity configurations. This again relies on the conditions of weak scattering.

Consequently, let us define three projection operators which act on the Hilbert space of the wavefunction $|\psi\rangle$ to project it into three different subspaces: a subspace with no electrons on the impurity, a subspace with one electron on the impurity, and a subspace in which the impurity is doubly occupied by electrons. For these projection operators, we adopt the notation [2, 4–6]:

$$P_0 = \text{projects out states in which the impurity contains no electrons,} \tag{3.170a}$$

$$P_1 = \text{projects out states in which the impurity contains one electron,} \tag{3.170b}$$

$$P_2 = \text{projects out states in which the impurity contains two electrons.} \tag{3.170c}$$

Operating on $|\psi\rangle$ it then follows that [2, 5, 6]

$$P_0|\psi\rangle = |\psi_0\rangle \text{ the part of the wavefunction containing no impurity electrons,} \tag{3.170d}$$

$$P_1|\psi\rangle = |\psi_1\rangle \text{ the part of the wavefunction containing one impurity electron,} \tag{3.170e}$$

$$P_2 |\psi\rangle = |\psi_2\rangle \text{the part of the wavefunction containing two impurity electrons.} \quad (3.170f)$$

These projections are important regarding our conditions in Eq. (3.168) as it is expected that $|\psi_1\rangle$ is a much larger component of the wavefunction $|\psi\rangle$ than either $|\psi_0\rangle$ or $|\psi_2\rangle$. We now set out to study the properties of $|\psi_1\rangle$ by rewriting the eigenvalue problem from Eq. (3.166) in terms of the eigenvalue problem of an effective Hamiltonian acting only on the eigenvector $|\psi_1\rangle$.

Within the basis $\{|\psi_0\rangle, |\psi_1\rangle, |\psi_2\rangle\}$ we write the wave function $|\psi\rangle$ as

$$|\psi\rangle = \sum_{i=0}^{2} |\psi_i\rangle, \quad (3.171a)$$

and the eigenvalue problem in Eq. (3.169) of the Hamiltonian in Eq. (3.166) then takes the matrix form [2, 5, 6]

$$\begin{vmatrix} H_{00} & H_{01} & 0 \\ H_{10} & H_{11} & H_{12} \\ 0 & H_{21} & H_{22} \end{vmatrix} \begin{vmatrix} |\psi_0\rangle \\ |\psi_1\rangle \\ |\psi_2\rangle \end{vmatrix} = E \begin{vmatrix} |\psi_0\rangle \\ |\psi_1\rangle \\ |\psi_2\rangle \end{vmatrix} \quad (3.171b)$$

where E is the energy eigenvalue of Eqs. (3.166) and (3.169), $H_{ij} = P_i H P_j$ for $i, j = 0,1,2$, and we have taken $H_{02} = H_{20} = 0$ as H only changes the electron impurity occupancy by one electron. From the top row of Eq. (3.171b) we find that $H_{00}|\psi_0\rangle + H_{01}|\psi_1\rangle = E|\psi_0\rangle$ so that, consequently, it follows that [2, 5, 6]

$$|\psi_0\rangle = \frac{1}{E - H_{00}} H_{01} |\psi_1\rangle, \quad (3.172a)$$

and from the bottom row of Eq. (3.171b) we find $H_{21}|\psi_1\rangle + H_{22}|\psi_2\rangle = E|\psi_2\rangle$ so that

$$|\psi_2\rangle = \frac{1}{E - H_{22}} H_{21} |\psi_1\rangle. \quad (3.172b)$$

From the middle row of Eq. (3.171b) we then find $H_{10}|\psi_0\rangle + H_{11}|\psi_1\rangle + H_{12}|\psi_2\rangle = E|\psi_1\rangle$ and applying Eqs. (3.172a) and (3.172b) to this gives [2, 5, 6]

$$\left\{ H_{10} \frac{1}{E - H_{00}} H_{01} + (H_{11} - E) + H_{12} \frac{1}{E - H_{22}} H_{21} \right\} |\psi_1\rangle = 0 \quad (3.172c)$$

or rewritten in the format of an eigenvalue problem

$$\left\{ H_{10} \frac{1}{E - H_{00}} H_{01} + H_{11} + H_{12} \frac{1}{E - H_{22}} H_{21} \right\} |\psi_1\rangle = E|\psi_1\rangle \quad (3.172d)$$

The resulting Eqs. (3.172c) and (3.172d) are now a reduced or effective eigenvalue problem for E and $|\psi_1\rangle$.

The operator form in the $\{\ \}$ of Eqs. (3.172c) and (3.172d) can be simplified using some operator identities. To this end, consider the first term in the brackets of Eq. (3.172c),

$$H_{10} \frac{1}{E - H_{00}} H_{01} |\psi_1\rangle, \quad (3.173)$$

and the process of its simplification. From Eq. (3.166) it follows that [2, 5, 6]

$$H_{01}|\psi_1\rangle = P\sum_{\vec{k},\sigma} v^*_{\vec{k}} c^+_{\vec{k},\sigma} c_{d,\sigma} |\psi_1\rangle \tag{3.174}$$

removes the one electron on $|\psi_1\rangle$ and places it in the electron gas mode $\vec{k},\sigma$. (Note in this regard that $\sum_{\vec{k},\sigma} v^*_{\vec{k}} c^+_{\vec{k},\sigma} c_{d,\sigma}$ is the only operator in Eq. (3.166) that reduces the number of electrons on the impurity by one). Consequently, the left side of Eq. (3.174) has no electrons on the impurity and P_0 is a bit redundant.

Since the perturbation is small, it changes the fermi surface very little from that of the zero-temperature fermi gas, so that we expect that the $\vec{k},\sigma$ state into which the electron on the impurity was sent is near the fermi surface, i.e., this accounts for a minimal change in the fermi surface. With this assumption, it then follows that [2, 5, 6]

$$\begin{aligned} H_{10}\frac{1}{E-H_{00}}H_{01}|\psi_1\rangle &= H_{10}P_0\frac{1}{E-H_{00}}\sum_{\vec{k},\sigma} v^*_{\vec{k}} c^+_{\vec{k},\sigma} c_{d,\sigma} |\psi_1\rangle \\ &\approx H_{10}\sum_{\vec{k},\sigma} v^*_{\vec{k}} c^+_{\vec{k},\sigma} c_{d,\sigma} \frac{1}{E-\left(H_{11}+\epsilon_{\vec{k},\sigma}-\epsilon_d\right)}|\psi_1\rangle \\ &\approx \left(\sum_{\vec{k}',\sigma'} v_{\vec{k}'} c^+_{d,\sigma'} c_{\vec{k}',\sigma'}\right)\sum_{\vec{k},\sigma} v^*_{\vec{k}} c^+_{\vec{k},\sigma} c_{d,\sigma} \frac{-1}{\epsilon_{\vec{k},\sigma}-\epsilon_d}|\psi_1\rangle \end{aligned} \tag{3.175}$$

Here, in the first approximation leading to the second line, we have used the anti-commutators in Eq. (3.165) to move H_{01} from the right to the left of $\frac{1}{E-H_{00}}$. In the course of this movement, we have also approximated $\frac{1}{E-H_{00}}$ using from Eq. (3.166) that

$$H_{11} = P_1\left[\epsilon_d + \sum_{\vec{k},\sigma} \epsilon_{\vec{k},\sigma} n_{\vec{k},\sigma}\right]P_1 \tag{3.176a}$$

and

$$H_{00} = P_0\sum_{\vec{k},\sigma} \epsilon_{\vec{k},\sigma} n_{\vec{k},\sigma} P_0. \tag{3.176b}$$

Consequently, it approximately follows that [2, 5, 6]

$$\frac{1}{E-H_{00}}\sum_{\vec{k},\sigma} v^*_{\vec{k}} c^+_{\vec{k},\sigma} c_{d,\sigma} |\psi_1\rangle \approx \sum_{\vec{k},\sigma} v^*_{\vec{k}} c^+_{\vec{k},\sigma} c_{d,\sigma} \frac{1}{E-\left(H_{11}-\epsilon_d+\epsilon_{\vec{k},\sigma}\right)}|\psi_1\rangle \tag{3.176c}$$

as the electron with energy ϵ_d on $|\psi_1\rangle$ is removed and placed in a free particle state $\vec{k},\sigma$ that is expected to be near the fermi surface. Notice, however, that [2, 5, 6]

$$E - H_{11} \approx 0 \tag{3.176d}$$

because, for the small perturbation, H_{11} approximates the energy of the unperturbed system and must be very close to the eigenvalue of the perturbed system. Applying this we arrive at the expression on the third line of Eq. (3.175).

Continuing, consider now the last line in Eq. (3.175). This can be rewritten as [2, 5, 6]

$$H_{10}\frac{1}{E-H_{00}}H_{01}\left|\psi_1\right\rangle=\left(\sum_{\vec{k}',\sigma'}v_{\vec{k}'}c^{+}_{d,\sigma'}c_{\vec{k}',\sigma'}\right)\sum_{\vec{k},\sigma}v^{*}_{\vec{k}}c^{+}_{\vec{k},\sigma}c_{d\sigma}\frac{-1}{\epsilon_{\vec{k},\sigma}-\epsilon_d}\left|\psi_1\right\rangle \tag{3.177}$$

$$=\sum_{\vec{k},\sigma}\sum_{\vec{p},\sigma'}\frac{v_p v^{*}_{\vec{k}}}{\epsilon_{\vec{k},\sigma}-\epsilon_d}c^{+}_{d,\sigma'}c^{+}_{\vec{k}\sigma}c_{\vec{p},\sigma'}c_{d,\sigma}\left|\psi_1\right\rangle$$

where, again, the scattering is largely about the fermi surface, and we have applied the anticommutation relations in Eq. (3.159). It is now helpful to introduce a new notation which allows us to write the scattering of the electrons in the gas by the impurity as arising from an interaction of the gas electrons with the spin on the impurity.

It is useful to rewrite Eq. (3.177) into an effective Hamiltonian form involving the spin operators of the electrons occupying the impurity. These are given by [2, 5, 6]

$$S^z=\frac{n_{d,\uparrow}-n_{d,\downarrow}}{2}, \tag{3.178a}$$

$$S^{+}=S^{x}+iS^{y}=c^{+}_{d,\uparrow}c_{d,\downarrow} \tag{3.178b}$$

and

$$S^{-}=S^{x}-iS^{y}=c^{+}_{d,\downarrow}c_{d,\uparrow} \tag{3.178c}$$

where S^z, S^+, S^- operate on the spin state of the impurity. The operator S^z gives the net z- component of the impurity spin in terms of the electron occupancy on the impurity, while S^+ raises the z-component of the impurity spin by one and S^- lowers the z- component of the impurity spin by one. In terms of these operators, Eq. (3.177) has the form [2, 5, 6]

$$H_{10}\frac{1}{E-H_{00}}H_{01}\left|\psi_1\right\rangle\approx \tag{3.179}$$

$$\approx\sum_{\vec{k}}\sum_{\vec{p}}\frac{v_{\vec{p}}v^{*}_{\vec{k}}}{\epsilon_{\vec{k}}-\epsilon_d}\left\{S^{-}c^{+}_{\vec{k},\uparrow}c_{\vec{p},\downarrow}+S^{+}c^{+}_{\vec{k},\downarrow}c_{\vec{p},\uparrow}+S^{z}\left(c^{+}_{\vec{k},\uparrow}c_{\vec{p},\uparrow}-c^{+}_{\vec{k},\downarrow}c_{\vec{p},\downarrow}\right)+\frac{1}{2}\sum_{\sigma}c^{+}_{\vec{k},\sigma}c_{\vec{p},\sigma}\right\}\left|\psi_1\right\rangle$$

where we have used the spin independence of the electron gas, i.e., $\epsilon_{\vec{k}}=\epsilon_{\vec{k},\sigma}$.

Proceeding by a similar consideration to that in Eqs. (3.173) through (3.179) we can also show that

$$H_{12}\frac{1}{E-H_{22}}H_{21}\left|\psi_1\right\rangle\approx \tag{3.180}$$

$$\approx\sum_{\vec{k}}\sum_{\vec{p}}\frac{v_{\vec{p}}v^{*}_{\vec{k}}}{U+\epsilon_d-\epsilon_{\vec{p}}}\left\{S^{-}c^{+}_{\vec{k},\uparrow}c_{\vec{p},\downarrow}+S^{+}c^{+}_{\vec{k},\downarrow}c_{\vec{p},\uparrow}+S^{z}\left(c^{+}_{\vec{k},\uparrow}c_{\vec{p},\uparrow}-c^{+}_{\vec{k},\downarrow}c_{\vec{p},\downarrow}\right)-\frac{1}{2}\sum_{\sigma}c^{+}_{\vec{k},\sigma}c_{\vec{p},\sigma}\right\}\left|\psi_1\right\rangle.$$

The reasoning involved in obtaining Eq. (3.180) is along the same lines as that in our discussions leading up to Eq. (3.179).

From Eqs. (3.172), (3.179), and (3.180) we then arrive at the effective Hamiltonian eigenvalue problem [2, 5, 6]

$$H_{eff}\left|\psi_1\right\rangle=E\left|\psi_1\right\rangle \tag{3.181}$$

where

$$H_{eff} = H_{11} + \sum_{\vec{k}} \sum_{\vec{p}} J_{\vec{k},\vec{p}} \left\{ S^- c^+_{\vec{k},\uparrow} c_{\vec{p},\downarrow} + S^+ c^+_{\vec{k},\downarrow} c_{\vec{p},\uparrow} + S^z \left(c^+_{\vec{k},\uparrow} c_{\vec{p},\uparrow} - c^+_{\vec{k},\downarrow} c_{\vec{p},\downarrow} \right) \right\} \tag{3.182}$$

$$+ \sum_{\vec{k}} \sum_{\vec{p}} K_{\vec{k},\vec{p}} \sum_{\sigma} c^+_{\vec{k},\sigma} c_{\vec{p},\sigma}$$

with

$$J_{\vec{k},\vec{p}} = \frac{v^*_{\vec{k}} v_{\vec{p}}}{\epsilon_{\vec{k}} - \epsilon_d} + \frac{v^*_{\vec{k}} v_{\vec{p}}}{U + \epsilon_d - \epsilon_{\vec{p}}} > 0 \tag{3.183a}$$

for an antiferromagnetic spin coupling between the spins of the conduction and impurity electrons, and [2, 5, 6]

$$K_{\vec{k},\vec{p}} = \frac{1}{2} \left(\frac{v^*_{\vec{k}} v_{\vec{p}}}{\epsilon_{\vec{k}} - \epsilon_d} - \frac{v^*_{\vec{k}} v_{\vec{p}}}{U + \epsilon_d - \epsilon_{\vec{p}}} \right). \tag{3.183b}$$

The effective Hamiltonian [5] in Eqs. (3.181) through (3.183) characterizes the effective interaction of the spins on the impurity with the states in the electron gas. This gives an effective Hamiltonian eigenvalue problem for $|\psi_1\rangle$ which is rewritten in terms of a consideration of the scattering of the electrons in the conduction band of the electron gas as they are scattered from the spin developed on the impurity. The Hamiltonian form in Eq. (3.182) is often referred to as the s-d Hamiltonian and represents a reformulation of the Anderson problem to handle the specific energy configuration in Eq. (3.168).

3.5.3 Application to the Kondo Effect: Resistivity Increase with Temperature Decrease

The low-temperature origins of the increase in resistivity with decreasing temperature can be understood in the context of a perturbation theory calculation based on the effective s-d Hamiltonian. Specifically, we consider an approach to determining the current flow in the electron gas based on the Boltzmann transport equation in a treatment involving a generalized form of the s-d Hamiltonian. The considerations begin with the study of the case of a single impurity characterized by the form [2, 5, 6]

$$H_{eff} = \sum_{\vec{k},\sigma} \epsilon_{\vec{k},\sigma} n_{\vec{k},\sigma} \tag{3.184}$$

$$+ \sum_{\vec{k}} \sum_{\vec{p}} J_{\vec{k},\vec{p}} \left\{ S^- c^+_{\vec{k},\uparrow} c_{\vec{p},\downarrow} + S^+ c^+_{\vec{k},\downarrow} c_{\vec{p},\uparrow} + S^z \left(c^+_{\vec{k},\uparrow} c_{\vec{p}\uparrow} - c^+_{\vec{k},\downarrow} c_{\vec{p},\downarrow} \right) \right\}$$

$$= \sum_{\vec{k},\sigma} \epsilon_{\vec{k},\sigma} n_{\vec{k},\sigma} + \mathcal{J} \sum_{\vec{k}} \sum_{\vec{p}} \left\{ S^- c^+_{\vec{k},\uparrow} c_{\vec{p},\downarrow} + S^+ c^+_{\vec{k},\downarrow} c_{\vec{p},\uparrow} + S^z \left(c^+_{\vec{k},\uparrow} c_{\vec{p}\uparrow} - c^+_{\vec{k},\downarrow} c_{\vec{p},\downarrow} \right) \right\}$$

where $\epsilon_{\vec{k},\sigma} = \frac{1}{2m} \left(\hbar \vec{k} \right)^2$, and as a simplification we take $J_{\vec{k},\vec{p}} = \mathcal{J} = \frac{J}{N_s}$ where N_s is the number of conduction electrons in the electron gas. As shall be seen in the course of the following development, the essential nature of the resistivity increase can be modeled based on the results from the scattering of the electron gas by the single s-d impurity modeled in Eq. (3.184). In this regard, the effects of the s-d impurities will enter the treatment as a dissipative term dependent only on the concentration of impurities in the system and arising from phase incoherent impurity scattering.

Our focus then becomes the determination of the current density of the conduction electrons subject to s-d impurity scattering. In this regard, the current density is defined as

$$\vec{j} = -en\langle \vec{v} \rangle \tag{3.185}$$

where $e > 0$ is the unit of electric charge, n is the density of conduction electrons, and $\langle \vec{v} \rangle$ is the thermal average of the electron velocity. In the Boltzmann transport approach, the average electron velocity of the gas is determined directly from a study of the time dependence of the thermal distribution of conduction electrons.

In the formulation of the Boltzmann transport equation, the equilibrium Fermi-Dirac distribution of electrons, in the absence of an applied electric field, is given by

$$f(\vec{k}) = \frac{1}{1 + e^{\beta(\epsilon_{\vec{k}} - \mu)}} \tag{3.186}$$

where $\epsilon_{\vec{k}} = \epsilon_{\vec{k},\sigma}$ is the spin-independent energy of the electron orbitals of the gas, μ is the chemical potential, $\beta = \frac{1}{k_B T}$, T is the temperature, and k_B is the Boltzmann constant. Notice that a homogeneous system is treated which does not depend on the position in space $\vec{r}$. The object is then to compute how the distribution of electrons in Eq. (3.186) changes in time with the application of an external electric field to the system. In this regard, if the system at time t evolved from its state at $t - dt$ where dt is an infinitesimal, then the system at time t is related to that at $t - dt$ by [2, 4–6]

$$f(\vec{k},t) = f\left(\vec{k} - \frac{\partial \vec{k}}{\partial t}dt, t - dt\right) + \left(\frac{d}{dt} f(\vec{k})\right)_{collisions} dt \tag{3.187a}$$

$$= f(\vec{k},t) - \frac{\partial \vec{k}}{\partial t} \cdot \nabla_{\vec{k}} f(\vec{k}) dt - \frac{\partial f(\vec{k},t)}{\partial t} dt + \left(\frac{d}{dt} f(\vec{k})\right)_{collisions} dt.$$

Here the first term in the first equality on the right describes the evolution due to the semi-classical dynamics of the Hamiltonian in the absence of impurity and phonon scattering, and the second term comes from a consideration of collisions with impurities and phonons in the system.

Specifically, if $f_{\vec{E}}(\vec{k})$ is the Fermi-Dirac distribution evolving in an external field $\vec{E}$, it follows that its dynamics is described as a process involving the acceleration of each of its momentum states $\vec{k}$ by the external force. From this consideration and for the steady state in which $\frac{\partial f_{\vec{E}}(\vec{k},t)}{\partial t} = 0$, we find from the far right side of Eq. (3.187a) that [2, 4–6]

$$\frac{\partial \vec{k}}{\partial t} \cdot \nabla_{\vec{k}} f_{\vec{E}}(\vec{k}) = \left(\frac{d}{dt} f_{\vec{E}}(\vec{k})\right)_{collisions} \tag{3.187b}$$

where the acceleration of the electron with momentum $\vec{k}$ due to $\vec{E}$ is given by the semi-classical expression

$$\frac{\partial \vec{k}}{\partial t} = -\frac{e\vec{E}}{\hbar}. \tag{3.187c}$$

The dynamics arising from the electric field then enters through $\frac{\partial \vec{k}}{\partial t}$ and, according to Eq. (3.187c), the acceleration of the distribution from this term is along the field direction.

To arrive at a model of a steady state conduction process, however, the rate of change in Eq. (3.187) must also take into account the relaxation of the system arising from electron collisions with impurities and with thermal disorder. These processes are accounted for on the right side of Eq. (3.187b). As we shall see, the nature of the relaxation in the system is then observed as a tendency to push the system back to the unperturbed state of the Fermi-Dirac distribution given by $f(\vec{k})$ from Eq. (3.186). A consequence of this is that when the applied electric field is switched off, the system eventually returns to its configuration $f(\vec{k})$ in the absence of the applied field.

An appropriate form describing such a relaxation is [1, 2, 4–6]

$$\left(\frac{d}{dt}f_{\vec{E}}(\vec{k})\right)_{collisions} = -\frac{f_{\vec{E}}(\vec{k})-f(\vec{k})}{\tau(\vec{k})} \tag{3.188}$$

where $\tau(\vec{k})$ is a characteristic relaxation time for excitations of wavevector $\vec{k}$. This is a commonly used phenomenological form in studies of the Boltzmann transport equation. Combining the dynamical terms in Eq. (3.187) with the relaxation terms in Eq. (3.188), we then find that

$$\frac{f_{\vec{E}}(\vec{k})-f(\vec{k})}{\tau(\vec{k})} = \frac{e\vec{E}}{\hbar}\cdot\nabla_{\vec{k}}f_{\vec{E}}(\vec{k}) \tag{3.189}$$

characterizes the steady state dynamics of our system. In our particular study, we shall see that the explanation of the Kondo effect resides in the determination of the relaxation time $\tau(\vec{k})$.

We now focus on the terms which determine the steady state transport solutions of Eq. (3.189). These are obtained by rewriting Eq. (3.189) into the form [1, 2, 4–6]

$$f_{\vec{E}}(\vec{k}) = f(\vec{k})+\tau(\vec{k})\frac{e\vec{E}}{\hbar}\cdot\nabla_{\vec{k}}f_{\vec{E}}(\vec{k}) \approx f(\vec{k})+\tau(\vec{k})\frac{e\vec{E}}{\hbar}\cdot\nabla_{\vec{k}}f(\vec{k}) \tag{3.190}$$

where $f_{\vec{E}}(\vec{k}) \approx f(\vec{k})$ has been approximated on the far right side of the equation. From Eq. (3.185) it then follows that

$$\vec{j} = -2e\int f_{\vec{E}}(\vec{k})\vec{v}\frac{d\vec{k}}{(2\pi)^3} = -2e\int \tau(\vec{k})\frac{e\vec{E}}{\hbar}\cdot\nabla_{\vec{k}}f(\vec{k})\vec{v}\frac{d\vec{k}}{(2\pi)^3} \tag{3.191}$$

$$= -2\frac{e^2\tau(\vec{k}_f)}{m}\vec{E}\cdot\int\left[\nabla_{\vec{k}}f(\vec{k})\right]\vec{k}\frac{d\vec{k}}{(2\pi)^3}$$

where the factor of 2 in the first equality comes from the spin degeneracy of the electron energy levels, in the second equality the first $f(\vec{k})$ term in Eq. (3.190) contributes no current to the system, for the third equality $\vec{v} = \frac{\hbar\vec{k}}{m}$, and $\tau(\vec{k}_f)$ is the relaxation time of the electrons at the fermi surface. (Notice in regard to the third equality, the gradient of the Fermi-Dirac distribution is highly peaked at the fermi surface.) Integrating by parts over the wavevector components on the far right of Eq. (3.191) then gives [1, 2]

$$\vec{j} = \frac{e^2\tau(\vec{k}_f)n}{m}\vec{E} \tag{3.192}$$

where n is the density of conduction electrons. The result in Eq. (3.192) is the standard Drude form for the electron conductivity, i.e., $\sigma = \frac{e^2\tau(\vec{k}_f)\, n}{m}$. Our focus next turns to the determination of the relaxation time $\tau(\vec{k}_f)$ which ultimately explains the Kondo observations.

In computing the current density from Eq. (3.192) we need to determine the relaxation time $\tau(\vec{k}_f)$ for an electron of wavevector $\vec{k}$ to be elastically scattered into a conduction electron state $\vec{k}' \neq \vec{k}$ near the fermi surface. These processes most contribute to the electron transport properties in the system. Let us then start by determining $\tau(\vec{k})$ for a single impurity in the gas within the context of the Born or golden rule approximation. This approximation represents the contribution to the relaxation from the lowest order scattering processes.

Considering the s-d Hamiltonian in Eq. (3.184), we find that the free particle electron gas states are described by $H_0 = \sum_{\vec{k},\sigma} \epsilon_{\vec{k},\sigma} n_{\vec{k},\sigma}$ and the scattering potential for the free particle states is given by the potential scattering term describing the interaction of the electron gas with the single impurity [2, 4–6], i.e.,

$$H_I = \mathcal{J} \sum_{\vec{k}} \sum_{\vec{p}} \left\{ S^- c^+_{\vec{k},\uparrow} c_{\vec{p},\downarrow} + S^+ c^+_{\vec{k},\downarrow} c_{\vec{p},\uparrow} + S^z \left(c^+_{\vec{k},\uparrow} c_{\vec{p},\uparrow} - c^+_{\vec{k},\downarrow} c_{\vec{p},\downarrow} \right) \right\}. \tag{3.193}$$

Of particular interest are matrix elements of H_I between the plane wave states of the gas which enter into the fermi golden rule for transitions from states of $\vec{k}$ into states of $\vec{k}'$. Specifically, at zero temperature the rate of such transitions $\frac{1}{\tau_0(\vec{k})}$ between wavevector states at the fermi surface is given by [1, 2, 4–6]

$$\frac{1}{\tau_0(\vec{k})} = \frac{2\pi}{\hbar} \sum_{\vec{k}'} \sum_{\sigma,\sigma'=\downarrow}^{\uparrow} \langle \vec{k},\sigma | H_I | \vec{k}',\sigma' \rangle \langle \vec{k}',\sigma' | H_I | \vec{k},\sigma \rangle \delta\left(\epsilon_{\vec{k}} - \epsilon_{\vec{k}'}\right). \tag{3.194a}$$

(Note: $\tau_0(\vec{k})$ differs from $\tau(\vec{k})$ in that $\tau_0(\vec{k})$ sums over the two spin polarizations of the wavevector states while $\tau(\vec{k})$ involves only a single electron polarization.) In this regard, our interest is in [4–6]

$$\sum_{\vec{k}'} \sum_{\sigma,\sigma'=\downarrow}^{\uparrow} \langle \vec{k},\sigma | H_I | \vec{k}',\sigma' \rangle \langle \vec{k}',\sigma' | H_I | \vec{k},\sigma \rangle \delta\left(\epsilon_{\vec{k}} - \epsilon_{\vec{k}'}\right) \tag{3.194b}$$

$$= \left[S^- S^+ + S^+ S^- + 2\left(S^z\right)^2 \right] \left(\frac{J}{N_s} \right)^2 \sum_{\vec{k}'} \delta\left(\epsilon_{\vec{k}} - \epsilon_{\vec{k}'}\right),$$

$$= 2S(S+1) \left(\frac{J}{N_s} \right)^2 \sum_{\vec{k}'} \delta\left(\epsilon_{\vec{k}} - \epsilon_{\vec{k}'}\right),$$

which occurs in the fermi golden rule for elastic scattering transition rates, and where we consider transitions involving all the degenerate spin polarizations. We shall focus on these contributions to $\tau(\vec{k})$ in the vicinity of the fermi surface and ignore other phonon scattering and nonmagnetic impurity scattering contributions to the relaxation. Such other processes are known to become small at low temperatures and do not contribute to the Kondo effect.

The matrix elements of Eq. (3.194b) enter into the fermi golden rule scattering at zero temperature. For the temperature dependent Kondo discussion, however, they need to be generalized to handle thermal effects. This is done by rewriting Eq. (3.194) accounting for the electron occupancies of the energy states of the system. These are given by the Fermi-Dirac distribution. Making such an accounting for the transitions in the thermal model and taking the continuum limit gives [1, 4–6]

$$\left(\frac{d}{dt} f_{\vec{E}}\left(\vec{k}\right)\right)_{collisions} = -\frac{f_{\vec{E}}\left(\vec{k}\right)-f\left(\vec{k}\right)}{\tau\left(\vec{k}\right)} \tag{3.195}$$

$$\approx -\frac{2\pi}{\hbar}\nu\int\frac{d\vec{k}'}{(2\pi)^3}\delta\left(\epsilon_{\vec{k}}-\epsilon_{\vec{k}'}\right)\frac{1}{2}\sum_{\sigma,\sigma'=\downarrow}^{\uparrow}\left\langle\vec{k},\sigma|H_I|\vec{k}',\sigma'\right\rangle\left\langle\vec{k}',\sigma'|H_I|\vec{k},\sigma\right\rangle f_{\vec{E}}\left(\vec{k}\right)\left[1-f_{\vec{E}}\left(\vec{k}'\right)\right]$$

$$+\frac{2\pi}{\hbar}\nu\int\frac{d\vec{k}'}{(2\pi)^3}\delta\left(\epsilon_{\vec{k}}-\epsilon_{\vec{k}'}\right)\frac{1}{2}\sum_{\sigma,\sigma'=\downarrow}^{\uparrow}\left\langle\vec{k},\sigma|H_I|\vec{k}',\sigma'\right\rangle\left\langle\vec{k}',\sigma'|H_I|\vec{k},\sigma\right\rangle f_{\vec{E}}\left(\vec{k}'\right)\left[1-f_{\vec{E}}\left(\vec{k}\right)\right]$$

$$= -\frac{2\pi}{\hbar}\nu\int\frac{d\vec{k}'}{(2\pi)^3}\delta\left(\epsilon_{\vec{k}}-\epsilon_{\vec{k}'}\right)\frac{1}{2}\sum_{\sigma,\sigma'=\downarrow}^{\uparrow}\left\langle\vec{k},\sigma|H_I|\vec{k}',\sigma'\right\rangle\left\langle\vec{k}',\sigma'|H_I|\vec{k},\sigma\right\rangle\left[f_{\vec{E}}\left(\vec{k}\right)-f_{\vec{E}}\left(\vec{k}'\right)\right]$$

where ν is the volume of the electron gas and the factor $\frac{1}{2}$ in front of the summation signs averages over the two polarization states of $\vec{k}$ so that $\tau\left(\vec{k}\right)$ is for a single electron. In Eq. (3.195) the first term on the right gives the transition rate for elastic scattering of $\vec{k}$ into $\vec{k}'$, where $f_{\vec{E}}\left(\vec{k}\right)$ is the probability that $\vec{k}$ is occupied and $1-f_{\vec{E}}\left(\vec{k}'\right)$ is the probability that $\vec{k}'$ is unoccupied. Similarly, the second term on the right gives the elastic transition rate for elastic scattering of $\vec{k}'$ into $\vec{k}$, where $f_{\vec{E}}\left(\vec{k}'\right)$ is the probability that $\vec{k}'$ is occupied and $1-f_{\vec{E}}\left(\vec{k}\right)$ is the probability that $\vec{k}$ is unoccupied. The two processes combined account for the net elastic scattering involving $\vec{k}$ and $\vec{k}'$ occurring near the fermi surface.

Next define

$$f_1\left(\vec{k}\right)=f_{\vec{E}}\left(\vec{k}\right)-f\left(\vec{k}\right) \tag{3.196a}$$

and

$$f_1\left(\vec{k}'\right)=f_{\vec{E}}\left(\vec{k}'\right)-f\left(\vec{k}'\right) \tag{3.196b}$$

so that it follows from the last equality in Eq. (3.195) that

$$\frac{1}{\tau\left(\vec{k}\right)}=\frac{2\pi}{\hbar}\nu\int\frac{d\vec{k}'}{(2\pi)^3}\delta\left(\epsilon_{\vec{k}}-\epsilon_{\vec{k}'}\right)\frac{1}{2}\sum_{\sigma,\sigma'=\downarrow}^{\uparrow}\left\langle\vec{k},\sigma|H_I|\vec{k}',\sigma'\right\rangle\left\langle\vec{k}',\sigma'|H_I|\vec{k},\sigma\right\rangle\left[1-\frac{f_1\left(\vec{k}'\right)}{f_1\left(\vec{k}\right)}\right]. \tag{3.197}$$

In the case of elastic scattering near the fermi surface we then have [1, 2, 4–6]

$$\frac{f_1\left(\vec{k}'\right)}{f_1\left(\vec{k}\right)}=\cos\theta_{k'} \tag{3.198}$$

where $\theta_{k'}$ is the angle between $\vec{k}'$ and $\vec{k}$. For s-wave scattering the cosine term in Eq. (3.197) from Eq. (3.198) drops out so that for relaxation near the fermi surface

$$\frac{1}{\tau(\vec{k})} = \frac{2\pi}{\hbar}\nu\int\frac{d\vec{k}'}{(2\pi)^3}\delta\left(\epsilon_{\vec{k}}-\epsilon_{\vec{k}'}\right)\frac{1}{2}\sum_{\sigma,\sigma'=\downarrow}^{\uparrow}\left\langle\vec{k},\sigma|H_I|\vec{k}',\sigma'\right\rangle\left\langle\vec{k}',\sigma'|H_I|\vec{k},\sigma\right\rangle \tag{3.199}$$

$$= \frac{2\pi}{\hbar}S(S+1)\left(\frac{J}{N_s}\right)^2\nu\int\frac{d\vec{k}'}{(2\pi)^3}\delta\left(\epsilon_{\vec{k}}-\epsilon_{\vec{k}'}\right)$$

$$= \frac{(2\pi)^2}{\hbar}S(S+1)\left(\frac{J}{N_s}\right)^2\nu\int\frac{(k')^2\,dk'}{(2\pi)^3}\sin\theta_{k'}d\theta_{k'}\delta\left(\epsilon_{\vec{k}}-\epsilon_{\vec{k}'}\right)$$

$$= \frac{(2\pi)^2 S(S+1)}{\hbar}\left(\frac{J}{N_s}\right)^2\frac{1}{2}\frac{D(\epsilon_f)}{4\pi}2 = \frac{3\pi S(S+1)N_s}{2\hbar\epsilon_f}\left(\frac{J}{N_s}\right)^2 = \frac{3\pi S(S+1)J^2}{2\hbar\epsilon_f N_s}$$

where $D(\epsilon_f) = \frac{3N_s}{2\epsilon_f}$ for the 3-dimensional electron gas. Here $\tau(\vec{k}) = \tau(\vec{k}_f)$ is for the scattering of a single electron at the fermi surface. It is spin independent even though it is the scattering rate for one of the two degenerate spin politization scattering processes.

At this point it is also important to note that the result in Eq. (3.199) is the rate for scattering from a single isolated impurity, not a uniform distribution of impurities. In the treatment of the Kondo effect, it is assumed that the impurity concentration is small enough that the rate of scattering from each impurity of the material is independently additive, i.e., phase interference effects between the impurity scattering of different impurities are less important than the incoherent sum of the scattering from different impurities. Consequently, the Kondo scattering rate $\frac{1}{\tau_{Kondo}(\vec{k})}$ is given from Eq. (3.199) by [4–6]

$$\frac{1}{\tau_{Kondo}(\vec{k})} = \frac{N_{imp}}{\tau(\vec{k})} = \frac{3\pi S(S+1)J^2}{2\hbar\epsilon_f}\frac{N_{imp}}{N_s} = c_{imp}\frac{3\pi S(S+1)J^2}{2\hbar\epsilon_f} \tag{3.200}$$

where N_{imp} is the number of magnetic impurities in the system, and c_{imp} is the fractional concentration of impurities.

Returning to Eq. (3.192), it now follows that the resistivity of the system is given from Eq. (3.200) by{4–6]

$$\rho_{Kondo} = \frac{1}{\sigma_{Kondo}} = \frac{m}{e^2 n}\frac{1}{\tau_{Kondo}(\vec{k})} = \frac{3\pi mJ^2 S(S+1)}{2e^2\hbar\epsilon_f}\frac{c_{imp}}{n}. \tag{3.201}$$

Notice that in this lowest order scattering from the magnetic impurities the contribution to the resistivity is made only to the residual ρ_0 (i.e., temperature independent) term in Eq. (3.156). Consequently, we have not yet arrived at an explanation for the Kondo effect. The explanation of the Kondo effect is obtained from a consideration of the next higher order terms in the single impurity scattering process.

To obtain the next order scattering terms of interest in the perturbation theory we continue to treat scattering from a single impurity site, going to the next order of single site impurity scattering. This is done by returning to the expression in Eq. (3.194a) for the transition rate from first order or Born scattering processes in H_I and including processes of second order in H_I for each of the single site impurities of concentration c_{imp} embedded in the metal. Specifically, if Eq. (3.200) is recomputed to include terms of order J^3 it is found that [4–6]

$$\frac{1}{\tau_{Kondo}(\vec{k})} = c_{imp}\frac{3\pi S(S+1)J^2}{2\hbar\epsilon_f}\left[1 - 2J\left(g(\epsilon_f) - g^*(\epsilon_f)\right)\right] \tag{3.202}$$

$$= c_{imp}\frac{3\pi S(S+1)J^2}{2\hbar\epsilon_f}\left[1 - \frac{2JD(\epsilon_f)}{\mathcal{V}}\ln\frac{k_B T}{D}\right]$$

where $D(\epsilon_f) = \frac{3N_s}{2\epsilon_f}$ and

$$g(\epsilon_f) = 2P\int_0^D \frac{f(\epsilon_{\vec{k}})}{\epsilon_{\vec{k}} - \epsilon_f}\frac{d\vec{k}}{(2\pi)^3}. \tag{3.203}$$

Here D is an energy cutoff parameter which is chosen to be of order of the bandwidth of the conduction electrons. The cutoff D is introduced to limit the logarithmic divergence of the integral which occurs in the unphysical limit that $D \to \infty$. Its presence is an indication that a renormalization of the perturbation theory series is needed. This involves summing an infinite set of divergent terms in the treatment of the full perturbation theory series. The reader is referred to the literature [5–6] for a discussion of this renormalization.

Using the result in Eq. (3.202) to recompute the resistivity in Eq. (3.201), it then follows that [4–6]

$$\rho_{Kondo} = \frac{m}{e^2 n}\frac{1}{\tau_{Kondo}(\vec{k})} = \frac{3\pi m J^2 S(S+1)}{2e^2\hbar\epsilon_f}\frac{c_{imp}}{n}\left[1 - \frac{2JD(\epsilon_f)}{\mathcal{V}}\ln\frac{k_B T}{D}\right]. \tag{3.204}$$

Here again phase-coherent processes involving the scattering from multiple different impurities in the system are ignored. Notice, in this regard, that it is seen from the result in Eq. (3.204) that the condition for the Kondo corrections to become dominant is that

$$1 = -\frac{2JD(\epsilon_f)}{\mathcal{V}}\ln\frac{k_B T}{D}. \tag{3.205}$$

This condition defines the so-called Kondo temperature [4–6]

$$k_B T_{Kondo} = De^{-\frac{\mathcal{V}}{D(\epsilon_f)J}} \tag{3.206}$$

which sets the temperature scale at which Kondo effects should become important.

In the $\frac{k_B T}{D} \ll 1$ limit it follows for $J > 0$ that the resistivity in Eq. (3.204) diverges. This is then the origin of the Kondo effect. Again, in Figure 3.6 a schematic plot is shown of a typical result of a study of resistivity versus temperature for Kondo materials. The temperature scale encompasses temperatures well below the Debye temperature.

As a final point: An interesting, experimentally observed result in general Kondo systems is obtained if we sum the resistivity from the phonon scattering in Eq. (3.156) with that for the Kondo scattering in Eq. (3.204). For this sum it is found, retaining only the temperature-dependent terms, that for constants a and b

$$\rho_{Total}(T) = \rho_{phonon}(T) + \rho_{Kondo}(T) = aT^5 - c_{imp}b\ln\frac{k_B T}{D}. \tag{3.207}$$

Consequently, it follows that the minimum of the resistivity occurs at the temperature [4–6]

$$T_{min} = \left[\frac{c_{imp} b}{5a} \right]^{\frac{1}{5}} \propto c_{imp}^{\frac{1}{5}}. \tag{3.208}$$

The temperature of the minimum resistivity in Kondo materials is generally observed to scale with the magnetic impurity concentration.

3.6 DISLOCATIONS: LINE DEFECTS AND PLASTICITY

We next consider a different type of imperfection in crystalline materials that is characterized by a more complex geometric structure than that of point or site impurities [7–13]. These new types of imperfections are so-called line defects which occur in the structure of the crystal lattice itself and which are commonly known as dislocations. Dislocations arise as local disruptions of the periodicity of the crystal lattice and are found as mismatches along lines passing through the space of the lattice. They tend to be present in all crystalline materials and are ultimately responsible for the plastic properties of materials associated with irreversible changes in the shape of crystals. They also enter as significant factors involved in the development of grain boundaries within materials composed of crystalline domains. In addition, dislocations play important roles as mechanisms involved in the condensation process during a gas-solid or liquid-solid transformation of a material into crystal form. In this regard, it is difficult to condense a material into a crystal lattice without the introduction into the lattice of these types of line imperfections [7–13].

Two common forms of dislocations found in crystals are the so-called edge and screw dislocations. The edge dislocation shows up in the crystal as an additional half plane of atoms introduced into the lattice. (See Figure 3.8a for a schematic of a simple form of edge dislocation.) If we consider a crystal as a layering of identical atomic planes, an edge dislocation can be thought of as being formed by replacing one of the atomic planes of the crystal with a half plane of atoms. Of course, in this process there is a certain amount of relaxation of the atoms near the half plane, particularly in the vicinity of the edge of the half plane in the bulk of the crystal. The edge of the half plane within the bulk region is known as the core of the dislocation and is very important in studies of the dynamics of the dislocation as it moves through the crystal. It is the line characterizing the dislocation as a line imperfection of the crystal lattice [7–12].

A screw dislocation, on the other hand, can be generated from a perfect crystal lattice. (See Figure 3.8b for a schematic of a simple form of screw dislocation.) To do this, one breaks the nearest-neighbor lattice bonds between two neighboring half planes of the lattice. The edge of the half plane region in the bulk (where the bond breaking between the two half planes ends) forms the axis of the dislocation. This is then the line defining the screw dislocation as a line defect and is known as the core of the screw dislocation. The dislocation is then formed in the region of cut bonds by shifting one edge of the cut upward (e.g., the U lip in Figure 3.8b) parallel to the axis of the cut and the other edge downward (e.g., the L lip in Figure 3.8b) parallel to the axis of the cut. This is done so that there is a relative displacement of the two original cut lattice regions by one lattice constant (i.e., a smallest distance of the periodicity of the lattice). The nearest-neighbor bonds are then reintroduced between the new sets of nearest-neighbor pairs between the upper and lower halves of the cut. Again, in this process there is a certain amount of relaxation of the atoms near the half planes of the cut, particularly in the vicinity of the axis of the dislocation in the bulk of the crystal. The region around the axis is known as the core of the screw dislocation and is very important in studies of the dynamics of the dislocation as it moves through the crystal. Notice that in the presence of the screw dislocation the lattice assumes a sort of spiral staircase structure about the axis of the dislocation [7–12].

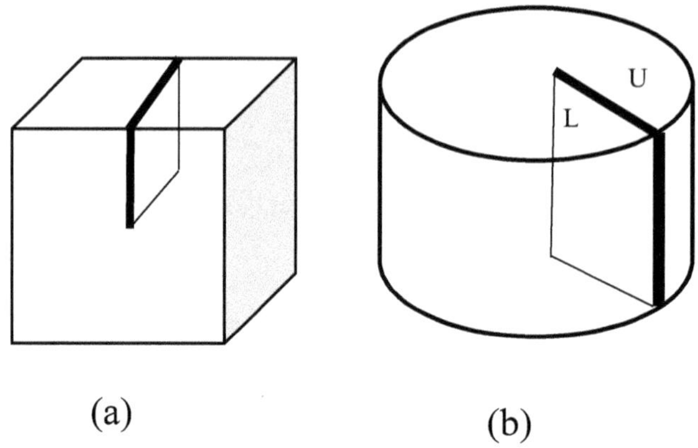

FIGURE 3.8 Schematic of: a) A cubic macroscopic piece of crystal containing an edge dislocation formed as an added half plane of atoms (indicated in black). The core of the dislocation is the edge of the added plane contained within the bulk of the crystal. b) A screw dislocation created in a bulk cylindrical piece of crystal. The dislocation is formed by breaking the bonds in the rectangular region forming the cut. The vertical light gray line in the bulk is the axis of the screw dislocation which is also the dislocation core, and the dislocation is formed by shifting the upper lip of the cut (denoted by U) and the lower lip of the cut (denoted L) relative to one another by one lattice constant in a direction parallel to the dislocation axis.

We see that the core region in both types of dislocations represents a distortion of the crystal lattice itself. In addition, the distortion of the core regions can be further generalized to other types of line dislocations exhibiting properties reminiscent of either edge or screw dislocations. For example, if the inserted half-plane of the edge dislocation is not rectangular in shape, one arrives at a more complex variety of line imperfection. For a discussion of these varieties and generalizations made from the edge and screw dislocations, the reader is referred to the literature [7–12]. Here we will address some of the basic properties of the edge and screw dislocations. In this regard, we next turn to a discussion of how dislocations determine the plasticity or irreversible deformation properties of crystals.

3.6.1 Plasticity

To begin, we estimate the plasticity of a material condensed into the form of a perfect crystalline lattice, i.e., we will study the stresses and strains required to make an irreversible change in the material shape [7]. For simplicity consider a hypothetical system composed of identical atoms arrayed on a cubic lattice having nearest neighbor atomic separations a. In Figure 3.9 a schematic is presented of a cube of side a in the lattice with atoms located at each of the vertices of the cube. If a shear stress τ_0 (defined in Figure 3.9a as the force per area acting in the x-direction in the $x-y$ plane) is applied between the top and bottom planes containing the top and bottom of the cube, the cube will be distorted by a simple shear strain σ_0 so that the top and bottom of the cube will be shifted relative to one another. Here, for the distorted cube in Figure 3.9b the simple shear strain is defined by $\sigma_0 = \frac{\delta a}{a}$.

A way of plastically and irreversibly deforming the crystal is to slide the top plane over the bottom plane by at least an atomic separation a. This would move the upper plane from one equilibrium position to its next nearest equilibrium position. The shear stress required to do this is approximately the shear stress to make the distortion $\delta a = \frac{1}{2}a$ which has the simple shear strain $\sigma_0 = \frac{\delta a}{a} = \frac{1}{2}$. In this regard, once the upper plane has slipped half an atomic separation over the

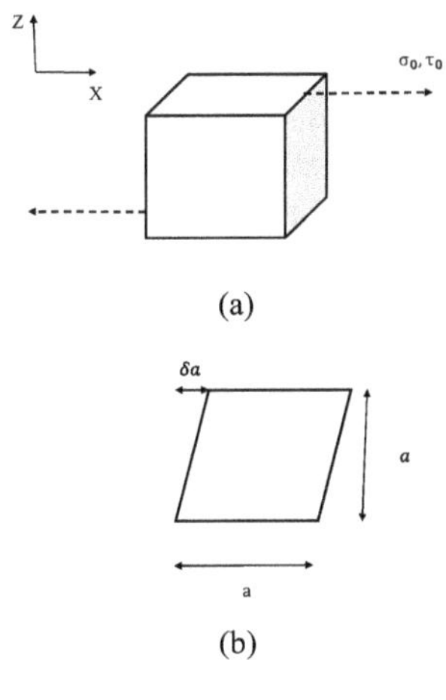

FIGURE 3.9 Schematic of: a) Eight nearest neighbor atoms on the vertices of a cubic crystal lattice. A shear stress τ_0 in the $x-y$ plane is applied in the x-direction between the top and bottom planes of the cube of nearest neighbors causing a simple strain σ_0 of the $x-y$ plane along the x – direction. b) A shear stress applied between the top and bottom planes in the x – direction causes a distortion δ_a in the x – direction of the top plane relative to the bottom plane and the nearest neighbor separation in the unstress crystal is a.

lower plane, by symmetry the stress required to finish the rest of the journey of length a should be the same. The stress and strain, however, are related by Hooke's law so that [7]

$$\tau_0 = \mu\sigma_0 \tag{3.209}$$

where the shear modulus μ is obtained in terms of the elastic constants of the medium. Consequently, from Eq. (3.209) the stress needed to cause a plastic deformation (i.e., an irreversible shape change in the crystal) is given by [7]

$$\tau_0 = \frac{1}{2}\mu. \tag{3.210}$$

For a metal such as magnesium (which has a hexagonal close-packed structure) $\tau_0 = \frac{2.5\times10^6}{2} = 1.25\times10^6\,psi$. This is about 10^4 times the observed required stress for an irreversible plastic deformation of the magnesium crystal [7].

The reason for the great disparity in the plasticity of magnesium between our theory and the experimental data is due to the mechanism we have used to explain the irreversible plastic deformation of the magnesium crystal. Specifically, the irreversible shape change of the crystal is not accomplished by the sliding of neighboring atomic planes over one another. This involves a significant distortion of the crystal lattice. Rather, it is made by moving a dislocation core through the lattice. This motion is accomplished by translating the core of the dislocation along a plane of the crystal lattice. In this regard, the core represents a compact localized region of greatest deviation of the dislocation from the periodicity of the lattice [7–12]. As we travel away from the core region of the dislocation, the periodicity of the crystal lattice is quickly regained so that the region outside the core has a smaller influence on the motion of the dislocation. In this sense, the motion of a plane and the motion of a dislocation are significantly different in terms of the facilitation of their movements within the lattice.

The plane on which the dislocation moves is known as a slip plane, and the motion along this plane arises through the application of a shear stress applied across the slip plane. In the mechanism of plasticity based on the motion of dislocations, the change in the crystal structure is made mainly through the atomic distortions in the dislocation core as it translates along the slip plane. Consequently, if an edge dislocation on one side of a crystal is moved along a slip plane to the other side of a crystal, an irreversible plastic shape change is introduced into the crystal. Similarly, if a screw dislocation on one side of a crystal is moved along a slip plane to the other side of a crystal, an irreversible plastic shape change is again introduced into the crystal. As demonstrated in computer simulation studies of the dislocation core, the core motion requires much less change in the crystal and much less stress than does the translation of an entire plane of atoms over another.

Another important property of dislocations is their role in the crystallization process [7]. In this regard, dislocations are naturally introduced into crystals during the process in which the solid is condensed out of its vapor or liquid state. During these processes, the atoms of the solid interface between the bulk solid and the liquid or gas are less stably bound in the solid than are those atoms deep in the condensed bulk of the crystal. Specifically, they have fewer solid bonds with their atomic neighbors, particularly with neighboring liquid state atoms. In this regard, the more solid bonds an atom can form, the more stably it will become attached to the condensing solid. For example, at a planar solid surface, an atom may develop a single solid bond attaching it to the surface. However, if the surface is not planar but has a 90° ledge or corner on the surface, it may be possible for an atom to bond at the bottom of the ledge or corner via multiple bonds. (See Figure 3.10 for a schematic.) This is a much stabler process which provides a natural way to form, for example, a screw dislocation. For the screw dislocation, due to the relative shift between the upper and lower lips in the cut region of its formation, a natural ledge or corner on the solid surfaces perpendicular to the

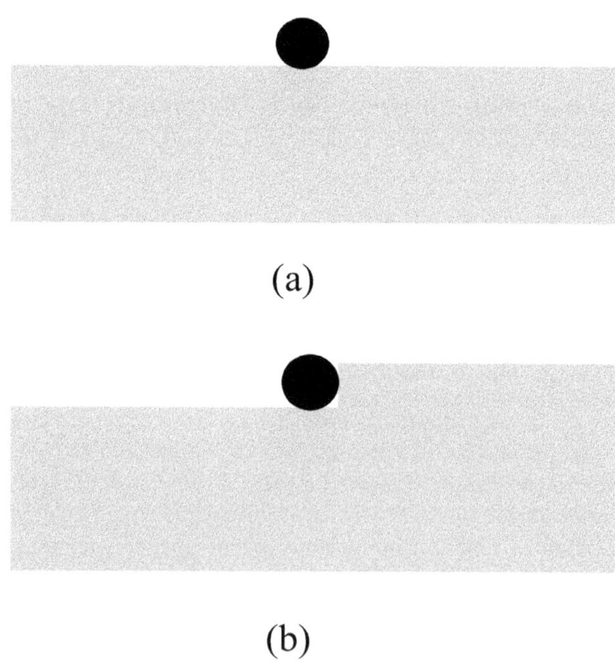

FIGURE 3.10 Schematic of: (a) The binding of a single atom on a planar surface between a solid (grey) and a liquid (white) region. Here a single bond is made by the atom to the surface. (b) The binding of a single atom on a surface with a A schematic with two panels compares atom binding at a planar solid surface and at a ledge, highlighting single and double bond formation. ledge between a solid (grey) and a liquid (white) region. Notice that here two bonds (made at A schematic with two panels compares atom binding at a planar solid surface and at a ledge, highlighting single and double bond formation. to one another) are made by the atom to the surface.

dislocation axis facilitates the formation process. Screw dislocations are generally observed on the surface of condensed metals as spiral edges of uneven crystal growth formed into ledges. Similarly, edge dislocations may enter in the condensation process, with their edges providing regions of attraction for unattached atoms entering into the process of solidification.

Consequently, dislocations occur in crystals as part of their natural condensation process, so that it is often difficult to avoid the formation of a crystal that does not contain dislocations. We shall next discuss the influence of dislocations on the elastic and plasticity processes observed in crystalline materials. The focus will be on how these properties arise directly from the dynamics of dislocations within the crystal lattice. To understand this, we next consider the stress-strain plot found in general crystalline materials and the contributions of dislocation to the features of this plot.

3.6.2 Stress-Strain Plot of Crystals

The elastic properties of a crystalline material are found to fall into a number of stress-strain regions. In a first region, involving small applied stresses, the system obeys Hooke's law (linear stress-strain laws) type of stress-strain relationship [7]. This is a region of reversible deformation as releasing the material from the applied stress causes it to return to its original shape. At some stress, known as the yield stress, the system begins to deform irreversibly. This is the second stress-strain region and involves a region in which the dislocations begin to move through the crystal [7]. The motion of dislocations occurring in this region causes irreversible shape changes in the crystal and the shape changes increase with increasing strain.

Since stationary dislocations (due to the distortions associated with their strain fields) tend to attract site impurities in the crystal, once the dislocations begin to move, they lose their associated impurity clouds. As a result, less stress than the yield stress is often required for the continued motion of the dislocations. This loss of restraint leads to a third region of the stress-strain curve[7]. Eventually, however, as the dislocations move, they tend to collide with one another and with other types of impurities. This gives rise to a fourth region which is one of so-called work hardening in which more stress is required for dislocation motion [7]. In Figure 3.11 is presented a hypothetical stress-strain plot for a typical system of dislocations indicating the variety of regions of elasticity

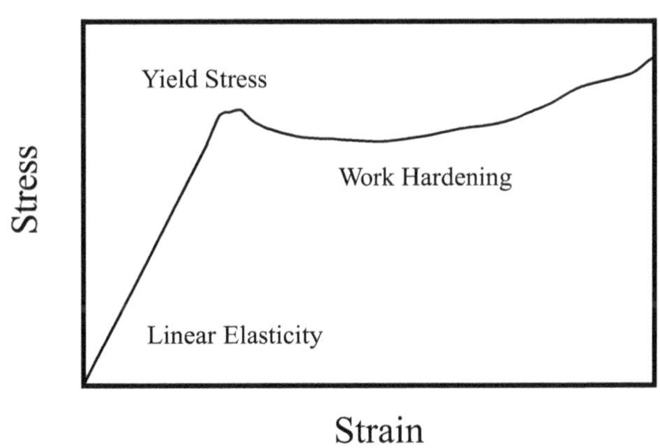

FIGURE 3.11 Schematic Stress versus Strain plot of a hypnotical crystalline solid. At the far left of the plot is a region of elastic behavior in which the crystal shape is restored upon releasing the stress on the crystal. Proceeding to the right, at the yield stress the crystal plastically and irreversibly deforms as the dislocations break away from their inertial restraints and begin motion along their slip planes. This is followed by a region of work hardening as one proceeds to the far right of the figure. In the work hardening region the dislocations entangle with one another and with impurities and grain boundaries in the system. This inhibits their motion through the crystal.

and plastic flow. Depending on the material the relative importance of the four different regions may be amplified or diminished from one another.

Another source of dislocations in crystals arises from stacking faults or in solids in which the condensation is into a solid phase composed as a collection of crystallites or crystal grains. Within each grain, the system is a crystal with crystal geometry, but the crystal axes of the grains are randomly oriented with respect to one another. In this regard, the randomly oriented grains are separated by grain boundaries from which dislocations can arise. These are another complicating feature of crystalline plasticity.

3.6.3 Grain Boundaries and Stacking Faults

Crystals often occur in composite structures, formed as a mixture of little crystallites [7]. Each crystallite, itself, is a small crystal, but the crystallites forming the composite may not share the same crystallographic axes. The regions between the crystallites are referred to as grain boundaries and have intermediate atomic structures mediating a transition between two different crystallites with different crystallographic axes. The grain boundaries can act as sources and sinks for dislocations introduced into the crystallites and in this way ultimately determine the plastic deformations of the total system. In this regard, the grain boundaries are important features contributing to the net plastic properties exhibited by crystals formed as composite structures.

Similar types of domain structures occur in ferromagnetic materials. In a ferromagnet, one finds it energetically stable to break the magnetization of the system into a composite of magnetic domains each having a different orientation of its permanent magnetization. This type of configuration is instrumental in reducing the total magnetic field generated by the permanent magnet domains, making for a lower energy state of the system [1, 2, 7].

Another type of transition region is that created by stacking faults [7]. Some crystals are formed by stacking a series of different planes of atoms on one another. For example, given a plane labeled as A, on top of A is placed a second different type of plane which we labeled B. In this way, the entire crystal may be formed as the layering ABABAB of these two types of planes, continuing this indefinitely to form an infinite crystal. In this way the layering sequence ABABAB... is observed in the generation of the hcp or hexagonally close-packed crystals. By interrupting the stacking sequence an imperfection (a stacking fault) can be created, e.g., consider the sequences ABABBBAB of repeated B's or ABABCBAB where C is a third different plane in the layering sequence. The stacking faults offer another type of separation between crystallites of a material and again enter into determining the plastic properties of the material.

In general, we see that the study of dislocations is a complicated many-body problem, often requiring a computer simulation. Some insight, however, can be obtained from an early treatment of dislocation properties given on the basis of continuum elasticity theory. In this regard, we conclude our study with the continuum treatment of edge and screw dislocations. This is followed by an interesting discussion of a simple one-dimensional lattice model for a dislocation, illustrating the importance of the detailed atomic core structure in a dislocation.

3.6.4 The Stress-Strain Fields of Dislocations

As an interesting problem in elasticity theory, we next look at the nature of the stress and strain fields for edge and screw dislocations created within a uniform isotropic elastic medium. This provides an illustration of the general features of these types of dislocations, though there are certain difficulties with the elasticity theory description involving regions of the dislocations in which the theory breaks down. The breakdown of the theory arises from the application of the ideas representing the atomic system as a continuum involving the small displacements of an elastic medium. In the breakdown regions of elasticity theory, singular elastic behavior occurs, and the continuum theory must be abandoned in favor of a treatment involving a discussion of the detailed atomic displacements.

An alternative description to elasticity theory is then developed in the context of atomic models of the materials, or by treating the system as a mixture of the continuum theory applied in one region of space matched to an atomic model in another region of space. Nevertheless, the elastic continuum treatment we describe here is valuable as an indication of the general properties of dislocations and often provides a starting point for more advanced considerations [7–12].

In elasticity theory the elastic medium is viewed as a mechanical continuum [7–12]. For an unstressed elastic medium at rest, the equilibrium points of the continuum have the space coordinates $\vec{r} = x_1\hat{i} + x_2\hat{j} + x_3\hat{k}$ which may alternatively be denoted by (x_1, x_2, x_3). In terms of the unstressed configuration, distortions of the medium from its equilibrium configuration are specified by giving the displacement vectors of the points of the medium from their original equilibrium configuration. Consequently, the displacement of the continuum which is initially located at an equilibrium point (x_1, x_2, x_3) is characterized by the displacement vector [7–12]

$$\vec{u}(x_1, x_2, x_3) = (u_1(x_1, x_2, x_3), u_2(x_1, x_2, x_3), u_3(x_1, x_2, x_3)) \tag{3.211}$$

and $\vec{u}(x_1, x_2, x_3) = 0$ represents the stationary unstressed continuum.

Coming from this description of the elastic medium in terms of displacement vectors, a set of elastic strains of the medium can be determined. These elastic strains characterize the deformations of the medium arising due to stresses acting on the material. They are defined in a way so that they are independent of non-elastic motions of the material such as translations and rotations of the medium as a whole.

In this sense, an elastic strain in the system is then characterized in terms of the displacement vectors by the strain tensor [1, 2, 7–12]

$$e_{i,j} = \frac{1}{2}\left(\frac{\partial u_i}{\partial x_j} + \frac{\partial u_j}{\partial x_i}\right) \tag{3.212}$$

where $i, j = 1,2,3$. This is a popular characterization as it is found to be independent of a constant translation or of a uniform rotation of the medium and only treats elastic deformations of the material. It represents a shape change in the medium due to the application of an external stress to the medium and is also ultimately related to the stress necessary to create an elastic change of shape. As a further simplification in our presentation, we shall also limit our discussions to a treatment of the simplest, most highly symmetric form of an elastic medium. This is known as an isotropic elastic medium and provides examples of elastic materials without the complex restrictions arising from the details of the crystal symmetry group. More complex systems can be discussed which have lower degrees of symmetry (i.e., cubic, rhombic, tetragonal symmetries, etc.) but these do not change the basic ideas developed in the study of isotropic materials. The isotropic media, in this regard, have the simplest relationship between the strains and stresses applied in the system.

For such an isotropic elastic medium, the Hooke's law relations for the stresses needed to deform the medium according to a particular strain tensor are given by [1, 2, 7–12]

$$\sigma_{i,i} = 2Ge_{i,i} + \lambda(e_{1,1} + e_{2,2} + e_{3,3}) \tag{3.213a}$$

where $i = 1,2,3$, and

$$\sigma_{i,j} = 2Ge_{i,j} \tag{3.213b}$$

where $i \neq j$ and $i, j = 1,2,3$. Here λ and the shear modulus G are the Lame constants characterizing the elasticity of the isotropic medium under consideration, and the stresses in the system $\sigma_{i,j}$ arising from the strains $e_{i,j}$ are related through Eq. (3.213). Given the stress-strain relationships of our material we can now develop the dynamical equations characterizing the elastic medium. This is done through a straightforward application of Newton's laws of motion.

If we consider the dynamics of an infinitesimal mass element dm contained within an infinitesimal volume element dV of the elastic medium, Newton's laws require that [1, 2, 7–12]

$$\vec{f}dV = \frac{\partial^2 \vec{u}}{\partial t^2}dm = \rho\frac{\partial^2 \vec{u}}{\partial t^2}dV \tag{3.214a}$$

where $\vec{f}$ is the force per volume acting on dm and ρ is the mass per volume of the element. In addition, the force per volume acting on the element is of the form

$$\vec{f} = \nabla\cdot\sigma^T + \vec{f}_{ext}. \tag{3.214b}$$

Here σ^T is the matrix transpose of the stress matrix and $\vec{f}_{ext}$ is the sum of the external forces acting on the medium (e.g., gravity, electrostatic interactions, etc.). In Eq. (3.214b) the first term on the right is obtained from the consideration of a surface integral of the stresses acting over the normal to the surface of dV followed by a subsequent application of the divergence theorem. It represents forces which elastically deform the medium. In our considerations, as a further simplification we will only treat systems for which $\vec{f}_{ext} = 0$ as these forces are not of interest in our basic treatment of the dislocations.

From Eq. (3.214) we then find that [1, 2, 7–12]

$$\rho\frac{\partial^2 \vec{u}}{\partial t^2} = \nabla\cdot\sigma^T. \tag{3.215}$$

In the case of the symmetric isotropic stress tensor in Eq. (3.213) and considering only small elastic displacements, it follows from Eqs. (3.212) through (3.214) that

$$\rho\frac{\partial^2 \vec{u}}{\partial t^2} = \mathrm{G}\nabla^2\vec{u} + (G+\lambda)\nabla(\nabla\cdot\vec{u}). \tag{3.216}$$

For the consideration of an elastostatic medium this in turn reduces to [1, 2, 7–12]

$$\mathrm{G}\nabla^2\vec{u} + (G+\lambda)\nabla(\nabla\cdot\vec{u}) = 0 \tag{3.217a}$$

and from Eq. (3.217a) it also follows for systems which are basically 2−dimensional in nature (i.e., our dislocation models) that [11]

$$\nabla^4\vec{u} = 0. \tag{3.217b}$$

The solutions of Eqs. (3.216) and (3.217) subject to appropriate boundary conditions then yield the displacement fields arising, respectively, from studies in elasticity theory for dynamical and elastostatic problems.

Our next focus is on the structure of the screw and edge dislocations, which are ultimately obtained as solutions of the elastostatic equations in Eq. (3.217).

3.6.5 Elastic Theory of Screw Dislocations

Let us now consider the two different dislocation types and study the relationship between their strain and stress fields within the elastic medium. Beginning with the screw dislocation, consider a dislocation with its axis coinciding with the $z-$ or x_3 $-$axis of space. The displacement vectors of such a screw dislocation are then given by [7–12]

$$u_1(x_1,x_2,x_3) = u_2(x_1,x_2,x_3) = 0 \tag{3.218a}$$

and

$$u_3(x_1,x_2,x_3) = \frac{a\theta}{2\pi} = \frac{a}{2\pi}\arctan\frac{x_2}{x_1} \tag{3.218b}$$

where θ is measured in the plane perpendicular to the dislocation axis which coincides with the polar axis. Notice that this displacement field is consistent with our earlier description of the screw dislocation and is also a solution of the elastostatic equation Eq. (3.217). Here the conditions in Eq. (3.218a) indicate that there is no distortion in the plane perpendicular to the dislocation axis while the condition in Eq. (3.218b) represents the shift by the constant a every time the dislocation axis is circled by 2π. This shift is the definition of the screw dislocation.

Applying Eq. (3.218) in Eq. (3.212) we find that [7–12]

$$e_{11} = e_{22} = e_{33} = e_{12} = e_{21} = 0, \tag{3.219a}$$

$$e_{13} = e_{31} = -\frac{a}{4\pi}\frac{x_2}{x_1^2+x_2^2} = -\frac{a}{4\pi}\frac{\sin\theta}{r}, \tag{3.219b}$$

and

$$e_{23} = e_{32} = \frac{a}{4\pi}\frac{x_1}{x_1^2+x_2^2} = \frac{a}{4\pi}\frac{\cos\theta}{r}. \tag{3.219c}$$

Notice that the lattice constant a, which is of order of the interatomic separations in the bulk unstrained medium, can be taken positive or negative depending on the sense of the dislocation. It is also seen from Eq. (3.219) that the elasticity theory fails at the $r = \sqrt{x_1^2 + x_2^2} = 0$ axis of the dislocation where the strain becomes singular, and otherwise the strain decreases like $\frac{1}{r}$ as one travels away from the axis of the dislocation. In this regard, from the dislocation strains in Eq. (3.219) it is found for $r \approx a$ that [2, 7–12]

$$e_{13} = e_{31} \approx -\frac{1}{4\pi} \tag{3.220a}$$

and

$$e_{23} = e_{32} \approx \frac{1}{4\pi}. \tag{3.220b}$$

These are approximately 10% strains of the medium and provide a rough measure of the radius of the core region in which elasticity theory fails. In this inner core region about the axis of dislocation, a detailed treatment of the individual atomic displacements must be made.

From Eq. (3.213) the stresses associated with the strain fields in Eq. (3.219) are given by

$$\sigma_{11} = \sigma_{22} = \sigma_{33} = \sigma_{12} = \sigma_{21} = 0, \tag{3.221a}$$

$$\sigma_{13} = \sigma_{31} = -\frac{Ga}{2\pi}\frac{x_2}{x_1^2 + x_2^2} = -\frac{Ga}{2\pi}\frac{\sin\theta}{r}, \tag{3.221b}$$

and

$$\sigma_{23} = \sigma_{32} = \frac{Ga}{2\pi}\frac{x_1}{x_1^2 + x_2^2} = \frac{Ga}{2\pi}\frac{\cos\theta}{r}. \tag{3.221b}$$

Again, it is found that the stresses within the core region of the dislocation are outside the limits required for an elastic theory treatment. As the core region is an essential component entering into a description of the dislocation dynamics, this accounts for the inability of elasticity theory to accurately determine the plasticity properties of crystalline materials.

3.6.6 Elastic Theory of Edge Dislocations

We next turn to a brief consideration of the elastic properties of the edge dislocation [11]. Here only an outline is presented, and the reader is referred to the literature [10, 11] for a more detailed treatment of the solution. As in the case with the screw dislocation, the edge dislocation can be formed in an isotropic elastic medium by cutting the medium along a half plane, shifting the two surfaces of the cut relative to one another, and reattaching them. The shift, however, is now made in a different direction relative to the axis of the cut than was made for the screw dislocation.

For the study of the edge dislocation, consider cutting the isotropic elastic medium in the $x_1 < 0$ region of the $x_1 - x_3$ plane, shifting the two surfaces of the cut in the x_1 –direction relative to one another by a, and bonding the adjusted surfaces back together. This is much the same process as was used to form the screw dislocation; however, now the surfaces of the cut are shifted in the x_1 – rather than the x_3 –directions. As in the case of the screw dislocation, the axis of the resulting edge dislocation is along the x_3 –axis of space. In addition, the x_3 –axis is also the axis of the core region of the edge dislocation. More formally, the displacement vectors of the edge dislocation in the $x_1 < 0$ region of the $x_1 - x_3$ half plane are given by the boundary conditions [11]

$$u_1(x_1, x_2 = \pm 0, x_3) = \pm\frac{1}{2}a \tag{3.222a}$$

and

$$u_2(x_1, x_2 = 0, x_3) = u_3(x_1, x_2 = 0, x_3) = 0. \tag{3.222b}$$

These conditions should be compared with the conditions in Eq. (3.218) for the screw dislocations. Notice, however, that unlike Eq. (3.218), which defined the displacement field of the dislocation in all space, the displacements in Eq. (3.222) are only defined over the $x_1 < 0$ region of the $x_1 - x_3$ half plane.

A formal solution over all space of the displacement field of the edge dislocation having a dislocation core along the x_3 –axis is obtained from the elastostatic equation in Eq. (3.217). This yields the displacement fields [11]

$$u_1(x_1, x_2, x_3) = \frac{a}{2\pi}\left[-\arctan\frac{x_1}{x_2} + \frac{x_1 x_2}{2(1-\upsilon)(x_1^2 + x_2^2)} + \frac{\pi}{2}sgn(x_2)\right], \tag{3.223a}$$

$$u_2(x_1, x_2, x_3) = \frac{a}{2\pi}\left[-\frac{1-2\upsilon}{4(1-\upsilon)}\ln(x_1^2 + x_2^2) + \frac{x_2^2}{2(1-\upsilon)(x_1^2 + x_2^2)}\right], \tag{3.223b}$$

and

$$u_3(x_1, x_2, x_3) = 0. \tag{3.223c}$$

Here υ is the Poisson ratio, which is related to the Young modulus E and the shear modulus G by the relationships $E = 2G(1+\upsilon)$ and $\lambda = \frac{E\upsilon}{(1+\upsilon)(1-2\upsilon)}$.

From the relationships in Eq. (3.223) the strain and the stress field tensors associated with the edge dislocation are directly obtained. For an isotropic elastic medium, the strain tensor is given by [11]

$$e_{11} = -\frac{a}{4\pi}\frac{1}{1-\upsilon}\frac{x_2}{\left(x_1^2+x_2^2\right)^2}\left[3x_1^2 + x_2^2 - 2\upsilon\left(x_1^2+x_2^2\right)\right], \tag{3.224a}$$

$$e_{22} = \frac{a}{4\pi}\frac{1}{1-\upsilon}\frac{x_2}{\left(x_1^2+x_2^2\right)^2}\left[x_1^2 - x_2^2 + 2\upsilon\left(x_1^2+x_2^2\right)\right], \tag{3.224b}$$

$$e_{33} = 0 \tag{3.224c}$$

$$e_{12} = e_{21} = -\frac{a}{4\pi}\frac{1}{1-\upsilon}\frac{x_1\left(x_1^2-x_2^2\right)}{\left(x_1^2+x_2^2\right)^2}, \tag{3.224d}$$

and

$$e_{13} = e_{31} = e_{23} = e_{32} = 0. \tag{3.224e}$$

Similarly, from Eq. (3.213) the stress tensor is [11]

$$\sigma_{11} = -Dx_2\frac{3x_1^2+x_2^2}{\left(x_1^2+x_2^2\right)^2}, \tag{3.225a}$$

$$\sigma_{22} = Dx_2\frac{x_1^2-x_2^2}{\left(x_1^2+x_2^2\right)^2}, \tag{3.225b}$$

$$\sigma_{33} = \upsilon\left(\sigma_{11}+\sigma_{22}\right), \tag{3.225c}$$

$$\sigma_{12} = \sigma_{21} = -Dx_1\frac{x_1^2-x_2^2}{\left(x_1^2+x_2^2\right)^2}, \tag{3.225d}$$

and

$$\sigma_{13} = \sigma_{31} = \sigma_{23} = \sigma_{32} = 0. \tag{3.225e}$$

Here [11]

$$D = \frac{Ga}{2\pi(1-\nu)} \tag{3.225f}$$

and

$$\nu = \frac{\lambda}{2(\lambda + G)}. \tag{3.225g}$$

Notice that for the edge dislocation a shear stress (denoted σ_{12}) is created in the $x_1 - x_3$ plane acting in x_1 –direction and a shear stress (denoted σ_{21}) is created in the $x_2 - x_3$ plane acting in x_2 –direction.

By similar considerations to those made for the screw dislocation, it follows that the condition

$$x_1^2 + x_2^2 \approx a^2 \tag{3.226}$$

offers an indicator of the size of the dislocation core region, i.e., the core tends to be in a region about the dislocation axis of radius $r = \sqrt{x_1^2 + x_2^2} \approx a$. In this regard, within the core region, the positions of the individual atoms of the material become important in determining the dynamical properties of the dislocation.

As with the screw dislocation, it is found that the stresses within the core region of the edge dislocation are outside the limits required for a treatment by elasticity theory. As the core region is an essential component entering into a description of the dislocation dynamics, this accounts for the inability of elasticity theory to accurately determine the plasticity of crystalline materials.

3.6.7 Dislocation Core: Frenkel-Kontorova Model

The stress needed to move a dislocation along its slip plane is determined by the detailed atomic structure and motions generated within the core region of the dislocation [2, 14]. An understanding of this usually requires a computer simulation modeling of the arrangements of the individual atoms forming the dislocations and their subsequent motions under a shear stress. However, the ideas of the mechanisms involved in the process can be developed by the consideration of a one-dimensional model of a chain of atoms simulating the properties of a dislocation. This is based on the Frenkel-Kontorova model [2, 14], and it is useful in seeing how the shear stress required to move a dislocation can be much less than that required to slide two entire atomic planes over one another. In addition, this is done in the context of an analytical treatment.

In the Frenkel-Kontorova model, we begin with the consideration of a chain of identical atoms which are coupled together by identical nearest neighbor spring forces. The atoms are located periodically along an x –axis in a chain with nearest neighbor equilibrium atomic separations a, and the atoms are labeled consecutively by integers $n = 0, \pm 1, \pm 2, \ldots, \pm N$ for $N \to \infty$ so that the position of the nth atom is denoted x_n. The potential energy arising from the spring forces between the atoms which are located along the chain at positions $\{x_n\}$ is then given by

$$U_{Harmonic} = \frac{1}{2} k \sum_{n=-\infty}^{\infty} \left[x_{n+1} - x_n - a \right]^2. \tag{3.227}$$

where k is the spring constant of the chain. (See Figure 3.12a for a schematic of the harmonically coupled chain of atoms.)

In addition, the atoms in the chain are also subject to a periodic potential along the x –axis originating from outside the chain. This new potential comes from the interaction of the atoms in the chain with other atoms of the crystal located outside the chain. In this way, for an atom of the chain located at the position x and positioned within the nth segment of the chain (denoted as the interval $na - \frac{a}{2} \le x \le na + \frac{a}{2}$) the new periodic interaction is defined by the potential energy form [2]

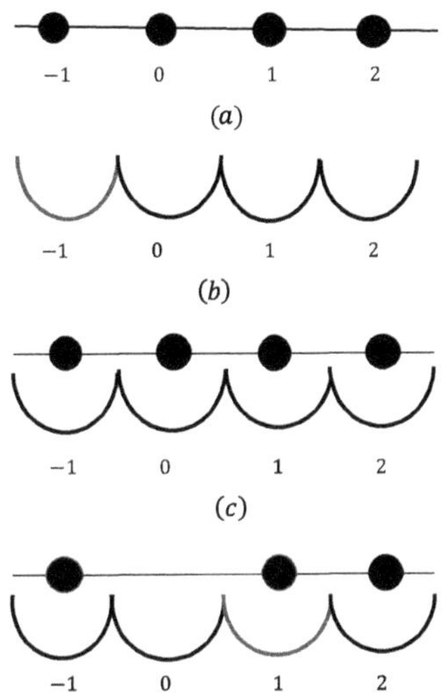

FIGURE 3.12 Schematic of a portion of: a) an infinite chain of atoms coupled together by nearest neighbor springs with equilibrium separation a, b) a periodic potential with periodicity a interacting with the atoms of the chain, c) the combination of the infinite chain of atoms in a) interacting with the periodic potential of b), and d) the combination of the infinite chain of atoms in a) interacting with the periodic potential of b) in which the chain is distorted so that no atom of the chain is located in the 0th well of the periodic potential.

$$U_{Periodic}(x) = \frac{1}{2}k_0\left[x - na\right]^2 \quad \text{for } na - \frac{a}{2} \le x \le na + \frac{a}{2} \tag{3.228}$$

where n is an integer. In Eq. (3.228), k_0 is an interaction constant characterizing the strength of the periodic potential. (See Figure 3.12b for a schematic of the periodic potential formed from the segmental potentials in Eq. (3.228).)

Finally, a third interaction potential is an external force exerted on each of the atoms of the chain. It has the form of an external force given by the potential energy form

$$U_{external}(x) = -fx \tag{3.229}$$

where x is the position of an atom of the chain and f is the constant force acting on the atom. The force f is introduced to simulate a shear stress pulling the atoms along the chain. In our later treatment, it will ultimately cause the motion of dislocations introduced into the atomic configuration of the chain. In this regard, the x-axis of the chain will play the role of a slip plane for the motion of dislocations.

First consider the chain in the absence of a dislocation so that each atom of the chain resides in one of the parabolic segments (given in Eq. (3.228)) of the chain. Summing the three potentials in Eqs. (3.227) through (3.229), we find for this configuration a total potential energy of the chain given by

$$U = \frac{1}{2}k\sum_{n=-\infty}^{\infty}\left[x_{n+1} - x_n - a\right]^2 + \frac{1}{2}k_0\sum_{n=-\infty}^{\infty}\left[x_n - na\right]^2 - \sum_{n=-\infty}^{\infty} fx_n \tag{3.230}$$

where each x_n is restricted to the segment $na - \frac{a}{2} \le x_n \le na + \frac{a}{2}$. In this regard, at equilibrium for $f = 0$ the nth atom of the chain is located in the nth segment of the periodic potential. (See Figure 3.12c for a schematic of the composite system.)

For the case that $f = 0$ the equilibrium configuration of the atoms in Eq. (3.230) occurs for the set $\{x_n\}$ given by

$$x_n = na, \tag{3.231}$$

where n is an integer. As $f > 0$ increases from zero the atoms in Eq. (3.230) continuously exhibit small uniform displacements from their $f = 0$ equilibrium positions in Eq. (3.231). These are obtained by minimizing the total potential energy so that [2]

$$-\frac{\partial U}{\partial x_n} = 0 = k\left[x_{n+1} + x_{n-1} - 2x_n\right] - k_0\left[x_n - na\right] + f. \tag{3.232}$$

The solution of Eq. (3.232) for the form $x_n = na + \delta$ yields the uniform displacement of the chain

$$\delta = \frac{f}{k_0}. \tag{3.233}$$

For our piecewise restrictions in Eq. (3.230), this is valid for $[\delta] < \frac{a}{2}$. Outside these limits, for $[\delta] > \frac{a}{2}$, the atoms travel into the neighboring segment of the periodic potential in Eq. (3.228) and plastic deformation results.

Returning to our considerations in Eqs. (3.209) and (3.210) for estimating the force needed in the absence of a dislocation to slide the chain over its substrate plane, we see that the condition to do this is that the chain displacement in the system is given by $[\delta] = \frac{a}{2}$. At this force the atom will enter the next segment of the periodic potential. From Eq. (3.233) it is then found that the required force for plastic deformation is given by

$$f = k_0\frac{a}{2}. \tag{3.234}$$

This result will now be compared with the force required to move a model dislocation which we introduce into the chain.

To introduce a dislocation into our model, let us make a modification regarding the piecewise nature of Eq. (3.228). In the piecewise condition in Eq. (3.228), each segment of the periodic potential contains an atom of the chain. We will now sequentially place atoms within each interval, except the interval $-\frac{a}{2} \le x \le \frac{a}{2}$ (i.e., the 0th segment) will be left unoccupied by an atom of the chain. This results in an imperfection, destroying the periodicity of the chain-periodic potential system and representing a dislocation introduced into our system. (See Figure 3.12d for a schematic of the composite system containing a dislocation.)

The potential energy of the imperfect system is modified from Eq. (3.230) and is now given by the form [2, 14]

$$U = \frac{1}{2}k\sum_{n=-\infty}^{\infty}\left[x_{n+1} - x_n - a\right]^2 + \frac{1}{2}k_0\sum_{n=1}^{\infty}\left[x_n - na\right]^2 \tag{3.235}$$

$$+\frac{1}{2}k_0\sum_{n=-\infty}^{0}\left[x_n - (n-1)a\right]^2 - \sum_{n=-\infty}^{\infty} f x_n$$

where for $n > 0$ each x_n is restricted by $na - \frac{a}{2} \le x_n \le na + \frac{a}{2}$ and for $n \le 0$ each x_n is restricted by $(n-1)a - \frac{a}{2} \le x_n \le (n-1)a + \frac{a}{2}$. Consequently, there is no atom of the chain in the interval $-\frac{a}{2} \le x \le \frac{a}{2}$ on the x-axis. (See Figure 3.12d for a schematic of the harmonically composite system with a dislocation.)

The resulting force difference equations for the chain are obtained from Eq. (3.235) by computing $f_n = -\frac{\partial U}{\partial x_n}$. In this way, we find for $n > 0$

$$f_n = k\left[x_{n+1} + x_{n-1} - 2x_n\right] - k_0\left[x_n - na\right] + f \tag{3.236a}$$

and for $n \le 0$

$$f_n = k\left[x_{n+1} + x_{n-1} - 2x_n\right] - k_0\left[x_n - (n-1)a\right] + f. \tag{3.236b}$$

The equilibrium solution of Eq. (3.236a) in the region $n > 0$ is of the form

$$x_n = \frac{f}{k_0} + na + A_r e^{-qn} \tag{3.237}$$

which for $f = 0$ is found to exhibit a displacement from the equilibrium configuration in Eq. (3.231) decaying exponentially as $n \to \infty$. From a substitution of Eq. (3.237) in Eq. (3.236a) it follows that

$$k\left[e^{-q} - 2 + e^{q}\right] - k_0 = 0. \tag{3.238}$$

This sets the spatial decay rate q as $n \to \infty$. On the other hand, the solution of Eq. (3.236b) in the region $n \le 0$ is of the form

$$x_n = \frac{f}{k_0} + (n-1)a + A_l e^{qn} \tag{3.239}$$

which for $f = 0$ exhibits a displacement from the equilibrium configuration in Eq. (3.231), decaying exponentially as $n \to -\infty$. From a substitution of Eq. (3.239) in Eq. (3.236b) the condition in Eq. (3.238) is again obtained, setting the same value of q for the rate of decay of the solution on both sides of $n = 0$. In addition, there is now no atom in the parabolic potential centered at the origin. This forms the core region of the lattice imperfection or dislocation.

The two solutions in Eqs. (3.237) and (3.239) must agree where they match up near the origin of the x-axis. Matching Eqs. (3.237) and (3.239) solutions at x_0 it follows that [2]

$$x_0 = \frac{f}{k_0} - a + A_l = \frac{f}{k_0} + A_r, \tag{3.240a}$$

and matching Eqs. (3.237) and (3.239) solutions at x_1 it follows that

$$x_1 = \frac{f}{k_0} + A_l e^{q} = \frac{f}{k_0} + a + A_r e^{-q}. \tag{3.240b}$$

The solutions of these conditions determine the values of A_l and A_r which are found to be

$$A_l = \frac{a}{e^{q} + 1} \text{ and } A_r = -\frac{a}{e^{-q} + 1}. \tag{3.241}$$

Substituting these values of A_l and A_r into Eq.(3.239) a condition for the motion of the dislocation can be obtained. Specifically, the equilibrium of the model will break down when the force f becomes large enough that $x_0 = -\frac{a}{2}$. This is the condition that the left edge of the dislocation will enter the 0th segment and the stability of the model is violated. It then follows from an examination of this condition applied in Eq. (3.239) that [2] for $x_0 = -\frac{a}{2} = \frac{f}{k_0} - a + A_l$ the force f for dislocation motion is given by

$$f = k_0 \frac{a}{2} \tanh \frac{q}{2} < k_0 \frac{a}{2}, \tag{3.242}$$

and since $\left|\tanh \frac{q}{2}\right| < 1$ we see that the force to move the dislocation is less than that in Eq. (3.234) to move the entire chain. The degree of this reduction is found from Eq. (3.238) to depend on the value of q which is obtained from the condition

$$\cosh q = \frac{k_0}{2k} + 1. \tag{3.243}$$

Summarizing, we find that for a dislocation solution to exist it is required that $\frac{k_0}{2k} + 1 > 1$, and as the ratio $\frac{k_0}{2k}$ of the spring constants is adjusted the reduction from the force in Eq. (3.234) is modulated.

As a final point, notice from the forms of Eqs. (3.237) and (3.239) that the lattice disruption of the dislocation is primarily around the origin of the lattice. As one goes to the limits $N \to \pm\infty$ the periodic nature of the chain is restored. Consequently, it is the local nature of the disruption of the atomic order in the chain at the dislocation core which is found to reduce the force required for dislocation motion.

REFERENCES

1. Ashcroft, N. W. and Mermin, N. D. 1976. *Solid State Physics* (Saunders College Publishing, Philadelphia).
2. Marder, M. P. 2000. *Condensed Matter Physics* (John Willey & Sons, Inc., New York).
3. Kittle, C. 1986. *Introduction to Solid State Physics,* 6th Edition (John Willey & Sons, Inc., New York).
4. Economou, E. N. 2006. *Green's Functions in Quantum Physics* (Springer-Verlag, Berlin).
5. Hewson, A. C. 1997. *The Kondo Problem to Heavy Fermions* (Cambridge University Press, Cambridge).
6. Kuramoto, Y. 2020. *Quantum Many-Body Physics: A Perspective on Strong Correlations* (Springe Japan, KK).
7. Reed-Hill, R. E. 1973. *Physical Metallurgy Principles,* 2nd Edition (D. Van Nostrand Company, New York).
8. Hull, D. and Bacon, D. J. 2001. *Introduction to Dislocations* (Elsevier Ltd., Amsterdam).
9. Nabarro, F. R. N. and Hirth, J. P. 2007. *Dislocations in Solids* (Elsevier, Amsterdam).
10. Mann, E. H. 1949. An Elastic Theory of Dislocations. *Proceedings of the Royal Society A* 199, 376–394.
11. Moss, W. C. and Hoover, W. G. 1978. Edge-Dislocation Displacements in an Elastic Strip. *Journal of Applied Physics* 49, 5449–5451.
12. Garcia, C. B., Endo, C. K., Chang, M., and Beltz, G. E. 1999. Dislocation Models as Teaching Aids. *Journal of Materials Education* 21, 149–156.
13. Arfken, G. B. and Weber, H. J. 2005. *Mathematical Methods for Physicists,* 6th Edition (Elsevier, Burlington).
14. Braun, O. M. and Kivshar, Y. S. 2004. *The Frenkel-Kontorova Model: Concepts, Methods, and Applications* (Springer, Berlin).

4 Entropy, Information, and Disorder

In this chapter the ideas of entropy as a measure of information and disorder are introduced [1–10]. In this regard, entropy has many applications. These include, for example, the development of statistical mechanics as a means of explaining from their microscopic structure the macroscopic properties displayed by physical systems [1–4]; the development of the ideas of both classical and quantum information as a measure that can be applied in the quantification, storage, and transmission of information [1–3, 4–10]; and the development of various statistical schemes for approximating the solutions of problems involving complex systems [1–10]. In general, entropy will be shown to represent a means of characterizing what can be most accurately stated about a statistical system of which we have an incomplete knowledge [1–4].

We begin by studying systems described by n possible different observations of a variable x, denoted by the set $\{x_1, x_2, \ldots, x_n\}$, where each observation is measured with the corresponding unknown probabilities $\{p_1, p_2, \ldots, p_n\}$ [1, 4]. The question of interest here is then: if a function $f(x)$ is known to have an average value $\langle f(x)\rangle = \sum_{i=1}^{n} p_i f(x_i)$, what is the most likely set of probabilities $\{p_1, p_2, \ldots, p_n\}$ statistically characterizing the system. Similarly, these questions are extended to the consideration: if a function $f(x)$ and $g(x)$ are known to have average values $\langle f(x)\rangle = \sum_{i=1}^{n} p_i f(x_i)$ and $\langle g(x)\rangle = \sum_{i=1}^{n} p_i g(x_i)$, what is the most likely set of probabilities $\{p_1, p_2, \ldots, p_n\}$ characterizing the system. The answers to these questions involve the quantification and evolution of the ideas of entropy [1, 4].

A natural application of the answers to the above problems is in statistical mechanics [1, 4]. Here physical systems are characterized by large numbers of microscopic degrees of freedom, having values that are unknown to us. However, macroscopically, the macroscopic variables of the system (e.g., energy, magnetization, etc., and the temperature and chemical potential) are known. The relationship between the macroscopic variables and the microscopic variables is developed from the ideas of entropy. These ideas of entropy and their applications, in turn, originate from the general considerations of the two problems outlined in the preceding paragraph. In this way, through the concept of entropy, we arrive at the canonical and grand canonical ensembles of statistical mechanics [4].

The next topic is the treatment of entropy as a characterization of information [1, 5–10]. Information arises as a general idea about probability distributions. Specifically, it is concerned with how much we learn about a system when we measure one of its variables that is described by a probability distribution [1, 5–10]. If the variable occurs with a high probability, then we learn little about the system by its measurement. We already had a high expectation that the system would be found in a highly probable state. On the other hand, if the variable occurs with a low probability, then we learn a great deal about the system. The result of measuring a low probability case is very surprising, and we learn a great deal about the system we are measuring. In this regard, the entropy of a probability distribution is formulated as a measure of the information content of a probability distribution.

DOI: 10.1201/9781032674919-4

An important application of entropy as a measure of information is in the storage, retrieval, and transmission of information [1, 5–10]. It enters into considerations of the efficiency of coding information symbolically and in the discussions of the efficient transmission or retrieval of stored information [1]. In this regard, entropy again becomes important in the treatment of errors due to random or systematic noise introduced during information handling [1, 5–10].

When handling classical systems, the entropy is defined in terms of classical variables and is known as the Shannon entropy [1, 5–10]. When dealing with quantum mechanical systems, the entropy is defined in terms of the density matrix formulation and is known as the von Neumann entropy [1, 6–10]. While both entropies provide measures of information, the von Neumann entropy also enters into considerations of quantum mechanical entanglement. In this regard, entanglement is a property specific to many-particle quantum mechanical systems. It arises in quantum mechanics from the focus on the quantum mechanical wavefunction as a probability amplitude. The probability distribution of a particle in space and time is determined as the modulus squared of its quantum wavefunction. This allows for quantum systems to exhibit space-time probability distributions which differ from those found in classical mechanics, and in many particle systems this difference is due to the entanglement of the system wavefunction. Examples of quantum probability distributions involving two particles with entanglement properties are those generated by the so-called Bell's states [1, 6–15]. These display probability distributions in space and time which are fundamentally different from those found in classical mechanical systems.

The final application of entropy is in the development of certain so-called maximum entropy methods of numerical and computer simulation work and in derivative simulation approaches based on search algorithms [16–26]. Examples of these are the development of the Metropolis-Hastings sampling methods [16–18] which allow for the simulation of thermodynamic properties and for the extraction of solutions to inverse problems by reverse Monte Carlo [20, 21] techniques, and in the methods of genetic algorithms [22–24] which are search routines based on ideas from biological evolution.

In the Metropolis-Hastings method [16–18] the thermodynamic properties of materials are simulated by randomly generating system configurations which are distributed statistically according to the Boltzmann distribution. From these configurations the variety of statistical averages of the system are approximated. This is a method that has been commonly applied to the study of both classical and quantum mechanical systems.

Another method of simulation based on statistical methods is that of genetic algorithms [22–26]. In genetic algorithms, search methods involving the application of principles from biological evolution are developed so as to optimize the solutions of physical problems. These methods are based on the introduction of the concepts of chromosomes, which represent possible solutions of optimized physical systems. Through the mutation of these chromosomes by means of genetic crossover and mutation methods and the application of natural selection to the property to be optimized, an optimized solution to a problem is evolved in the set of chromosomes.

In the following, first the ideas of entropy are introduced as a measure of the information contained within a probability distribution. This is followed by applications in statistical mechanics, information theory, quantum entanglement, Monte Carlo simulation methods, and methods of genetic algorithms.

4.1 ENTROPY

The idea of entropy is to establish a measure of the amount of disorder in a system [1–3]. In addition, it will be seen during the discussions that the entropy is ultimately related to the information obtained upon performing a measurement on a system. These are two seemingly different processes, disorder and information, that are found nevertheless to complement each other. Both aspects of entropy are found to be useful in understanding and quantifying a wide variety of statistical and information processes in several seemingly unrelated statistical systems. Later, when applied to statistical mechanics, these results will be seen eventually to lead to the important formulation in physics known as the canonical ensemble.

First, the idea of entropy is introduced in the context of providing an indicator of disorder [1–4]. This is done by designing an entropy function which numerically quantifies the statistical disorder present in a set of randomly occurring events of a system. In this view, under a measurement, a system exhibits one of n possible different outcomes from the set of possible measurements [1–4]

$$\{x_1, x_2, \ldots\ldots, x_n\} \tag{4.1a}$$

and the elements of this set are measured, respectively, with probabilities

$$\{p_1, p_2, \ldots\ldots, p_n\}. \tag{4.1b}$$

Consequently, the outcome x_1 is found to occur with probability p_1, the outcome x_2 is found with probability p_2, etc. In such a system, Shannon found that a unique numerical measure of the disorder of the probability distribution $\{p_1, p_2, \ldots\ldots, p_n\}$ is given by a so-called entropy function that operates directly on the elements of the distribution. This function is defined under a broad set of general statistical guidelines that are discussed now in the course of this chapter.

In this regard, the entropy of a distribution is shown to be defined solely in terms of the elements of the probability distribution $\{p_1, p_2, \ldots\ldots, p_n\}$ and is ultimately related to the relative uncertainties in observing the measured outcomes $x_1, x_2, \ldots\ldots, x_n$ in the systems [1–4]. It gives us a sense of how much we learn about the system as measurements are made on it. For example, if the probability distribution is narrowly peaked about an outcome x_i so that the other outcomes are far less frequently measured, the probability distribution is characterized as one with a small disorder. We often measure the system in the state x_i and in fact on any given measurement would tend to expect this outcome. On the other hand, if the probability distribution is broadly developed over all the possible random variables, it is characterized as one with a large disorder. Now measurements on the system exhibit a significant scatter in the measured events. We are less expectant of a particular outcome. In accord with our ideas of entropy as a measure of disorder, in the first example the entropy of the system must be small while in the second the entropy of the system must be large. Consequently, in the first system the measurement of x_i tells us less about the system as we expect the system to be in the state x_i. However, this is not the case in the second system where our prior ability to guess a given measured outcome is less likely than in the first system. Consequently, we learn more from a measurement on the second system than from a measurement on the first. This provides a relation between disorder and information content.

In the early development of information theory, Shannon formulated the numerical measure of the entropy [1, 3, 5–10]. Specifically, he found that in terms of the probability distribution $\{p_1, p_2, \ldots\ldots, p_n\}$ a unique representation of the entropy of the distribution is given by the positive value of the function defined by [1, 3, 5–10]

$$H(p_1, p_2, \ldots\ldots, p_n) \equiv -K \sum_{i=1}^{n} p_i \ln p_i \tag{4.2}$$

where $K > 0$ is a constant which sets the scale of the entropy. Defined in this manner, the entropy function is shown to be an increasing function of the statistical disorder present in a system with the relative disorder between two different realizations of the probability distribution $\{p_1, p_2, \ldots\ldots, p_n\}$ defined on $\{x_1, x_2, \ldots\ldots, x_n\}$ indicated by the relative numerical magnitudes of the respective $H(p_1, p_2, \ldots\ldots, p_n)$ values between the two distributions. For a distribution peaked about a specific x_i the entropy is small, whereas the entropy becomes larger in a broader distribution defined on the same random outcomes. In addition, the entropy function is invariant under the permutation of the subscripts on the p_i's so that the function does not, a priori, favor one outcome of the system over another. All outcomes are treated equally by the form of the entropy function.

As an example, consider the case in which the outcome of a measure on the system is certain. The probability distribution of the measurement has only one certain outcome represented by the delta-function distribution [1–10]

$$p_i = \delta_{i,j} \tag{4.3}$$

(i.e., $p_i = 1$ for $i = j$, but is otherwise zero), and it follows from Eq. (4.2) that

$$H(p_1, p_2, \ldots.., p_n) = 0. \tag{4.4}$$

Since there is no disorder in the system, the outcome of a measurement on the system is certain, and the entropy function is zero. A measurement on such a system tells us nothing, as we already know the outcome of the measurement. The entropy function, however, immediately becomes positive nonzero with the introduction of disorder into the system.

With increasing disorder, the entropy increases from zero, ultimately approaching a maximum value for a completely disordered system [1–10]. In this regard, we expect that the greatest disorder in a system is when all the possible outcomes are equally probable [1–10], and

$$p_1 = p_2 = \ldots = p_n = \frac{1}{n}. \tag{4.5}$$

In this case, all the outcomes $x_1, x_2, \ldots.., x_n$ are equally likely, and no single x_j is expected to be more likely than any of the other outcomes. From Eqs. (4.2) and (4.5), it is seen that for this case

$$H\left(p_1 = \frac{1}{n}, p_2 = \frac{1}{n}, \ldots.., p_n = \frac{1}{n}\right) \equiv -K\sum_{i=1}^{n} p_i \ln p_i = -K \ln\frac{1}{n} = K \ln n, \tag{4.6}$$

and the greatest entropy in the systems is directly related to the number of elements in the random distribution of possible occurrences.

From Eqs. (4.4) and (4.6) we see that the entropy function for the distribution $\{p_1, p_2, \ldots.., p_n\}$ satisfies

$$0 \le H(p_1, p_2, \ldots.., p_n) \le K \ln n. \tag{4.7}$$

Furthermore, from Eq. (4.6) it also follows that between two different probability distributions defined, respectively, on n possible different outcomes with a distribution $\{p_1, p_2, \ldots.., p_n\}$ and $n+1$ possible different outcomes with a distribution $\{p_1, p_2, \ldots.., p_{n+1}\}$ the relative entropy maxima are related by [1]

$$H\left(p_1 = \frac{1}{n+1}, p_2 = \frac{1}{n+1}, \ldots.., p_{n+1} = \frac{1}{n+1}\right) > H\left(p_1 = \frac{1}{n}, p_2 = \frac{1}{n}, \ldots.., p_n = \frac{1}{n}\right) \tag{4.8a}$$

and for a system with only one possible event or outcome

$$H\left(p_1 = \frac{1}{1}\right) \equiv 0. \tag{4.8b}$$

As an example of Eq. (4.2) for a nontrivial probability distribution, consider two events $\{x_1, x_2\}$ that occur with the probability distribution $\{p_1 = p, p_2 = 1 - p\}$. Here the outcome of the system is definite when $p = 0,1$ and the probability distribution is most broadly defined for $p = \frac{1}{2}$. From Eq. (4.2) it then follows that the entropy of this system is given by the analytic form [1]

$$H(p_1, p_2) \equiv -K\sum_{i=1}^{2} p_i \ln p_i = -K\left[p\ln p + (1-p)\ln(1-p)\right] \tag{4.9}$$

and this form offers a basic illustration of the symmetries of the entropy. This symmetry is seen in Figure 4.1 which yields a plot of $H(p_1, p_2)$ versus p over the range of $0 \le p \le 1$. It is interesting to note that the entropy is zero at $p = 0$ and 1 where the system is known with certainty while the entropy is maximum at $p = \frac{1}{2}$ with $H\left(\frac{1}{2}, \frac{1}{2}\right) = K\ln 2$ for the system with maximum disorder. At all other probabilities the distribution is below $K\ln 2$.

Additional important properties used in the development of the entropy function $H(p_1, p_2, \ldots, p_n)$ require that it be invariant under a partitioning of the random events $\{x_1, x_2, \ldots, x_n\}$ amongst subsets [1]. Such a partitioning should not change the nature of the original unpartitioned distributions $\{p_1, p_2, \ldots, p_n\}$, and, in addition, should determine the probability distribution of a specific subset to contain the outcome of a measurement. As we shall soon see, this is an important property in defining the form of $H(p_1, p_2, \ldots, p_n)$ and is basic in arriving at the analytic form of the entropy in Eq. (4.2).

As another important symmetry of the entropy function, consider a partitioning of the n elements of $\{x_1, x_2, \ldots, x_n\}$ into $m \le n$ nonempty subsets. This is done so that no two subsets contain the same element, and each element is found in one of the subsets. Specifically, consider the partition [1, 5–10]:

1. $Y_1 = \{x_1, x_2, \ldots, x_{k_1}\}$ with the corresponding probabilities $\{p_1, p_2, \ldots, p_{k_1}\}$,
2. $Y_2 = \{x_{k_1+1}, x_{k_1+2}, \ldots, x_{k_2}\}$ with the corresponding probabilities $\{p_{k_1+1}, p_{k_1+2}, \ldots, p_{k_2}\}$,
3. $Y_i = \{x_{k_{i-1}+1}, x_{k_{i-1}+2}, \ldots, x_{k_i}\}$ with the corresponding probabilities $\{p_{k_{i-1}+1}, p_{k_{i-1}+2}, \ldots, p_{k_i}\}$,
4. $Y_m = \{x_{k_{m-1}+1}, x_{k_{m-1}+2}, \ldots, x_{k_m=n}\}$ with the corresponding probabilities $\{p_{k_{m-1}+1}, p_{k_{m-1}+2}, \ldots, p_{k_m=n}\}$.

For this partitioning the probability of an outcome being found in the subset Y_1 is $w_1 = p_1 + p_2 + \ldots + p_{k_1}$ and in general of being found in Y_i is $w_i = p_{k_{i-1}+1} + p_{k_{i-1}+2} + \ldots + p_{k_i}$.

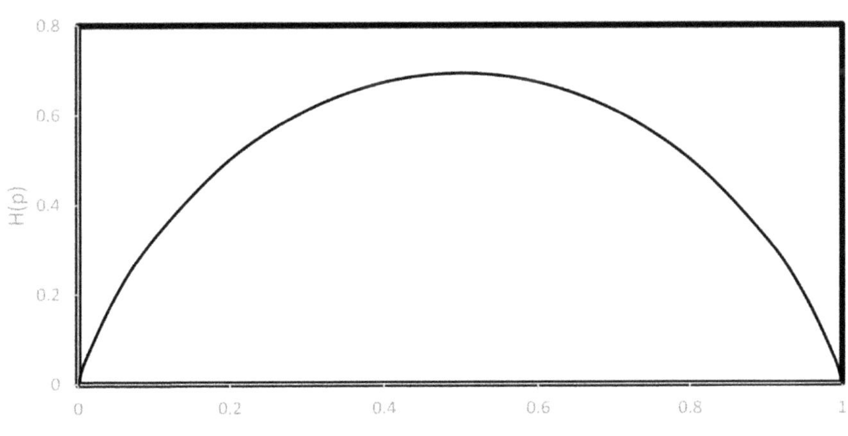

FIGURE 4.1 Plot of $H(p) \equiv H(p_1 = p, p_2 = 1-p)$ in units of K versus p over the interval $0 \le p \le 1$.

The resulting partitioning is important and leads to a useful relationship between the partitioned and non-partitioned system. This relationship allows for the entropy function to be re-expressed in terms of what we know about $\{p_1, p_2, \ldots.., p_n\}$ and the distribution of nonempty subsets $\{w_1, w_2, \ldots.., w_m\}$ where $m \leq n$. It appears as a limiting statement of restrictions on the functional form of the entropy function.

The entropy function in Eq. (4.2) can now be expressed in terms of the subset partitioning and written in terms of both the p_i's and w_i's. In this regard, from the rules of probabilities it follows that [1, 5–10]

$$H(p_1, p_2, \ldots.., p_n) = H(w_1, w_2, \ldots.., w_m) + w_1 H\left(\frac{p_1}{w_1}, \frac{p_2}{w_1}, \ldots.., \frac{p_{k_1}}{w_1}\right) + \ldots$$

$$+ w_m H\left(\frac{p_{k_{m-1}+1}}{w_m}, \frac{p_{k_{m-1}+2}}{w_m}, \ldots.., \frac{p_{k_m}}{w_m}\right) =$$

$$= H(w_1, w_2, \ldots.., w_m) + \sum_{i=1}^{m} w_i H\left(\frac{p_{k_{i-1}+1}}{w_i}, \frac{p_{k_{i-1}+2}}{w_i}, \ldots.., \frac{p_{k_i}}{w_i}\right). \tag{4.10}$$

In this expression $H(w_1, w_2, \ldots.., w_m)$ is the entropy associated with choosing a subset during a measurement, and the remaining terms are the entropy contributions associated with choosing a specific outcome x_j within that chosen subset. Consequently, once an outcome x_i is found in a given subset $Y_j = \{x_{k_{j-1}+1}, \ldots x_i \ldots x_{k_j}\}$ with the subset occurrence probability w_j and in which the outcomes $x_{k_{j-1}+1}, \ldots x_i \ldots x_{k_j}$ of the subset are found with the associated probabilities $\{p_{k_{j-1}+1}, \ldots p_i \ldots p_{k_j}\}$, then the probability of that x_i being amongst the subset elements of Y_j is $\frac{p_i}{w_j}$. The resulting subset entropy arising from the $\frac{p_i}{w_j}$ probability distribution within the subset Y_j is $H\left(\frac{p_{k_{i-1}+1}}{w_i}, \frac{p_{k_{i-1}+2}}{w_i}, \ldots.., \frac{p_{k_i}}{w_i}\right)$ and occurs in the sum in Eq. (4.10) weighted by the subset probability w_j of its contribution to the entropy $H(p_1, p_2, \ldots.., p_n)$. The resulting entropy function in Eq. (4.2) was found by Shannon to satisfy the condition in Eq. (4.10).

To see an example of the application of Eq. (4.10), consider the case of a system with maximum disorder in which $p_1 = p_2 = \ldots = p_n = \frac{1}{n}$ which is partitioned into m subsets containing equal numbers of elements q so that $w_1 = w_2 = \ldots w_m = \frac{1}{m}$ and $n = mq$. From Eq. (4.10) it follows that under these conditions [1, 5–10]

$$H\left(p_1 = p_2 = \ldots = p_n = \frac{1}{n}\right) = H\left(w_1 = w_2 = \ldots w_m = \frac{1}{m}\right) + \sum_{i=1}^{m} \frac{1}{m} H\left(\frac{1}{q} = \frac{1}{q} = \ldots = \frac{1}{q}\right). \tag{4.11}$$

Defining the functions

$$A(n) = H\left(p_1 = p_2 = \ldots = p_n = \frac{1}{n}\right), \tag{4.12a}$$

$$A(m) = H\left(w_1 = w_2 = \ldots = w_m = \frac{1}{m}\right), \tag{4.12b}$$

and

$$A(q) = H\left(p_1 = p_2 = \ldots = p_q = \frac{1}{q}\right), \tag{4.12c}$$

it is then found from Eq. (4.11) that

$$A(n) = A(m) + A(q) = A(mq). \tag{4.13}$$

The relationship in Eq. (4.13) is seen to be satisfied by the general functional form of a logarithm written as [1]

$$A(r) = K \ln r. \tag{4.14}$$

This is directly from Eq. (4.2) evaluated for the case in which $p_1 = p_2 = \ldots = p_r = \frac{1}{r}$, i.e.,

$$H\left(p_1 = p_2 = \ldots p_r = \frac{1}{r}\right) \equiv -K \sum_{i=1}^{r} p_i \ln p_i \to K \ln r. \tag{4.15}$$

4.2 PROBABILITY

An interesting application of entropy is in the context of the following problem. Consider that we have a set of possible outcomes $\{x_1, x_2, \ldots.., x_n\}$ but do not know the associated probabilities $\{p_1, p_2, \ldots.., p_n\}$ with which they occur. We are then given a function $f(x)$ where x runs over the set $\{x_1, x_2, \ldots.., x_n\}$ so that the average value of $f(x)$ is obtained from the sum [1, 2, 5–10]

$$\langle f(x) \rangle \equiv \sum_{i=1}^{n} p_i f(x_i). \tag{4.16}$$

(Note: Here the notation $\langle \ \rangle$ indicates an average over $\{p_1, p_2, \ldots.., p_n\}$ of the formula contained within the left and right brackets.) A question of interest is, given $\{x_1, x_2, \ldots.., x_n\}$ and the average $\langle f(x) \rangle$, what is the most general statement we can make regarding the unknown probability distribution $\{p_1, p_2, \ldots.., p_n\}$. In this regard, we would expect that the distribution should be best represented by the state of most disorder in $\{p_1, p_2, \ldots.., p_n\}$ which also reproduces the average value in Eq. (4.16) and for which the distribution sums to unity. Another way of putting this is that the entropy of the probability distribution $H(p_1, p_2, \ldots.., p_n)$ expressed in Eq. (4.2) should be maximized subject to the constraint that the average in Eq. (4.16) is a known value in the normalized probability distribution [1, 2]. This is the idea of the so-called maximum entropy method applied in the determination of the p_i's.

To perform this maximization of Eq. (4.2) subject to constraints, we use the method of Lagrange multipliers. In this method, the maximization of $H(p_1, p_2, \ldots.., p_n)$ subject to the stated constraints on the system is done by considering the function [1, 2]

$$\mathcal{H}(p_1, p_2, \ldots.., p_n, \lambda, \mu) = H(p_1, p_2, \ldots.., p_n) - \lambda\left[\sum_{i=1}^{n} p_i f(x_i) - \langle f(x) \rangle\right] - \mu\left[\sum_{i=1}^{n} p_i - 1\right] =$$

$$= -K\sum_{i=1}^{n} p_i \ln p_i - \lambda\left[\sum_{i=1}^{n} p_i f(x_i) - \langle f(x)\rangle\right] - \mu\left[\sum_{i=1}^{n} p_i - 1\right]. \quad (4.17)$$

Here $\langle f(x)\rangle$ is the known average value of $f(x)$ in the distribution of states or outcomes of the system, and the two terms added to $H(p_1, p_2, \ldots, p_n)$ are statements of the constraints from $\langle f(x)\rangle$ and the normalization of the probability distribution. The parameters λ and μ (e.g., the Lagrange multipliers) are then new variables introduced as a means of including the conditions of constraint into $\mathcal{H}(p_1, p_2, \ldots, p_n, \lambda, \mu)$. In this way, by studying the extrema of the new function, $\mathcal{H}(p_1, p_2, \ldots, p_n, \lambda, \mu)$, a solution will be generated to our problem for maximizing the entropy subject to the constraints [1, 2].

To see this let us take the extremum of $\mathcal{H}(p_1, p_2, \ldots, p_n, \lambda, \mu)$ in the variables $p_1, p_2, \ldots, p_n, \lambda, \mu$. We find that [1, 2]

$$\frac{\partial \mathcal{H}}{\partial p_i} = -K\left[\ln p_i + 1\right] - \lambda f(x_i) - \mu = 0 \text{ for } 1 \le i \le n, \quad (4.18a)$$

$$\frac{\partial \mathcal{H}}{\partial \lambda} = \langle f(x)\rangle - \sum_{i=1}^{n} p_i f(x_i) = 0, \quad (4.18b)$$

and

$$\frac{\partial \mathcal{H}}{\partial \mu} = 1 - \sum_{i=1}^{n} p_i = 0. \quad (4.18c)$$

From Eq. (4.18a) it then follows that the probabilities have become functions of K, λ, μ, $f(x_i)$, and are given by the general form [1, 2]

$$p_i = \exp\left[-\frac{\lambda}{K} f(x_i) - \frac{\mu + K}{K}\right]. \quad (4.19)$$

In addition, the conditions obtained in Eqs. (4.18b) and (4.18c) are just the constraints earlier stated on the problem. Eqs. (4.18b) and (4.18c) then constitute two equations in the known variables $K, f(x_i)$, allowing for the determination of the new unknown variables λ and μ introduced into the problem. For these values of λ and μ it follows that the constraining conditions in Eq. (4.17) are set to zero and drop out of the problem.

From Eqs. (4.18a) and (4.19) it then follows that [1, 2]

$$\langle f(x)\rangle = \sum_{i=1}^{n} p_i f(x_i) = \sum_{i=1}^{n} f(x_i) \exp\left[-\frac{\lambda}{K} f(x_i) - \frac{\mu + K}{K}\right] \quad (4.20a)$$

and from Eq. (4.18c) that

$$1 = \sum_{i=1}^{n} p_i = \sum_{i=1}^{n} \exp\left[-\frac{\lambda}{K} f(x_i) - \frac{\mu + K}{K}\right]. \quad (4.20b)$$

Applying the values of λ and μ obtained from the solutions of Eq. (4.20), Eq. (4.17) reduces to

$$\mathcal{H}(p_1, p_2, \ldots, p_n, \lambda, \mu) = -K\sum_{i=1}^{n} p_i \ln p_i. \quad (4.21)$$

As a result, the remaining term on the right of Eq. (4.21) is maximized subject to the constraining conditions that [1, 2]

$$\langle f(x)\rangle = \sum_{i=1}^{n} p_i f(x_i) \tag{4.22a}$$

and that

$$\sum_{i=1}^{n} p_i = 1 \tag{4.22b}$$

from our solved values of λ and μ.

Considering these expressions in greater detail, we see from Eq. (4.20b) that it then follows that [1, 2]

$$\exp\left[\frac{\mu+K}{K}\right] = \sum_{i=1}^{n} \exp\left[-\frac{\lambda}{K} f(x_i)\right] \equiv Z\left(\frac{\lambda}{K}\right) \tag{4.23}$$

where the function $Z(\alpha) = \sum_{i=1}^{n} \exp\left[-\alpha f(x_i)\right]$ is defined as the partition function of the problem. As shall be seen in the course of this chapter, the partition function plays an important role in the applications of maximum entropy methods. To begin with, the present notation is simplified by writing results in terms of the partition function, and it is found from Eq. (4.20a) that [1, 2]

$$\langle f\rangle = \sum_{i=1}^{n} p_i f(x_i) = \frac{\sum_{i=1}^{n} f(x_i)\exp\left[-\frac{\lambda}{K} f(x_i)\right]}{\sum_{i=1}^{n} \exp\left[-\frac{\lambda}{K} f(x_i)\right]} = -\frac{d\ln Z\left(\frac{\lambda}{K}\right)}{d\left(\frac{\lambda}{K}\right)}. \tag{4.24a}$$

In addition, it follows that for a general function $g(x)$, its average in $\{x_1, x_2, \ldots\ldots, x_n\}$ is obtained from the form [1, 2]

$$\langle g\rangle = \sum_{i=1}^{n} p_i f(x_i) = \frac{\sum_{i=1}^{n} g(x_i)\exp\left[-\frac{\lambda}{K} f(x_i)\right]}{\sum_{i=1}^{n} \exp\left[-\frac{\lambda}{K} f(x_i)\right]} = \frac{\sum_{i=1}^{n} g(x_i)\exp\left[-\frac{\lambda}{K} f(x_i)\right]}{Z\left(\frac{\lambda}{K}\right)}. \tag{4.24b}$$

so that the average properties of the system at maximum entropy subject to the Eq. (4.22) constraints are determined from averages involving the weighting factors or probabilities

$$p_i = \frac{\exp\left[-\frac{\lambda}{K} f(x_i)\right]}{\sum_{i=1}^{n} \exp\left[-\frac{\lambda}{K} f(x_i)\right]} = \frac{\exp\left[-\frac{\lambda}{K} f(x_i)\right]}{Z\left(\frac{\lambda}{K}\right)}. \tag{4.25}$$

Returning to a consideration of the entropy, from Eqs. (4.21) and (4.25) the entropy of the constrained system is obtained as

$$S = \mathcal{H}(p_1, p_2, \ldots\ldots, p_n, \lambda, \mu) = -K\sum_{i=1}^{n} p_i \ln p_i = \tag{4.26}$$

$$= K \ln Z\left(\frac{\lambda}{K}\right) - \lambda \frac{d \ln Z\left(\frac{\lambda}{K}\right)}{d\left(\frac{\lambda}{K}\right)} = \lambda_0 + \lambda \langle f \rangle$$

where S is the entropy and $\lambda_0 \equiv K \ln Z\left(\frac{\lambda}{K}\right)$. This offers a measure of the disorder in the constrained systems. In addition, taking another derivative in Eq. (4.24a) we also find the important statistical quantity given by [1, 2]

$$\frac{d^2 \ln Z\left(\frac{\lambda}{K}\right)}{d\left(\frac{\lambda}{K}\right)^2} = \langle f^2 \rangle - \langle f \rangle^2. \tag{4.27}$$

This is a measure of the statistical fluctuations present in the measurement of $\langle f \rangle$ in the constrained system.

4.3 EXAMPLE: THERMODYNAMICS

An interesting application of the ideas of maximum entropy is in the area of statistical mechanics. In this problem, a system with energy levels $\{\epsilon_1, \epsilon_2, \ldots\ldots, \epsilon_n\}$ is found to occur with unknown probabilities $\{p_1, p_2, \ldots\ldots, p_n\}$ subject to the constraints that the average energy, U, of the system is given by [1, 2, 4]

$$U = \sum_{i=1}^{n} p_i \epsilon_i, \tag{4.28a}$$

and

$$\sum_{i=1}^{n} p_i = 1. \tag{4.28b}$$

Note that with the identification $\langle f \rangle \leftrightarrow U$ and $f(x_i) \leftrightarrow \epsilon_i$, Eqs. (4.16) and (4.28a) transform into each other so that the thermodynamic problem is isomorphic to the study given in the previous two sections.

The results for the statistical mechanic problem can, consequently, be obtained from the studies in the previous two sections. Upon this consideration, it is found that in the thermodynamic system the entropy in Eq. (4.26) takes the form [1, 2, 4]

$$S = -K \sum_{i=1}^{n} p_i \ln p_i = -k_B \sum_{i=1}^{n} p_i \ln p_i \tag{4.29}$$

where the scaling constant K is set as the Boltzmann constant, k_B. From Eq. (4.25), taking $K = k_B$ and $\lambda = \frac{1}{T}$ for temperature T and making the identification $f(x_i) \leftrightarrow \epsilon_i$, the probabilities of the system become

$$p_i = \frac{\exp\left[-\frac{\lambda}{K} f(x_i)\right]}{\sum_{i=1}^{n} \exp\left[-\frac{\lambda}{K} f(x_i)\right]} = \frac{\exp\left[-\frac{\lambda}{K} f(x_i)\right]}{Z\left(\frac{\lambda}{K}\right)} \tag{4.30}$$

$$= \frac{\exp\left[-\frac{1}{k_B T}\epsilon_i\right]}{\sum_{i=1}^{n} \exp\left[-\frac{1}{k_B T}\epsilon_i\right]} = \frac{\exp\left[-\frac{1}{k_B T}\epsilon_i\right]}{Z\left(\frac{1}{k_B T}\right)}.$$

The result in Eq. (4.30) is the standard Boltzmann factors from statistical mechanics. Substituting this in Eq. (4.29), we then find [1, 2, 4]

$$S = -k_B \sum_{i=1}^{n} p_i \ln p_i = \frac{U}{T} + k_B \ln Z\left(\frac{1}{k_B T}\right). \tag{4.31}$$

In statistical mechanics, from the entropy it is also useful to define the free energy as

$$F = U - TS = U - T\left[\frac{U}{T} + k_B \ln Z\left(\frac{1}{k_B T}\right)\right] = -k_B T \ln Z\left(\frac{1}{k_B T}\right). \tag{4.32}$$

This is applied for generating thermodynamic averages when the energy of the system depends on a variety of external fields, i.e., volume, gravitational fields, applied electric fields, applied magnetic fields, etc.

For example, if the system interacts with an externally applied field denoted by α, the free energy becomes a function of α. Specifically, the free energy becomes of the general form $F = F(\alpha)$. From this field interaction then arises thermodynamic forces β which are given from [4]

$$\beta = k_B T \frac{\partial}{\partial \alpha} \ln Z\left(\frac{1}{k_B T}\right) \tag{4.33}$$

and are proportional to a variety of measurable quantities including pressures, electric and magnetic moments, etc.

4.3.1 An Example: Classical Ideal Gas

As an example of a partition function consider the classical ideal gas of N atoms. In this model the energy of the ideal gas is given by [1, 2, 4]

$$E\left(\overrightarrow{p_1}, \overrightarrow{p_2}, \ldots, \overrightarrow{p_N}\right) = \sum_{i=1}^{N} \frac{p_i^2}{2m} \tag{4.34}$$

where $\frac{p_i^2}{2m}$ is the energy of the ith classical atom which has mass m and momentum $\overrightarrow{p_i}$. The canonical partition function for the system of N atoms is then given (from Eq. (4.23) and our discussions in Section 4.3) in terms of the energy by [1, 2, 4]

$$Z_N\left(\frac{1}{k_B T}\right) = \frac{1}{h^3 N!} \int \prod_{i=1}^{N} \left(dV_i d^3 p_i\right) e^{-\frac{1}{k_B T} E\left(\overrightarrow{p_1}, \overrightarrow{p_2}, \ldots, \overrightarrow{p_N}\right)} \tag{4.35}$$

$$= \frac{1}{h^3 N!}\left[\int_V \int_{-\infty}^{\infty} dp_x \int_{-\infty}^{\infty} dp_y \int_{-\infty}^{\infty} dp_z e^{-\frac{1}{k_BT}\frac{p^2}{2m}}\right]^N = \frac{V^N}{N!}\left[\frac{2\pi m k_B T}{h^2}\right]^{3N/2}$$

where h is Planck's constant, dV_i is the volume integral for the ith atom, d^3p_j is the momentum integral for the jth atom, and V is the volume of the gas. (Note that the factors of h make the partition function dimensionless but do not affect the values of the thermodynamic quantities determined from the partition function, and the factor $N!$ accounts for the $N!$ ways that the particles labels $1, 2, \ldots, N$ can be assigned to the same configuration of identical particles.) From this partition function and Eqs. (4.31) through (4.33) the thermodynamics of the gas follows.

Specifically, from Eq. (4.31) the entropy of the gas is obtained as [1, 2, 4]

$$S = \frac{U}{T} + k_B \ln Z_N\left(\frac{1}{k_BT}\right) = \frac{3}{2}Nk_B + k_B\left[\frac{3N}{2}\ln\frac{2\pi m k_B T}{h^2} + N\ln V - \ln N!\right] \tag{4.36}$$

where we have used the result $U = \frac{3}{2}Nk_BT$.

4.4 PROBABILITIES WITH MORE CONSTRAINS

An important generalization of our earlier treatment is to extend the maximum entropy discussions to treat systems in which we know the averages of two different functions (e.g., $\langle f(x)\rangle$ and $\langle h(x)\rangle$) instead of the average of just one (e.g., $\langle f(x)\rangle$). This is now considered by making an adjustment to the earlier discussions for a system with a single know average [1, 2, 4]. The results from the discussion when applied to statistical mechanics lead to the generation of another important formulation in physics known as the grand canonical ensemble.

To perform the new maximization of Eq. (4.2) subject to the two constraining averages, we again use the method of Lagrange multipliers. Specifically, consider again the case of a system in which the set of outcomes $\{x_1, x_2, \ldots.., x_n\}$ occurs with probabilities $\{p_1, p_2, \ldots.., p_n\}$. For this system, we are given as a first constraint that the function $f(x)$ has the known average value [1, 2, 4]

$$\langle f(x)\rangle \equiv \sum_{i=1}^{n} p_i f(x_i), \tag{4.37a}$$

and, additionally, as a second constraint that the function $h(x)$ has the known average value [1, 2, 4]

$$\langle h(x)\rangle \equiv \sum_{i=1}^{n} p_i h(x_i). \tag{4.37b}$$

The question now is what the maximum entropy of the system obtained from Eq. (4.2) is, subject to these two different constraints.

In this problem, the maximization of $H(p_1, p_2, \ldots.., p_n)$ restricted by the two constraints is done by considering the function [1, 2, 4]

$$\mathcal{H}(p_1, p_2, \ldots.., p_n, \lambda, \lambda_1, \mu) = -K\sum_{i=1}^{n} p_i \ln p_i - \lambda\left[\sum_{i=1}^{n} p_i f(x_i) - \langle f(x)\rangle\right]$$

$$-\lambda_1\left[\sum_{i=1}^{n} p_i h(x_i) - \langle h(x)\rangle\right] - \mu\left[\sum_{i=1}^{n} p_i - 1\right] \tag{4.38}$$

Here $\langle f(x)\rangle$ is the known average value of $f(x)$, $\langle h(x)\rangle$ is the known average value of $h(x)$, and the three terms added to $H(p_1,p_2,\ldots..,p_n)$ are statements of the constraints from $\langle f(x)\rangle$ and $\langle h(x)\rangle$ and the normalization of the probability distribution. The Lagrange multipliers λ, λ_1, and μ introduce the constraints into $\mathcal{H}(p_1,p_2,\ldots..,p_n,\lambda,\lambda_1,\mu)$ so that by studying the extrema of $\mathcal{H}(p_1,p_2,\ldots..,p_n,\lambda,\lambda_1,\mu)$ a solution is arrived at for the constrained maximum entropy.

To see this, we take the extremum of $\mathcal{H}(p_1,p_2,\ldots..,p_n,\lambda,\lambda_1,\mu)$ in the variables $p_1,p_2,\ldots..,p_n,\lambda,\lambda_1,\mu$. In this regard, it is required that [1, 2, 4]

$$\frac{\partial\mathcal{H}}{\partial p_i}=-K\left[\ln p_i+1\right]-\lambda f(x_i)-\lambda_1 h(x_i)-\mu=0 \text{ for } 1\le i\le n, \tag{4.39a}$$

$$\frac{\partial\mathcal{H}}{\partial\lambda}=\langle f(x)\rangle-\sum_{i=1}^{n}p_i f(x_i)=0, \tag{4.39b}$$

$$\frac{\partial\mathcal{H}}{\partial\lambda_1}=\langle h(x)\rangle-\sum_{i=1}^{n}p_i h(x_i)=0, \tag{4.39c}$$

and

$$\frac{\partial\mathcal{H}}{\partial\mu}=1-\sum_{i=1}^{n}p_i=0. \tag{4.39d}$$

It follows from Eq. (4.39a) that the probabilities are now functions of K, λ, λ_1, μ, $f(x_i)$, $h(x_i)$ given by the general form

$$p_i=\exp\left[-\frac{\lambda}{K}f(x_i)-\frac{\lambda_1}{K}h(x_i)-\frac{\mu+K}{K}\right], \tag{4.40}$$

and the conditions obtained in Eqs. (4.39b) through (4.39d) are just the earlier stated constraints on the problem. Eqs. (4.39b) through (4.39d) then constitute three equations in the known variables $K, f(x_i)$, $h(x_i)$ for the determination of the new unknown Lagrange multipliers λ, λ_1, and μ. For these values of λ, λ_1, and μ it follows that the constraining conditions in Eq. (4.38) are set to zero.

From Eqs. (4.39b), (4.39c), and (4.40) it then follows that [1, 2, 4]

$$\langle f(x)\rangle=\sum_{i=1}^{n}p_i f(x_i)=\sum_{i=1}^{n}f(x_i)\exp\left[-\frac{\lambda}{K}f(x_i)-\frac{\lambda_1}{K}h(x_i)-\frac{\mu+K}{K}\right] \tag{4.41a}$$

$$\langle h(x)\rangle=\sum_{i=1}^{n}p_i h(x_i)=\sum_{i=1}^{n}h(x_i)\exp\left[-\frac{\lambda}{K}f(x_i)-\frac{\lambda_1}{K}h(x_i)-\frac{\mu+K}{K}\right] \tag{4.41b}$$

and from Eqs. (4.39d) and (4.40) that

$$1=\sum_{i=1}^{n}p_i=\sum_{i=1}^{n}\exp\left[-\frac{\lambda}{K}f(x_i)-\frac{\lambda_1}{K}h(x_i)-\frac{\mu+K}{K}\right]. \tag{4.41c}$$

Applying the values of λ, λ_1, and μ obtained from the solutions of Eq. (4.41), Eq. (4.38) reduces to

$$\mathcal{H}(p_1,p_2,\ldots..,p_n,\lambda,\lambda_1,\mu)=-K\sum_{i=1}^{n}p_i\ln p_i, \tag{4.42}$$

maximized subject to the constraining conditions

$$\langle f(x)\rangle = \sum_{i=1}^{n} p_i f(x_i), \tag{4.43a}$$

$$\langle h(x)\rangle = \sum_{i=1}^{n} p_i h(x_i), \tag{4.43b}$$

and

$$\sum_{i=1}^{n} p_i = 1. \tag{4.43c}$$

We see from Eq. (4.41c) that it follows that

$$\exp\left[\frac{\mu+K}{K}\right] = \sum_{i=1}^{n} \exp\left[-\frac{\lambda}{K} f(x_i) - \frac{\lambda_1}{K} h(x_i)\right] \equiv Z\left(\frac{\lambda}{K}, \frac{\lambda_1}{K}\right) \tag{4.44a}$$

where the function $Z(\alpha,\alpha_1) = \sum_{i=1}^{n} \exp\left[-\alpha f(x_i) - \alpha_1 h(x_i)\right]$ is defined as the partition function of the problem. As in the single constraint problem, a variety of important averages are obtained from $Z(\alpha,\alpha_1)$. In particular, it follows from Eq. (4.40) that the probability of finding the system in the state of x_i with $f(x_i)$, $h(x_i)$ is given by [1, 2, 4]

$$p_i = \frac{\exp\left[-\frac{\lambda}{K} f(x_i) - \frac{\lambda_1}{K} h(x_i)\right]}{Z\left(\frac{\lambda}{K}, \frac{\lambda_1}{K}\right)}. \tag{4.44b}$$

4.5 SYSTEMS WITH VARIABLE NUMBERS OF PARTICLES: CHEMICAL POTENTIALS

Returning to the ideas of statistical mechanics, we now consider systems which have variable numbers of particles [1, 2, 4]. During the treatment, the formulation of statistical physics known as the grand canonical ensemble is ultimately generated. Specifically, for this type of problem a system having energy levels $\{\epsilon_1, \epsilon_2, \ldots, \epsilon_n\}$ with the corresponding number of particles $\{n_1, n_2, \ldots, n_n\}$ is found to occur with the unknown probabilities $\{p_1, p_2, \ldots, p_n\}$ subject to the constraints that the average energy, U, of the system is given by [1, 2, 4]

$$U = \sum_{i=1}^{n} p_i \, \epsilon_i, \tag{4.45a}$$

the average number of particles, N, of the system is given by

$$N = \sum_{i=1}^{n} p_i n_i, \tag{4.45b}$$

and the probability distribution is normalized so that

$$\sum_{i=1}^{n} p_i = 1. \tag{4.46}$$

This is just the two average problems we have just considered. In this regard, we note, with reference to our discussions in the previous section of systems constrained by known $\langle f(x)\rangle$ and $\langle h(x)\rangle$, that under the identification $\langle f(x)\rangle \leftrightarrow U$, $f(x_i) \leftrightarrow \epsilon_i$, and $\langle h(x)\rangle \leftrightarrow N$, $h(x_i) \leftrightarrow n_i$, Eqs. (4.37) and (4.45a) and (4.45b) transform into each other.

The results for this problem are obtained from those of the previous section. In the process of transforming the results in Section 4.4 into those for the statistical mechanical system, we again take the constant K to be the Boltzmann constant, k_B. This sets the scale of the entropy. In addition, in Eq. (4.44) we make the identifications $K = k_B$, $\lambda = \frac{1}{T}$, $\lambda_1 = -\frac{\mu_c}{T}$ for temperature T and chemical potential μ_c while making the identification $f(x_i) \leftrightarrow \epsilon_i$ and $h(x_i) \leftrightarrow n_i$. Under these conditions the probabilities, $\{p_1, p_2, \ldots, p_n\}$, of the system take the general form

$$p_i = \frac{\exp\left[-\frac{1}{k_BT}\epsilon_i + \frac{\mu_c}{k_BT}n_i\right]}{Z\left(\frac{1}{k_BT}, -\frac{\mu_c}{k_BT}\right)} \tag{4.47}$$

and now are functions of the temperature T and the chemical potential μ_c. Notice that here the chemical potential, μ_c, is introduced in order to set the number of particles, N, in the system.

The result in Eq. (4.47) is the standard Boltzmann factor accounting for systems with variable numbers of particles. Substituting this in Eq. (4.42), we find the expression for the entropy as [1, 2, 4]

$$S = -k_B\sum_{i=1}^{n} p_i \ln p_i = \frac{U}{T} - \frac{\mu_c N}{T} + k_B \ln Z\left(\frac{1}{k_BT}, -\frac{\mu_c}{k_BT}\right) \tag{4.48a}$$

The free energy of the system then follows and is given by

$$F = U - TS = U - T\left[\frac{U}{T} - \frac{\mu_c N}{T} + k_B \ln Z\left(\frac{1}{k_BT}, -\frac{\mu_c}{k_BT}\right)\right] = \mu_c N - k_BT \ln Z\left(\frac{1}{k_BT}, -\frac{\mu_c}{k_BT}\right). \tag{4.48b}$$

In these expressions it follows from Eq. (4.44) that [1, 2, 4]

$$Z\left(\frac{1}{k_BT}, -\frac{\mu_c}{k_BT}\right) = \sum_{i=1}^{n} \exp\left[-\frac{1}{k_BT}\epsilon_i + \frac{\mu_c}{k_BT}n_i\right] = \sum_{m=1}^{n} \exp\left[\frac{\mu_c}{k_BT}m\right] Z_m\left(\frac{1}{k_BT}\right) \tag{4.48c}$$

where $Z_m\left(\frac{1}{k_BT}\right)$ is the canonical partition function for a system with m atoms.

The free energy in Eq. (4.48) is applied for generating thermodynamic averages in systems with variable numbers of particles when the energy of the system depends on a variety of external fields, i.e., volume, gravitational fields, applied electric fields, applied magnetic fields, etc. For example, if the system interacts with an externally applied field denoted by α, the free energy becomes a function of α. Specifically, from this field interaction arise thermodynamic forces β which are given from [1, 2, 4]

$$\beta = k_B T \frac{\partial}{\partial \alpha} \ln Z\left(\frac{1}{k_B T}, -\frac{\mu_c}{k_B T}\right) \tag{4.49a}$$

and are proportional to a variety of measurable quantities including pressures, electric and magnetic moments, etc. In addition, now that the number of particles is uncertain the average number of particles N is found to be obtained from [1, 2, 4]

$$N = \frac{\partial F}{\partial \mu_c}. \tag{4.49b}$$

4.5.1 An Example: Classical Ideal Gas

As an example of a grand canonical partition function consider the classical ideal gas of atoms. In this model the energy of the ideal gas of N atoms is given by [1, 2, 4]

$$E\left(\vec{p_1}, \vec{p_2}, \ldots, \vec{p_N}\right) = \sum_{i=1}^{N} \frac{p_i^2}{2m} \tag{4.50}$$

where $\frac{p_i^2}{2m}$ is the energy of the ith classical atom which has mass m and momentum $\vec{p_i}$. The grand canonical partition function in a gas with variable number of particles is then given in terms of the energy and chemical potential by

$$Z\left(\frac{1}{k_B T}, -\frac{\mu_c}{k_B T}\right) = \sum_{n=1}^{\infty} e^{\frac{\mu_c}{k_B T} n} Z_n\left(\frac{1}{k_B T}\right) = \sum_{n=1}^{\infty} e^{\frac{\mu_c}{k_B T} n} \frac{V^n}{n!}\left[\frac{2\pi m k_B T}{h^2}\right]^{3n/2}$$

$$= \sum_{n=1}^{\infty} \frac{1}{n!}\left\{V e^{\frac{\mu_c}{k_B T}}\left[\frac{2\pi m k_B T}{h^2}\right]^{3/2}\right\}^n = \exp\left(V e^{\frac{\mu_c}{k_B T}}\left[\frac{2\pi m k_B T}{h^2}\right]^{3/2}\right) \tag{4.51}$$

where μ_c is the chemical potential and $Z_n\left(\frac{1}{k_B T}\right)$ is given in Eq. (4.35). From this partition function all the thermodynamics of the gas follows.

4.6 CONTINUUM PROBABILITY DENSITIES

The earlier discussions focused on discrete probability distributions $\{p_1, p_2, \ldots\ldots, p_n\}$ defined over a set of discrete variables $\{x_1, x_2, \ldots\ldots, x_n\}$ and subject to the condition $\sum_{i=1}^{n} p_i = 1$. These can be extended to a consideration of continuum probability densities $p(x)$ defined over a set of continuum variable $x \in [a, b]$ and subject to the condition $\int_a^b p(x)dx = 1$. For our purposes, this extension is made by replacing the entropy definition of the discrete set, which for $K = 1$ is given by the form [1, 2, 4]

$$H(p_1, p_2, \ldots\ldots, p_n) \equiv -\sum_{i=1}^{n} p_i \ln p_i, \tag{4.52a}$$

by the continuum generalization of Eq. (4.52a) given by the Riemann integral

$$H\left[p(x)\right] \equiv -\sum_{i=1}^{n} p(x_i)\Delta x \ln\left[p(x_i)\Delta x\right], \tag{4.52b}$$

where $\Delta x \to 0$ as $n \to \infty$. (Notice that here the Δx is introduced to convert the probability density into a discretized probability distribution.) The integral defined by Eq. (4.52b) then reduces to

$$H\left[p(x)\right] = -\sum_{i=1}^{n} p(x_i)\Delta x \ln\left[p(x_i)\right] - \ln \Delta x \tag{4.52c}$$

where the second term on the right is an infinite constant, independent of the form of $p(x)$, which will be ignored in the following discussions. In this regard, we redefine the entropy of the continuum system as [1, 2, 4]

$$H\left[p(x)\right] = -\int_a^b p(x)\ln p(x)dx. \tag{4.53}$$

(Note that this is often referred to as the relative entropy of the now continuum probability density.) Using the continuum generalization of the entropy defined in Eq. (4.53), the maximum entropy principle can now be used to define the Poisson distribution and Gaussian distributions commonly found in the study of statistical physics.

4.6.1 Poisson Distribution

First, we apply the maximum entropy principle to generate the Poisson distribution. In this treatment, the entropy is defined over the interval $x \in \left[0, \infty\right)$ so that Eq. (4.53) takes the form [1, 2, 4]

$$H\left[p(x)\right] = -\int_0^\infty p(x)\ln p(x)dx. \tag{4.54}$$

To arrive at the Poisson distribution, the entropy in Eq. (4.54) is maximized subject to conditions characterizing the basic properties of the distribution $p(x)$. In this regard, a basic feature of the Poisson distribution is the specification of its first moment given by the formula

$$\mu = \langle x \rangle = \int_0^\infty xp(x)dx \tag{4.55}$$

where the notation $\langle\ \rangle$ indications an average using $p(x)$. This one restriction placed on the maximization of the entropy will be seen to lead to the result.

The application of the maximum entropy principle subject to Eqs. (4.54) and (4.55) then leads to a consideration of the maximization of the form

$$J\left[p(x)\right] = -\int_0^\infty p(x)\ln p(x)dx + \lambda_0\left(\int_0^\infty p(x)dx - 1\right) + \lambda_1\left(\int_0^\infty xp(x)dx - \mu\right). \tag{4.56}$$

Here the entropy is subject to the conditions for the normalization of the distribution

$$\int_0^\infty p(x)dx = 1 \tag{4.57a}$$

and the specification of the first moment μ by

$$\int_0^\infty xp(x)\,dx = \mu. \tag{4.57b}$$

These requirements are introduced into the problem through the Lagrange multipliers λ_0 and λ_1.

To treat the maximization of Eq. (4.56), we consider the variation of the probability distribution given by the form

$$p(x) \to p(x) + \delta p(x). \tag{4.58}$$

Introducing the variation in Eq. (4.58) into Eq. (4.56) gives

$$J\left[p(x)+\delta p(x)\right] = -\int_0^\infty \left[p(x)+\delta p(x)\right]\ln\left[p(x)+\delta p(x)\right]dx + \tag{4.59}$$

$$+\lambda_0\left(\int_0^\infty \left[p(x)+\delta p(x)\right]dx - 1\right) + \lambda_1\left(\int_0^\infty x\left[p(x)+\delta p(x)\right]dx - \mu\right).$$

Collecting terms of order $\delta p(x)$ we find at the maximum of $J\left[p(x)\right]$ that

$$-\int_0^\infty dx\left[\ln p(x) + 1 - \lambda_0 - \lambda_1 x\right]\delta p(x) = \int_0^\infty \frac{\partial J\left[p(x)\right]}{\partial p(x)}\delta p(x) = 0 \tag{4.60a}$$

or

$$\frac{\partial J\left[p(x)\right]}{\partial p(x)} = -\left[\ln p(x) + 1 - \lambda_0 - \lambda_1 x\right] = 0. \tag{4.60b}$$

In addition, from the terms of order $\left[\delta p(x)\right]^2$ we find

$$-\int_0^\infty \frac{1}{p(x)}\left[\delta p(x)\right]^2 = \int_0^\infty \frac{\partial^2 J\left[p(x)\right]}{\partial\left[p(x)\right]^2}\left[\delta p(x)\right]^2 \tag{4.60c}$$

or

$$\frac{\partial^2 J\left[p(x)\right]}{\partial\left[p(x)\right]^2} = -\frac{1}{p(x)} < 0. \tag{4.60d}$$

Here the condition in Eq. (4.60d) requires that the extremum is a maximum as from the nature of the probability distribution $p(x) > 1$.

From Eq. (4.60b) the form of $p(x)$ is given by [1, 2, 4]

$$p(x) = e^{\lambda_0 - 1}e^{\lambda_1 x}. \tag{4.61}$$

For the conditions in Eq. (4.57) to be met requires that

$$e^{\lambda_0 - 1} = -\lambda_1 \tag{4.62a}$$

and

$$\lambda_1 = -\frac{1}{\mu}. \tag{4.62b}$$

It then follows that

$$p(x) = \frac{1}{\mu} e^{-\frac{x}{\mu}} \tag{4.63}$$

which is the standard form of the Poisson distribution.

4.6.2 Gaussian Distribution

Next, we apply the maximum entropy principle to generate the Gaussian distribution. The entropy is now defined over the interval $x \in (-\infty, \infty)$ so that Eq. (4.53) has the form [1, 2, 4]

$$H[p(x)] = -\int_{-\infty}^{\infty} p(x) \ln p(x) dx. \tag{4.64}$$

The entropy in Eq. (4.64) is again maximized subject to conditions characterizing the basic properties of the distribution $p(x)$. For the Gaussian distribution, we specify both the first moment of the distribution, given by the formula [1, 2, 4]

$$\mu = \langle x \rangle = \int_{-\infty}^{\infty} x p(x) dx, \tag{4.65a}$$

and the second moment of the distribution, given by the formula

$$\langle x^2 \rangle = \int_{-\infty}^{\infty} x^2 p(x) dx. \tag{4.65b}$$

Here the notation $\langle\ \rangle$ indications an average using $p(x)$. Also, note that now two restrictions are placed on the maximization of the entropy in order to lead to the result.

The application of the maximum entropy principle subject to the two restrictions in Eqs. (4.65a) and (4.65b) and the normalization then leads to a consideration of the maximization of the form

$$J[p(x)] = -\int_{-\infty}^{\infty} p(x) \ln p(x) dx + \lambda_0 \left(\int_{-\infty}^{\infty} p(x) dx - 1 \right) + \lambda_1 \left(\int_{-\infty}^{\infty} x p(x) dx - \mu \right) \tag{4.66}$$
$$+ \lambda_2 \left(\int_{-\infty}^{\infty} x^2 p(x) dx - \langle x^2 \rangle \right).$$

In Eq. (4.66) the entropy is subject to the conditions for the normalization of the distribution

$$\int_{-\infty}^{\infty} p(x) dx = 1 \tag{4.67a}$$

and the specification of the first moment μ by

$$\int_{-\infty}^{\infty} x p(x) dx = \mu \tag{4.67b}$$

and the specification of the second moment $\langle x^2 \rangle$ by

$$\int_{-\infty}^{\infty} x^2 p(x)dx = \langle x^2 \rangle. \tag{4.67c}$$

These requirements are introduced into the problem through the Lagrange multipliers λ_0, λ_1, and λ_2.

To treat the maximization of Eq. (4.66), we consider the variation of the probability distribution as represented by the form

$$p(x) \to p(x) + \delta p(x). \tag{4.68}$$

Proceeding as in the case of the Poisson distribution we find

$$\frac{\partial J[p(x)]}{\partial p(x)} = -\left[\ln p(x) + 1 - \lambda_0 - \lambda_1 x - \lambda_2 x^2\right] = 0. \tag{4.69a}$$

and

$$\frac{\partial^2 J[p(x)]}{\partial [p(x)]^2} = -\frac{1}{p(x)} < 0. \tag{4.69b}$$

Here again the curvature condition in Eq. (4.69b) requires that the extremum is a maximum as from the nature of the probability distribution $p(x) > 1$.

From Eq. (4.69a) the form of $p(x)$ is given by [1, 2, 4]

$$p(x) = e^{\lambda_0 - 1} e^{\lambda_1 x + \lambda_2 x^2} \tag{4.70}$$

where with the normalization it follows that $e^{\lambda_0 - 1} = \sqrt{\frac{-\lambda_2}{\pi}} e^{\frac{\lambda_1^2}{4\lambda_2}}$. For the conditions in Eq. (4.67b) to be met, it is then required that

$$\mu = -\frac{1}{2}\frac{\lambda_1}{\lambda_2} \tag{4.71a}$$

and for the condition in Eq. (4.67c) to be met it is required that

$$\langle x^2 \rangle = \frac{1}{4\lambda_2^2}\left(\lambda_1^2 - 2\lambda_2\right). \tag{4.71b}$$

Defining the variance as $\sigma^2 = \langle x^2 \rangle - \langle x^2 \rangle$, it then follows from Eqs. (4.70) and (4.71) that

$$p(x) = \frac{1}{\sqrt{2\pi\sigma^2}} e^{-\frac{1}{2\sigma^2}(x-\mu)^2}. \tag{4.72}$$

This is the standard form of the Gaussian distribution written in terms of the position μ of the distribution peak and the variance σ^2 of the distribution about its mean.

We see that placing a condition on the first moment of the distribution leads to the Poisson distribution while placing an additional restriction on the second moment of the distribution leads to the

Gaussian distribution. For a comparison, it is interesting to look at the system in the case that none of the moments enter as restrictions. In Eq. (4.56) this amounts to taking $\lambda_1 = 0$ while in Eq. (4.66) we take $\lambda_1 = \lambda_2 = 0$. Treating the resulting system defined on the interval $x\epsilon\left[a,b\right]$ where $b > a$ leads us to a consideration of the form [1, 2, 4]

$$J\left[p(x)\right] = -\int_a^b p(x)\ln p(x)dx + \lambda_0\left(\int_a^b p(x)dx - 1\right). \tag{4.73}$$

This is maximized by

$$\frac{\partial J\left[p(x)\right]}{\partial p(x)} = -\left[\ln p(x) + 1 - \lambda_0\right] = 0. \tag{4.74a}$$

and

$$\frac{\partial^2 J\left[p(x)\right]}{\partial\left[p(x)\right]^2} = -\frac{1}{p(x)} < 0, \tag{4.74b}$$

leading to the form

$$p(x) = e^{\lambda_0 - 1} \tag{4.75}$$

subject to the condition

$$\int_{-\infty}^{\infty} p(x)dx = 1. \tag{4.76}$$

It follows that the resulting distribution is

$$p(x) = \frac{1}{b-a}, \text{ for } x \in \left[a,b\right] \tag{4.77}$$

$$p(x) = 0, \text{ otherwise.}$$

Consequently, a normalized distribution with maximum entropy is just a uniform constant distribution.

4.7 INFORMATION

Another aspect of entropy involves the quantification and coding of information [1, 3, 5–10]. This is important in the development of an understanding of the storage and retrieval of information, and in the processes of information transmission. It has entered into many of the considerations of the representation of information to be sent along noisy transmission lines and in the error correction of transmitted and stored messages. Here only a brief discussion of the most elementary aspects of the subject is presented.

An important application of the idea of entropy is involved in the description of information and its transmission and storage. In such studies, it is common to define the measure of information using logarithms to the base two (i.e., the logarithm of x is denoted as $Log_2(x)$ so that $x = 2^{Log_2(x)}$) rather than in the natural logarithm (i.e., the logarithm of x is denoted as $\ln(x)$ so that $x = e^{ln(x)}$) or the logarithm to the base ten (i.e., the logarithm of x is denoted as $Log_{10}(x)$). Note that between these defining conventions, it is found that [1, 3, 5–10]

$$Log_2(x) = \frac{\ln(x)}{\ln(2)} \text{ and } Log_{10}(x) = \frac{ln(x)}{ln(10)} \tag{4.78a}$$

so that the corresponding entropies and information in these three systems are proportional to one another.

In base two, the informational content in selecting a random variable x_i which is expected with probability p_i (i.e., how much information is obtained in the selection of x_i) is given by

$$-Log_2(p_i) \tag{4.78b}$$

where the unit of the information is the bit. In this scheme, when $p_i = 1$ the information content of the event is 0 bits as nothing is learned from its measurement. On the other hand, when $p_i = 0$ the information content of the event is infinite, it is completely unexpected. Specifically, a lot is learned from selecting a measurement for which $p_i = 0$. For the other probabilities, $0 < p_i < 1$, the information falls between these limits, with the selection of less probable events providing more information about the system than is provided by selecting more probable events.

Similarly, these considerations can be extended to entropies and information defined using the other logarithms so that in the base ten system the information related to an event of probability p_i is [1, 3, 5–10]

$$-Log_{10}(p_i) \tag{4.78c}$$

(being measured in information units known as Harleys) and the information

$$-ln(p_i), \tag{4.78d}$$

measured in the natural logarithm, is given in information units known as nets. In both of these logarithmic systems it is again found that for $0 < p_i < 1$ the logarithms range between infinity and zero. Note, for the following discussions of information systems, we will stick to the base 2 logarithms and the units of bits discussed earlier. These are the most commonly used units in the measure of information in communication systems.

Given a complete set of events $\{x_1, x_2, ..x_N\}$ with corresponding probabilities $\{p_1, p_2, ..p_N\}$ the average informational content of the entire system is then defined by [1, 3, 5–10]

$$H(p_1, p_2, .., p_N) = -\sum_{i=1}^{N} p_i Log_2(p_i). \tag{4.79}$$

This is the standard entropy in base two of a set of probabilities as we have discussed earlier in our studies of statistics and statistical mechanics.

As an example of the use of the information and entropy measures in Eqs. (4.78) and (4.79), consider a simple probability system of coin tosses. For a single coin toss, the probability of an event (i.e., heads or tails) is $\frac{1}{2}$ so that the information received from the toss is

$$-Log_2\left(\frac{1}{2}\right) = 1 \tag{4.80a}$$

i.e., one bit of information. Note that this is also the entropy of the system. If we are given a biased coin so that in the single coin toss heads occurred with probability $p = \frac{3}{4}$ and tails with probability $1 - p = \frac{1}{4}$ the outcome has an average entropy or information content given as

$$-\frac{3}{4} Log_s\left(\frac{3}{4}\right) - \frac{1}{4} Log_2\left(\frac{1}{4}\right) = 0.81 \text{ bits.} \tag{4.80b}$$

We see that the information is less due to the bias in the coin. In this regard, for the extreme case of a biased coin toss in which heads occurs with probability $p = 1$ and tails with probability $1 - p = 0$, the outcome of the single toss is certain. In this case, the information received is

$$-Log_s(1) - 0Log_2(0) = 0 \text{ bits.} \tag{4.80c}$$

These three exams illustrate the basic nature of the information and entropy functions considering only a single coin toss.

Next consider a sequence of coin tosses with heads occurring with probability p and tails with probability $1 - p$. For a sequence of N such coin tosses with heads probability p and tails probability $1 - p$ the average information entropy is from Eq. (4.79) given by

$$H(p) = -p Log_2(p) - (1 - p) Log_2(1 - p). \tag{4.80d}$$

The information received from the N tosses, $NH(p)$, is then directly proportional to the number of tosses made.

Communication systems are generally complex and require not just probabilities of events associated with single outcomes for their description but also various joint and conditional probabilities involving the outcomes of two distinct events [1, 3, 5–10]. In such systems, the joint and conditional probabilities characterize the probabilities associated with the simultaneous occurrence of two or more events, e.g., the sending and receiving of a message. This brings us to discussions of entropies involving the operation of sources of information and of the resultant received messages. To begin, we look at some of the information processes of messages in a transmission line between the source and the receiver.

4.7.1 Systems for the Transmission of Information

A simple example of such an information process and the application of entropy ideas in its description is developed by considering a highly idealized information system. Our idealized model of an information system involves the treatment of messages composed of sequences of 0's and 1's as in a type of binary computer code. To simplify the treatment, then, the possible messages in the system are limited to arrays of $N \gg 1$ sites that are randomly occupied by 0's or 1's, and associated with each of the strings of length N is a unique message. For example, one array of length N may correspond to the message "I'd like chocolate ice cream," while another different array of length N may correspond to the message "I like red cars." In this way, the system can in principle handle 2^N different statements which occur in various types of storage or transmission processes.

Let us assume that in our system there is a source S which generates messages of length $N \gg 1$ and that during the generation process the digit 0 is generated with probability p while the digit 1 is generated with probability $1 - p$. The messages from the source are then distributed in a binomial distribution so that the probability of S emitting a message with m zeros is [1, 3, 5–10]

$$P(m) = \frac{N!}{m!(N - m)!} p^m (1 - p)^{N - m}. \tag{4.81}$$

For large N this is a highly peaked binomial distribution, and in the limit that $N \to \infty$ the average number of zeros in the typical message of the system is given by $m_{average} = pN$. In this sense, the typical state generated by the source will overwhelmingly have $m_{average}$ zeros, and we shall assume in the following that all the meaningful messages of our interest are included in this set of messages.

In fact, the number of such states of $m_{average}$ zeros (denoted as N_S) is given from the prefactor in Eq. (4.81) by [1, 3, 5–10]

$$N_S = \frac{N!}{m_{average}!\left(N - m_{average}\right)!} \tag{4.82}$$

$$= \frac{N!}{(pN)!\left[(1-p)N\right]!}.$$

For large N Eq. (4.82) can be simplified through an application of Stirling's approximation for large factorials. This approximation states that for large M the logarithm of $M!$ approaches the limiting form [1]

$$log_2\left(M!\right) \to M\,log_2\left(M\right) - \frac{M}{\ln 2} \tag{4.83}$$

where $log_2(\ )$ is the logarithm to the base 2. (Note that Stirling's approximation can be rewritten to handle logarithms in general bases. For example, in the natural logarithm, Stirling's approximation takes the form $\ln\left(M!\right) \to M \ln\left(M\right) - M$. In our discussion, the base 2 form has been given as it is useful to discuss binary coding in base 2.) Applying Eq. (4.83) to $log_2\left(N_S\right)$ from Eq. (4.82) then gives [1]

$$log_2\left(N_S\right) = N\,log_2\left(N\right) - \frac{N}{\ln 2} - pN\,log_2\left(pN\right) + \frac{pN}{\ln 2} - \tag{4.84}$$

$$-(1-p)N\,log_2\left[(1-p)N\right] + \frac{(1-p)N}{\ln 2}$$

$$= N\left[-p\,log_2(p) - (1-p)log_2(1-p)\right]$$

$$= NH_s\left(p\right)$$

where only the terms in log_2 remain.

Notice in Eq. (4.84) that $H_s\left(p\right) = -p\,log_2\left(p\right) - (1-p)log_2\left(1-p\right)$ has the form of the entropy $H\left(p\right)$ of the binary probabilities p and $\left(1-p\right)$. This is given from Eq. (4.9) in the base 2 logarithms by

$$H\left(p\right) = -p\,log_2\left(p\right) - \left(1-p\right)log_2\left(1-p\right). \tag{4.85}$$

Consequently, from Eqs. (4.84) and (4.85) it follows that [1, 3, 5–10]

$$N_S = 2^{NH_s(p)} = 2^{NH(p)}, \tag{4.86}$$

and the number of states generated by the source is ultimately related to the entropy $H\left(p\right)$ of the p and $1-p$ probabilities generating the binomial distribution. In this regard, it can be shown that

Eq. (4.86) represents the minimum information content which the source can launch into the system to be transmitted to a receiver.

In many cases the messages generated by the source are processed by a device, and the results are communicated to a receiver R as a final message. If it is assumed that the message received by the receiver R is again of length N and composed of 0's and 1's occurring with probabilities p and $(1-p)$, then, using the same arguments as in our treatment of the source messages, the number of received messages must be of the form [1, 3, 5–10]

$$N_R = 2^{NH_R(p)} = 2^{NH(p)}, \tag{4.87}$$

where $H_R(p) = H(p)$ is again given in Eq. (4.85). (Here we have assumed for simplicity that the messaging properties of the source channel introducing the messages into the device are the same as those of the receiver channel passing the processed messages to the receiver.) Similar to our discussion of the source signal, it can be shown that Eq. (4.87) represents the minimum information content which the receiver can accept from the source.

Let us next look, for some cases of interest, at the process of generating a message at the source and receiving it at the receiver. This involves combining the two processes we have just treated as they are mediated by a transmission channel. The fundamental question in the generation and retrieval of messages is how faithfully they are transmitted so that information is not lost from the system due to noise or other error mechanisms. This topic requires the development of a measure for correlating the information content in the received and sent messages as they are processed by the system. We begin developing such a measure from the consideration of some examples of messages sent and received along a transmission channel.

First, consider the case in which there is no correlation between the sent and received messages, i.e., the sent message does not determine the message received. In these processes, the device arbitrarily generates a message to send to the receiver in response to any message it receives from the source. Consequently, the total possible number of sent and received message pairs, N_T, is [1, 3, 5 –10]

$$N_T = N_S N_R = 2^{NH_s(p)} 2^{NH_R(p)} = 2^{N\left[H_s(p)+H_R(p)\right]} = 2^{2NH(p)}. \tag{4.88a}$$

It follows that the total entropy for the transmission of messages of length N in the absence of correlation between the sent and received signals is [1, 3, 5–10]

$$2H(p). \tag{4.88b}$$

This entropy indicates a high degree of disorder in the transmission process.

On the other hand, next consider the case in which the received message is completely determined from the sent message, i.e., the received message is exactly the same as the sent message. Now the total possible number of sent and received messages in the system is [1, 3, 5–10]

$$N_T = N_S N_R = 2^{NH_s(p)} 2^{N0} = 2^{NH_s(p)} = 2^{NH(p)}. \tag{4.89a}$$

In this regard, the resulting entropy of the transmission,

$$H(p), \tag{4.89b}$$

is half that in the totally uncorrelated system.

Between these two limits of totally uncorrelated transmissions and totally faithful transmissions are a variation of partial transmissions of information. These partial transmissions of information may originate due to noise or through a conscious modification of the message. The next step in the discussions is to extend our ideas of probability to include probabilities related to the occurrence of two events. This in turn allows for the development of a measure quantifying the degree of correlation between the sent and received messages. Ultimately, we are brought to some discussions involving entropies based on conditional or joint probabilities.

4.7.2 Joint Probabilities in Transmission Processes

Intermediate between the two limiting behaviors of a message being generated at the source and the reception of the message at the receiver is the channel transmission. During this transmission, changes in the message signal can occur so that the sent and received messages are only probabilistically related to one another. In this type of system, there is then a focus on the determination of the joint probabilities quantifying the extent that the sent and received messages are related to one another, i.e., given a sent message, what is the probability that it will be received in a specific form which may differ from the original sent message. The changes in the sent message may arise from random noise or some other systematic process.

Specifically, given the set of possible sent messages $S = \{s_1, s_2, \ldots\ldots\ldots\}$ each of length N and the set of received messages $R = \{r_1, r_2, \ldots\ldots\ldots\}$ each of length N, define the function [1, 3, 5–10]

$$P(r_i \mid s_j) \tag{4.90a}$$

as the probability that the message r_i is received when the message s_j is sent. In such a system, for example, the received message may have been altered from the sent message by a noisy transmission or by the processing of some intermediate device. From Eq. (4.90a) it then follows that the chance that both r_i and s_j occur together in the system is given by

$$P(r_i, s_j) = P(r_i \mid s_j) P(s_j) \tag{4.90b}$$

where $P(s_j)$ is the probability of s_j. The probability $P(r_i, s_j)$ is often referred to as the joint probability and the probability $P(r_i \mid s_j)$ is known as the conditional probability.

Associated with the set of joint probabilities $\{P(r_i, s_j)\}$ for the transmissions, one can develop the transmission entropy function given by [1, 3, 5–10]

$$H(R,S) = -\sum_{i=1}^{N}\sum_{j=1}^{N} P(r_i, s_j) Log_2 P(r_i, s_j) \tag{4.91}$$

defined on the sets $R \equiv \{r_1, r_2, \ldots\ldots\ldots\}$ of received messages and $S \equiv \{s_1, s_2, \ldots\ldots\ldots\}$ of sent messages. This entropy characterizes the amount of information within the system of sent and received messages, and, in turn, can be used to develop a measure of the correlation between the elements of the sets R and S in their probabilistic linkage with one another. These correlations are important in estimating the degree of faithfulness with which the received messages represent the sent messages.

Specifically, the quantification of the correlations in the sent and received messages is done by defining the correlation function [1, 3, 5–10]

$$I(R:S) = \sum_{i=1}^{N}\sum_{j=1}^{N} P(r_i, s_j) Log_2 \frac{P(r_i, s_j)}{P(r_i) P(s_j)} \tag{4.92a}$$

where

$$P(r_i)=\sum_{j=1}^{N}P(r_i,s_j), \tag{4.92b}$$

and

$$P(s_j)=\sum_{i=1}^{N}P(r_i,s_j)=\sum_{i=1}^{N}P(s_j,r_i) \tag{4.92c}$$

represent, respectively, the probability of the occurrence of r_i and s_j. (Note that the probabilities in Eqs. (4.92b) and (4.92c) which involve only a single variable are often referred to as marginal probabilities.) The correlator defined in Eq. (4.92a) is termed the average mutual information as it is commonly used as a measure of the efficiency of the transmission of information from a source S to a receiver R. In this regard, it is a measure of the mutual dependence of the elements in S and the elements in R.

From Eq. (4.92a) it follows that in terms of the entropies of the marginal probabilities of the individual elements and joint probabilities defined on the sets R and S the correlation function becomes [1, 3, 5–10]

$$I(R:S)=\sum_{i=1}^{N}\sum_{j=1}^{N}P(r_i,s_j)Log_2\frac{P(r_i,s_j)}{P(r_i)P(s_j)} \tag{4.93}$$

$$=\sum_{i=1}^{N}\sum_{j=1}^{N}P(r_i,s_j)Log_2P(r_i,s_j)-\sum_{i=1}^{N}\sum_{j=1}^{N}P(r_i,s_j)Log_2P(r_i)$$

$$-\sum_{i=1}^{N}\sum_{j=1}^{N}P(r_i,s_j)Log_2P(s_j)$$

$$=-H(R,S)+H(R)+H(S).$$

Here the probabilities $\{P(r_i)\}$ and $\{P(s_j)\}$ are the marginal probabilities of the r_i and s_j variables, and from the definition it follows that $I(R:S)=I(S:R)$. It is seen that the average mutual information is ultimately expressed in terms of the entropies $H(R)$, $H(S)$, and $H(R,S)$.

From Eq. (4.93) we find that when the elements in R and S occur independently of one another so that $P(r_i,s_j)=P(r_i)P(s_j)$, the average mutual information in Eq. (4.93) yields [1, 3, 5–10]

$$I(R:S)=0. \tag{4.94}$$

In this limit, there is no correlation in the system because the sent and received messages do not affect one another. On the other hand, consider a completely correlated system in which the source messages show up as the received message. Specifically,

$$s_i=r_i \text{ for all } i \tag{4.95a}$$

and

$$P(r_i, s_j) = P(r_i) = P(s_i) \text{ for } i = j \tag{4.95b}$$

$$= 0, \text{ otherwise.}$$

Under these conditions it follows from Eq. (4.93) that [1, 3, 5–10]

$$I(R:S) = H(R) = H(S) > 0 \tag{4.96}$$

where $H(R)$ is the entropy of the received messages and $H(S)$ is the entropy of the sent messages. It is seen that Eq. (4.94) is in agreement with the result in Eq. (4.88b) and Eq. (4.96) is in agreement with Eq. (4.89b).

We now consider a model which interpolates between the limits of no correlation and complete correlation. It illustrates a range in values of the average mutual information which quantifies the efficiency of sending and accurately receiving messages.

4.7.3 An Example of Signal Change during Transmission

A simple, idealized model of a noisy communication system which generates, transmits, and receives messages is now treated [1, 3, 5–10]. The proposed model is of interest as it can be solved exactly and provides an illustration of many of the information properties of correlation in communication systems. In the simple system, source messages are generated from a set of two possible messages $S \equiv \{s_1 = 0, s_2 = 1\}$ and these are eventually received as one of a set of two received messages from the set $R \equiv \{r_1 = 0, r_2 = 1\}$. The probabilistic process for this transmission is described by a set of nontrivial joint probabilities defined between the elements of S and R so that the value of the sent signal digits of S probabilistically determines those received in the set R. This describes a system in which the sent signal S interacts with a noisy transmission channel before it arrives as the received signal in R. Consequently, the source signal may undergo a change due to the presence of the noise.

In the model, the values of R are directly determined only by those of S. We begin with a consideration of the generation of messages at the source: in the messages generated by the source S, the variable $s_1 = 0$ is generated with marginal probability

$$P(s_1 = 0) = s \tag{4.97a}$$

and the variable $s_2 = 1$ is generated with marginal probability

$$P(s_2 = 1) = 1 - s. \tag{4.97b}$$

During the transmission process from S to R the message is affected by a noisy transmission channel. This transmission to R is described probabilistically by a set of conditional probabilities so that [1, 3, 5–10]:

1. A sent signal with $s_1 = 0$ is received as $r_1 = 0$ with conditional probability $P(r_1 = 0 \mid s_1 = 0) = 1 - p$,
2. A sent signal with $s_1 = 0$ is received as $r_2 = 1$ with conditional probability $P(r_2 = 1 \mid s_1 = 0) = p$,
3. A sent signal $s_2 = 1$ is received as $r_1 = 0$ with conditional probability $P(r_1 = 0 \mid s_2 = 1) = p$,
4. A sent signal $s_2 = 1$ is received as $r_2 = 1$ with conditional probability $P(r_2 = 1 \mid s_2 = 1) = 1 - p$.

(See Figure 4.2 for a schematic diagram of this transmission process of transmission between S and R.) Here we have introduced the conditional probability $P(r_j \mid s_i)$ which gives the probability of r_j

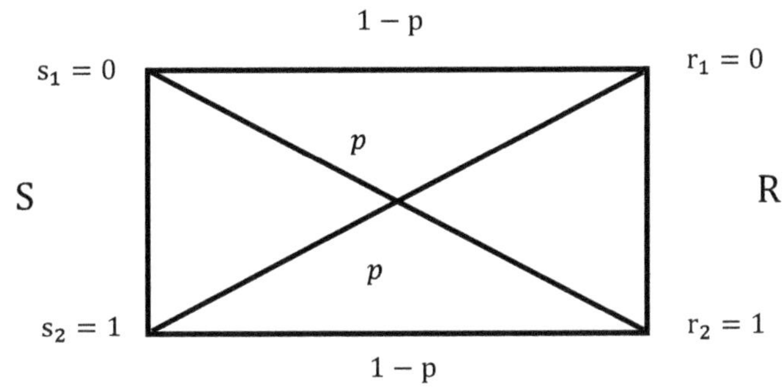

FIGURE 4.2 Schematic drawing for signal transmission between the source S and the receiver R. At the source, two signals s_1 and s_2 may be sent. At the receiver two signals r_1 and r_2 may be received. The conditional probabilities $P(r_j \mid s_i)$ for $i = 1,2$ and $j = 1,2$ are indicated on the lines connecting the vertices $\{s_1, s_2\}$ of the source signals and $\{r_1, r_2\}$ of the received signals.

given the certainty of s_i. Consequently, the join, conditional, and marginal probabilities are related by $P(r_j, s_i) = P(r_j \mid s_i)P(s_i)$.

As a consequence of the transmission process in Figure 4.2, it is found that the variables received by R assumes the value 0 with marginal probability [1, 3, 5–10]

$$P(r_1 = 0) = P(s_1 = 0)(1-p) + P(s_2 = 1)p = s(1-p) + (1-s)p \tag{4.98a}$$

and the value 1 with marginal probability

$$P(r_2 = 1) = P(s_2 = 1)(1-p) + P(s_1 = 0)p = (1-s)(1-p) + sp. \tag{4.98b}$$

Consequently, the reception probabilities of the two messages depend both on their marginal probability of generation at the source and their conditional probability of transmission to the receiver.

From the discussions above, the resultant joint probabilities for the transmission of s_i to r_j are then given by [1, 3, 5–10]

$$P(r_1 = 0, s_1 = 0) = P(s_i = 0)(1-p) = s(1-p), \tag{4.99a}$$

$$P(r_1 = 0, s_2 = 1) = P(s_2 = 1)p = (1-s)p, \tag{4.99b}$$

$$P(r_2 = 1, s_1 = 0) = P(s_1 = 0)p = sp, \tag{4.99c}$$

and

$$P(r_2 = 1, s_2 = 1) = P(s_2 = 1)(1-p) = (1-s)(1-p). \tag{4.99d}$$

The joint probabilities in Eq. (4.99) and the marginal probabilities in Eqs. (4.97) and (4.98) are now applied in Eqs. (4.92a) and (4.93) to determine the entropy of the total process and the average mutual information. This is now considered.

Specifically, from Eq. (4.92a) the average mutual information developed between the elements of our sets $S=\{s_1=0,s_2=1\}$ and $R=\{r_1=0,r_2=1\}$ by the process in Figure 4.2 follows from

$$I(R:S)=\sum_{i=1}^{2}\sum_{j=1}^{2}P(r_i,s_j)Log_2\frac{P(r_i,s_j)}{P(r_i)P(s_j)}. \tag{4.100}$$

Upon the substitution of Eqs. (4.97) through (4.99) in Eq. (4.100), it then yields [1, 3, 5–10]

$$I(R:S)=(1-p)s\ Log_2\left(\frac{(1-p)s}{s\left[(1-p)s+(1-s)p\right]}\right)+ps\ Log_2\left(\frac{ps}{s\left[ps+(1-p)(1-s)\right]}\right) \tag{4.101}$$

$$+p(1-s)Log_2\left(\frac{p(1-s)}{(1-s)\left[(1-p)s+(1-s)p\right]}\right)+(1-p)(1-s)Log_2\frac{(1-p)(1-s)}{(1-s)\left[ps+(1-p)(1-s)\right]}.$$

Following some algebra, the expression in Eq. (4.101) reduces to the form

$$I(R:S)=(1-p)Log_2(1-p)+p\ Log_2p \tag{4.102}$$

$$-\left[s(1-p)+(1-s)p\right]Log_2\left[s(1-p)+(1-s)p\right]$$

$$-\left[sp+(1-s)(1-p)\right]Log_2\left[sp+(1-s)(1-p)\right].$$

This determines the general transmission characteristics between the source and the receiver for a general signal of our system. Of interest, however, is the maximum value of $I(R:S)$ which is the ultimate possible efficiency of the transmission channel.

The quantity of interest regarding the channel transmission is the channel capacity. This is defined as the values of the source marginal probabilities (in our case s) that make Eq. (4.102) a maximum. Given the rate of signal generation at the source, this gives the maximum rate at which information is transmitted in the system. Specifically, to obtain this for our system we determine [1, 3, 5–10]

$$\frac{\partial I(R:S)}{\partial s}=0=(1-2p)Log_2\left[\frac{sp+(1-s)(1-p)}{s(1-p)+(1-s)p}\right] \tag{4.103}$$

which is satisfied by

$$s=\frac{1}{2}. \tag{4.104}$$

For $s=\frac{1}{2}$ from Eq. (4.102), the maximum is then given by

$$I(R:S)_{Max}=1+(1-p)Log_2(1-p)+p\ Log_2(p). \tag{4.105}$$

It is found that for $p=0$ or 1 the correlation in the system is the greatest and

$$I(R:S)_{Max}=1. \tag{4.106a}$$

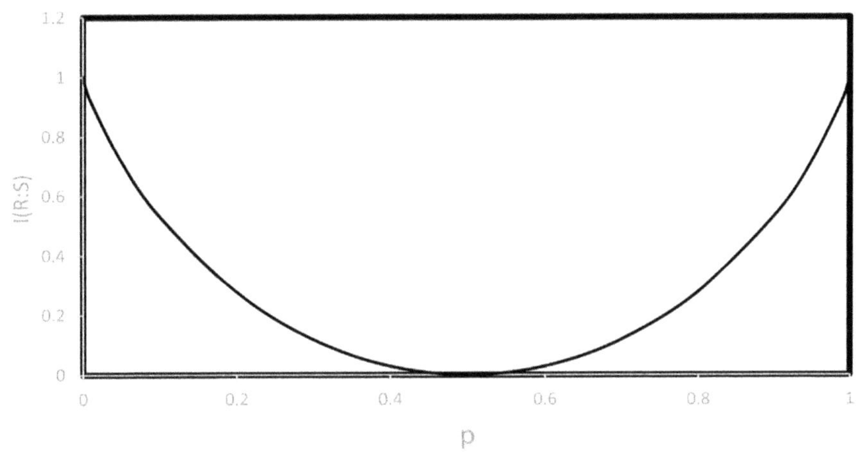

FIGURE 4.3 Plot of the channel capacity $I(R:S)=I(R:S)_{Max}$ versus p.

On the other hand, for $p=\frac{1}{2}$ the correlation in the system is the least and

$$I(R:S)_{Max}=0. \tag{4.106b}$$

From the Figure 4.3 plot of the channel capacity $I(R:S)_{Max}$ versus p we see that the remaining values of $I(R:S)_{Max}$ fall between the two limits in Eq. (4.106).

From Figure 4.3 it is seen that when $p=0$ the signal generated at S is received unchanged at R, while when $p=1$ the signal generated at S is flipped so that at R a sent 0 is received as a 1 and a sent 1 is received as a 0. In either of these cases, there is no uncertainty as to the intent of the original messages. However, for increasing $0<p<\frac{1}{2}$ the nature of the original message from S as received at R is increasingly uncertain. At $p=\frac{1}{2}$ the uncertainty at R in the nature of the sent messages from S is the greatest. In the next interval, for increasing $\frac{1}{2}<p<1$ the nature of the original message from S is increasingly more certain as it is received at R. In this sense, it is found that the correlator $I(R:S)_{Max}$ offers a relative indicator of the efficiency and fidelity of the transmission process in the transmitting channel between S and R.

4.7.4 An Example of Signal Loss during Transmission

Another simple model of a noisy communication system is now treated, which illustrates another type of error process [1, 3, 5–10]. Of interest in the present system is that during the transmission process, information can be lost from the system during the transmission. Again, the proposed model can be solved exactly, providing an illustration of the information properties of a type of lossy communication system. In the system, source messages are generated from a set of two possible messages $S\equiv\{s_1=0,s_2=1\}$ and these may be eventually received as one of a set of two received messages from the set $R\equiv\{r_1=0,r_2=1\}$. In addition, however, the signals in S can be lost to an eraser bin denoted as E. Signals that arrive at the eraser bin are considered to be lost from the system. The probabilistic process for this transmission is described by a set of nontrivial conditional probabilities defined between the elements of S and R and E so that the value of the sent source messages of S probabilistically determines the messages received in the set R and the eraser bin E.

In the model the messages received by R and the losses to E are directly determined by the messages sent from S. (See Figure 4.4 for a schematic diagram representing the proposed transmission

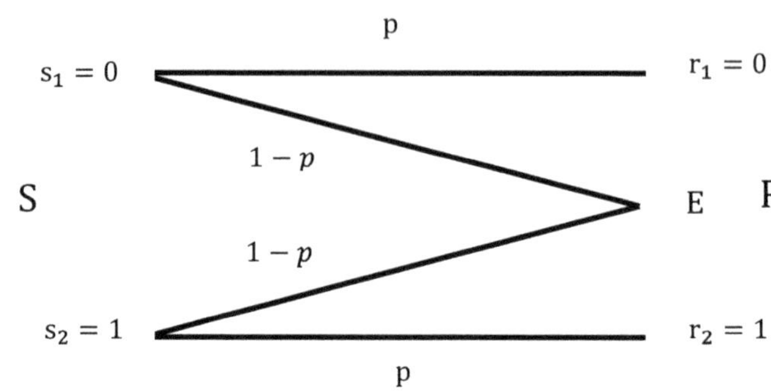

FIGURE 4.4 Schematic drawing for signal transmission between the source S, the receiver R, and an eraser bin E which accounts for losses in the system. At the source, two signals s_1 and s_2 may be sent. At the receiver two signals r_1 and r_2 may be received. In addition, some of the source signals are lost to the eraser bin. The conditional probabilities $P(r_i \mid s_i)$ for $i = 1,2$ are indicated on the lines connecting the vertices $\{s_1, r_1\}$ and the vertices $\{s_2, r_2\}$. In addition, the condition probabilities $P(E \mid s_i)$ for $i = 1,2$ indicating losses to the eraser bin are indicated on the lines connecting the vertices $\{s_1, E\}$ and the vertices $\{s_2, E\}$.

process.) For the messages generated by the source S, the message $s_1 = 0$ is generated with the marginal probability

$$P(s_1 = 0) = s \tag{4.107a}$$

and the message $s_2 = 1$ is generated with the marginal probability

$$P(s_2 = 1) = 1 - s. \tag{4.107b}$$

During the transmission process from S to R the message is affected by a noisy transmission channel which causes some messages to be lost to E. The transmission to R is described probabilistically by the conditional probabilities so that:

1. A source message with $s_1 = 0$ is received as $r_1 = 0$ with the conditional probability $P(r_1 = 0 \mid s_1 = 0) = p$,
2. A source message $s_2 = 1$ is received as $r_2 = 1$ with the conditional probability $P(r_2 = 1 \mid s_2 = 1) = p$,
3. A source message with $s_1 = 0$ is received in the eraser bin E with the conditional probability $P(E \mid s_1 = 0) = 1 - p$,
4. A source message $s_2 = 1$ is received in the eraser bin E with the conditional probability $P(E \mid s_2 = 1) = 1 - p$,

Figure 4.4, again, offers a schematic labeling the various transmission processes between S and R and E.

As a consequence of this transmission process, it is found that the messages received by R assume the value 0 with marginal probability [1, 3, 5–10]

$$P(r_1 = 0) = P(s_1 = 0)p = sp \tag{4.108a}$$

and the value 1 with marginal probability

$$P(r_2=1)=P(s_2=1)p=(1-s)p. \tag{4.108b}$$

Consequently, the reception probabilities of the two messages depend both on their marginal probability of generation at the source and their conditional probability of transmission to the receiver. In addition, the messages can be lost or erased during transmission. The probability of either of the source messages ending in the eraser bin is then given by

$$P(E)=s(1-p)+(1-s)(1-p)=1-p. \tag{4.108c}$$

From the above discussions, the resultant joint probabilities for the transmission of s_i to r_j and s_i to E are then given by [1, 3, 5–10]

$$P(r_1=0,s_1=0)=P(s_i=0)p=sp, \tag{4.109a}$$

$$P(r_1=0,s_2=1)=0, \tag{4.109b}$$

$$P(r_2=1,s_1=0)=0, \tag{4.109c}$$

$$P(r_2=1,s_2=1)=P(s_2=1)p=(1-s)p, \tag{4.109d}$$

$$P(E,s_1=0)=P(s_i=0)(1-p)=s(1-p), \tag{4.109e}$$

and

$$P(E,s_2=1)=P(s_2=1)(1-p)=(1-s)(1-p) \tag{4.109f}$$

The joint probabilities in Eq. (4.109) and the marginal probabilities in Eqs. (4.107) and (4.108) can now be applied in Eqs. (4.92a) and (4.93) to determine the entropy of the total transmission process and its average mutual information.

In the present problem, from Eq. (4.92a) the average mutual information developed between the elements of our sets $S=\{0,1\}$ and $R=\{0,1\}$ and E by the processes in Figure 4.4 follows from [1, 3, 5–10]

$$I(R,E:S)=\sum_{i=1}^{2}\sum_{j=1}^{2}P(r_i,s_j)Log_2\frac{P(r_i,s_j)}{P(r_i)P(s_j)} \tag{4.110}$$

$$+\sum_{i=1}^{2}P(r_i,E)Log_2\frac{P(r_i,E)}{P(r_i)P(E)}.$$

Upon the substitution of Eqs. (4.107) through (4.109) in Eq. (4.110) it then yields

$$I(R,E:S)=sp\ Log_2\left(\frac{sp}{s[sp]}\right)+(1-s)p\ Log_2\left(\frac{(1-s)p}{(1-s)[(1-s)p]}\right) \tag{4.111}$$

$$+s(1-p)Log_2\left(\frac{s(1-p)}{s(1-p)}\right)+(1-s)(1-p)Log_2\frac{(1-s)(1-p)}{(1-s)(1-p)}.$$

Following some algebra, this reduces to the form

$$I(R,E:S) = -p\left[s\,Log_2 s + (1-s)Log_2(1-s)\right] = pH(s) \tag{4.112}$$

involving the entropy $H(s) = -\left[s\,Log_2 s + (1-s)Log_2(1-s)\right]$. This determines the transmission characteristics between the source and the receiver for a general signal. Of particular interest is the maximum value of $I(R,E:S)$ which is the ultimate possible efficiency or capacity of the channel.

The quantity of interest regarding the channel transmission is the channel capacity obtained from the source marginal probabilities (i.e., the value of s) that makes Eq. (4.112) a maximum. Specifically, for our study, it is obtained from considering [1, 3, 5–10]

$$\frac{\partial I(R:S)}{\partial s} = 0 = -p\,Log_2\left[\frac{s}{(1-s)}\right] \tag{4.113}$$

which is satisfied by

$$s = \frac{1}{2}. \tag{4.114}$$

For $s = \frac{1}{2}$, from Eq. (4.112) the maximum is given by

$$I(R,E:S)_{Max} = p\,Log_2\,2 = p. \tag{4.115}$$

For $p = 1$ the correlation in the system is the greatest and

$$I(R,E:S)_{Max} = 1. \tag{4.116a}$$

On the other hand, for $p = 0$ the correlation in the system is the least and

$$I(R,E:S)_{Max} = 0. \tag{4.116b}$$

From the Figure 4.5 plot of the channel capacity $I(R,E:S)_{Max}$ versus p we see that the remaining values of $I(R,E:S)_{Max}$ fall between these limits.

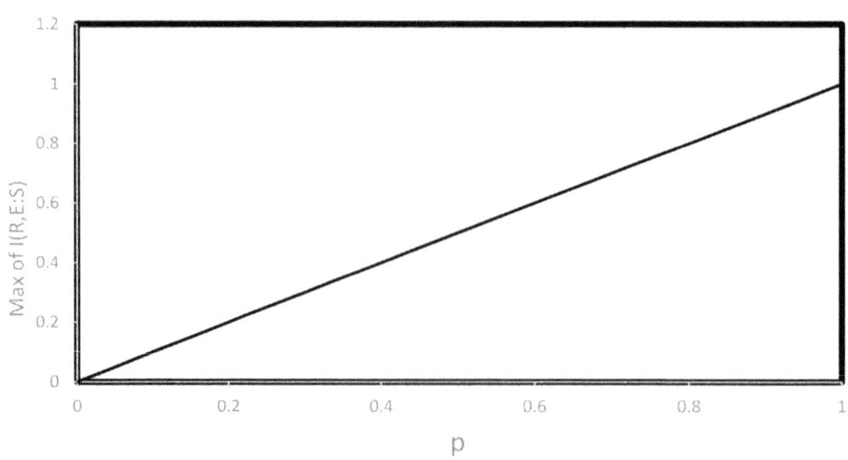

FIGURE 4.5 Plot of the channel capacity $Max\,of\,I(R,E:S) = I(R,E:S)_{Max}$ versus p.

From Figure 4.5 it is found that when $p = 0$ the signal generated at S is never received at R, while when $p = 1$ the signal generated at S is always received undistorted at R. As p is decreased from 1, the signal is increasingly erased from the system during the transmission process.

7.4.5 Example of a Noisy System with Greater Numbers of Messages

Next, we consider a system with a greater number of possible sent and received messages [1, 3, 5–10]. This example shows how to extend the network transmissions to more complex messaging, while also extending the entropy considerations quantifying the information involved in such processes. Specifically, a model is treated in which there are three possible source messages $S = \{s_1, s_2, s_3\}$ which show up at the receiver as one of the three possible received messages $R = \{r_1, r_2, r_3\}$. The study modifies the first noisy transmission example to include the basic features of most multiple message communications.

The network studied is shown diagrammatically in Figure 4.6. Here during the transmission process from S to R the messages $S = \{s_1, s_2, s_3\}$ are respectively sent by the source with marginal probabilities $\{P(s_1), P(s_2), P(s_3)\}$ and received as one of the set $R = \{r_1, r_2, r_3\}$ with marginal probabilities $\{P(r_1), P(r_2), P(r_3)\}$. The transmissions received at R are again described probabilistically by conditional probabilities so that if s_i is sent by the source, then r_j is received at the receiver with the conditional probability $P(r_j \mid s_i)$. In terms of these conditional probabilities, the probability that both s_i is sent and r_j is received is then given by $P(r_j, s_i) = P(r_j \mid s_i)P(s_i) = P(s_j \mid r_i)P(r_i) = P(s_j, r_i)$. From these relationships the average mutual information and the entropy of the communication are computed for the transmission process.

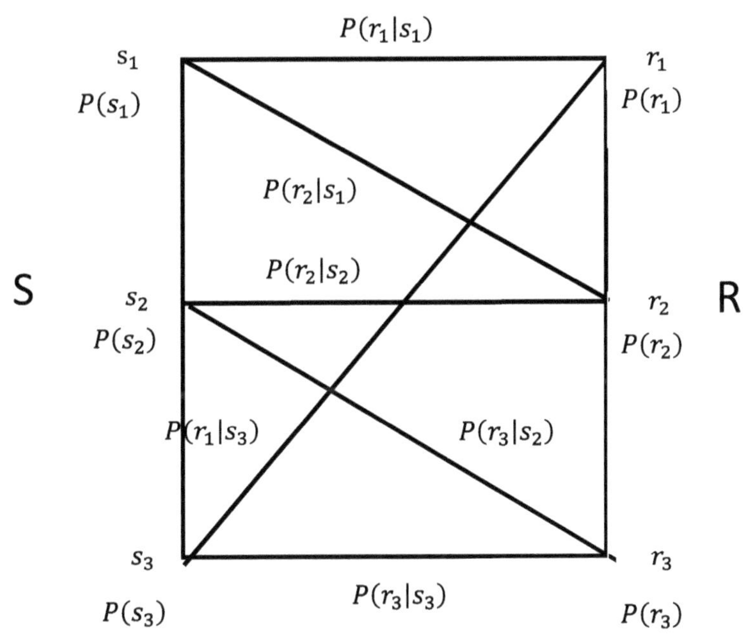

FIGURE 4.6 Schematic drawing for signal transmission between the source S, and the receiver R. At the source three signals s_1 and s_2 and s_3 may be sent. At the receiver, three signals r_1 and r_2 and r_3 may be received. The conditional probabilities $P(r_j \mid s_i)$ for $i = 1,2,3$ and $j = 1,2,3$ are indicated on the lines connecting the vertices $\{s_1, s_2, s_3\}$ and the vertices $\{r_1, r_2, r_3\}$.

For a general set of marginal probabilities and joint probabilities between the sets $S = \{s_1, s_2, s_3\}$ and $R = \{r_1, r_2, r_3\}$ the average mutual information is given from Eq. (4.93) by [1, 3, 5–10]

$$I(R:S) = \sum_{i=1}^{3}\sum_{j=1}^{3} P(r_i, s_j) Log_2 \frac{P(r_i, s_j)}{P(r_i)P(s_j)} \tag{4.117}$$

$$= \sum_{i=1}^{3}\sum_{j=1}^{3} P(r_i, s_j) Log_2 \frac{P(r_i \mid s_j)P(s_j)}{P(r_i)P(s_j)}$$

$$= \sum_{i=1}^{3}\sum_{j=1}^{3} P(r_i \mid s_j)P(s_j) Log_2 P(r_i \mid s_j) - \sum_{i=1}^{3}\sum_{j=1}^{3} P(r_i \mid s_j)P(s_j) Log_2 P(r_j).$$

This is conveniently rewritten into the form

$$I(R:S) = -\sum_{i=1}^{3} P(s_i)P(r_i \mid s_i) Log_2 \sum_{k=1}^{3} P(s_k)P(r_i \mid s_k) \tag{4.118a}$$

$$-\sum_{i=1 \neq j=1}^{3} P(s_i)P(r_j \mid s_i) Log_2 \sum_{k=1}^{3} P(s_k)P(r_j \mid s_k)$$

$$+\sum_{i=1}^{3} P(s_i)P(r_i \mid s_i) Log_2\, P(r_i \mid s_i)$$

$$+\sum_{i=1 \neq j=1}^{3} P(s_i)P(r_j \mid s_i) Log_2\, P(r_j \mid s_i)$$

which separates the $i = j$ from the $i \neq j$ terms.

As per our previous discussions, $I(R:S)$ ultimately provides a measure of the fidelity with which the launched signal is received upon its passage through the system. Similarly, the entropy of the transmission process is given by

$$H(R,S) = -\sum_{i=1}^{3}\sum_{j=1}^{3} P(r_i, s_j) Log_2\, P(r_i, s_j) \tag{4.118b}$$

and quantifies the amount of information transmission in the system when signals are passed and received. A further understanding of the nature of $I(R:S)$ and $H(R,S)$ in this regard can be developed by a consideration of some particular cases.

To begin, consider a particular set of probabilities and conditional probabilities. A simple system is composed of the network in Figure 4.6 if it is assumed that the conditional probabilities have the form

$$P(r_i \mid s_i) = \frac{1}{2}, \tag{4.119a}$$

$$P(r_{i+1} \mid s_i) = \frac{1}{2} \text{ for } i = 1,2, \tag{4.119b}$$

$$P(r_1 \mid s_3) = \frac{1}{2}, \tag{4.119c}$$

and

$$P(r_j \mid s_i) = 0, \text{otherwise.} \tag{4.119d}$$

Here each of the s_i and r_i are connected to one another, while s_i can also be received as an alternative signal r_j with $i \neq j$. From the configuration defined in Eq. (4.119) it follows from Eq. (4.118a) that [1, 3, 5–10]

$$\begin{aligned} I(R:S) = &-\frac{1}{2}\left[P(s_1)+P(s_3)\right]Log_2\left(\frac{1}{2}\left[P(s_1)+P(s_3)\right]\right) \\ &-\frac{1}{2}\left[P(s_2)+P(s_1)\right]Log_2\left(\frac{1}{2}\left[P(s_2)+P(s_1)\right]\right) \\ &-\frac{1}{2}\left[P(s_2)+P(s_3)\right]Log_2\left(\frac{1}{2}\left[P(s_2)+P(s_3)\right]\right)-1 \end{aligned} \tag{4.120}$$

where we have used $Log_2 \frac{1}{2} = -1$ to obtain the final form of Eq. (4.120).

As an illustration, consider Eq. (4.120) for three interesting cases of the set of source marginal probabilities $\{P(s_1), P(s_2), P(s_3)\}$ characterizing the source message generation in the system. In the case that $P(s_1) = P(s_2) = P(s_3) = \frac{1}{3}$ (so that all the possible messages are sent with equal probability), it follows that

$$I(R:S) = Log_2 3 - 1 = 0.585 \text{ bits.} \tag{4.121a}$$

Similarly, in the case that $P(s_1) = P(s_2) = \frac{1}{2}$ and $P(s_3) = 0$ (so that the s_1 and s_2 are sent with equal probability but s_3 is never sent) it follows that

$$I(R:S) = \frac{1}{2} \text{ bits.} \tag{4.121b}$$

Reducing the message generation further so that $P(s_1) = 1$, $P(s_2) = 0$, and $P(s_3) = 0$ (only one signal is ever sent) we find

$$I(R:S) = 0. \tag{4.121c}$$

The greatest signal mutual information, as expected, occurs for $P(s_1) = P(s_2) = P(s_3) = \frac{1}{3}$, and the mutual information is reduced as the possible messages generated by the source are increasingly restricted.

On the other hand, looking at the informational entropy for these three configurations, we find a similar trend in the information of the system. From Eq. (4.118b), it is found that for $P(s_1) = P(s_2) = P(s_3) = \frac{1}{3}$

$$H(R,S) = Log_2 6 = 2.585 \text{ bits} \tag{4.122a}$$

while for $P(s_1) = P(s_2) = \frac{1}{2}$ and $P(s_3) = 0$

$$H(R,S) = Log_2 4 = 2.0 \text{ bits.} \tag{4.122b}$$

Finally, for the case $P(s_1)=1,\ P(s_2)=0$, and $P(s_3)=0$ again from Eq. (4.118b) it follows that

$$H(R,S)=Log_2\,2=1.0\,\text{bits}. \tag{4.122c}$$

It is found that the more general the messages that are possible in the system, the more informational entropy is contained in the system process. In this regard, more information is learned from a message received in the $P(s_1)=P(s_2)=P(s_3)=\frac{1}{3}$ system than from a message received in the $P(s_1)=P(s_2)=\frac{1}{2}$ and $P(s_3)=0$ system. Also, more information is learned in receiving a message in the $P(s_1)=P(s_2)=\frac{1}{2}$ and $P(s_3)=0$ system than is learned from a message received in the $P(s_1)=1, P(s_2)=0$ and $P(s_3)=0$ system.

From both the mutual information and entropy, it is observed that restricting the possible messages generated at the source limits the information handled by the transmission.

Next, as a final consideration of the properties of information entropy, we will consider some of the ideas of error correction. Specifically, how can information transmitted or stored in a system be corrected if errors are introduced into it before it is received or read from the system.

4.7.6 Error Correction

We have seen that in processing information in a system it is possible to have certain random processes occur which introduce noise into the received message [1, 3, 5–10]. One of the goals of information theory is to determine the conditions under which it might be possible to correct poorly transmitted messages and the characteristics required in the transmission line for such corrections to be made possible.

To discuss the basis of error correction let us go back to the treatment in Section 7.1 and give an alternative approach to the ideas of message transmission in the absence of noise. Consider messages composed as strings of $N \gg 1$ sites with each site occupied by a 0 or a 1. The possible messages from the source are random sequences with 0's occurring at each message site with probability p and 1's occurring at each message site with probability $1-p$. In this system for $N \to \infty$ the most probable messages sent from the source occur with probability [1]

$$P = p^{Np}\left(1-p\right)^{N(1-p)}. \tag{4.123a}$$

Taking the logarithm of both sides of Eq. (4.123a), we find

$$Log_2\,P = N\left[p\,Log_2\,p+\left(1-p\right)Log_2\left(1-p\right)\right] \tag{4.123b}$$

$$=-NH\left(p\right)$$

where $H(p)=-\left[p\,Log_2\,p+(1-p)Log_2(1-p)\right]$ is the information entropy. Consequently,

$$P=2^{-NH(p)} \tag{4.123c}$$

where we have again used the definition of information entropy in Eq. (4.85), and

$$NH\left(p\right) \tag{4.123d}$$

is seen to represent the information carried by the message measured in bits [1].

Consequently, assuming the source and receiver have the same entropy characteristics, from Eq. (4.86) it follows for the source that we have [1]

$$P_S = \frac{1}{N_S} = 2^{-NH(p)} \tag{4.124a}$$

and from Eq. (4.87) it follows for the receiver that

$$P_R = \frac{1}{N_R} = 2^{-NH(p)} \tag{4.124b}$$

where P_S and P_R are, respectively, the probabilities of one particular of the most probable source and receiver messages occurring. This alternative approach can now be extended to handle a noisy system which introduces errors in a source message during its transmission to the receiver.

Next, we treat a system in which, between the generation of a message at the source and its reception at the receiver, an amount of random error is introduced into it. Specifically, given a message composed of N sites randomly occupied by 0's or 1's, let an error be randomly introduced during its transmission to the receiver. This is done by randomly replacing some of its 0's by 1's and some of its 1's by 0's. A question of interest for such a system is how much information is required to arrive at the receiver in order to reassemble the correct message as it was originally sent by the source.

To understand the message correction, let us start with a message containing $N_0 = NH(p)$ bits of information which is launched from the source. As it moves through the transmission channel, random errors are introduced into it, and somehow, we need to send some additional information to the receiver in order to let the receiver staff correct the message.

Assume that the probability of an error being introduced at any one of the N_0 sites in the original message is q and that the errors are introduced into the message randomly and independently of one another. Consequently, $N_0 \gg 1$ the number of errors in a received message should on average be qN_0. The most probable configuration of errors in a message received at the receiver occurs then with probability [1, 3, 5–10]

$$P_e = q^{N_0 q}\left(1-q\right)^{N_0(1-q)}. \tag{4.125a}$$

Taking the logarithm of Eq. (4.125a), it therefore follows that

$$Log_2 P_e = N_0\left[q\, Log_2\, q + \left(1-q\right) Log_2\left(1-q\right)\right] \tag{4.125b}$$

$$= -N_0 H\left(q\right)$$

where $H\left(q\right) = -\left[q\, Log_2\, q + \left(1-q\right) Log_2\left(1-q\right)\right]$ is the entropy associated with the random errors. As a result, a most probable error configuration occurs with the error probability

$$P_e = 2^{-N_0 H(q)}. \tag{4.125c}$$

The probability P_e is, consequently, found to be related to the information entropy $H\left(q\right)$ where q is the probability that an error is introduced at any given site of the message. Here we have again used the definition of information entropy in Eq. (4.85), and

$$N_0 H\left(q\right) \tag{4.125d}$$

represents the bits of information associated with the error introduced into the message. The information in Eq. (4.125d) is lost in the noise of the system and somehow needs to be sent on to the receiver in order for the imperfect received message to be corrected.

To correct the errors introduced into the original message of bit length N_0 would require transmitting an additional $N_0 H(q)$ bit of information, represented in Eq. (4.125d), to the receiver. (That is, at a minimum, the lost information needs to be forwarded on to the receiver.) To account for this correction, the original message of N_0 bits must at a minimum be augmented to include the error correction information. The resulting message, including the minimal correction, would have an information content of

$$N_{c1} = N_0\left[1 + H(q)\right]\text{bits} \tag{4.126}$$

where N_{c1} is the corrected message bits.

However, note that the $N_0 H(q)$ bits of correction would be subject to $qN_0 H(q)$ errors just as the original signal of N_0 bits was subject to qN_0 errors. To account for these additional $qN_0 H(q)$ errors, an additional information of $N_0\left[H(q)\right]^2$ bits must be tacked on to the result in Eq. (4.126). This would result in a message of corrected bit length [1, 3, 5–10]

$$N_{c2} = N_0\left[1 + H(q) + \left[H(q)\right]^2\right]\text{bits} \tag{4.127}$$

where N_{c2} is the new corrected message bit length Again, the tacked-on correction would require correct and so on, etc.

Proceeding in this manner by correcting each error in the transmitting channel, we find that the total minimal number of bits required in the completely corrected message would be [1]

$$N_c = N_0\left[1 + H(q) + \left[H(q)\right]^2 + \ldots\right] = \frac{N_0}{1 - H(q)}\text{bits.} \tag{4.128}$$

Here N_c is the totally corrected message bits needed to reassemble the originally sent message of N_0 bits.

This concludes the treatment of classical information theory. Next, we give a brief introduction to some of the new features arising in the treatment of the information carried in quantum mechanical systems.

4.8 QUANTUM INFORMATION

Now we consider the reformulation of the ideas of information theory to deal with quantum mechanical systems [1, 6–15]. This involves the development of an expression for the information entropy in the context of a density matrix expressed in the quantum mechanical eigenmodes of the system. The resulting entropy is known as the von Neumann entropy. As shall be seen, the von Neumann entropy is of particular interest in the study of entanglement, which is often present in quantum mechanical systems. Entanglement is a property that can only be found in quantum mechanical systems and arises from the particular probabilistic nature of quantum theory. Specifically, its origin is in the role of the quantum mechanical wavefunction as a probability amplitude from which the probability distribution of a particle or field in space and time is obtained in terms of the modulus squared of the wavefunction.

To develop the ideas of the von Neumann entropy, consider a quantum mechanical system characterized by a set of orthonormal eigenfunctions [1, 6–15]

$$\{|\psi_0\rangle, |\psi_1\rangle, |\psi_2\rangle, \ldots\} \tag{4.129a}$$

with the corresponding energy eigenvalues

$$\{\epsilon_0, \epsilon_1, \epsilon_2, \ldots\}. \tag{4.129b}$$

In addition, it is assumed that in the description of the system the occupation of the eigenfunctions and their corresponding energy eigenvalues is also characterized by a corresponding set of probabilities

$$\{\rho_0, \rho_1, \rho_2, \ldots\}. \tag{4.129c}$$

These indicate the occupation probabilities of the states in the system arising from an incomplete knowledge of the system. (For example, the probabilities $\{\rho_0, \rho_1, \rho_2, \ldots\}$ may arise from statistical mechanics or noisy disorder considerations in the system studied. They account for an uncertainty in the description of the system over that arising from its quantum nature.) In terms of the elements in Eq. (4.129), the density matrix of the system is then defined as [1, 6–15]

$$P = \sum_{i=0}^{\infty} |\psi_i\rangle \rho_i \langle \psi_i | \tag{4.130}$$

where $\sum_{i=0}^{\infty} \rho_i = 1$. From this operator, the macroscopic physical properties of the system are determined.

Specifically, in the considerations of an operator of the form [1, 6–15]

$$A = \sum_{i=1}^{\infty} \sum_{j=1}^{\infty} |\psi_i\rangle A_{i,j} \langle \psi_j |, \tag{4.131}$$

the average of A in the system (denoted as $\langle A \rangle$) is given in terms of Eqs. (4.130) and (4.131) by

$$\langle A \rangle = Tr\{PA\}, \tag{4.132}$$

where in terms of the operator B the trace is defined by

$$Tr\{B\} = \sum_{i=1}^{\infty} \langle \psi_i | B | \psi_i \rangle. \tag{4.133}$$

Applying Eqs. (4.130) and (4.131) in Eq. (4.132), it then follows that the average of A in the quantum system is obtained as

$$\langle A \rangle = \sum_{i=0}^{\infty} \rho_i A_{i,i}. \tag{4.134}$$

so that Eq. (4.134) is completely specified once the set of probabilities $\{\rho_i\}$, external to the quantum states, is specified.

The density function is particularly useful in studying the average macroscopic properties of a system and is often applied to introduce the Boltzmann weighting into the thermodynamic treatment of quantum systems. In this regard, for the case of statistical mechanics

$$\rho_i = \frac{e^{-\beta\epsilon_i}}{Z} \tag{4.135}$$

where $\beta = \frac{1}{k_B T}$, T is the temperature, and $Z = \sum_{i=0}^{\infty} e^{-\beta\epsilon_i}$ is the partition function. Applying Eqs. (4.131) and (4.135) in Eq. (4.134), it follows that the thermal average of A in the quantum system is obtained as

$$\langle A \rangle = \sum_{i=0}^{\infty} \rho_i A_{i,i} = \sum_{i=0}^{\infty} \frac{e^{-\beta\epsilon_i}}{Z} A_{i,i}. \tag{4.136}$$

Returning to an example of an average based on the general form of the density matrix in Eq. (4.130), if we take $A = P$ in Eqs. (4.132) and (4.134) we find that [1, 6–15]

$$\langle P \rangle = Tr\{PP\} = \sum_{i=0}^{\infty} \rho_i^2 \le \left(\sum_{i=0}^{\infty} \rho_i \right)^2 = 1. \tag{4.137}$$

In this expression for the average probability, it is seen that the $\le$ is replaced by $=$ only when all the ρ_i's are zero except for one $\rho_j = 1$. This is the case when the system is in a pure quantum state, and the density matrix formulation is not necessary for the description of the system.

As a case of Eqs. (4.130) and (4.132) which is of particular interest in the study of information, a generalization of the information entropy in Eq. (4.79) can be made to quantum systems involving the various quantum modes and their density matrix. The generalization, known as the von Neumann entropy, is defined by taking $A = -\ln P$ in Eq. (4.132) to obtain the entropy expressed as

$$H = \langle -\ln P \rangle = -Tr\{P \ln P\} = -\sum_{i=0}^{\infty} \rho_i \ln \rho_i \tag{4.138}$$

where we have used the natural logarithm in place of Log_2 and the orthogonality of the eigenvectors $\{|\psi_i\rangle\}$.

Again, similar to in the classical information entropy formulation, in the case of a very narrow probability distribution in which

$$\rho_i = 1 \text{ for } i = n \tag{4.139}$$

$$\rho_i = 0 \text{ otherwise,}$$

it follows from Eq. (4.138) that

$$H = -\sum_{i=0}^{\infty} \rho_i \ln \rho_i = 0. \tag{4.140}$$

In this limit, the entropy is zero as no new information is obtained from studying the system. On the other hand, when there are m eigenmodes in the system which are each uniformly occupied, i.e.,

$$\rho_i = \frac{1}{m} \text{ for } i = 0,1,2,\ldots,m-1, \tag{4.141}$$

the von Neumann entropy becomes

$$H = -\sum_{i=0}^{\infty} \rho_i \ln \rho_i = -\ln \frac{1}{m} = \ln m. \tag{4.142}$$

This is the distribution in which the most is learned from the system of m states. It again has the same result as that obtained in a uniformly occupied system in the context of the classical information entropy.

An interesting property of the averages in Eqs. (4.132) through (4.138) is that they are all invariant under a unitary transformation from the orthonormal basis $\{|\psi_0\rangle, |\psi_1\rangle, |\psi_2\rangle, \ldots\}$ to another orthonormal basis $\{|\varphi_0\rangle, |\varphi_1\rangle, |\varphi_2\rangle, \ldots\}$. This is a basic property of the $Tr\{\ \}$ operation and is often useful in simplifying the evaluation of the average properties of a system described in the density matrix formulation [1, 6–15].

As a last example, consider a set of n degenerate orthonormal eigenmodes, $\{|\varphi_i\rangle\}$, all with energy ϵ. A unitary transformation can be made to another basis of n degenerate eigenmodes, $\{|\mu_i\rangle\}$, in the same subspace. Consequently, a general wavefunction in the new basis of the subspace has the form [1, 6–15]

$$|\mu_j\rangle = \sum_{i=0}^{n-1} c_{j,i} |\varphi_i\rangle \tag{4.143}$$

where $c_{j,i}$ are the elements of the unitary transformation linking the two bases, and $\langle \mu_j | \mu_j \rangle = \sum_{i=0}^{n-1} |c_{j,i}|^2 = 1$. The density matrix of the single pure state $|\mu_j\rangle$ is represented as

$$P = |\mu_j\rangle\langle\mu_j| = \sum_{i=0}^{n-1}\sum_{k=0}^{n-1} c_{j,i} c_{j,k}^* |\varphi_i\rangle\langle\varphi_k|. \tag{4.144}$$

Specifically, Eq. (4.144) describes the system when it is in the state $|\mu_j\rangle$. Applying this in Eq. (4.138), it is then found that the entropy of the system in the state $|\mu_j\rangle$ is given by

$$H = -Tr\{P \ln P\} = -\left\langle \mu_j \left| \left\{ |\mu_j\rangle\langle\mu_j| \ln\left(|\mu_j\rangle\langle\mu_j|\right)\right\} \right| \mu_j \right\rangle = -\left\langle \mu_j \left| \ln\left(|\mu_j\rangle\langle\mu_j|\right)\right| \mu_j \right\rangle \tag{4.145}$$

$$= -\ln 1 = 0.$$

Here the trace is made first over the states $\{|\mu_l\rangle\}$ and for the second to the last expression over the states $\{|\varphi_l\rangle\}$.

The ideas of quantum information entropy are next applied to study the entanglement properties of bipartite quantum systems.

4.8.1 Entanglement of Bipartite Systems

The von Neumann entropy allows us to determine the entanglement properties of quantum mechanical states in some multiparticle systems [1, 11–15]. This is done by determining the form of the wavefunction as the entanglement arises from the specific way in which multiparticle states are composed from their single particle constituents. Consequently, we start the discussions with a brief introduction to the ideas associated with quantum entanglement and the nature of entangled wavefunctions. This is followed by an example illustrating the application of von Neumann entropy in determining the entanglement properties of bipartite wavefunctions. In this regard, entanglement can be found in many types of many-particle systems, but only in bipartite wavefunctions have effective, systematic measures been developed to treat the identification of entanglement properties of wavefunctions.

We begin with a study of bipartite wavefunctions and the nature of entangled states composed within their context [1, 11–15]. Consider a two-particle wavefunction defined on particles denoted as A and B. Particle A is characterized in terms of two degenerate wavefunctions from the set $\{|0_A\rangle, |1_A\rangle\}$, while particle B is characterized in the basis of two degenerate wavefunctions from the set $\{|0_B\rangle, |1_B\rangle\}$. Within these two sets, the wavefunctions are orthonormal such that $\langle i_A \mid j_A\rangle = \delta_{i_A, j_A}$ for $i_A, j_A = 0_A, 1_A$ and $\langle i_B \mid j_B\rangle = \delta_{i_B, j_B}$ for $i_B, j_B = 0_B, 1_B$. An example of such states of two particles is provided by a system of two identical particles characterized as states of spin up and spin down, in which the energy of the particles is independent of their spins.

The wave function of the two particles considered together is then represented in the basis set of four energy degenerate direct product states [1, 11–15]

$$\{|0_A\rangle |0_B\rangle, |0_A\rangle |1_B\rangle, |1_A\rangle |0_B\rangle, |1_A\rangle |1_B\rangle\} \tag{4.146}$$

so that the general wave function is of the form

$$|\psi_{AB}\rangle = \sum_{i_A=0_A}^{1_A} \sum_{j_B=0_B}^{1_B} c_{i_A, j_B} |i_A\rangle |j_B\rangle. \tag{4.147}$$

Here the c_{i_A, j_B} are complex coefficients such that

$$\langle \psi_{AB} \mid \psi_{AB}\rangle = 1 \tag{4.148}$$

for a normalized two-particle state.

In the particular case that

$$c_{i_A, j_B} = c_{i_A} c_{j_B}, \tag{4.149}$$

where $\sum_{i_A=0_A}^{1_A} |c_{i_A}|^2 = 1$ and $\sum_{j_B=0_B}^{1_B} |c_{j_B}|^2 = 1$, the wavefunction in Eq. (4.147) is said to be separable or not to be entangled. For this system, substituting Eq. (4.149) into Eqs. (4.147) and (4.148), we find that [1, 11–15]

$$\begin{aligned} |\psi_{AB}\rangle &= \sum_{i_A=0_A}^{1_A} \sum_{j_B=0_B}^{1_B} c_{i_A} c_{j_B} |i_A\rangle |j_B\rangle \\ &= \left(\sum_{i_A=0_A}^{1_A} c_{i_A} |i_A\rangle \right) \left(\sum_{j_B=0_B}^{1_B} c_{j_B} |j_B\rangle \right) \\ &= |\psi_A\rangle |\psi_B\rangle, \end{aligned} \tag{4.150}$$

where

$$|\psi_A\rangle = \sum_{i_A=0_A}^{1_A} c_{i_A} |i_A\rangle \tag{4.151a}$$

$$|\psi_B\rangle = \sum_{j_B=0_B}^{1_B} c_{j_B} \,|\, j_B\rangle \tag{4.151b}$$

are eigenfunctions of the separate A and B single particle modes with

$$\langle\psi_A \,|\, \psi_A\rangle = \langle\psi_B \,|\, \psi_B\rangle = 1. \tag{4.151c}$$

The wavefunction is, consequently, found to separate into the product of an A single particle eigenfunction and a B single particle eigenfunction.

Similarly, in the absence of entanglement, the density matrix [1, 11–15]

$$\rho_{AB} = |\psi_{AB}\rangle\langle\psi_{AB}| = |\psi_A\rangle|\psi_B\rangle\langle\psi_B|\langle\psi_A| = |\psi_A\rangle\langle\psi_A||\psi_B\rangle\langle\psi_B| = \rho_A\rho_B \tag{4.152}$$

where $\rho_A = |\psi_A\rangle\langle\psi_A|$ and $\rho_B = |\psi_B\rangle\langle\psi_B|$ separates into the product of two separate A and B single particle density matrices.

Consider now the following sequence of operations performed on Eq. (4.152). First, begin by tracing over the B single particle states; in this way, it is found that

$$Tr_B\{\rho_{AB}\} = \rho_A Tr_B\{\rho_B\} = \rho_A. \tag{4.153}$$

Next from the discussions in Eqs. (4.143) through (4.145) the von Neumann entropy of the matrix $Tr_B\{\rho_{AB}\} = \rho_A$ is determined. It follows from Eq. (4.153) and the discussions in Eqs. (4.143) through (4.145) that [1, 11–15]

$$Tr_A\{\rho_A \ln \rho_A\} = 0, \tag{4.154}$$

and the von Neumann entropy is seen to be zero in the absence of entanglement.

Consequently, the von Neumann entropy of the matrix operator $Tr_B\{\rho_{AB}\}$ is zero in the case of a wavefunction that is not entangled. Note that in the same way a similar result follows for $Tr_A\{\rho_{AB}\} = \rho_B$ and the determination of its von Neumann entropy.

On the other hand, for entangled states

$$c_{i_A,j_B} \neq c_{i_A} c_{j_B} \text{ for some } i_A \text{ and } j_B \tag{4.155}$$

and the eigenfunctions do not factorize as in the case of Eq. (4.150). Similarly, the separation of the partition function in Eq. (4.152) cannot be made. In fact, one can show in general that the von Neumann entropy computed for $Tr_B\{\rho_{AB}\}$ or $Tr_A\{\rho_{AB}\}$ of an entangled state will be nonzero. These results of the von Neumann entropy are now illustrated by an example involving a simple set of bipartite wavefunctions.

4.8.2 An Example

The properties of the von Neumann entropy in bipartite systems are illustrated by a commonly given example. Consider the four degenerate basis states in Eq. (4.146). From these the state

$$|\psi_{AB}\rangle = |0_A\rangle|0_B\rangle \tag{4.156}$$

will be shown to be separable or not entangled, while the four states (known as the Bell states) of the form [1, 11–15]

$$|\psi_+\rangle = \frac{1}{\sqrt{2}}\left[|0_A\rangle|0_B\rangle + |1_A\rangle|1_B\rangle\right], \tag{4.157a}$$

$$|\psi_-\rangle = \frac{1}{\sqrt{2}}\left[|0_A\rangle|0_B\rangle - |1_A\rangle|1_B\rangle\right], \tag{4.157b}$$

$$|\varphi_+\rangle = \frac{1}{\sqrt{2}}\left[|0_A\rangle|1_B\rangle + |1_A\rangle|0_B\rangle\right], \tag{4.157c}$$

and

$$|\varphi_-\rangle = \frac{1}{\sqrt{2}}\left[|0_A\rangle|1_B\rangle - |1_A\rangle|0_B\rangle\right] \tag{4.157d}$$

are shown to be entangled states. Specifically, while the wavefunction in Eq. (4.156) separates into the product of two separate $|0_A\rangle$ and $|0_B\rangle$ single particle states, the A and B particle states in Eq. (4.157) cannot be factored into a product of terms each only A or B single particle states.

In the case of Eq. (4.156), the considerations of Eqs. (4.152) through (4.154) apply and the von Neumann entropy is found to be zero. This, as we have seen, is an indication that Eq. (4.156) is not an entangled state. The Bell states in Eq. (4.157), however, are now shown to exhibit nonzero von Neumann entropy which indicates their entangled natures [1, 11–15].

For the consideration of the Bell states we begin with Eqs. (4.157a) and (4.157b). The density matrix of these states has the general form [1, 11–15]

$$|\psi_\pm\rangle\langle\psi_\pm| = \frac{1}{\sqrt{2}}\left[|0_A\rangle|0_B\rangle \pm |1_A\rangle|1_B\rangle\right]\frac{1}{\sqrt{2}}\left[\langle 0_B|\langle 0_A| \pm \langle 1_B|\langle 1_A|\right]. \tag{4.158a}$$

Tracing over the B states in Eq. (4.158a) gives

$$\begin{aligned} Tr_B\{|\psi_\pm\rangle\langle\psi_\pm|\} &= \langle 0_B|\frac{1}{\sqrt{2}}\left[|0_A\rangle|0_B\rangle \pm |1_A\rangle|1_B\rangle\right]\frac{1}{\sqrt{2}}\left[\langle 0_B|\langle 0_A| \pm \langle 1_B|\langle 1_A|\right]|0_B\rangle \\ &+ \langle 1_B|\frac{1}{\sqrt{2}}\left[|0_A\rangle|0_B\rangle \pm |1_A\rangle|1_B\rangle\right]\frac{1}{\sqrt{2}}\left[\langle 0_B|\langle 0_A| \pm \langle 1_B|\langle 1_A|\right]|1_B\rangle \\ &= \frac{1}{2}|0_A\rangle\langle 0_A| + \frac{1}{2}|1_A\rangle\langle 1_A|, \end{aligned} \tag{4.158b}$$

and from this the von Neumann entropy is determined to be

$$\begin{aligned} -Tr_A&\left\{\left[\frac{1}{2}|0_A\rangle\langle 0_A| + \frac{1}{2}|1_A\rangle\langle 1_A|\right]\ln\left[\frac{1}{2}|0_A\rangle\langle 0_A| + \frac{1}{2}|1_A\rangle\langle 1_A|\right]\right\} \\ &= -\frac{1}{2}\ln\frac{1}{2} - \frac{1}{2}\ln\frac{1}{2} = -\ln\frac{1}{2} = \ln 2. \end{aligned} \tag{4.159}$$

Similarly, the Bell state operators $|\varphi_\pm\rangle\langle\varphi_\pm|$ are evaluated in the same way, giving the same $\ln 2$ results for the von Neumann entropy. We see that the Bell states exhibit the greatest von Neumann entropy of states which do not factorize and are thus known as maximally entangled states.

As a final consideration, we treat a more general case of the wavefunctions in Eq. (4.157). Specifically, consider a wavefunction of the form [1, 11–15]

$$|\psi_\alpha\rangle = \frac{\alpha|0_A\rangle|0_B\rangle + (1-\alpha)|1_A\rangle|1_B\rangle}{\sqrt{\alpha^2 + (1-\alpha)^2}}, \tag{4.160}$$

where $0 \le \alpha \le 1$. This wavefunction interpolates between the non-entangled wavefunction in Eq. (4.156) (i.e., $\alpha = 1$), the maximally entangled Bell state in Eq. (4.157a) (i.e., $\alpha = \frac{1}{2}$), and the non-entangled state $|1_A\rangle|1_B\rangle$ (i.e., $\alpha = 0$). From Eq. (4.160) the density matrix obtained from Eq. (4.160) has the form

$$|\psi_\alpha\rangle\langle\psi_\alpha| = \frac{\left[\alpha|0_A\rangle|0_B\rangle \pm (1-\alpha)|1_A\rangle|1_B\rangle\right]\left[\alpha\langle 0_B|\langle 0_A| \pm (1-\alpha)\langle 1_B|\langle 1_A|\right]}{\alpha^2 + (1-\alpha)^2}. \tag{4.161}$$

Now, tracing over the B states gives

$$Tr_B\left\{|\psi_\alpha\rangle\langle\psi_\alpha|\right\} = \tag{4.162}$$

$$= \frac{\langle 0_B|\left[\alpha|0_A\rangle|0_B\rangle \pm (1-\alpha)|1_A\rangle|1_B\rangle\right]\left[\alpha\langle 0_B|0_A| \pm (1-\alpha)\langle 1_B|\langle 1_A|\right]|0_B\rangle}{\alpha^2 + (1-\alpha)^2}$$

$$+ \frac{\langle 1_B|\left[\alpha|0_A\rangle|0_B\rangle \pm (1-\alpha)|1_A\rangle|1_B\rangle\right]\left[\alpha\langle 0_B|\langle 0_A| \pm (1-\alpha)\langle 1_B|\langle 1_A|\right]|1_B\rangle}{\alpha^2 + (1-\alpha)^2}$$

$$= \frac{\alpha^2}{\alpha^2 + (1-\alpha)^2}|0_A\rangle\langle 0_A| + \frac{(1-\alpha)^2}{\alpha^2 + (1-\alpha)^2}|1_A\rangle\langle 1_A|,$$

and the von Neumann entropy is then given from Eq. (4.162) by

$$-Tr_A\left\{\left[\frac{\alpha^2}{\alpha^2+(1-\alpha)^2}|0_A\rangle\langle 0_A| + \frac{(1-\alpha)^2}{\alpha^2+(1-\alpha)^2}|1_A\rangle\langle 1_A|\right]\ln\left[\frac{\alpha^2}{\alpha^2+(1-\alpha)^2}|0_A\rangle\langle 0_A| + \frac{(1-\alpha)^2}{\alpha^2+(1-\alpha)^2}|1_A\rangle\langle 1_A|\right]\right\}$$

$$= -\frac{\alpha^2}{\alpha^2+(1-\alpha)^2}\ln\frac{\alpha^2}{\alpha^2+(1-\alpha)^2} - \frac{(1-\alpha)^2}{\alpha^2+(1-\alpha)^2}\ln\frac{(1-\alpha)^2}{\alpha^2+(1-\alpha)^2}. \tag{4.163}$$

This is the result for the von Neumann entropy computed for Eq. (4.162). An identical result is obtained from Eq. (4.161) by first tracing over the A states and then evaluating the von Neumann entropy of the resulting matrix in the B states.

For $\alpha = 0,1$ the von Neumann entropy in Eq. (4.163) is zero and the system is found to be in a state which is not entangled. In the case that $\alpha = \frac{1}{2}$ in Eq. (4.163) the von Neumann entropy is at its largest, the state is maximally entangled. Between these two limits, the entropy is nonzero but less than its maximum value $\ln 2$. This indicates for $\alpha \ne 0, \frac{1}{2}, 1$ a degree of entanglement, but the entanglement is less than in its maximum entangled state.

In Figure 4.7 a plot of the entropy in Eq. (4.163) is made versus α. In this plot, the value of the entropy indicates the relative degree of entanglement of the wavefunction between 0 and $\ln 2$.

4.8.3 Tripartite Entangled States

The ideas of entanglement are extended to higher multiparticle states than those involving just two particles [1, 11–15]. For example, the wavefunction in Eq. (4.147) can be generalized to three particles where it has the form

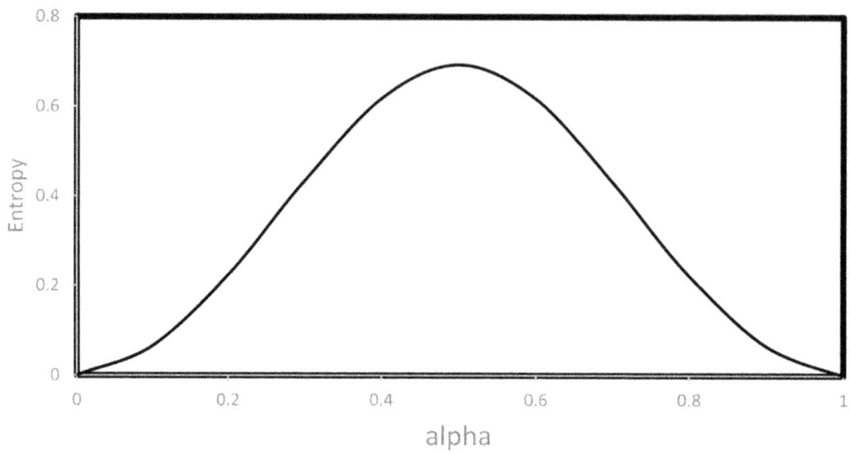

FIGURE 4.7 A plot of the entropy in Eq. (4.163) versus α.

$$|\psi_{ABC}\rangle = \sum_{i_A=0_A}^{1_A} \sum_{j_B=0_B}^{1_B} \sum_{k_C=0_C}^{1_C} c_{i_A,j_B,k_C} \, |i_A\rangle \, |j_B\rangle \, |k_C\rangle. \tag{4.164}$$

Here the c_{i_A,j_B,k_C} coefficients are again complex and defined such that

$$\langle \psi_{ABC} \mid \psi_{ABC} \rangle = 1 \tag{4.165}$$

for a normalized three-particle state.

A separable or non-entangled state is obtained from Eq. (4.164) when c_{i_A,j_B,k_C} is of the form

$$c_{i_A,j_B,k_C} = c_{i_A} c_{j_B} c_{k_C}, \tag{4.166}$$

where $\sum_{i_A=0_A}^{1_A} |c_{i_A}|^2 = 1$, $\sum_{j_B=0_B}^{1_B} |c_{j_B}|^2 = 1$, and $\sum_{k_C=0_C}^{1_C} |c_{j_B}|^2 = 1$. Specifically, under these conditions it is found that

$$\begin{aligned} |\psi_{ABC}\rangle &= \left(\sum_{i_A=0_A}^{1_A} c_{i_A} \, |i_A\rangle \right) \left(\sum_{j_B=0_B}^{1_B} c_{j_B} \, |j_B\rangle \right) \left(\sum_{k_C=0_C}^{1_C} c_{k_C} \, |k_C\rangle \right) \\ &= |\psi_A\rangle \, |\psi_B\rangle \, |\psi_C\rangle, \end{aligned} \tag{4.167}$$

where

$$|\psi_A\rangle = \sum_{i_A=0_A}^{1_A} c_{i_A} \, |i_A\rangle \tag{4.168a}$$

$$|\psi_B\rangle = \sum_{j_B=0_B}^{1_B} c_{j_B} \, |j_B\rangle \tag{4.168b}$$

$$|\psi_C\rangle = \sum_{k_C=0_C}^{1_C} c_{k_C} \, |k_C\rangle \tag{4.168c}$$

and

$$\langle \psi_A \mid \psi_A \rangle = \langle \psi_B \mid \psi_B \rangle = \langle \psi_C \mid \psi_C \rangle = 1. \tag{4.168d}$$

However, in the case that [1, 11–15]

$$c_{i_A, j_B, k_C} \neq c_{i_A} c_{j_B} c_{k_C} \tag{4.169}$$

for some set of i_A, j_B, and k_C, the state exhibits a certain degree of entanglement. In this regard, two highly entangled three particle states that have been of technological importance are the so-called GHZ (Greenberger-Horne-Zeilinger) state and the W-state. These have found applications in quantum computing, information, cryptography, and teleportation [1, 11–15].

The GHZ state is of the form

$$|\psi_{GHZ}\rangle = \frac{1}{\sqrt{2}}\left(|0_A\rangle|0_B\rangle|0_C\rangle + |1_A\rangle|1_B\rangle|1_C\rangle\right). \tag{4.170}$$

It has a wavefunction which does not factorize into the product of an A state times a B state times a C state wavefunction. Nevertheless, it can be written in terms of the two-particle entangled Bell states in Eq. (4.157a) and (4.157b) times a third particle wavefunction so that

$$|\psi_{GHZ}\rangle = \frac{1}{\sqrt{2}}|\psi_+\rangle\frac{1}{\sqrt{2}}\left[|0_C\rangle + |1_c\rangle\right] + \frac{1}{\sqrt{2}}|\psi_-\rangle\frac{1}{\sqrt{2}}\left[|0_C\rangle - |1_c\rangle\right], \tag{4.171}$$

giving a type of partial factorization [1, 11–15].

The W-state is also entangled, with the form given by [1, 11–15]

$$|\psi_W\rangle = \frac{1}{\sqrt{3}}\left[|0_A\rangle|0_B\rangle|1_C\rangle + |0_A\rangle|1_B\rangle|0_C\rangle + |1_A\rangle|0_B\rangle|0_C\rangle\right]. \tag{4.172}$$

It represents, in the three particle states, structurally a different type of entanglement than that found in the GHZ state.

These two states involve fundamentally different forms of entanglement. This is seen by considering the probability amplitudes for the measurement of one of the three particles. Consider measuring the amplitude for finding the $|0_C\rangle$ and $|1_C\rangle$ states in $|\psi_{GHZ}\rangle$. From Eq. (4.170) we find

$$\langle 0_C \mid \psi_{GHZ} \rangle = \frac{1}{\sqrt{2}}|0_A\rangle|0_B\rangle \tag{4.173a}$$

and

$$\langle 1_C \mid \psi_{GHZ} \rangle = \frac{1}{\sqrt{2}}|1_A\rangle|1_B\rangle. \tag{4.173b}$$

The states after the measurement are each separable, non-entangled states. In this regard, tracing the density matrix $|\psi_{GHZ}\rangle\langle\psi_{GHZ}|$ over the C particle states gives

$$Tr_C\left\{|\psi_{GHZ}\rangle\langle\psi_{GHZ}|\right\} = \frac{1}{2}|0_A\rangle|0_B\rangle\langle 0_B|\langle 0_A| + \frac{1}{2}|1_A\rangle|1_B\rangle\langle 1_B|\langle 1_A| \tag{4.174}$$

indicating that the resulting system is a mixed state of the $|0_A\rangle|0_B\rangle$ and $|1_A\rangle|1_B\rangle$ eigenmodes.

Next, consider measuring the amplitude for finding the $|0_C\rangle$ and $|1_C\rangle$ states in $|\psi_W\rangle$. From Eq. (4.172) we find [1, 11–15]

$$\langle 0_C | \psi_W \rangle = \frac{1}{\sqrt{3}}\left[|0_A\rangle|1_B\rangle + |1_A\rangle|0_B\rangle\right] \tag{4.175a}$$

which is an entangled two-particle state and

$$\langle 1_C | \psi_W \rangle = \frac{1}{\sqrt{3}}|0_A\rangle|0_B\rangle \tag{4.175b}$$

which is separable. After the measurement, the state in Eq. (4.175a) contains an element of entanglement of the A and B particles while that in Eq. (4.175b) is not entangled. In this regard, tracing the density matrix $|\psi_W\rangle\langle\psi_W|$ over the C particle states gives

$$Tr_C\left\{|\psi_W\rangle\langle\psi_W|\right\} = \frac{1}{\sqrt{3}}\left[|0_A\rangle|1_B\rangle + |1_A\rangle|0_B\rangle\right]\frac{1}{\sqrt{3}}\left[\langle 0_A|\langle 1_B| + \langle 1_A|\langle 0_B|\right] \tag{4.176}$$

$$+\frac{1}{3}|0_A\rangle|0_B\rangle\langle 0_B|\langle 0_A|$$

indication of a mixed state composed of the entangled $|0_A\rangle|1_B\rangle$ and $|1_A\rangle|0_B\rangle$ eigenmodes and a separable state of $|0_A\rangle|0_B\rangle$.

4.9 MONTE CARLO AND REVERSE MONTE CARLO SIMULATION

Another application of entropy involves the development of the methods of Monte Carlo and reverse Monte Carlo simulation [16–21]. The Monte Carlo simulation is a means of numerically estimating the thermodynamic properties of physical systems through the generation of configurations of its microscopic variables using an algorithm involving random numbers. The generated configurations are then used to compute the average values of the macroscopic thermodynamic variables of the system. Similarly, the reverse Monte Carlo method uses random numbers to generate configurations of a system which approximate the scattering characteristics of a material. In this regard, it focuses on determining possible solutions to the inverse scattering problem. We begin with a discussion of the ideas of Monte Carlo simulation for the determination of the thermodynamic properties of a system.

4.9.1 Monte Carlo Simulation

Consider a classical mechanical system characterized by a set of canonical coordinates $\{x_1, x_2, .. p_1, p_2, \ldots\}$. These may, for example, be the microscopic positions and momenta of a set of interacting particles. From the various possible configurations of the system, we find a set of energy levels $\{\epsilon_0, \epsilon_1, \epsilon_2, \ldots\}$ associated with each member of the set of configurations, i.e., each of the set $S = \left\{\{x_1, x_2, .. p_1, p_2, \ldots\}_0, \{x_1, x_2, .. p_1, p_2, \ldots\}_1, \{x_1, x_2, .. p_1, p_2, \ldots\}_2, \ldots\right\}$ where the integer subscripts label the different configurations of the system. (Here, specifically, note that the configuration $\{x_1, x_2, .. p_1, p_2, \ldots\}_n$ has the energy ϵ_n.) Ultimately the probability of finding a particular energy state ϵ_n corresponding to the nth configuration $\{x_1, x_2, .. p_1, p_2, \ldots\}_n$ in the system is found by the maximum entropy principle to be obtained from the Boltzmann probability

$$P_n = \frac{e^{-\beta\epsilon_n}}{Z} \tag{4.177a}$$

where

$$Z = \sum_{n=1}^{\infty} e^{-\beta\epsilon_n} \tag{4.177b}$$

is the partition function, $\beta = \frac{1}{k_B T}$, and T is the temperature of the system. The problem of the Monte Carlo simulation is to randomly generate a set S composed of many different random configurations $\{x_1, x_2, .. p_1, p_2, \ldots\}_n$ so that these configurations occur in S distribution according to the Boltzmann probability [16–21]. In this regard, if the nth configuration $\{x_1, x_2, .. p_1, p_2, \ldots\}_n \in S$ corresponds to the energy ϵ_n, its fractional occurrence in S is given by $P_n = \frac{e^{-\beta\epsilon_n}}{Z}$.

Consider then that we have generated the set S containing $N \gg 1$ different configurations, each of which is the form $\{x_1, x_2, .. p_1, p_2, \ldots\}_n$ and that their presence is accurately represented in S by the Boltzmann probability $P_n = \frac{e^{-\beta\epsilon_n}}{Z}$. The elements of the set S can then be used to calculate the averages of thermodynamic quantities which are dependent on the coordinates $\{x_1, x_2, .. p_1, p_2, \ldots\}$. In this regard, given an operator of the form $A(x_1, x_2, .. p_1, p_2, \ldots)$, its thermal average $\langle A(x_1, x_2, .. p_1, p_2, \ldots)\rangle$ is approximated in the set S by [16–21]

$$\langle A(x_1, x_2, .. p_1, p_2, \ldots)\rangle \cong \frac{1}{N} \sum_{\{x_1, x_2, .. p_1, p_2, \ldots\}_n \in S} A(x_1, x_2, .. p_1, p_2, \ldots) \tag{4.178}$$

where the sum is over all the $\{x_1, x_2, .. p_1, p_2, \ldots\}_n$ contained in S. Consequently, the object of a computer simulation is to develop an algorithm for generating S and then use it to determine the thermodynamic averages of the system. A commonly used algorithm for this process is known as the Metropolis Algorithm. It is now discussed.

The Metropolis Algorithm starts with the statement of detailed balance [16]. Considering a transition between two energy levels ϵ_m and ϵ_p in a thermodynamic system, the rate of transition from ϵ_m to ϵ_p is given by [16]

$$T(p, m) P_m \tag{4.179a}$$

where P_m is the probability of ϵ_m being occupied and $T(p, m)$ is the rate of transition from an ϵ_m to an ϵ_p state. Similarly, the rate of transition from ϵ_p to ϵ_m is given by

$$T(m, p) P_p \tag{4.179a}$$

where P_p is the probability of ϵ_p being occupied and $T(m, p)$ is the rate of transition from an ϵ_p to an ϵ_m state. For a system in thermodynamic equilibrium, it follows that we must have [16]

$$T(p, m) P_m = T(m, p) P_p. \tag{4.180}$$

Otherwise, there will be a net change with time in the occupancy of the ϵ_m and ϵ_p states due to the differing transition rates.

From our earlier entropy considerations, the equilibrium configuration of the thermodynamic system has the form

$$P_m = \frac{e^{-\beta\epsilon_m}}{Z} \text{ and } P_p = \frac{e^{-\beta\epsilon_p}}{Z}. \tag{4.181}$$

Consequently, from Eq. (4.180) it follows that [16]

$$T(p,m)e^{-\beta\epsilon_m} = T(m,p)e^{-\beta\epsilon_p} \tag{4.182a}$$

or

$$\frac{T(p,m)}{T(m,p)} = e^{-\beta(\epsilon_p - \epsilon_m)}. \tag{4.182b}$$

The Boltzmann factors then set the relative magnitude of the transition rates between the ϵ_m and ϵ_p energy levels. It is, consequently, necessary to develop a sampling algorithm which simulates the detailed balance transition rate given in Eqs. (4.180) through (4.182) in generating the set S. A commonly used algorithm for this was formulated by Metropolis based on a sampling which mimics Eq. (4.182b).

In the Metropolis sampling we generate a sequence $l = 0,1,2,\ldots,N-1$ of configurations of the system to create the set S. In this generation, the jth configuration of the system is of the form $\{x_1, x_2,..p_1, p_2,\ldots\}_j$ with an associated energy denoted ϵ_j, and it is generated directly from the $(j-1)^{th}$ configuration of the sequence [16]. The generation process results in a set S of configurations labeled $l = 0,1,2,\ldots,i,\ i+1,\ldots,N-1$ with their corresponding energies $\epsilon_0,\epsilon_1,\ldots\epsilon_i,\epsilon_{i+1},\ldots\epsilon_{N-1}$. In this regard, the greater N becomes, the longer is the sequence of S and the more accurately Eq. (4.178) represents the average value of the operator $A(x_1, x_2,..p_1, p_2,\ldots)$ of the thermodynamic system being simulated. In the following, we introduce the sequential Metropolis sampling process for generating the $\{x_1, x_2,..p_1, p_2,\ldots\}_{i+1}$ configuration from the $\{x_1, x_2,..p_1, p_2,\ldots\}_i$ configuration in the set S.

The generation [16] of the $(i+1)^{th}$ configuration with energy ϵ_{i+1} is made by making a small random change in the ith configuration (i.e., the system configuration $\{x_1, x_2,..p_1, p_2,\ldots\}_i$) which has an associated energy ϵ_i. This applied change generates a new configuration $\{x_1, x_2,..p_1, p_2,\ldots\}_{i+1}$ with a new energy ϵ_{i+1}. Once the change to $\{x_1, x_2,..p_1, p_2,\ldots\}_i$ is made, an acceptance criterion is applied which determines the probability of the acceptance of the generated $\{x_1, x_2,..p_1, p_2,\ldots\}_{i+1}$ as the new $(i+1)^{th}$ configuration in S with the new energy ϵ_{i+1}. On the other hand, if the newly generated state is not accepted, then a copy of the ith state, $\{x_1, x_2,..p_1, p_2,\ldots\}_i$, is retained as the $(i+1)^{th}$ configuration of the sequence with energy $\epsilon_{i+1} = \epsilon_i$, i.e., the ith state becomes repeated in the sequence. Note that in order to start the process, the first configuration of the sequence $\{x_1, x_2,..p_1, p_2,\ldots\}_0$ can be chosen arbitrarily. Often, however, considerations specific to the study of a particular thermodynamic system are helpful in formulating a guess for the first configuration.

In the Metropolis formulation, the acceptance probability P_a for determining whether a state generated from $\{x_1, x_2,..p_1, p_2,\ldots\}_s$ will be accepted as the next configuration $\{x_1, x_2,..p_1, p_2,\ldots\}_r$ is given by [16]

$$\begin{aligned} P_a &= 1 \qquad \text{if } \epsilon_r \le \epsilon_s \\ &= e^{-\beta(\epsilon_r - \epsilon_s)} \text{ otherwise.} \end{aligned} \tag{4.183}$$

Here ϵ_r is the new energy of the newly generated configuration from the state of energy ϵ_s. If the new configuration is not accepted, a copy of the old configuration with energy ϵ_s is retained as the new configuration corresponding to the rth configuration with energy $\epsilon_r = \epsilon_s$.

To see that this, indeed, satisfies the conditions of detailed balance, let us treat the following transitions between the energy levels ϵ_i and ϵ_{i+1} in the generated sequence of states in S. First consider the case in which $\epsilon_{i+1} \leq \epsilon_i$ so that applying Eq. (4.183) we find for the transition from ϵ_i to ϵ_{i+1}

$$T(i+1,i) = \alpha P_a = \alpha \tag{4.184a}$$

where α is a constant of proportionality associated with the rate of acceptance. (Here in applying Eq. (4.183) we have taken $\epsilon_s = \epsilon_i$ and $\epsilon_r = \epsilon_{i+1}$). On the other hand, the transition $T(i,i+1)$ in which the system goes from an energy ϵ_{i+1} to ϵ_i is also obtained from Eq. (4.183) giving

$$T(i,i+1) = \alpha P_a = \alpha\left[e^{-\beta(\epsilon_i - \epsilon_{i+1})}\right]. \tag{4.184b}$$

(Here in applying Eq. (4.183) we have now taken $\epsilon_s = \epsilon_{i+1}$ and $\epsilon_r = \epsilon_i$). Again α is a constant of proportionality associated with the rate of acceptance.

Consequently, collecting the results in Eq. (4.184), we see that they satisfy the detailed balance condition in Eq. (4.182b), which is given by [16]

$$\frac{T(i+1,i)}{T(i,i+1)} = e^{-\beta(\epsilon_{i+1} - \epsilon_i)}. \tag{4.185}$$

The conditions of detailed balance are, therefore, met by the application of the Metropolis criterion in Eq. (4.183) for the case $\epsilon_{i+1} \leq \epsilon_i$.

Next consider the case $\epsilon_i \leq \epsilon_{i+1}$ so that applying Eq. (4.183) we find [16]

$$T(i+1,i) = \alpha P_a = \alpha\left[e^{-\beta(\epsilon_{i+1} - \epsilon_i)}\right] \tag{4.186a}$$

where α is a constant of proportionality. (Here in applying Eq. (4.183) we have taken $\epsilon_s = \epsilon_i$ and $\epsilon_r = \epsilon_{i+1}$). From Eq. (4.182b) it also follows that

$$T(i,i+1) = \alpha P_a = \alpha. \tag{4.186b}$$

(Here in applying Eq. (4.183) we have now taken $\epsilon_s = \epsilon_{i+1}$ and $\epsilon_r = \epsilon_i$). Consequently, it again follows that

$$\frac{T(i+1,i)}{T(i,i+1)} = e^{-\beta(\epsilon_{i+1} - \epsilon_i)} \tag{4.187}$$

so that, by the application of the conditions in Eq. (4.187), the conditions of detained balance in Eq. (4.182b) are again met for the case $\epsilon_i \leq \epsilon_{i+1}$.

4.9.2 An Example

As a highly simplified example of the Metropolis Algorithm, consider the following problem: Treat a single two-level system which can exist in a ground state with energy ϵ_g and an excited state with energy ϵ_e. The average energy of the system can be solved exactly in statistical mechanics giving the average energy $\langle \epsilon \rangle$ as

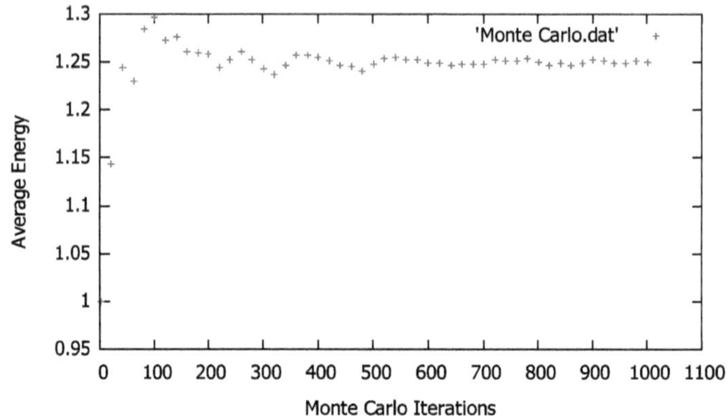

FIGURE 4.8 A plot of the average energy obtained by averaging over the first l energies of the generated sequence versus l for the case that $\frac{\epsilon_e}{\epsilon_g} = 2$ and $\beta\epsilon_g = 1$. In the plot, aside from the point at $l = 1$, the points are displayed at every 20th iteration of the simulation.

$$\langle \epsilon \rangle = \epsilon_g \frac{1 + \frac{\epsilon_e}{\epsilon_g} e^{-\beta(\epsilon_e - \epsilon_g)}}{1 + e^{-\beta(\epsilon_e - \epsilon_g)}}. \tag{4.188}$$

Now let us obtain this result using a Monte Carlo simulation based on the Metropolis Algorithm.

In the application of the Metropolis formulation to the two-level system, an initial i^{th} state (i.e., the initial state being $\epsilon_i = \epsilon_g$ or ϵ_e) is modified to a proposed $(i+1)^{th}$ state (i.e., a proposed state being $\epsilon_{i+1} = \epsilon_g$ or ϵ_e) of the system. The acceptance probability P_a for retaining the proposed state as the new $(i+1)^{th}$ state in the Metropolis sequence is [16]

$$\begin{aligned} P_a &= 1 \qquad \text{if } \epsilon_{i+1} \le \epsilon_i \\ &= e^{-\beta(\epsilon_{i+1} - \epsilon_i)} \text{ otherwise.} \end{aligned} \tag{4.189}$$

If these conditions are not met, then the ith state is copied as the new $(i+1)^{th}$ state so that $\epsilon_{i+1} = \epsilon_i$.

Starting from the initial 0th state of the sequence in S and taking $\epsilon_0 = \epsilon_g$ we apply the Metropolis Algorithm to generate the sequence of states shown in Figure 4.8. Here we plot the average energy obtained by averaging over the first l energies of the generated sequence versus l for the case that $\frac{\epsilon_e}{\epsilon_g} = 2$ and $\beta\epsilon_g = 1$. In the plot, aside from the point at $l = 1$, the points are displayed at every 20th iteration of the simulation. It is seen that, following some initial scatter, the results eventually approach the thermodynamic limit, $\langle \epsilon \rangle = \epsilon_g \frac{1 + 2e^{-1}}{1 + e^{-1}} \cong 1.269\ \epsilon_g$, obtained from Eq. (4.188).

4.9.3 Quantum Monte Carlo

The Monte Carlo methods can be extended to treat quantum mechanical systems [18, 19]. This is done in a number of ways, but one method which has been a focus is by treating the dynamics of the quantum mechanical system in its path-integral formulation [19]. This allows for a treatment of the problem through the evaluation of a functional integral written in terms of real and complex variables. Such a formulation becomes accessible to numerical computer simulations.

In the path-integral formulation, for example, the statistical physics of a 3-dimensional quantum mechanical system is rewritten in terms of a path-integral in 4-dimensions involving complex variables [18, 19]. In this regard, the time evolution of a dynamical system governed by a Hamiltonian H is closely related to its statistical mechanics. Specifically, the dynamics is generated by the time translation operator

$$e^{-\frac{i}{\hbar}H(t-t_0)} \tag{4.190}$$

where t_0 is an initial time and t is the current time. Upon making the replacement [19]

$$\frac{i}{\hbar}(t-t_0) \to \beta = \frac{1}{k_B T} \tag{4.191}$$

Eq. (4.190) becomes

$$e^{-\beta H} \tag{4.192}$$

which is proportional to the Boltzmann weight. In this respect, it is not surprising that the path-integral responsible for the dynamics of a quantum mechanical system is transformed into a functional integral describing the statistical mechanics of the same system. The resulting functional integral is readily available to the application of Monte Carlo simulation methods.

For the details of quantum Monte Carlo, the reader is referred to the Refs. 18 and 19 and to the last chapter in this text.

4.9.4 Reverse Monte Carlo

To conclude the discussions of Monte Carlo simulation methods, an introduction to the basic techniques of reverse Monte Carlo is given [20, 21]. This is a method that has found applications in the study of inverse problems arising from scattering measurements. It, specifically, is involved with the generation of a model of a scattering medium which is consistent with the scattering measured from the medium. Examples are the problem of simulating the atomic-molecular scale structure of amorphous materials, fluids, or disordered media, and the technique has been applied in both X-ray, neutron scattering, and optics [20, 21].

In the reverse Monte Carlo, an initial guess is made of the structure of a medium, and the scattering properties of the initial guess are computed. A comparison of the scattering from the guessed medium with experimental scattering data is made using the form [20, 21]

$$\chi^2 = \sum_i \frac{\left(s_{obsi} - s_{calci}\right)^2}{\sigma_i^2} \tag{4.193}$$

where i runs over the points of the scattering data, s_{obsi} is the observed scattering property, s_{calci} is the calculated scattering property from the model, and σ_i accounts for the experimental uncertainty of the observation. It is seen that the smaller χ^2 the better is the agreement between the observed and calculated data. In addition, it should be noted that certain restrictions can also be made on the system when guessing the configuration. These restrictions may arise from a priori understanding of the nature of the systems being modeled.

In the next step of the reverse Monte Carlo, a small change is made to the initially guessed configurations and a new χ^2 is computed with the newly generated configuration. Again, it should be noted that certain a priori restrictions on the nature of the small changes in the system may apply.

The new configuration is then accepted or rejected as a better configuration of the system using the principles of Metropolis sampling.

This is done repeatedly so that at some point in the process a $(n+1)^{th}$ model is generated from a nth model which has been accepted during the generation process. At this point a measure of the change in suitability between the nth and the $(n+1)^{th}$ model of the scattering medium is made. Specifically, we compute [20, 21]

$$\Delta\chi_{n+1}^2 = \chi_{n+1}^2 - \chi_n^2 \tag{4.194}$$

where χ_n^2 is the measure in Eq. (4.193) for the nth model and χ_{n+1}^2 is the measure in Eq. (4.193) for the $(n+1)^{th}$ model. The value of $\Delta\chi_{n+1}^2$ is now used to apply a Metropolis sampling criterion to determine the suitability of the $(n+1)^{th}$ model as a representation of the system being studied.

In the Metropolis formulation for the reverse Monte Carlo, the acceptance probability P_a for determining whether the $(n+1)^{th}$ model will be accepted as the next configuration representing the system is given by

$$\begin{aligned} P_a &= 1 \qquad \text{if } \chi_{n+1}^2 \leq \chi_n^2 \\ &= e^{-\frac{1}{2}\Delta\chi_{n+1}^2} \quad \text{otherwise.} \end{aligned} \tag{4.195}$$

If the new configuration is not accepted, a copy of the old configuration with χ_n^2 is retained as the new configuration corresponding to the $(n+1)^{th}$ model. Following this process eventually a $(n+1)^{th}$ model is generated which represents the observed scattering to the desired accuracy.

While the reverse Monte Carlo does not necessarily generate that exact configuration of a system, it does generate a configuration which is consistent, within the experimental error, with reproducing the scattering data. This is often of use for determining the atomic correlation functions and other statistical properties of the material.

As an example of the results of such a reverse Monte Carlo process applied to optics, in Figure 4.9 are presented results of a reverse Monte Carlo study applied to elastic optical scattering to retrieve the surface profile of a one-dimensional random grating.

4.9.5 Genetic Algorithms

Genetic algorithms [22–26] are closely related to the methods applied in developing the Monte Carlo and reversed Monte Carlo, but, in addition, incorporate ideas of biological evolution. They are often applied to problems involving finding optimal solutions through the performance of statistical searches. This occurs, for example, in searches for the maxima or minima of functions. In this regard, genetic algorithms allow for the determination of the global or absolute optimized solution in a system which can also exhibit a number of additional local optimized solutions.

In genetic algorithms, samplings of system configurations are made in accord with the ideas of genetics and the theory of natural selection so as to arrive at an optimized solution to a physics or engineering problem [22–26]. In these approaches, one typically starts with a collection of guessed solutions, and these solutions are termed chromosomes. Each chromosome is a proposed solution from which the properties of the system it represents are computed. The members of the chromosome set are then each tested for their abilities to yield an optimized value of the property of interest, and poor representations are discarded or selected out. A list of best guesses (i.e., good chromosomes) is made, and some of the chromosomes from this list are acted upon using evolutionary principles to produce a next generation of chromosomes offered as optimized solutions. The

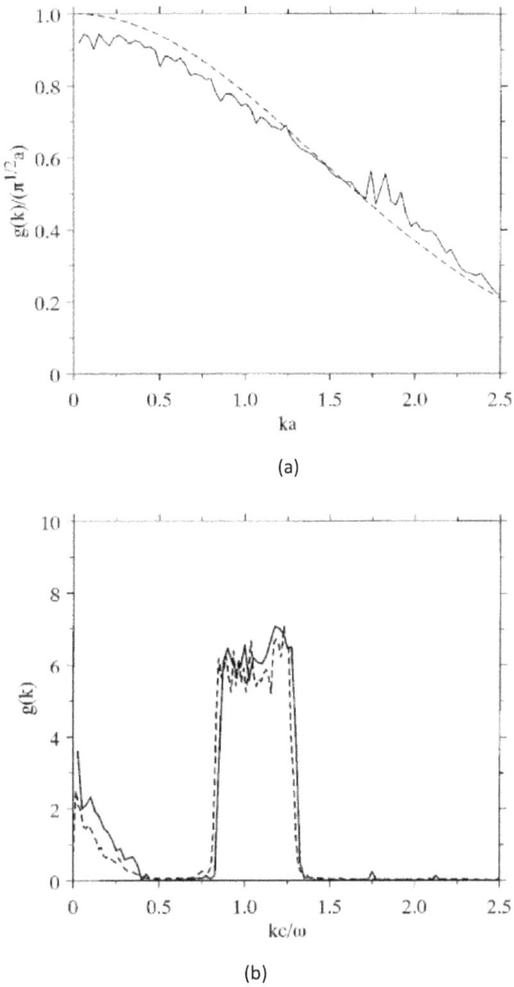

FIGURE 4.9 Plot from Ref. 21 of the power spectrum of the surface profile function in wavenumber space of a one-dimensional grating with Gaussian random disorder. The experimental curve from profile data is the dashed curve and the RMC extraction from the diffuse scattering is the solid curve. Reprint with permission from Ref. [21], Optical Society of America.

evolutionary principles most often employed involve ideas of genetic crossover and genetic mutation of the chromosomes. In this way, through application of genetic evolution on a number of successive generations of chromosomes, an optimized result is ultimately generated.

An illustration of the rationale behind genetic algorithms arises in the study of the maximization of functions which exhibit a number of local maxima in addition to an overall global maximum. Conventional methods of optimizing a solution of such a problem often lead to solutions which are stuck in local maxima. This would be an unsatisfactory result in the process of locating the greatest maximum value of a system. The jolts given to the system by crossover and mutation processes are meant to allow the solution to pass through the local maximum and ultimately arrive at the one global maximum in the system. In this regard, the algorithm is a type of search routine.

Genetic algorithms begin by encoding information for the system in chromosomes which contain numerical information from which the properties of the system are determined [22–26]. The coding can be done in a number of ways, but one technique that is often used in physics and engineering is binary coding. This represents numbers in the basis 2. Specifically, in the base 2 the number 01011 (composed of five digits) has a value in the base 10 given by

$$01011 = (1)2^0 + (1)2^1 + (0)2^2 + (1)2^3 + (0)2^4 = 17 \tag{4.196a}$$

so that for a general number of the form $i_1i_2i_3i_4i_5$ where $i_j = 1$ or 0 for $j = 1,2,3,4,5$

$$i_1i_2i_3i_4i_5 = i_5 2^0 + i_4 2^1 + i_3 2^2 + i_2 2^3 + i_1 2^4. \tag{4.196b}$$

In turn, these ideas generalize to general numbers of the form $i_1i_2i_3i_4i_5.....i_n$ with $i_j = 1$ or 0 for $n > 5$.

Similarly, for the decimal 0.010101... its value in base 10 is given by

$$0.010101.. = (0)\left(\frac{1}{2}\right)^1 + (1)\left(\frac{1}{2}\right)^2 + (0)\left(\frac{1}{2}\right)^3 + (1)\left(\frac{1}{2}\right)^4 + (0)\left(\frac{1}{2}\right)^5 + (1)\left(\frac{1}{2}\right)^6 + \ldots \tag{4.197a}$$

$$= \frac{1}{4} + \frac{1}{16} + \frac{1}{64} + \ldots = \frac{1}{3}$$

so that for a general number $0.i_1i_2i_3i_4i_5$ where $i_j = 1$ or 0 for $j = 1,2,3,4,5$

$$0.i_1i_2i_3i_4i_5 = i_1\left(\frac{1}{2}\right)^1 + i_2\left(\frac{1}{2}\right)^2 + i_3\left(\frac{1}{2}\right)^3 + i_4\left(\frac{1}{2}\right)^4 + i_5\left(\frac{1}{2}\right)^5. \tag{4.197b}$$

In turn, this idea again generalizes to general numbers of the form $0.i_1i_2i_3i_4i_5.....i_n$ with $i_j = 1$ or 0 for $n > 5$.

Consider now discussing genetic algorithm methods by means of explaining how they would be applied to a simple maximization problem. Specifically, let us develop the method as it could be used to find the maximum of a function of the form [22–26]

$$y = f(x) \tag{4.198}$$

evaluated over the integers in the interval $0 \le x \le 31$.

Within the interval $0 \le x \le 31$ the integers are represented by binary numbers of the form $i_1i_2i_3i_4i_5$ where $i_j = 1$ or 0 for $j = 1,2,3,4,5$. Consequently, $x = 0$ in base 10 is $x = 00000$ in base 2, $x = 15$ in base 10 is $x = 01111$ in base 2, and $x = 31$ in base 10 is $x = 11111$ in base 2, etc. In the following the base 10 values of x are useful in evaluating the function $y = f(x)$ so that

$$\begin{aligned} y_0 &= f(x = 0), \\ y_1 &= f(x = 1), \\ y_2 &= f(x = 2), \\ &\ldots\ldots\ldots\ldots, \\ y_{31} &= f(x = 31) \end{aligned} \tag{4.199}$$

give the values of the function (i.e., y_0, y_1, ..., y_{31}) in base 10. On the other hand, the binary representations of x are the chromosomes and are useful for executing genetic operations. This will presently be seen when we discuss the genetic operations applied to the optimization process.

The optimization then begins by randomly choosing a set of chromosomes to represent a type of genetic population to be mated and evolved into the next generations of chromosomes. For an illustration, let us consider a population of four chromosomes. (In dealing with a real problem, this population would, of course, be taken to contain many more chromosomes.)

In our example, let us assume that the four chromosomes chosen to represent the set of potential solutions are associated with the base 10 solutions in the set

$$\{y_5 = f(5), y_{11} = f(11), y_{15} = f(15), y_{20} = f(20)\}. \tag{4.200}$$

(Note, here the base 10 notation in Eq. (4.199) is used.) The members of the set of possible solutions in Eq. (4.200) are then associated, respectively, with the set of their x base 10 values $\{5,11,15,20\}$ which are in turn represented in binary form by

$$\{00101, 01011, 01111, 10100\}. \tag{4.201}$$

where 00101,01011,01111, and 10100 are now the chromosomes for the problem. The binary set in Eq. (4.201) contains the chromosomes representing, respectively, each of the proposed potential solutions $\{y_5 = f(5), y_{11} = f(11), y_{15} = f(15), y_{20} = f(20)\}$. We shall now operate on the chromosomes in Eq. (4.201) to generate an evolved next generation set of chromosomes.

First, let us summarize our discussions so far. This is made in Table 4.1.

In Table 4.1 is collected a summary of our treatment to this point. The values of x to the base 10 are in the first column and the corresponding chromosomes are listed in column two. The associated values of the function $y = f(x)$ in base 10 are in the third column. The fourth column gives the fractional weight of y_i in base 10 relative to the total of the generated y_j's from the chromosomes treated in the first generation of the population [22–26].

The object of our study then is to find the maximum of the function $y = f(x)$. Consequently, we compare the solution base 10 values $\{y_5, y_{11}, y_{15}, y_{20}\}$ from our four selected chromosomes to find the most suitable ones to mate, i.e., those that have the potential to generate an increase in $y = f(x)$. These chosen chromosomes will mate in order to maximize the y_i characteristic. To this end in Table 4.1 the values p_i compare y_i to the sum of all the y_i's in the set of chromosomes being treated. This indicates the relative suitability of each chromosome to the others in the set under consideration. On the basis of the p_i, the chromosomes in column two of the Table 4.1 are genetically modified to generate the next population of chromosomes.

TABLE 4.1
Summary of Data for the Maximization of $y = f(x)$

x base 10	x base 2	$y_i = f(x_i)$ base 10	p_i base 10
5	00101	$y_5 = f(x_5)$	$\dfrac{y_5}{y_5 + y_{11} + y_{15} + y_{20}}$
11	01011	$y_{11} = f(x_{11})$	$\dfrac{y_{11}}{y_5 + y_{11} + y_{15} + y_{20}}$
15	01111	$y_{15} = f(x_{15})$	$\dfrac{y_{15}}{y_5 + y_{11} + y_{15} + y_{20}}$
20	10100	$y_{20} = f(x_{20})$	$\dfrac{y_{20}}{y_5 + y_{11} + y_{15} + y_{20}}$

The chromosomes $\{00101, 01011, 01111, 10100\}$ are now mated to form the next generation of chromosomes. In this process, say that we have made a comparison of the p_i members of the set $\{y_5, y_{11}, y_{15}, y_{20}\}$ under consideration and have decided to mate the set of chromosomes $\{y_{11}, y_{20}\}$ and $\{y_{11}, y_{15}\}$ to create two new chromosomes to model the system. For this mating, we will use two genetic processes known as crossover and mutation in order to arrive at the members of the next generation of chromosomes.

Consider crossover applied to the mating of the chromosomes in the set $\{y_{11}, y_{20}\}$. The chromosome for y_{11} is 01011 and the chromosome for y_{20} is 10100. In crossover, we exchange segments of chromosomes between each of the pairs considered. For example, let us exchange the chromosome segments following the second digits from the left end of each of the mating chromosomes. In this process

$$01011 \text{ becomes } 01100 \tag{4.202a}$$

and

$$10100 \text{ becomes } 10011. \tag{4.202b}$$

The new chromosomes from Eq. (4.199) represent the new solutions

$$01100 \text{ corresponding to } y_{12} \tag{4.203a}$$

and

$$10011 \text{ corresponding to } y_{19}. \tag{4.203b}$$

These new chromosomes are now available for comparison and mating to make a new chromosome population in the subsequent generation.

Note that this process can be extended to treat more complex forms of exchanges of chromosome segments between pairs of chromosomes, i.e., segments of different lengths and defined by different end points within the chromosomes. In all such exchanges, however, it is also important to note that the number of digits in the final chromosome pairs remains the same as those in the beginning pairs. In this way, we maintain the genetic population to represent chromosomes for x within the interval $0 \le x \le 31$.

Another process that is commonly used in chromosome evolution is the process of mutation. In this process, a small number of random changes are made in a chromosome, e.g., changing a single 1 to 0 or a single 0 to 1. In such a process, for example

$$01011 \text{ could become } 01111 \tag{4.204}$$

by replacing the middle 0 with a 1. Here 01111 represents y_{15} which now must be compared with the other values in the considerations to determine whether or not a more suitable maximum is obtained in the mutation process.

These methods are repeated, generating successive new populations until a solution for the global maximum is confidently settled upon.

For further details and examples of the applications of genetic algorithms, the reader is referred to the literature [22–26].

REFERENCES

1. Barnett, S. M. 2009. *Quantum Information* (Oxford University Press, New York).
2. Alencar, M. S. 2014. *Information Theory* (Momentum Press, New York).
3. Shannon, C. E. 1948. A Mathematical Theory of Communication. *The Bell System Technical Journal* 27, 379–423.
4. Jaynes, E. T. 1957. Information Theory and Statistical Mechanics. *Physical Review* 106, 620–630.
5. Cover, T. M. and Thomas, J. A. 1991. *Elements of Information Theory* (John Wiley & Sons, Inc, Chichester, Great Britain).
6. Vedral, V. 2007. *Introduction to Quantum Information Science* (Oxford University Press, New York).
7. Hayashi, M. 2006. *Quantum Information: An Introduction* (Springer, Berlin).
8. Shepard, A. 2021. *Let's Calculate Bach: Applying Information Theory and Statistics to Numbers in Music* (Springer Nature, Switzerland).
9. Petz, D. 2008. *Quantum Information Theory and Quantum Statistics* (Springer-Verlag, Berlin).
10. Bergou, J. A. and Hillery, M. 2013. *Introduction to the Theory of Quantum Information Processing* (Springer, New York).
11. Wu, J.-Q. 2019. *AdS3/CFT2 and Holographic Entanglement Entropy* (Springer Nature, Singapore).
12. Fickler, R. 2015. *Quantum Entanglement of Complex Structures of Photons* (Springer, Cham).
13. Elsert, J., Cramer, M., and Plenio, M. B. 2010. Area Laws for the Entanglement Entropy-A Review. *Reviews of Modern Physics* 82, 277–305.
14. Duarte, F. J. 2019. *Fundamentals of Quantum Entanglement* (IOP Publishing Ltd, Bristol).
15. Cunha, M. M., Fonseca, A., and Silva, E. O. 2019. Tripartite Entanglement: Foundations and Applications. *Universe* 5, 209–242.
16. Landau, D. P. and Binder, K. 2015. *A Guide to Monte Carlo Simulations in Statistical Physics, Fourth Edition* (Cambridge University Press, Cambridge).
17. Barbu, A. and Zhu, S.-C. 2020. *Monte Carlo Methods* (Springer Nature, Singapore).
18. Pang, T. 2016. *An Introduction to Quantum Monte Carlo Methods* (Morgan & Claypool Publishers, San Rafael).
19. Feynman, R., Hibbs, A. R., and Styer, D. F. 2010. *Quantum Mechanics and Path Integrals* (Dover Publications Inc., Mineola).
20. McGreevy, R. L. and Pusztai, L. 1988. Reverse Monte Carlo Simulation: A New Technique for the Determination of Disordered Structures. *Molecular Simulation* 1, 359–367.
21. Malyshkin, V., Simeonov, S., McGurn, A. R., and Maradudin, A. A. 1997. Determination of Surface Profile Statistics from Electromagnetic Scattering Data. *Optics Letters* 22, 58–60.
22. Vose, M. D. 2023. *The Simple Genetic Algorithm: Foundations and Theory* (EBSCO Pubilishing, Cambridge, MA).
23. Bajpai, P. and Kumar, M. 2010. Genetic Algorithms-an Approach to Solve Global Optimization Problems. *Indian Journal of Computer Science and Engineering* 1, 199–206.
24. Kramer, O. 2017. *Genetic Algorithm Essentials* (Springer International Publishing, AG).
25. Chibante, R. 2019. *Simulated Annealing: Theory with Applications* (INTECH d.o.o., Rijeka).
26. Tsuzuki, M. 2019. *Simulated Annealing: Single and Multiple Objective Problems* (INTECH d.o.o., Rijeka).

5 Amorphous Materials
Polymers, Glasses, Amorphous Networks, Quasicrystals, and Tilings

In this chapter we consider solid state systems which do not exhibit the translational symmetries associated with crystalline materials or the approximate (i.e., approximately on average) translational symmetries of crystals containing small amounts of impurities. These are generally classified as amorphous materials [1–16] and have properties dominated by local atomic order. This local order is then found to disappear relatively quickly with distance as one moves spatially from point to point in the amorphous material. In this regard, whereas crystals are noted for their long-range order, amorphous materials do not display long-range order. Consequently, with only the presence of short-range order, amorphous materials display properties heavily dependent on considerations of the local atomic environment and the nature of the local chemical bonding [1–16].

Examples of such short-ranged dependencies are those properties arising from the coordination of the individual atomic bonds and the nature of the bond formation through their specific geometric configurations [1–16]. These are important determiners of the properties exhibited by the macroscopic material, especially in the absence of the translational symmetry of a crystal lattice. As examples of the local determination of the general system properties, we will focus on the discussion of four important types of amorphous material properties and the techniques developed for their study. These are included in the topics of the elastic properties of polymers forming rubber and the origins of these properties in the entropy [1–4]; the randomness encountered in the solidification of glasses and the nature of the glass transition [3–5]; the residual entropy of ice [6–10]; and the study of the electronic and vibrational excitations of amorphous networks [11–16]. Some of these properties will also be further discussed in a later chapter in the context of our treatment of Anderson localization.

An introduction will also be given to a new class of recently discovered materials known as quasicrystals [17–21]. These are materials displaying properties intermediate between crystalline and amorphous materials. Quasicrystals exhibit a long-range order but with an absence of translational invariance symmetry [17–21]. In this regard, the long-range order of quasicrystals is quite different from that associated with the translational symmetry of crystals. Nonetheless, quasicrystals are similar to but not the same as amorphous materials due to their absence of translational symmetry and, consequently, an increased importance on the local atomic environment. An example of this difference is their ability to exhibit point group symmetries not found in crystals. As we shall find, the topic of quasicrystals is also related to the very interesting mathematical study of the tiling of surfaces and spaces [17–21].

To begin our study, in a first set of discussions in this chapter, a treatment of the elastic properties of rubber is given [1–4]. This is a material formed from polymers and may occur in arrays of loosely correlated long linear chain molecules of repeated monomer units or as higher dimensional networks of interconnected polymer chains. In both the linear chain and network models, the systems can be considered as being composed of chains of monomers which are free to arrange themselves in a large number of spatial conformations. In this regard, each different spatial arrangement of the chain is referred to as a conformation of the chain, and the diversity of chain conformations is an example of the absence of long-range spatial ordering in the material. In the formation of the

DOI: 10.1201/9781032674919-5

conformations, however, the nature of the bonds between monomers can significantly affect the overall nature of the distribution of polymer conformations. Also, the spatial exclusion of the chain from itself (i.e., the chain cannot occupy the same space twice) is an important restriction on conformation formation.

Whether the chains exist independently in space or bound as part of a network, a consequence of the large number of conformations is that a large entropy is associated with all the possible chain conformations [1–4]. By pulling on the chains or network of chains one tends to straighten the chains out, reducing the number of conformations of the chains, and, consequently, lowering the entropy of the system. This reduction of the entropy can be shown thermodynamically to manifest itself in setting the elastic force required to stretch the material. As an additional thermodynamic consequence, associated with the stretching or contraction of the material, a change in the temperature of the material is also observed [1–4].

The resulting process of elastic deformation turns out to be quite different from that observed in crystalline materials [1–4]. This is because the mechanisms of elastic deformation in crystalline solids and rubbers are completely different between the two systems. In crystalline materials, a deformation of the material requires that the atomic separations in the material change, and this is a process involving changes in the potential energy of the atomic interactions of the material [1–4]. This is part of the work done during the mechanical deformation. On the other hand, in polymer systems, the force arises from the change in the number of conformations of the chains in the material and the subsequent change in entropy of the material. The elastic change in this case does not necessarily involve the energetics of changing the energy associated with the atomic bonds in the material. It is a thermodynamic process involving the entropy [1–4].

In a related problem, discussions are given of some basics of the glass transition [3–5]. This is the transition from a liquid state to what is termed a glass state. The glass state is a nonequilibrium configuration of the system which, nevertheless, may be considered as a type of solid material. The glass transition is a slowing down of the system dynamics so that the solid becomes a metastable state in which the motion of the material is an extremely slow flow [3–5]. While the state is not in thermodynamic equilibrium, the relaxation of the material to thermodynamic equilibrium is so long that the material appears unchanged in most practical temporal considerations. In this respect, the glass transition is displayed as an inflection in some of the physical properties of the material considered over a range of temperatures as the material goes from the liquid to the glass state. The glass transition temperature T_g is usually defined within the temperature range of this inflectional change but is not a critical temperature in the sense that the system undergoes a real phase transition. In this regard, T_g is less well-defined than the critical temperature at a phase transition [3–5].

The glass state is a solid state in which there is an absence of long-range order. The absence of long-range order is observed in the elastic scattering from the material. In crystalline materials, sharp well-defined Bragg diffraction peaks are observed [1–5]. This is typical of the long-range periodic ordering of crystalline materials that are formed on periodic lattices. On the other hand, in amorphous or glass systems, such long-range order is absent, and the material exhibits instead a diffuse scattering. The absence of long-range order shows up as the absence of sharply defined Bragg diffraction peaks at specific characteristic angles and the presence, instead, of a general diffuse elastic scattering from the materials at all angles. This diffuse scattering in the amorphous state may be modulated by broad characteristic peaks of the material studied. In this regard, the diffuse scattering remains highly dependent on the presence of short-ranged correlations existent in the atomic system.

The glass transition is a characteristic of many types of amorphous materials that are of interest to physicists [1–5]. An important example of this is amorphous silicon [3–5]. Amorphous silicon is a material of great interest to electronics technology, where it finds applications in a large variety of devices. In its crystalline form, silicon is an important semiconductor which is formed of Si –atoms making tetragonal bonds with each other to condense into a diamond cubic structure. However, silicon also exists in an amorphous metastable state in which the atoms do not occupy a lattice structure but rather occur on an amorphous network of Si –atoms [3–5]. In the network, there can exist dangling bonds which are unsatisfied by other atoms; however, each Si –atom essentially seeks to form tetragonal bonds. These bonds, nevertheless, can be highly distorted by the lack of periodicity of the atomic network. As in all amorphous materials, amorphous silicon has no long-range order but tends to exhibit properties arising from the underlying nature of the tetragonal bonding. This shows up, for example, in its electronic density of states.

The absence of periodicity is also manifest in the nature of the excitations of the amorphous system, and this difference is found in the study of the density of states of the excitations of the material [1–16]. In crystalline materials, the modes tend to be characterized by wavevectors associated with the translation symmetry of the lattice. As a result, the excitations in crystals exhibit dispersion relations and lifetimes determined as a function of their wavevectors and wavefunctions of the form of plane waves times a periodic function defined over the lattice. Amorphous materials, on the other hand, have excitations which are determined to a greater extent by the short-range order of the system. The properties of these excitations arise from the local characteristics of the bonding between the atoms in the material and have wavefunctions no longer given by simple Bloch wave forms [1 –16]. This difference in the nature of excitations in amorphous materials requires an alternative approach to the study of excitations than that commonly applied to crystalline materials. We shall consider some of the methods developed for such studies of amorphous materials.

Both crystalline and amorphous material density of states are determined by the Green's functions of the modal excitations [11–16]. In the case of amorphous materials, these Green's functions are determined using so-called recursive methods [11, 12]. Recursive methods are new techniques which begin by emphasizing the local properties of the network structure of the amorphous material. Starting from the study of a tight-binding model defined on an amorphous network, the recursive method [11, 12] is used to generate a new basis of orthonormal basis states and a new representation of the system's Hamiltonian in terms of a tridiagonal matrix. By studying the Green's function of the Hamiltonian in the representation of the tridiagonal matrix, a method for determining the local density of states at each site of the network of the amorphous material is developed. This summary is essentially an outline of the development and formulation of the recursive method as applied to amorphous materials [11–16]. The method, however, is general in nature and can be applied to study crystals, amorphous materials, or atomic clusters [11, 12].

Another problem of nonmetallic amorphous materials, illustrating the difference in these systems from their crystalline counterparts, is that of the so-called two-level states problem [13–16]. This is focused on the differences in the vibrational properties of amorphous and crystalline materials at temperatures well below their Debye temperatures. Specifically, in this temperature range, the temperature dependence of the phononic heat capacity and thermal conductivity of amorphous materials is observed to be quite different from those found in crystalline materials. This difference is surprising as the vibrational excitations in both crystalline and amorphous materials in this low-temperature range are expected to come from long wavelength phonons with wavelengths extending over many atoms composing the systems [13–16]. These types of long wavelength excitations should be relatively insensitive to the details of the atomic structure of the material, so that the material should approach the behavior of an elastic continuum. Consequently, the specific heat and thermal conductivities for both types of materials would be expected to be those from an elastic continuum model. In the case of amorphous materials, however, we shall find that another excitation mechanism is also at play in the material.

The anomalous low-temperature behavior of the amorphous materials from that of the continuum behavior is associated with small structural changes in the geometry of the amorphous network [13–16]. These are tunneling transitions rather than phononic excitations. In this regard, the network is found to tunnel locally between two slightly different network configurations. Such tunneling between different geometric configurations does not occur in rigid crystalline systems [13–16]. However, in the absence of long-range order, local distortions of the amorphous network allow for tunneling between different metastable network geometries [13–16]. The tunneling can be thought of as an introduction of two energy levels into the network associated with the two distinct local geometries.

Consequently, as the network system tunnels between its two network configurations, it moves between two different energy states of the network configuration. In this way, the occupancy of the two-level states of the network shows up as an additional contribution to the heat capacity and the thermal conductivity of the material. The amorphous materials then begin to exhibit characteristics of the general thermodynamic behavior found in a system consisting of numerous two-level energy scattering states found distributed within its amorphous network. This local tunneling of the many two-level states within the network accounts for the anomalous behavior of the amorphous materials [13–16].

For a final topic, we shall look at the related interesting problems of quasiperiodic crystals, quasicrystals, tilings, and systems with $5-$fold rotational symmetry [17–24]. These are, again, noncrystalline materials but, nevertheless, they exhibit an interesting patterning in their formation. While they represent materials that do not exhibit translational symmetry in space, they, nevertheless, exhibit a long-range spatial order. This long-range order is evidenced as a sequence of bright spots found in their Bragg diffraction patterns. Consequently, they are not crystals but due to the presence of long-range order, they are not amorphous materials. In this regard, perhaps they are intermediate structures between true crystalline materials and amorphous materials.

An interesting feature of quasicrystal materials is that some of them exhibit electron and X-ray diffraction patterns which have been shown both theoretically and experimentally to exhibit $5-$fold rotational symmetry [17–21, 24]. This is important as $5-$fold rotational symmetry axes are known to be excluded from crystalline materials. In addition, recently, patterns exhibiting $8-, 10-$ or $12-$fold axes of rotational symmetry have also been commonly found. These axes of rotational symmetry are also not allowed in a true crystal lattice [17–24].

We begin our studies with a discussion of polymer conformations and how these form a basis for understanding the elastic properties of rubber [1–4].

5.1 POLYMER CONFORMATIONS OF LINEAR CHAIN POLYMERS: RANDOM WALK MODEL

In a simplistic model of polymer conformations, the macromolecule is treated as a random walk through $3-$dimensional space [1–4]. For this representation, a cubic lattice is introduced into space with a separation between nearest-neighbor points on the lattice denoted by the distance a, and the individual monomers forming the polymer occur as steps between two nearest-neighbor sites of the lattice. The configuration of the polymer is then defined by specifying a connected sequence of nearest-neighbor steps on the lattice, beginning at one end of the molecule and going to the other end of the molecule. (See Figure 5.1 for a schematic of this process.)

In this regard, the conformation of the polymer is formed by a random walk involving a disordered process of connected nearest-neighbor moves [1–4]. As we shall see, the model is a rather crude model of how macromolecules occupy space because it ignores many basic constraints on the molecule other than those of the monomer length, and it involves an arbitrary limitation of the motion of the walker to a cubic lattice. Nevertheless, it offers a general idea of the basic properties and mechanisms found in some polymer materials approximated by simple chain molecules.

We begin our study by considering the set of all polymer conformations for molecules formed of n consecutive connected segments whose geometry is defined on a cubic lattice [1–4]. In this model, each monomer segment of the molecule occupies the region between two nearest-neighboring points on the lattice and is the length of the lattice constant or nearest-neighbor separation a on the lattice. The geometry of the molecule in space is then treated as being laid out in space by the n steps of a walker, randomly moving between nearest-neighbor sites on the lattice, and the statistical properties of the molecular geometry are determined from the sequence of these motions. In this process, one end of the molecule is located at $(0,0,0)$ and after n steps or segments the other end is located, for example, at the lattice coordinates (i, j, k).

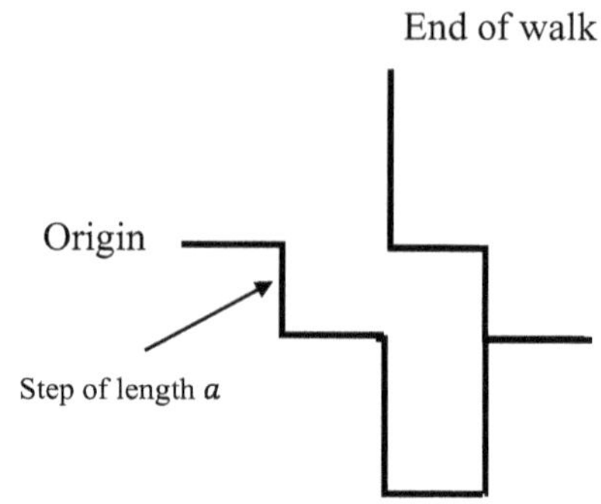

FIGURE 5.1 Schematic of a random walk polymer chain formed of steps of length a on a cubic lattice. As a simplification, the representation of the specific walk shown here is for a walk made in the $x-y$ plane of the cubic lattice.

For this system we define the probability [1–4]

$$P(n:i,j,k) \tag{5.1}$$

that starting at the $(0,0,0)$ end of the molecule after n steps the walker is found at the other end of the molecule located at the lattice point (i,j,k).

Notice, regarding the probability in Eq. (5.1) that $P(n:i,j,k)$ is related to the probabilities that a molecule composed of $n-1$ steps or segments ends on one of the nearest neighbor sites of the lattice site (i,j,k). Specifically, $P(n:i,j,k)$ can be written in terms of the probabilities $P(n-1:i\pm1,j,k)$, $P(n-1:i,j\pm1,k)$, and $P(n-1:i,j,k\pm1)$ as from these ends a single-step results in an n segment molecule ending at (i,j,k). It then follows that [1–4]

$$\begin{aligned}&P(n:i,j,k)=\\&\frac{1}{6}\{[P(n-1:i-1,j,k)+P(n-1:i+1,j,k)]+[P(n-1:i,j-1,k)+P(n-1:i,j+1,k)]\\&+[P(n-1:i,j,k-1)+P(n-1:i,j,k+1)]\}\end{aligned} \tag{5.2}$$

This is a statement that the probability $P(n:i,j,k)$ of a molecule of n segments having an end located at (i,j,k) is given by $\frac{1}{6}$ times the sum of the probabilities of $P(n-1:i-1,j,k), P(n-1:i+1,j,k)$, $P(n-1:i,j-1,k), P(n-1:i,j+1,k)$, $P(n-1:i,j,k-1), P(n-1:i,j,k+1)$ of a molecule of $n-1$ segments having an end occupying one of the nearest-neighbor sites surrounding (i,j,k). In this respect, the factor $\frac{1}{6}$ is the probability that the nth segment coming from each of the nearest-neighbor sites of (i,j,k) will go to the site (i,j,k).

The result in Eq. (5.2) can next be considered in a continuum limit approximation in which its physics is represented in terms of a diffusive process characterized by a differential diffusion equation [1–5]. This continuum formulation will facilitate approximating the statistical properties characterizing the geometry of the macromolecules and becomes increasingly valid in the limit that $n \gg 1$.

In the continuum limit the relationship in Eq. (5.2) is rewritten as a partial differential equation having the form of a diffusion equation [1–5]. To obtain the continuum limit we assume that $P(n:i,j,k)$ is a slowly varying function of the variables n,i,j,k and that $P(n:i,j,k)$ can be analytically continued by treating n,i,j,k as continuous real variables. For this process we begin by first subtracting $P(n-1:i,j,k)$ from both sides of Eq. (5.2).

Making this subtraction on the left side of Eq. (5.2) and Taylor expanding the result gives

$$P(n:i,j,k)-P(n-1:i,j,k)\approx P(n:i,j,k)-\left[P(n:i,j,k)-\frac{\partial}{\partial n}P(n:i,j,k)\right] \tag{5.3a}$$

$$=\frac{\partial}{\partial n}P(n:i,j,k)=\frac{\partial}{\partial n}P(n:x,y,z)$$

where in the first approximation on the right a Taylor expansion is made about n of the function $P(n-1:i,j,k)$ which is slowly varying over the integers n, and where the new real variables (x,y,z) on the far right side are evaluated at $(x,y,z)=(ia,ja,ka)$. Similarly, on the right-hand side of Eq. (5.2), $P(n-1:i,j,k)$ is subtracted and distributed equally between each of the three $[\ldots]$ terms in the brackets $\{\ \}$. In the first, $[\ldots]$ we Taylor expand about i to find [1–4]

$$P(n-1:i-1,j,k)-2P(n-1:i,j,k)+P(n-1:i+1,j,k)\approx a^2\frac{\partial^2}{\partial x^2}P(n:x,y,z) \tag{5.3b}$$

where the new real variables (x,y,z) are evaluated at (ia,ja,ka). Similarly, Taylor expanding about j in the second $[\ldots]$ we find

$$P(n-1:i,j-1,k)-2P(n-1:i,j,k)+P(n-1:i,j+1,k)\approx a^2\frac{\partial^2}{\partial y^2}P(n:x,y,z) \tag{5.3c}$$

where (x,y,z) are evaluated at (ia,ja,ka), and finally we Taylor expand about k in the third $[\ldots]$ to obtain

$$P(n-1:i,j,k-1)-2P(n-1:i,j,k)+P(n-1:i,j,k+1)\approx a^2\frac{\partial^2}{\partial z^2}P(n:x,y,z) \tag{5.3d}$$

where (x,y,z) are evaluated at (ia,ja,ka). These results can then be assembled in Eq. (5.2) to form a differential diffusion equation for the probability density $P(n:x,y,z)$ in terms of the continuous variables (x,y,z).

In this way Eq. (5.2) reduces an equation given by [1–5]

$$\frac{\partial}{\partial n}P(n:x,y,z)=\frac{a^2}{6}\nabla^2 P(n:x,y,z). \tag{5.4}$$

Note that Eq. (5.4) is in the form of a diffusion equation and can be evaluated by using standard techniques. In this way, the solution for the probability density $P(n:x,y,z)$ is given by

$$P(n:x,y,z)=\left(\frac{3}{2\pi}\right)^{3/2}\frac{1}{\left(a\sqrt{n}\right)^3}\exp\left[-\frac{3}{2a^2n}\left(x^2+y^2+z^2\right)\right] \tag{5.5}$$

which is normalized such that

$$\int_{-\infty}^{\infty}\int_{-\infty}^{\infty}\int_{-\infty}^{\infty}dxdydzP(n:x,y,z)=1. \tag{5.6}$$

This normalization is a statement that a molecule of length n has a unit probability of ending somewhere in $3-$dimensional space. The result in Eq. (5.5) gives an approximate representation within the random walk model of the probability of finding a polymer molecule composed of $n \gg 1$

monomer units distributed in space with one end located at $(0,0,0)$ and the other at (x,y,z). It is now used to study the geometric properties of the various conformations of the polymer molecule.

From Eq. (5.6) it then follows, considering the symmetry of the distribution $P(n:x,y,z)$ about the origin, that the average position of the ends of the polymers is located at

$$\int_{-\infty}^{\infty}\int_{-\infty}^{\infty}\int_{-\infty}^{\infty} dxdydzP(n:x,y,z)\left[x\hat{i}+y\hat{j}+z\hat{k}\right]=\left[0\hat{i}+0\hat{j}+0\hat{k}\right], \tag{5.7a}$$

i.e., at the origin of coordinates. The mean squared distance between the ends of the polymers can be similarly determined as

$$\int_{-\infty}^{\infty}\int_{-\infty}^{\infty}\int_{-\infty}^{\infty} dxdydzP(n:x,y,z)\left[x^2+y^2+z^2\right]=na^2 \tag{5.7b}$$

which is found to be nonzero and related to the monomer length and the number of monomers in the molecule.

Rewriting Eqs. (5.5) and (5.6) in terms of polar coordinates, we also obtain [1–4]

$$P(n:r,\theta,\varphi)dr\sin\theta d\theta d\varphi=\left(\frac{3}{2\pi}\right)^{3/2}\frac{1}{\left(a\sqrt{n}\right)^3}\exp\left[-\frac{3}{2a^2n}r^2\right]r^2dr\sin\theta d\theta d\varphi, \tag{5.8a}$$

and integrating over the angular variables yields the radial distribution of the polymer end about the origin in space. This is given by

$$P(n:r)dr=\left(\frac{3}{2\pi}\right)^{3/2}\frac{1}{\left(a\sqrt{n}\right)^3}\exp\left[-\frac{3}{2a^2n}r^2\right]4\pi r^2dr. \tag{5.8b}$$

and represents the probability distribution that if one end of the polymer is at the origin the other end occurs in the radial interval $[r,r+dr]$. Consequently, we see from this that the most probable radial separation of the two ends of the molecule is given by $\frac{d}{dr}P(n:r)=0$ and is found to be located at a radial separation

$$r=\sqrt{\frac{2}{3}na^2}=\sqrt{\frac{2}{3}n}a. \tag{5.9}$$

Next, we consider the entropy properties of the probability density $P(n:r)$ in Eq. (5.8b). This is solely a measure of the geometric disorder of the spatial configurations or conformations of the polymers. In this regard, we have not associated any energies in connection with the changes of the polymer conformations, and, consequently, the molecular conformations do not self-interact. From Eq. (5.8b) it then follows that the radial probabilities are related to the numbers of configurations in the system so that [1–5]

$$P(n:r)dr=\frac{N(n,r)}{N_{total}(n)} \tag{5.10a}$$

where $N_{total}(n)$ are all the possible conformations of polymers composed of n monomer units, and

$$N(n,r)=N_{total}(n)P(n:r)dr \tag{5.10b}$$

is the number of conformations with a separation r between their ends contained within the radial interval $[r, r+dr]$.

Consequently, the probability p_i of a particular conformation labeled i with end separation r being chosen from the set of all conformations with end separations in the interval $[r, r+dr]$ is given by [1–4]

$$p_i = \frac{1}{N(n,r)} \tag{5.11}$$

and is the same for each distinct individual conformation within the interval $[r, r+dr]$. The entropy associated with the system of n monomers within the radial interval $[r, r+dr]$ is then given in terms of these $N(n,r)$ identical probabilities by

$$S = -k_B \sum_{i=1}^{N(n,r)} p_i \ln p_i = -k_B \sum_{i=1}^{N(n,r)} \frac{1}{N(n,r)} \ln \frac{1}{N(n,r)} = k_B \ln N(n,r) \tag{5.12}$$

$$= k_B \ln\left[N_{total}(n) P(n:r) dr\right] \approx k_B \ln C - \frac{3k_B r^2}{2na^2}.$$

Here C is a constant, we have used Eqs. (5.8b) and (5.10) in the last two equalities on the right, and we have dropped a term weakly dependent on r (i.e., linear in $\ln r$) in the final expression on the right. It is seen from Eq. (5.12) that for a fixed number of monomer units n, the entropy of the system is decreased with an increasing radial separation between the ends of the polymer. As we shall see later, this has important implications in the context of the elasticity properties exhibited by polymers.

In this regard, we next consider a basic treatment of the elasticity properties of linear chain polymers. In particular, it is shown that the nature of the polymer elasticity is dominated by entropy considerations associated with the large number of different conformations available to the polymer molecules. This is unlike the elastic properties of crystalline materials in which the elasticity arises from the nature of the chemical bonds acting between the atoms and the potential energy associated with changes in the atomic separations during elastic distortion.

5.2 ELASTICITY OF LINEAR CHAIN POLYMERS

A rough model for the elasticity of polymers can now be presented in the context of the entropy discussions given in Eqs. (5.10) through (5.12). The model is formulated following some basic properties in thermodynamics as applied to the entropy form in Eq. (5.12) combined with a number of general experimental observations on polymeric materials exhibiting elasticity [1–4]. These observations were historically put forward early in the study of various types of stretching processes involving polymer materials [1–4]. Specifically, it was observed that as the polymer chains of a material are stretched, the number of their different conformations is reduced, causing a decrease in the disorder of their conformational structure. Ultimately, the resulting decrease in the entropy from the conformational disorder shows up as an elastic restoring force arising solely due to the entropy changes of the material.

The thermodynamic discussion begins with a consideration of the Helmholtz free energy potential of the polymer material and is concerned with the work done on the system during an isothermal elastic stretching process. This is basically a natural focus of the definition and the formal development of the Helmholtz potential theory. In this regard, the Helmholtz potential as a function of the material volume V and temperature T is defined by the form $F(V,T) = E(V,S) - TS$ where $E(V,S)$

is the energy of the elastic system and S is the entropy of the system [1–4]. It then follows that for an isothermal process the change of the Helmholtz potential is characterized by the differential form

$$dF = dE - TdS \tag{5.13}$$

where for a consideration of the linear polymer chain the entropy S is given by Eq. (5.12).

For a first consideration, we study the effects on the system of the elongation of the polymers composing the elastic material. This is done by studying the length change dl obtained through the application of a stretching force f on the molecules. The resulting molecular elongations ultimately show up as a general stretching of the elastic material, composed as a conformal system of molecular chains. From thermodynamic considerations, the work done during the elastic extension dl of the molecules in an isothermal process is obtained in terms of the Helmholtz function by [1–4]

$$fdl = dF \tag{5.14}$$

and represents the work fdl done on the molecule by the external stretching force f as the system is extended.

From Eqs. (5.13) and (5.14), it consequently follows that [1–4]

$$f = \left(\frac{dE}{dl}\right)_T - T\left(\frac{dS}{dl}\right)_T \tag{5.15}$$

so that the elastic force would appear to come from the energy of the system as well as from its entropy. However, general experiments made on elastic materials have shown that, to a reasonable approximation, in polymer materials $\left(\frac{dE}{dl}\right)_T \approx 0$ so that

$$f = -T\left(\frac{dS}{dl}\right)_T . \tag{5.16}$$

This represents a great simplification of the theory, expressing the elastic force as arising from a change in entropy of the material and is at odds with crystalline materials in which the elastic response of the material to an applied stress arises from the atomic bonds in the material. In this respect, as a crystalline material is distorted its potential energy is changed, making the first energy term in Eq. (5.15) important. Consequently, the force of deformation in crystals arises through the energy of the atomic interactions rather than directly from conformational entropy considerations.

We next consider an elastic material composed of polymers of length l (i.e., l is the distance between the ends of the polymers) composed of n monomer units each of length a. An approximation for the entropy of the molecules under these considerations has been obtained from Eq. (5.12) and is written with reference to an individual molecule in the system as [1–4]

$$S = k_B \ln C - \frac{3k_B l^2}{2na^2}. \tag{5.17}$$

This ultimately provides an approximation for the dependence of the entropy on the changing length between the ends of a polymer molecule.

The expression in Eq. (5.16) directly relates the elastic force to the change in entropy with changing length dl. Applying the result in Eq. (5.17) in Eq. (5.16), we find for a single polymer chain that [1–4]

$$f = \frac{3k_B Tl}{na^2} \tag{5.18}$$

as the elastic force causing the extension of a molecule. It is seen from this expression that the force required to elastically deform a polymer material is proportional to both the extended length of the polymers l and to the temperature T of the material it composes. Consequently, increasing either the length (i.e., the separation between the polymer ends) of the polymer or its temperature results in an increase in the elastic force f needed to extend the material.

The elasticity results presented here are highly simplistic, generated under the assumptions of a system composed of independent linear chain polymers. These types of chain polymers are only a small class of molecules that are classified as polymers, and many more complex geometric 2 – and 3 –dimensional structures are found existing as polymer materials. In addition, as another weakness, the random walk theory of entropy of polymer conformations applied to our linear chain model is also a gross approximation. This aspect of our treatment requires a more sophisticated approach, including self-avoiding geometries and more complex modeling of the bonds formed between neighboring monomers. We next present some considerations of a model which improves upon our present discussions.

5.2.1 Elasticity of Linear Chain Polymers: A Refinement

To improve the simple considerations given above, we generalize them to a treatment of networks of more complex geometry than is exhibited by a single polymer chain. In a more realistic model of rubber, long polymer chains are interconnected into a 3 –dimensional network of chains so that the material is actually better represented as a type of loose fishing net of interconnecting polymer chains. Stretching this type of net changes the entropy of the entire network by stretching each of the chains composing it.

In this way, upon stretching, the entropy of the conformations of the individual chains forming the network is reduced per our earlier discussions, and this reduces the entropy of the entire network of interconnected chains. (See Figure 5.2 for a schematic of such a network.) As is seen from the figure, the increase in polymer complexity can be accomplished by introducing various branchings and crosslinking segments which loosely connect chains to one another to form a network of chains [1–4] .The bonding between chains, in fact, also keeps the chains of the network from sliding past one another and aids in the stretching process which reduces the entropy of the total configuration. Nevertheless, the important conformational disorder is contained in the chain regions occurring between the chain branchings and bonding points found between different chains composing the network.

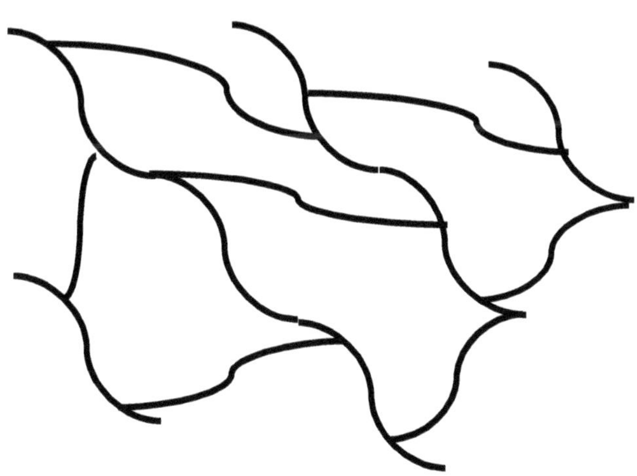

FIGURE 5.2 Schematic "fishnet" of interconnected polymer chains. The lines represent various long-chain polymer molecules which attach to each other to form a network.

Let us assume then that the system is composed of N chains of length l and these are connected together to make a 3−dimensional network such that the individual chains of the network occur randomly oriented about space. For this network the entropy (now of the 3−dimensional system of interconnected polymer molecules) is roughly approximated from Eq. (5.17) by [1–4]

$$S = k_B N \ln C - \frac{3k_B N}{2na^2}\left(x^2 + y^2 + z^2\right) \tag{5.19a}$$

where N are the number of chains of length l in the network, and we have written $l^2 = x^2 + y^2 + z^2$ explicitly in terms of the separation of the chain ends in (x, y, z) coordinates. (Note here that (x, y, z) represents a typical displacement of the polymer ends which satisfy $l^2 = x^2 + y^2 + z^2$ and aside from this constraint is randomly oriented in space.) In the unstretched state of the rubber, let us take the chain coordinates as (x_0, y_0, z_0) so that for this limit the entropy in Eq. (5.19a) is

$$S_0 = k_B N \ln C - \frac{3k_B N}{2na^2}\left(x_0^2 + y_0^2 + z_0^2\right) \tag{5.19b}$$

where $x_0^2 + y_0^2 + z_0^2 = l_0^2$ with the unstretched polymer length l_0 being defined between the ends of the chain.

Next consider stretching the material so that (x_0, y_0, z_0) of length l_0 changes to (x, y, z) of length l such that [1–4]

$$x = \lambda_1 x_0, y = \lambda_2 y_0, z = \lambda_3 z_0 \tag{5.20a}$$

represent the new coordinates of the stretched system. Here the parameters λ_1, λ_2, λ_3 characterize the stretching process of the material, and depending on the scale under consideration, can be written as

$$\lambda_1 = \frac{x}{x_0} \approx \frac{L_x}{L_{x0}}, \lambda_2 = \frac{y}{y_0} \approx \frac{L_y}{L_{y0}}, \lambda_3 = \frac{z}{z_0} \approx \frac{L_z}{L_{z0}} \tag{5.20b}$$

where L_{x0}, L_{y0}, L_{z0} are the unstretched lengths of the sample of material being studied and L_x, L_y, L_z are the stretched lengths of the sample of material being studied. (Notice that in this representation we assume that the length changes on the molecular level are scaled with those on the level of the dimensions of the bulk material.) Consequently, it follows for the lengths of the sample of elastic material that

$$L_{x0}^2 + L_{y0}^2 + L_{z0}^2 = L_0^2 \text{ and } L_x^2 + L_y^2 + L_z^2 = L^2 \tag{5.20c}$$

so that $L_x^2 + L_y^2 + L_z^2 = \left(\lambda_1^2 L_{x0}^2 + \lambda_2^2 L_{y0}^2 + \lambda_3^2 L_{z0}^2\right)$ as characterized by the ratios in Eq. (5.20b).

In terms of the λ −parameters, from Eq. (5.20a) the new entropy of the stretched material is given by [1–4]

$$S = k_B N \ln C - \frac{3k_B N}{2na^2}\left(\lambda_1^2 x_0^2 + \lambda_2^2 y_0^2 + \lambda_3^2 z_0^2\right) \tag{5.21a}$$

and for a symmetric material in which on average $x_0^2 = y_0^2 = z_0^2 = \frac{1}{3}na^2 = \frac{1}{3}l_0^2$ the entropy takes the form

$$S = k_B N \ln C - \frac{k_B N}{2}\left(\lambda_1^2 + \lambda_2^2 + \lambda_3^2\right). \tag{5.21b}$$

Similarly, in the unstressed system Eq. (5.21a) reduces to

$$S_0 = k_B N \ln C - 3\frac{k_B N}{2} \tag{5.21c}$$

for the material in the absence of stretching [1–4].

In the case that the system is stretched along the x –direction and the total volume of the material remains unchanged (Note: the approximate conservation of volume is again an experimental characteristic of rubber) we have

$$\lambda_1 \lambda_2 \lambda_3 = 1 \tag{5.22}$$

so that in an isotropic system $\lambda_2 = \lambda_3 = \frac{1}{\sqrt{\lambda_1}}$. From Eq. (5.13) in the case that $dE = 0$ and for an isothermal process, it then follows that [1–4]

$$dF = -TdS = -T\left(S - S_0\right) = \frac{1}{2} N k_B T\left(\lambda_1^2 + \lambda_2^2 + \lambda_3^2 - 3\right) \tag{5.23a}$$

$$= \frac{1}{2} N k_B T\left(\lambda_1^2 + \frac{2}{\lambda_1} - 3\right).$$

Accounting for the approximate relationships $l \approx \lambda_1 l_0$ and $L \approx \lambda_1 L_0$ where $\lambda_1 \approx \frac{l}{l_0} = \frac{L}{L_0}$, we obtain from Eq. (5.23a) that the consideration of the isothermal work $dF = fdL$ done on the macroscopic piece of material yields the relationships

$$f = \frac{dF}{dL} = -T\left(\frac{dS}{dL}\right)_T = -T\left(\frac{dS}{d\lambda_1}\right)_T \frac{\partial}{\partial L}\left(\frac{L}{L_0}\right) = \frac{N k_B T}{L_0}\left(\lambda_1 - \frac{1}{\lambda_1^2}\right) = \sigma A_0. \tag{5.23b}$$

where L_0 is for the undeformed system, A_0 is the cross-sectional area of the sample in the absence of stretching, and σ is the stress applied to the elastic medium. In a final form we then have [1–4]

$$\sigma = \frac{N k_B T}{A_0 L_0}\left(\lambda_1 - \frac{1}{\lambda_1^2}\right) = n_0 k_B T\left(\lambda_1 - \frac{1}{\lambda_1^2}\right) \tag{5.24}$$

where n_0 is the number of chain segments per volume. This provides a simple form for the dependence of the applied stress on the extension $L \approx \lambda_1 L_0$ of the elastic material.

As a final point, notice that in our discussions of the random walk model of polymer conformations we have disregarded energy considerations in choosing the conformations of the system. This is consistent with our assumption in this subsection that $dE = 0$ during the elastic processes treated. Consequently, the first law of thermodynamics for the case of our elastic material can be written in the form [1–4]

$$dE = dQ + fdL \tag{5.25}$$

where dQ is the heat energy flowing into the system, fdL is the work done on the system by the stretching force f, and dE is the change in total energy of the material. In the case of $dE = 0$ it then follows that $dQ = -fdL$ so that when work is done on the system $dQ = -fdL < 0$ and there is a net flow of heat out of the system, i.e., the system heats up and heat travels out of it. Similarly, in the case of $dE = 0$ it follows that $dQ = -fdL$ so that when work is done by the system $dQ = -fdL > 0$ and there is a net flow of heat into the system, i.e., the system cools down so that heat travels into it.

To conclude the discussions on polymers, we next give an illustration of a treatment which is meant to refine the general ideas presented up to now. Such treatments ultimately contribute to improvements in the further considerations of polymer elasticity. We begin with a theory which provides a more realistic assessment of the conformations of linear chain polymers. This is then followed by a linkage of the ideas of polymer conformations with the ideas of Brownian motion in diffusive processes.

5.3 A MORE DETAILED MODEL FOR THE STUDY OF LINEAR CHAIN POLYMER CONFORMATIONS

A problem with the random walk model is that many linear chain polymers have restricted bond angles between their monomers which are set by the nature of the chemical [1–4]. These type of restrictions are not correctly treated in the random walk models of our earlier considerations. In this regard, an interesting model of a polymer is obtained by fixing the bond angles between nearest-neighbor monomers along the chain. This provides an example of an incremental generalization of the problem of polymer conformations to treat more complex models.

We begin with a consideration of this type of problem by making a reformulation of the treatment of the polymer conformations. In this reformulation, the monomers of the system are treated as identical chemical units of length a, and each monomer is represented by a vector $\vec{r}_i$ where $|\vec{r}_i| = a$ and the subscript i indicates that $\vec{r}_i$ represent the ith monomer forming the polymer as a position vector going from one end of the monomer to the other. Again, the polymer is generated by a walk composed of successive $\vec{r}_i$. Consequently, starting at one end of the polymer located at the origin in space, the other end is found at [1–5]

$$\vec{R} = \sum_{i=1}^{n} \vec{r}_i \tag{5.26a}$$

where $\{\vec{r}_i\}$ are the successive steps along the chain. In our considerations we also assume that for this walk

$$\langle \vec{R} \rangle = \left\langle \sum_{i=1}^{n} \vec{r}_i \right\rangle = 0 \tag{5.26b}$$

where $\langle \ \rangle$ represents an average over all possible conformations of $\vec{R}$ and that

$$\langle \vec{R} \cdot \vec{R} \rangle = \sum_{i=1}^{n} \sum_{j=1}^{n} \langle \vec{r}_i \cdot \vec{r}_j \rangle. \tag{5.26c}$$

From Eq. (5.26b) it then follows that

$$\langle \vec{r} \rangle_i = 0 \tag{5.27a}$$

and, similarly, in Eq. (5.26c) we have [1–5]

$$\langle \vec{r}_i \cdot \vec{r}_j \rangle = a^2 \langle \cos\theta_{i,j} \rangle \tag{5.27b}$$

where $\theta_{i,j}$ is the angle between $\vec{r}_i$ and $\vec{r}_j$. Consequently, the angle of the chemical bond formed along the chain between the monomers $\vec{r}_i$ and $\vec{r}_{i+1}$ is given by $\theta_{i,i+1}$.

In this reformulation the results for the random walk model of our earlier considerations are obtained in the limiting case that [1–4]

$$\langle \vec{r}_i \cdot \vec{r}_j \rangle = \langle \vec{r}_i \rangle \cdot \langle \vec{r}_j \rangle = 0 \text{ for } i \neq j \tag{5.28a}$$

and

$$\langle \vec{r}_i \cdot \vec{r}_i \rangle = a^2. \tag{5.28b}$$

It then follows from Eq. (5.26) that

$$\langle \vec{R} \rangle = \left\langle \sum_{i=1}^{n} \vec{r}_i \right\rangle = 0 \tag{5.29a}$$

and

$$\langle \vec{R} \cdot \vec{R} \rangle = \sum_{i=1}^{n} \sum_{j=1}^{n} \langle \vec{r}_i \cdot \vec{r}_j \rangle = \sum_{i=1}^{n} \langle \vec{r}_i \cdot \vec{r}_i \rangle = na^2 \tag{5.29b}$$

which are again the results for the earlier consideration of the random walk problem obtained in Eq. (5.7). Consequently, the earlier result is again obtained in our new format of the problem.

The generalized formulation developed in Eqs. (5.26) and (5.27) is now shown to be helpful in developing more restrictive models than that involving the completely random walk considered earlier. We shall next briefly make a consideration of one of these types of restrictive models.

5.3.1 Chains with Fixed Bond Angles

Another popular model of linear chain polymers is one in which consecutive bonds between nearest-neighbor polymers are set at a fixed bond angle $\theta_{i,i+1} = \theta$, i.e., $\vec{r}_i \cdot \vec{r}_{i+1} = a^2 \cos\theta$ between neighboring segments along the entire polymer chain [1–4]. (See Figure 5.3a for a schematic.) In spite of the fixed angle of the chemical bonds, however, the monomers described by $\vec{r}_i$ and $\vec{r}_{i+1}$ continue to exhibit flexibility. Specifically, they can rotate 360° about each other taken as an axis of rotation, e.g., $\vec{r}_{i+1}$ rotates freely through 360° about the axis defined by $\vec{r}_i$ while maintaining the dot product $\vec{r}_i \cdot \vec{r}_{i+1} = a^2 \cos\theta$. The statistical properties of this restricted model can be generated by similar methods used to study the random walk in Eqs. (5.24) through (5.29). In the present case, however, instead of a total randomness in the trajectories of the walks the walks are restricted on successive steps to bond angles θ along the lengths of nearest-neighbor monomers.

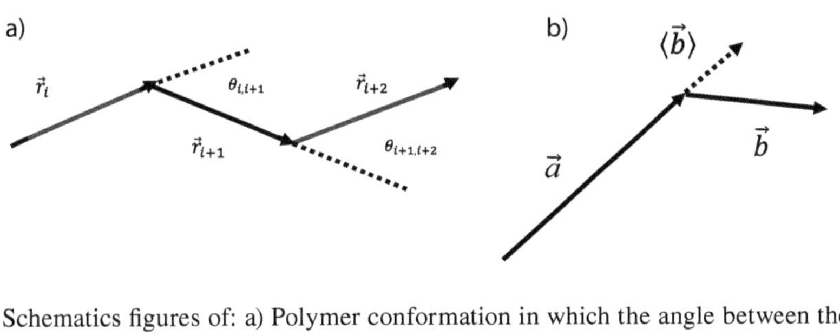

FIGURE 5.3 Schematics figures of: a) Polymer conformation in which the angle between the vector $\vec{r}_i$ and $\vec{r}_{i+1}$ for i running over the chain is $\theta_{i,i+1} = \theta$, and the length of each $\vec{r}_i$ is the same over the chain. Otherwise, neighboring segments of the polymer are free to rotate in space about the dotted axes. b) The vectors $\vec{a}$ and $\vec{b}$ and the projection of $\vec{b}$ on the axis of $\vec{a}$ denoted as $\langle \vec{b} \rangle$.

To understand the generation of the polymer conformations in this model, we begin by looking at the geometry of two random vectors which are attached to each other head to tail. Specifically, consider the two vectors $\vec{a}$ and $\vec{b}$ shown attached head to tail in Figure 5.3b. Assume now that these vectors satisfy the relationship [1–4]

$$\vec{a}\cdot\vec{b} = \mathrm{ab}\cos\theta \tag{5.30}$$

so that if we keep $\vec{a}$ fixed in space and rotate $\vec{b}$ by 360° about the axis of rotation defined by $\vec{a}$, we find upon averaging over the rotations of $\vec{b}$ that

$$\langle \vec{b} \rangle = \left(\frac{b}{a}\cos\theta\right)\vec{a}. \tag{5.31}$$

Here $\langle\ \rangle$ indicates the average of $\vec{b}$ as it is rotated by 360° about the axis of rotation defined by $\vec{a}$. In the following discussion, we shall find this general relationship helpful in discussing vector averages of the form

$$\langle \vec{r}_i\cdot\vec{r}_j \rangle \tag{5.32}$$

where $\langle\ \rangle$ indicates an average over the 360° rotation of $\vec{r}_{l+1}$ about the axis defined by $\vec{r}_l$ for $i \le l < j$.

Consider next a polymer chain with a conformation defined by the consecutive monomer displacement vectors $\{\vec{r}_i\}$ where $i = 1,2,3,\ldots,n$; such that $|\vec{r}_i| = a$ for each i; and where $\vec{r}_i\cdot\vec{r}_{i+1} = a^2\cos\theta$ for $i = 1,2,3,\ldots,n-1$. (See Figure 5.3a for a schematic.) Using Eq. (5.31) to compute $\langle \vec{r}_i\cdot\vec{r}_{i+1}\rangle$ for $1 \le i \le n-1$, we find by a straightforward application that [1–4]

$$\langle \vec{r}_i\cdot\vec{r}_{i+1} \rangle = \cos\theta\langle \vec{r}_i\cdot\vec{r}_i \rangle = \cos\theta a^2. \tag{5.33}$$

Similarly, if we consider $\langle \vec{r}_i\cdot\vec{r}_{i+2}\rangle$ for $1 \le i \le n-2$, we obtain

$$\langle \vec{r}_i\cdot\vec{r}_{i+2} \rangle = \cos\theta\langle \vec{r}_i\cdot\vec{r}_{i+1} \rangle = (\cos\theta)^2\langle \vec{r}_i\cdot\vec{r}_i \rangle = (\cos\theta)^2 a^2. \tag{5.34}$$

Notice that in generating Eq. (5.34), the result in Eq. (5.31) is applied between $\vec{r}_{i+2}$ and $\vec{r}_{i+1}$ in the first equality and then between $\vec{r}_{i+1}$ and $\vec{r}_i$ in the second equality. In this manner, the iterative procedure, applied successively along the chain, can similarly be generalized to show that

$$\langle \vec{r}_i\cdot\vec{r}_j \rangle = (\cos\theta)^{|j-i|} a^2 \tag{5.35}$$

for i and j contained within the set $\{1,2,3,\ldots,n\}$. In this way, the set of averages needed to evaluate Eqs. (5.26) and (5.27) is generated.

Specifically, from Eqs. (5.26) and (5.27) we find that [1]

$$\langle \vec{R} \rangle = \left\langle \sum_{i=1}^{n}\vec{r}_i \right\rangle = 0 \tag{5.36a}$$

and [1]

$$\langle \vec{R}\cdot\vec{R} \rangle = \sum_{i=1}^{n}\sum_{j=1}^{n}\langle \vec{r}_i\cdot\vec{r}_j \rangle = a^2\sum_{i=1}^{n}\sum_{j=1}^{n}(\cos\theta)^{|j-i|} \tag{5.36b}$$

$$= na^2 + 2a^2 \sum_{i=1}^{n-1} \sum_{j=i+1}^{n} (\cos\theta)^{j-i}$$

where in the last equality we have used the fact that $(\cos\theta)^{|j-i|} = (\cos\theta)^{|i-j|}$. If we next write $k = j - i$ and sum over the diagonals of the 2-dimensional (i, j) space representing fixed k we find [1]

$$\langle \vec{R} \cdot \vec{R} \rangle = na^2 + 2a^2 \sum_{i=1}^{n-1} \sum_{j=i+1}^{n} (\cos\theta)^{j-i} = na^2 \left[1 + \frac{2}{n} \sum_{k=1}^{n-1} (n-k)(\cos\theta)^k \right] \tag{5.37a}$$

$$= na^2 \left[1 + 2\sum_{k=1}^{n-1} (\cos\theta)^k - \frac{2}{n} \sum_{k=1}^{n-1} k(\cos\theta)^k \right]$$

$$= na^2 \left\{ 1 + 2\frac{(\cos\theta - \cos\theta^n)}{1-\cos\theta} - \frac{2}{n}\left[\frac{\cos\theta(1-\cos\theta)^n}{(1-\cos\theta)^2} - n\frac{(\cos\theta)^n}{1-\cos\theta} \right] \right\}$$

$$= na^2 \left\{ 1 + 2\frac{\cos\theta}{1-\cos\theta} - \frac{2}{n} \frac{\cos\theta(1-\cos\theta)^n}{(1-\cos\theta)^2} \right\}.$$

Here in the fourth equality, we have used standard power series identities for summing the finite series. Notice that a simplification of Eq. (5.37a) occurs in the limit that $n \gg 1$. In particular, for this limit it is found that [1]

$$\langle \vec{R} \cdot \vec{R} \rangle \approx na^2 \left\{ 1 + 2\frac{\cos\theta}{1-\cos\theta} \right\} = na^2 \left\{ \frac{1+\cos\theta}{1-\cos\theta} \right\}. \tag{5.37b}$$

Notice that for $\cos\theta = 0$ the result in Eq. (5.37b) reduces to our result $\langle \vec{R} \cdot \vec{R} \rangle = na^2$ for the random walk considered earlier. On the other hand, for $\cos\theta > 0$ we find from Eq. (5.37b) that $\langle \vec{R} \cdot \vec{R} \rangle > na^2$ while for $\cos\theta < 0$ we find that $\langle \vec{R} \cdot \vec{R} \rangle < na^2$.

Other limitations can be imposed to clean up the result of the totally random walk model. These will not be considered here, but the reader is referred to the literature for further discussions of these [1–4]. We next end our considerations with a discussion of the relationship of our problem of polymer conformations to that of the important problem of Brownian motion.

5.4 BROWNIAN MOTION

A model related to our treatment of the random walk polymer conformations is that applied to the study of Brownian motion [1–5]. Brownian motion is observed in fluids as the diffusive process involved in the motion of atoms of the fluid. It is characterized by apparently random space-time trajectories of the atoms generated from a set of disorienting collisions. For example, the study of the statistical properties of the motion of an impurity atom dissolved within a fluid appears roughly as a set of random spatial displacements. In this respect, a first treatment of this motion can regard the impurity atom as exhibiting a series of randomly generated spatial steps characterized by a few basic statistical constraints. Whereas polymer conformations involve connected monomer segments studied as a function of their sequence of occurrence in the chain of a polymer, Brownian motion involves a connected sequence of space displacements studied as a function of time steps in their dynamics.

To begin consider the Brownian motion of an atom [1–4] as evolving in a sequence of nearest-neighbor spatial steps of time interval Δt on a cubic space lattice composed of nearest-neighbor separations of length a. In this representation during the time Δt a diffusing atom can move a single step of length a along the $x-$, $y-$, or $z-$axis or it can stay on the same lattice site it occupied at the beginning of the time step. This last option of remaining on the same lattice site is a difference from the random walks in the polymer conformation problem of our earlier considerations. It leads to small variations in the statistical properties between the two models after accounting for the difference in space evolution in time or of evolution in segmental sequencing [1–4].

If for the Brownian motion problem we define [1–5]

$$P(x,y,z,t) \tag{5.38}$$

as the probability at time t of a particular diffusing particle being located at (x,y,z), the process of its development as a random walk is characterized by the relationship

$$\begin{aligned} P(x,y,z,t+\Delta t) = {} & s\left[P(x+a,y,z,t)+P(x-a,y,z,t)\right] \\ & +s\left[P(x,y+a,z,t)+P(x,y-a,z,t)\right] \\ & +s\left[P(x,y,z+a,t)+P(x,y,z-a,t)\right] \\ & +(1-6s)P(x,y,z,t) \end{aligned} \tag{5.39}$$

where s is the probability of making a step between nearest neighbor sites at time t onto the site (x,y,z) at time $t+\Delta t$, and the last term $(1-6s)P(x,y,z,t)$ represents the case in which the particle at the beginning and end of the Δt step continues to be located at (x,y,z). (Note: in our earlier considerations in Eqs. (5.1) through (5.4) we have considered the particular case in which $s=\frac{1}{6}$ and for this case $(1-6s)P(x,y,z,t)=0$. This is characteristic of a sequential array of polymer segments.) Assuming that $P(x,y,z,t)$ is slowly varying over the cubic lattice, we Taylor expand Eq. (5.39) in Δt and a to find that [1–4]

$$\Delta t\frac{\partial}{\partial t}P(x,y,z,t) = sa^2\nabla^2P(x,y,z,t) \tag{5.40}$$

or

$$\frac{\partial}{\partial t}P(x,y,z,t) = D\nabla^2P(x,y,z,t) \tag{5.41a}$$

where the diffusion constant is given by

$$D=\frac{sa^2}{\Delta t}. \tag{5.41b}$$

The result in Eq. (5.41) is the standard form of a space-time diffusion equation.

For a diffusing particle beginning its motion at the origin of the space coordinates, a solution of particle probability distributions from Eq. (5.41) is

$$P(x,y,z,t)=\left[\frac{1}{4\pi Dt}\right]^{3/2} e^{-\frac{1}{4Dt}\left(x^2+y^2+z^2\right)}. \tag{5.42}$$

The result in Eq. (5.42) exhibits a variety of statistical properties of the diffusive trajectories of the particle. In this way, for example, we find that

$$\int_{-\infty}^{\infty}\int_{-\infty}^{\infty}\int_{-\infty}^{\infty} dxdydz\left(x\hat{i} + y\hat{j} + z\hat{k}\right)P(x,y,z,t) = 0 \tag{5.43a}$$

so that the diffusive process is symmetric about the origin. The mean square displacement between the atom injection at the origin of the system at $t = 0$ and its location in space at time t is

$$\langle r^2 \rangle = \int_{-\infty}^{\infty}\int_{-\infty}^{\infty}\int_{-\infty}^{\infty} dxdydz\left(x^2 + y^2 + z^2\right)P(x,y,z,t) = \frac{D}{s}t = a^2 n \tag{5.43b}$$

where $n = \frac{t}{\Delta t}$ are the number of time steps in the time interval from $t = 0$ to time t. We see that if each time step on the trajectory is sequentially considered, Eq. (5.43b) retrieves the earlier random conformation result discussed in the context of polymer conformations.

In addition, from Eq. (5.43b) we obtain the result [1–4]

$$\left\langle r^2 \right\rangle = \frac{D}{s}t \tag{5.44a}$$

which for $s = \frac{1}{6}$ reduces to

$$\left\langle r^2 \right\rangle = 6Dt. \tag{5.44b}$$

This characterizes, in a fundamental way, the time development of the diffusive process.

5.5 RESIDUAL ENTROPY

An interesting problem of random disorder involving entropy considerations is that of residual entropy [6–10]. This comes from the observation that some disordered systems exhibit a nonzero entropy as they approach absolute zero of temperature. Early consideration of this problem was made by Pauling [6, 7] in the context of the study of the residual entropy of ice. In ice, it was noticed that the third law of thermodynamics was apparently not satisfied. Specifically, as absolute zero of temperature was approached, the entropy of ice did not go to zero but rather exhibited a so-called residual (i.e., nonzero) entropy. This behavior would appear to violate the third law of thermodynamics, which requires that in systems at thermodynamic equilibrium, the entropy goes to zero at the temperature $T = 0^\circ K$.

Pauling [6, 7] showed that the residual entropy in ice resulted from the existence of a large number of metastable degenerate energy states coming from the disorder associated with the nature of the bonds between the water molecules [6–10]. In this regard, at low-temperature ice forms into degenerate bonding configurations characterized as metastable states, and the associated entropy of these various degenerate energy configurations is reminiscent of the conformational entropy of the degenerate polymer conformations treated in the earlier considerations of rubber elasticity.

In ice, the system solidifies into a crystalline lattice structure [6, 7]. However, disorder enters the system from randomness associated with the nature of the bonds between the hydrogen and oxygen atoms composing the structure. Ice is a molecular solid in which water molecules, formed of hydrogen atoms strongly bonded to an oxygen atom, are arrayed in the crystal by means of weak hydrogen bonds, associated with the dipolar nature of the water molecules. The hydrogen bonds formed between molecules are essentially weak electrostatic couplings between the hydrogen ion of a water molecule and an oxygen ion in a neighboring water molecule. Consequently, there are two types of bonding of hydrogen in ice, one weak and one strong, and the degree of randomness

$$O = H - O \text{ or } O - H = O$$

FIGURE 5.4 The two possible bonding configurations of H and its two nearest neighboring O. Here = represents the chemical bond of the water molecule and – represents a hydrogen bond between water molecules.

in the occurrence of these two types of strong and weak bonds in the ice structure accounts for the disorder introduced into the system [6, 7]. This is the origin of the residual entropy [6–10].

In the solid form of ice, the oxygen atoms are positioned on a crystal lattice and form four bonds with nearest neighbor oxygen atoms on the lattice mediated by hydrogen atoms. If an oxygen atom in the lattice is considered, it is found to form two strong chemical bonds with hydrogen, arising from the original bond to hydrogen in the water molecule, and two weak hydrogen bonds from the dipole nature of the neighboring water molecules. Considering then the bond between two nearest neighbor oxygen atoms on the lattice as mediated by a hydrogen atom, one of the oxygen atoms will be strongly bonded to the hydrogen atom and the other atom will be weakly bonded to it by a hydrogen bond. (See the schematic in Figure 5.4.) In this way, there are two different types of bonds that can form between the nearest neighbor oxygen atoms, depending on which oxygen is strongly bonded to the hydrogen and which oxygen is weakly attached to the hydrogen atom. The entropy of the system arises from the disorder of the bonds in the solid ice in respect to the position of the strong and weak hydrogen bonds. This uncertainty allows for a large number of degenerate energy configurations in the system [6–10].

Pauling offered a rough computation of the entropy of the disordered system using the following reasoning. Consider a system of ice composed of N oxygen atoms and begin with a focus on the bonds between nearest neighbor oxygen atoms. These bonds are mediated by hydrogen atoms and associated with each of the N oxygen atoms there are 4 such bonds. However, each bond is shared by 2 oxygen atoms so there are $2N$ nearest neighbor oxygen bonds in the system of N oxygen atoms. This means that, since there are two possible types of bonds between nearest neighbor oxygen atoms (See Figure 5.4.), if these bonds occurred with equal probability and completely at random in the system there are [6, 7]

$$2^{2N} \tag{5.45}$$

possible bond configurations associated with the bonds in the system. This counting, however, is made in the absence of restrictions on how many strong or weak bonds can exist at a given oxygen atom in the lattice. This restriction must now be reintroduced to correct for the overcounting represented by Eq. (5.45).

Under the restrictions involving the nearest neighbor oxygen bonds mediated by hydrogen, each oxygen atom in the system must form two strong and two weak bonds with hydrogen. As a result, the count in Eq. (5.45) must be an overcount. This is a consequence of the fact that the configurations in Eq. (5.45) do not distinguish between the positioning of weak or strong bonds at each oxygen but consider all bonds as randomly occurring [6–10].

To factor for the overcounting, consider one of the oxygens on the lattice and its 4 associated bonds. Two of these must involve strong bonds to hydrogen and two must involve weak bonds to hydrogen atoms. Consequently, there are 6 possible combinations of the 4 bonds associated with the single oxygen having this pattern of bonding at the oxygen. In the absence of bonding preference, however, there are 16 ways of occupying the 4 bonds by the 2 different bond types. Consequently, we can scale the result in Eq. (5.45) by $\frac{6}{16}$ for each oxygen atom in the system in order to correct for the overcount. This scaling gives the approximate counting [6, 7]

$$\left(\frac{6}{16}\right)^N 2^{2N} = \left(\frac{3}{2}\right)^N \tag{5.46}$$

for the possible different configurations of the system. The result in Eq. (5.46) is approximately consistent with the bond restrictions associated with an oxygen atom in the system.

The entropy of the system of ice bond configurations is then given by [6, 7]

$$S = -k_B \sum_{i=1}^{\left(\frac{3}{2}\right)^N} p_i \ln p_i = -k_B \sum_{i=1}^{\left(\frac{3}{2}\right)^N} \frac{1}{\left(\frac{3}{2}\right)^N} \ln \frac{1}{\left(\frac{3}{2}\right)^N} = k_B \ln \left(\frac{3}{2}\right)^N \tag{5.47}$$

$$= Nk_B \ln \frac{3}{2} = 0.405 Nk_B = n\left(0.806\, cal\, mol^{-1} K^{-1}\right)$$

where $p_i = \frac{1}{\left(\frac{3}{2}\right)^N}$ is the probability of the ith configuration of bonds being found in the system, and n is the number of moles in the material. This compares to the experimental molar value of $S = 0.85\, cal\, mol^{-1} K^{-1}$.

The residual entropy problem has been treated in a number of other systems; CO being another commonly treated system and glass another. It has also generated the topic in statistical physics dealing with a variety of statistical models known as ice and vertex models [6, 7].

5.6 THE GLASS TRANSITION

One of the ways of making an amorphous material is by having the liquid or elastic state of the material undergo a glass transition [3–5]. This is a transition in which a material, initially in a liquid or easily deformable elastic state, is rapidly cooled to a transitional state exhibiting many of the properties of a solid material except that it is not in a true equilibrium state of the system. Such a type of solid is known as a glass. However, the glass state is not an absolute equilibrium state and does not display the long-range atomic and molecular correlations of a crystalline solid. Rather the glass transition is made to a disordered state of short-range order in which the rate of transition of the material has become so small that an equilibrium configuration of the system is never realized within a practical time scale. In this sense, the changes occurring in the glass state are made slowly enough that the material exists in an apparently permanent metastable state rather than reaching a true permanently stable absolute equilibrium configuration [3–5].

Associated with the glass transition is a glass transition temperature T_g which, depending on the nature of the system, may or may not be simply defined [3–5]. Specifically, the glass transition can be thought of as occurring over a range of temperatures, characteristically occurring in a neighborhood of temperatures surrounding the glass transition temperature T_g. As an example, T_g in some cases it can be arbitrarily defined as the temperature at which the system exhibits a certain level of viscosity or value of another property of the material. In other cases, the glass transition temperature may be defined relative to an inflection point in the plot of a specific material property versus temperature. Above the glass temperature the material exhibits behavior typified as that of a fluid or rubber elastic medium, with short characteristic relaxation times. Below the glass temperature, on the other hand, the material exhibits behaviors characteristic of solid materials, but typically with characteristically long relaxations times [3–5].

In this regard, it is seen that the glass transition is not a phase transition as it does not represent a transition between two different equilibrium states of a material, with a transition point uniquely defined between the two phases. Nevertheless, for applications, the glass state can be regarded as a metastable state of the solid materials. As a transition to a metastable state of the system, the glass transition temperature T_g then generally occurs at a temperature less than the temperature T_c at which the system undergoes a phase transition to a true equilibrium solid [3–5]. The glass transition is made rapidly enough, however, that the atoms of the material do not have enough time to finish completely rearranging themselves into a long-range ordered solid phase [3–5].

5.7 DENSITY OF ENERGY STATES: RECURSIVE METHODS

In Chapter 3, discussions were given for the determination of the density of states of the excitations in crystalline materials [11, 12]. It was shown there that the density of states was directly related to the Green's functions of the modal excitations which, due to the translational symmetry of the crystal lattice, had particularly simple forms. Specifically, the density of states was found to be obtained from the imaginary part of the Green's function expressed in terms of the Fourier transform of an integrand composed of simple complex frequency poles. In this context, the integral involved an evaluation in the complex plane, with the problem reduced to the determination of integrals using the calculus of residues. To provide a simple example, the particular model treated during the discussions was that of a basic tight-binding Hamiltonian composed of atomic lattice sites characterized by an on-site excitation energy and a hopping interaction between nearest-neighbor sites which uniquely characterized the jump of an excitation between two lattice sites.

In amorphous materials, on the other hand, the problem of studying the density of states is more complex than that encountered in the study of crystalline materials [1–16]. There is no translational symmetry in the amorphous network, and because of this, the problem of calculating the Green's functions and their density of states is of a less systematic nature. In this respect, the material is fundamentally disordered so that there is no simplification in approaching its study based on the general analytic forms of the wavefunctions of the eigenmodes and their symmetry properties [11, 12]. With the absence of translational symmetry as a fundamental aspect of the modal solutions of the material, another approach needs to be applied to determine the Green's functions and density of states of the amorphous system. The focus of the new approach introduced here is on the local behavior of the medium rather than on its global translational symmetry properties. In this regard, it is found that the local properties of a non-crystalline medium are more important in determining the characteristics of the density of states of the amorphous material than is the nature of its long-range order [11, 12].

The new approach we develop is based on a recursive method [11, 12]. In this approach, we start with a consideration of the wavefunction of the system at a site of the amorphous medium and apply dynamical considerations to generate new basis states of the system, taking into account the system properties at increasing ranges from that of the initial site wavefunction. This is done by developing a recursive equation based on the original Hamiltonian problem of the system and characterized by a new set of recursive coefficients and a new set of basis states.

In this way, a Green's function of the original Hamiltonian is ultimately generated in terms of the coefficients of the recursive equations and the new basis set [11, 12]. It appears in the form of an infinite continued fraction. As in the case of our study of the periodic system in Chapter 3, the density of states is eventually obtained in terms of the Green's function so that the expression for the density of states is also represented as an infinite continued fraction. The evaluation of the continued fraction representation to obtain accurate numerical results for the amorphous system is then our final consideration [11, 12].

In the following, this new recursive method will again be developed for a simple tight-binding Hamiltonian H composed of atomic network sites characterized by a single on-site excitation energy and a hopping interaction between close-neighbor sites which uniquely characterizes the jumps between two network sites. While this crude "toy" model glosses over many of the details involved in the quantum chemistry describing an actual amorphous material, the model provides a simple system illustrating the basic methodology at work in the application of a recursive treatment for the density of states of an amorphous material [11, 12]. For a more detailed consideration of the recursive techniques applied to actual materials, the reader is referred to the literature [11, 12].

We begin our treatment of H by recursively rewriting its eigenvalue problem into the form of a tridiagonal matrix of recursive coefficients. This development is initially focused on one particular site of the amorphous network of the tight-binding model and is ultimately applied to the determination of local density of states at the chosen site.

5.7.1 Development of Recursive Difference Equations from the Model Hamiltonian

Consider that the Hamiltonian of an amorphous medium is given by H which, for example, is a linear tight-binding model describing a system of noninteracting single particle excitations, i.e., there are no particle-particle interactions [11, 12]. The system is a collection of atoms which are bonded together, and the particles of the system move from atom to atom. In this regard, the system could be composed of a periodic lattice as in Chapter 3, an amorphous network as in our present case, or a finite cluster of atomic sites. The present method is, consequently, a generalization made to include less ordered media than studied earlier, and, as well, it could be applied to a periodic system. Our objective will be to calculate the energy density of states at each site of the network.

For simplicity let us begin by considering an amorphous network of identical atoms which are bound together in a dynamical system [11, 12]. In the network we focus on a single atom which we denote as 0 and which is characterized by a basis state denoted by $|0\rangle$ which is localized at the atomic site. The considerations will focus on evaluating the density of states at the 0 atomic site arising from the modal eigenvalues in the system as determined by the network geometry and the Hamiltonian defined upon it.

Begin by computing $H|0\rangle$ where we assume that the site state $|0\rangle$ is not an eigenmode of the Hamiltonian. (Notice, $|0\rangle$ may, for example, be a mode of the isolated 0 atom when the other atoms of the network are removed.) As a consequence, it follows that [11, 12]

$$H|0\rangle = \alpha_0 |0\rangle + \beta_1 |1\rangle \tag{5.48}$$

where the separation on the right has been made so that $\langle 0|1\rangle = \langle 1|0\rangle = 0$ and $\langle 0|0\rangle = 1|\langle 1|1\rangle$. Notice here that the term $\beta_1 |1\rangle$ arises because $|0\rangle$ is not an eigenmode of the Hamiltonian so that $\beta_1 |1\rangle$ is representative of the degree to which $|0\rangle$ is not an eigenstate. In this regard, while $|0\rangle$ is associated with the atomic site 0, $|1\rangle$ need not be associated with a particular atom of the network but may be some linear combination of wavefunctions of the atomic neighbors of the atom 0. In this regard, the separation in Eq. (5.48) can be made using the techniques of Gram-Schmidt orthogonalization applied to $H|0\rangle$.

Consequently, it follows from Eq. (5.48) and the orthonormality of the set $\{|0\rangle, |1\rangle\}$ that [11, 12]

$$\langle 0|H|0\rangle = \alpha_0 \tag{5.49a}$$

and

$$\langle 1|H|0\rangle = \beta_1. \tag{5.49b}$$

It then follows from Eq. (5.48) and the results in Eq. (5.49) that

$$|1\rangle = \frac{1}{\beta_1}(H - \alpha_0)|0\rangle. \tag{5.49c}$$

Consequently, the coefficients α_0 and β_1 are given in Eqs. (5.49a) and (5.49b), and these allow us to determine $|1\rangle$ as given in Eq. (5.49c).

Next consider the result of the application of H to $|1\rangle$. From this it then follows that [11, 12]

$$H|1\rangle = \alpha_1 |1\rangle + \beta_2 |2\rangle + \beta_1 |0\rangle \tag{5.50}$$

where $|2\rangle$ is separated out so that the set $\{|0\rangle, |1\rangle, |2\rangle\}$ are mutually orthonormal. Notice again that the term $\beta_2 |2\rangle$ arises because $|1\rangle$ is not an eigenmode of the Hamiltonian so that $\beta_2 |2\rangle$ and $\beta_1 |0\rangle$ are representative of the degree to which $|1\rangle$ is not an eigenstate. Consequently, applying the orthonormality of the members of the set $\{|0\rangle, |1\rangle, |2\rangle\}$ we find

$$\langle 1 | H | 1 \rangle = \alpha_1, \tag{5.51a}$$

$$\langle 0 | H | 1 \rangle = \beta_1 = \langle 1 | H | 0 \rangle, \tag{5.51b}$$

and

$$\langle 2 | H | 1 \rangle = \beta_2. \tag{5.51c}$$

From Eq. (5.50) it then follows that

$$|2\rangle = \frac{(H - \alpha_1)|1\rangle - \beta_1 |0\rangle}{\beta_2}, \tag{5.52}$$

and it is seen that the results in Eqs. (5.51a), (5.51c), and (5.52) advance us into the space of our approximation by defining α_1, β_2, and $|2\rangle$.

We next generalize our consideration to the treatment of the $(n+1)^{th}$ generated state $|n+1\rangle$ for $n \geq 1$. This follows from the application of H to the state $|n\rangle$. The result of the application of H to $|n\rangle$ then gives [11, 12]

$$H|n\rangle = \alpha_n |n\rangle + \beta_{n+1} |n+1\rangle + \beta_n |n-1\rangle \tag{5.53}$$

where $|n\rangle$ is separated out so that the set $\{|n-1\rangle, |n\rangle, |n+1\rangle\}$ are mutually orthonormal. Applying the orthonormality of $\{|n-1\rangle, |n\rangle, |n+1\rangle\}$ we find

$$\langle n | H | n \rangle = \alpha_n, \tag{5.54a}$$

$$\langle n-1 | H | n \rangle = \beta_n = \langle n | H | n-1 \rangle, \tag{5.54b}$$

and

$$\langle n+1 | H | n \rangle = \beta_{n+1}. \tag{5.54c}$$

From Eq. (5.53) it then follows that

$$|n+1\rangle = \frac{(H - \alpha_n)|n\rangle - \beta_n |n-1\rangle}{\beta_{n+1}}, \tag{5.55}$$

and it is seen that the results in Eqs. (5.54a), (5.54c), and (5.55) advance us into the space of factors controlling our approximation by defining α_n, β_{n+1}, and $|n+1\rangle$. In the above discussions it is, in general, found that for Hermitian H, the basis set $\{|p\rangle\}$ for $p = 0,1,2,\ldots,n+1$, the set of coefficients $\{\alpha_i\}$ for $i = 0,1,\ldots,n$ and the set of coefficients $\{\beta_i\}$ for $i = 0,1,\ldots,n+1$ can be all chosen to be real. In each case, the new basis state $|n+1\rangle$ is obtained from $H|n\rangle$ through a process of Gram-Schmidt orthogonalization.

As a final point in the development of the new orthonormal basis set $\{|p\rangle\}$ for $p = 0,1,2,\ldots,n+1$, we show that all of the generated basis states are in fact orthonormal to one another [11, 12]. To this end, first note that the states $\{|0\rangle,|1\rangle,|2\rangle\}$ generated in Eqs. (5.48) through (5.52) are all mutually orthonormal by their construction, and, similarly, the states $\{|n-1\rangle,|n\rangle,|n+1\rangle\}$ generated by Eqs. (5.53) through (5.55) are mutually orthonormal for the same reason. We need to assure ourselves, however, that the states between and including these sets are also mutually orthonormal.

Let us start by assuming that the set of generated states $\{|q\rangle\}$ for $q = 0,1,2,\ldots,n$ are mutually orthogonal and that we have just generated $|n+1\rangle$ by the process described in Eqs. (5.53) through (5.55). The object next is to show that the generated $|n+1\rangle$ states are in fact orthogonal to a general state of the set $\{|m\rangle\}$ for $m = 0,1,2,\ldots,n-2$. In this regard, we already know by construction in Eqs. (5.54) and (5.55) that $|n+1\rangle$ is orthonormal to $|n\rangle$ and $|n-1\rangle$.

To continue with these considerations, from Eq. (5.55) consider the form [11, 12]

$$\beta_{n+1}|n+1\rangle = (H-\alpha_n)|n\rangle - \beta_n|n-1\rangle, \tag{5.56}$$

and for $m = 0,1,2,\ldots,n-2$ the form

$$\beta_{n+1}\langle m \mid n+1\rangle = \langle m \mid H-\alpha_n \mid n\rangle - \beta_n\langle m \mid n-1\rangle \tag{5.57}$$

$$= \langle m \mid H \mid n\rangle$$

where we have used the orthogonality properties $\langle m \mid n\rangle = 0$ and $\langle m \mid n-1\rangle = 0$ on the far right-hand side of Eq. (5.57). From Eq. (5.56) evaluated for $n \leftrightarrow m$ it then follows that

$$\beta_{m+1}|m+1\rangle = (H-\alpha_m)|m\rangle - \beta_m|m-1\rangle, \tag{5.58}$$

and, consequently, from this that

$$H|m\rangle = \alpha_m|m\rangle + \beta_m|m-1\rangle + \beta_{m+1}|m+1\rangle \tag{5.59a}$$

or in an alternative adjoint form

$$\langle m|H = \alpha_m\langle m| + \beta_m\langle m-1| + \beta_{m+1}\langle m+1|. \tag{5.59b}$$

As a result of this, for $m = 0,1,2,\ldots,n-2$ we find from the orthonormality relations that [11, 12]

$$\langle m \mid H \mid n\rangle = \left[\alpha_m\langle m| + \beta_m\langle m-1| + \beta_{m+1}\langle m+1|\right]|n\rangle \tag{5.60}$$

$$= \alpha_m\langle m \mid n\rangle + \beta_m\langle m-1 \mid n\rangle + \beta_{m+1}\langle m+1 \mid n\rangle = 0$$

and using this in Eq. (5.57) it follows that

$$\langle m \mid n+1 \rangle = 0, \tag{5.61}$$

i.e., $|n+1\rangle$ is orthonormal to the previously generated states of the system.

Having generated an orthonormal basis including $|n+1\rangle$, we turn to the development of the Green's functions describing the nature of the modal excitations in the system. These Green's functions are naturally developed in the format of a continued fraction representation.

5.7.2 Continued Fraction Representation of the Green's Function

Let us begin the study of the Green's function of the Hamiltonian H by considering the associated eigenvalue problem expressed in terms of the new basis we have just generated. This problem is written as [11, 12]

$$H\sum_{n=0}^{\infty} P_n |n\rangle = E\sum_{n=0}^{\infty} P_n |n\rangle \tag{5.62}$$

where the set $\{|n\rangle\}$ for $n = 0,1,\ldots$ is our orthonormal basis, the Hamiltonian can be formally written as $H = \sum_{m=0}^{\infty}\sum_{l=0}^{\infty} |m\rangle\langle m \mid H \mid l\rangle\langle l|$, and $\{P_n\}$ and E are to be determined. In terms of the difference equation results in Eq. (5.56) the eigenvalue equation in Eq. (5.62) takes the form of a symmetric tridiagonal matrix. This is expressed by [11, 12]

$$\begin{vmatrix} E-\alpha_0 & -\beta_1 & 0 & 0 & 0 & \cdots \\ -\beta_1 & E-\alpha_1 & -\beta_2 & 0 & 0 & \cdots \\ 0 & -\beta_2 & E-\alpha_2 & -\beta_3 & 0 & \cdots \\ 0 & 0 & -\beta_2 & E-\alpha_3 & -\beta_4 & \cdots \\ \vdots & \vdots & \vdots & \vdots & \vdots & \vdots \end{vmatrix} \begin{vmatrix} P_0 \\ P_1 \\ P_2 \\ P_3 \\ \vdots \end{vmatrix} = 0. \tag{5.63}$$

The density of states of the amorphous system at the 0 atomic site, then, follows directly from the solutions of Eq. (5.63) or from a study of the Green's functions of the tridiagonal matrix based on Eq. (5.63). We shall continue our study using the method of Green's functions so that, consequently, the treatment at this stage directly parallels that from our earlier study in Chapter 3 of the density of states of periodic tight-binding models defined on periodic lattices.

Specifically, our focus is on computing the density of states at the atomic site 0 by studying the standard Green's function problem [11, 12]

$$[E-H]G(E)=1. \tag{5.64}$$

Here 1 is the unit diagonal matrix and $G(E)$ is of the form

$$\begin{vmatrix} G_{0,0}(E) & G_{0,1}(E) & G_{0,2}(E) & G_{0,3}(E) & G_{0,4}(E) & \cdots \\ G_{1,0}(E) & G_{1,1}(E) & G_{1,2}(E) & G_{1,3}(E) & G_{1,4}(E) & \cdots \\ G_{2,0}(E) & G_{2,1}(E) & G_{2,2}(E) & G_{2,3}(E) & G_{2,4}(E) & \cdots \\ G_{3,0}(E) & G_{3,1}(E) & G_{3,2}(E) & G_{3,3}(E) & G_{3,4}(E) & \cdots \\ \vdots & \vdots & \vdots & \vdots & \vdots & \vdots \end{vmatrix}. \tag{5.65}$$

The density of states that we compute is then obtained from $G_{0,0}(E)$, and we consider this to generate the site density of states at the 0 network site used in defining the $|0\rangle$ basis element. (Note, from our earlier construction, $|0\rangle$ is defined as a localized atomic state about one of the network atomic sites. It is often natural to take this as that site's atomic site wavefunction.)

The evaluation of the Green's function problem in Eq. (5.64) can be further simplified if we study the components of the Green's function matrix of the form $G_{n,0}(E)$ for $n = 0,1,2,\ldots$. The problem then reduces to that of computing [11, 12]

$$\begin{vmatrix} E-\alpha_0 & -\beta_1 & 0 & 0 & 0 & \cdots \\ -\beta_1 & E-\alpha_1 & -\beta_2 & 0 & 0 & \cdots \\ 0 & -\beta_2 & E-\alpha_2 & -\beta_3 & 0 & \cdots \\ 0 & 0 & -\beta_2 & E-\alpha_3 & -\beta_4 & \cdots \\ \vdots & \vdots & \vdots & \vdots & \vdots & \vdots \end{vmatrix} G(E) = \begin{vmatrix} 1 \\ 0 \\ 0 \\ 0 \\ \vdots \end{vmatrix} \tag{5.66a}$$

where

$$G(E) = \begin{vmatrix} G_{0,0}(E) \\ G_{1,0}(E) \\ G_{2,0}(E) \\ G_{3,0}(E) \\ \vdots \end{vmatrix}. \tag{5.66b}$$

Note that here we have restricted our Green's problem to involve only the first column of the $n \times n \to \infty \times \infty$ Green's function matrix. Similarly, only the first column of the identity matrix is considered on the right side of Eq. (5.66a). For this system, the site density of states is next determined from the Green's function element $G_{0,0}(E)$ of the solution of Eq. (5.66).

If $D_n(E)$ is defined as the determinant of the matrix on the left side of Eq. (5.66a) with the first n rows and columns deleted, then $G(E)$ in Eq. (5.66b) is obtained by a standard linear algebra treatment of Eq. (5.66). Specifically, to determine $G_{0,0}(E)$ we find that [11, 12]

$$G_{0,0}(E) = \frac{D_1(E)}{D_0(E)} = \frac{D_1(E)}{(E-a_0)D_1(E) - b_1^2 D_2(e)} = \frac{1}{E - a_0 - b_1^2 \dfrac{D_2(E)}{D_1(E)}} \tag{5.67}$$

$$= \frac{1}{E - a_0 - b_1^2 \dfrac{1}{E - a_1 - b_2^2 \dfrac{D_3(E)}{D_2(E)}}} = \frac{1}{E - a_0 - b_1^2 \dfrac{1}{E - a_1 - b_2^2 \dfrac{1}{E - \alpha_2 - \ldots}}}.$$

This is seen to generate an infinite continued fraction for the site Green's function, which offers an exact formal solution to our problem.

Once Eq. (5.67) is obtained, then, following our earlier discussions in Chapter 3, the 0 site density of states $n_0(E)$ as a function of E is obtained from $G_{0,0}(E)$ through the standard relationship [11, 12]

$$n_0(E) = -\frac{1}{\pi} Im\left[G_{0,0}(E)\right]. \tag{5.68}$$

It is seen that since $G_{0,0}(E)$ is an infinite continued fraction, so is $n_0(E)$. While the continued fraction represents a formal solution of the density of states, it also presents problems in regard to its accurate numerical evaluation. Specifically, how is Eq. (5.68) numerically evaluated in the most accurate fashion.

In practice, in the evaluation of Eq. (5.68) a terminating approximation of some sort must be made to the infinite continued fraction in order to generate a numerical result. A number of methods have been used to accomplish this, depending on the specifics of the problem under consideration, and some mention of these will be made later. For now, however, we consider a simple example of obtaining a density of states from a continued fraction.

5.7.3 An Example

As an example of the formalism in Eqs. (5.62) through (5.68), consider the treatment of the simplistic system in which $\alpha_i = 0$ and $\beta_i = \beta$ for $i = 0,1,2,\ldots$. Applying these coefficients in Eq. (6.67), it then follows that the continued fraction assumes the form [11, 12]

$$G_{0,0}(E) = \frac{1}{E - b^2 G_{0,0}(E)}. \tag{5.69a}$$

This in turn is found to reduce directly to the exact result

$$G_{0,0}(E) = \frac{E \pm \sqrt{E^2 - 4b^2}}{2b^2} \tag{5.69b}$$

where the choice of sign in Eq. (5.69b) is determined by the analytic properties of $G_{0,0}(E)$. In this regard, $G_{0,0}(E)$ as a function of E is known to have an asymmetric real part and a symmetric imaginary part.

Considering the density of states $n_0(E) = -\frac{1}{\pi} Im\, G_{0,0}(E)$, it then is found from Eq. (5.69) that [11, 12]

$$n_0(E) = -\frac{1}{\pi} Im\, G_{0,0}(E) = \frac{1}{\pi b}\sqrt{1 - \left(\frac{E}{2b}\right)^2} \tag{5.70}$$

which is a nonzero, symmetric, function defined in the region $-2b < E < 2b$ with a maximum value of $n_0(E=0) = \frac{1}{\pi b}$. As a point of interest, notice that if instead of taking $\alpha_i = 0$ for $i = 0,1,2,\ldots.$ in Eq. (5.69a), we had taken $\alpha_i = \alpha$ for $i = 0,1,2,\ldots.$, then E in Eq. (5.70) would have been replaced by $E - \alpha$. This is just an energy shift of the density of states.

As a variant of this model, one can study a system in which α_i and β_i are arbitrarily specified for $i = 0,1,2,\ldots,m$, but $\alpha_i = \alpha$ and $\beta_i = \beta$ for $i = m+1, m+2, m+3,\ldots$. This case of the model is also exactly solvable. The solution, however, becomes an increasingly more difficult process for increasing $m > 0$.

Note, as an important point, that this last result can be helpful as a method of terminating an infinite continued fraction [11, 12]. In practical applications, continued fractions are numerically evaluated for hundreds of α, β's, after which the continued fraction is terminated by an approximation. The nature of some of the possible termination schemes is now discussed.

5.7.4 Termination

In applications one needs to find a method of termination of the continued fraction in order to obtain numerical results [11, 12]. This is a difficult process which depends to a large extent on the nature of the system being treated. In general, however, most approaches for the determination of the Green's function and density of states focus on the asymptotic behaviors of the coefficients α_n, β_n as $n \to \infty$ and the analytic properties of the density of states as a function of energy. In this regard, a focus in many methods is on approximating the analytic properties of [11, 12]

$$\lim_{n\to\infty} \frac{\beta_n}{\alpha_n}. \tag{5.71}$$

Our earlier discussions in Eqs. (5.69) and (5.70) are an example of the importance of this limiting form.

As a final point of our discussion, note that we have focused on the density of states of the 0 site of the network [11, 12]. Especially in the case of amorphous media, it is important to remember that the total density of states of the material is the average of the site density of states over all the network sites. In this regard, an application must be made to a consideration of a range of sites in the amorphous network.

We next turn to the treatment of some of the low-temperature properties of amorphous insulating solids. These serve as important examples of systems in which the density of states is required for an understanding of the behaviors of their low-temperature heat capacities and thermal conductivities. In particular, it will be seen that, in addition to the modal excitations of the amorphous network, the dynamics of structural changes in the network itself are also required for a complete understanding of the systems. This introduces a new aspect to the thermodynamics of amorphous materials which is not encountered in the thermodynamics of crystalline solids.

For these discussions we shall consider an amorphous network of vibrating atoms, i.e., the focus is on the phonon modes of an amorphous network of atoms. The dynamics of this network is characterized by a type of tight-binding Hamiltonian, but we shall see that additional considerations must be made in order to completely describe the vibrational heat capacity and thermal conductivity of the system. While the phonon contribution to the vibrational properties can be studied in the context of the recursive methods just discussed, an additional contribution to the low-temperature properties of the system is found to arise from geometric changes in the network geometry. These additional changes in the network are treated by the two-level states theory, which is now the focus of our considerations.

5.8 TWO-LEVEL STATES SYSTEMS

Nonmetallic amorphous materials exhibit anomalous heat capacities and thermal conductivities at temperatures well below the Debye temperature as one approaches absolute zero of temperature [13–16]. These anomalies, arising as significant qualitative and quantitative differences from the behaviors observed in crystalline materials, come from the ability of the network to tunnel between various local geometric differences in the metastable network geometry. Specifically, the amorphous network displays different geometric structures which it can tunnel between in a manner similar to the tunneling of an atom in a potential well exhibiting two local minima. (See Figure 5.5 for a schematic representation of this.) The tunneling between the metastable configurations of the amorphous network can be characterized by a collection of two-level energy state models distributed throughout the network [13–16]. These two-level energy states ultimately supply a low-temperature contribution to the heat capacity and thermal conductivities of the amorphous material. Their contribution represents an additional effect from those coming solely from the phonon excitations in the material. Unlike the amorphous network geometry, the single stable energy minimum

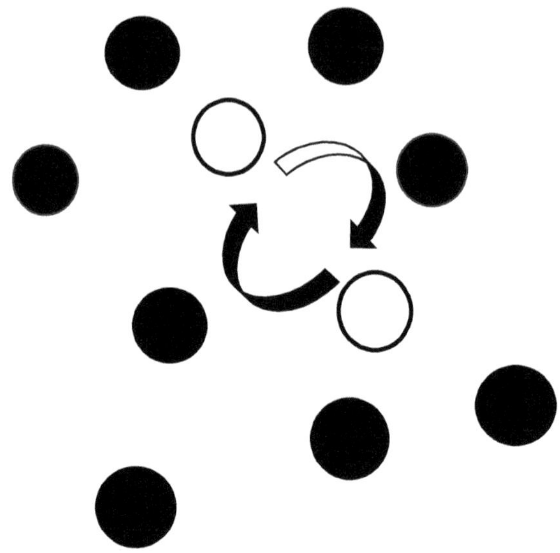

FIGURE 5.5 A portion of an amorphous material designed to indicate a possible tunneling between states of the material. The black circles represent stationary atoms in the material, and the set of two open circles represents a single atom which tunnels between metastable states of the system. The tunneling paths are represented by arrows. The figure is meant to indicate a type of tunneling that could give rise to the two-level mechanism, and many others are possible in the system.

of the lattice structure of a crystalline material does not exhibit such tunneling between metastable states of its geometry. Consequently, in insulating crystalline materials, the low-temperature properties arise solely from the phonons.

At low temperatures in a nonmetallic crystalline material, it is known from the Debye approximation that the material exhibits a heat capacity C characterized by the temperature dependence

$$C \sim T^3 \tag{5.72a}$$

and a thermal conductivity $\kappa = \frac{1}{3} C v_s l$ again characterized by the temperature dependence

$$\kappa \sim T^3 \tag{5.72b}$$

where T is the temperature in Kelvin, v_s is the speed of sound, and l is the mean free path for phonon scattering. In typical crystalline materials, these behaviors are generally found at temperatures well below the Debye temperature as one approaches absolute zero, e.g., in crystalline silicon at $20^\circ K$ and less. Notice in this regard that at low temperatures, the long wavelength speed of sound v_s is a constant, and the mean free path for phonon scattering l approaches a constant determined by the size of the sample. Consequently, the temperature dependence of the thermal conductivity at low temperatures is essentially set by the heat capacity.

In amorphous materials the low-temperature behaviors of the heat capacity and thermal conductivity differ from those in Eq. (5.72) for the crystalline materials. In a nonmetallic amorphous material at low temperatures approaching zero degrees Kelvin, the material displays a heat capacity C characterized by the temperature dependence [13–16]

$$C \sim T, \tag{5.73a}$$

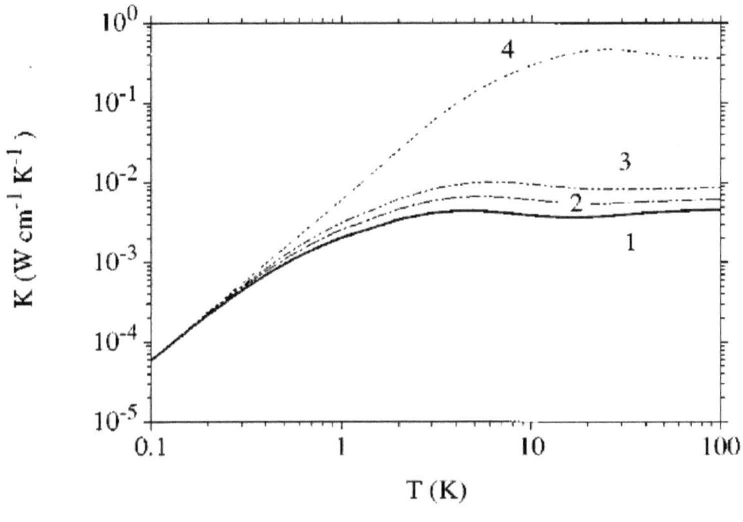

FIGURE 5.6 The thermal conductivity versus temperature of amorphous materials determined over the region $0^\circ K \le T \le 100^\circ K$ by a computer simulation is shown for different amorphous materials. The three characteristic temperature regions as discussed in the text are observed for each material [16]. Note that the phonon contributions to the thermal conductivity in these simulations were determined using recursive methods. (Reprinted figure with permission from [16]. Copyright 1996, American Physical Society.)

e.g., this behavior occurs in amorphous silicon at $20^\circ K$ and less. In addition, an amorphous material exhibits a thermal conductivity [13–16]

$$\kappa = \frac{1}{3} C v_s l \tag{5.73b}$$

which is now characterized by a complicated temperature dependence. (See Figure 5.6 for a schematic representation of some thermal conductivity behaviors.) Specifically, at very low temperatures, the thermal conductivity has the temperature dependence [13–16]

$$\kappa \sim T^{\gamma} \tag{5.73c}$$

with $1 \le \gamma \le 2$ for $T \le 1^\circ K$. As the temperature is increased, a plateau or nearly constant region of thermal conductivity is found for $1^\circ K \le T \le 10^\circ K$, and this is followed by a monotonic increase in the thermal conductivity with increasing T for $10^\circ K \le T \le 100^\circ K$. (Notice that in Eq. (5.73) T is the temperature in Kelvin, v_s is the speed of sound, and l is the mean free path for phonon scattering.)

The difference between the behaviors in Eqs. (5.72) and (5.73) is accounted for by a new mechanism which contributes to the heat capacity and thermal conductivity in amorphous materials but is not present in crystalline materials. To understand the additional contributions to the heat capacity and thermal conductivity coming from the geometric tunneling in an amorphous material, let us begin by considering the thermodynamics of two-level states systems which offer a simple model for the tunneling of a system between two different configurations. This model of tunneling will be seen to explain the differences in the properties of crystalline and amorphous materials [13–16].

To understand the thermodynamics of the two-level states contribution to amorphous materials, consider the one-dimensional problem of atomic tunneling between two local minima. (See Figure 5.7 for a schematic representation of this problem.) The Hamiltonian of the atomic tunneling problem is given by [13–15]

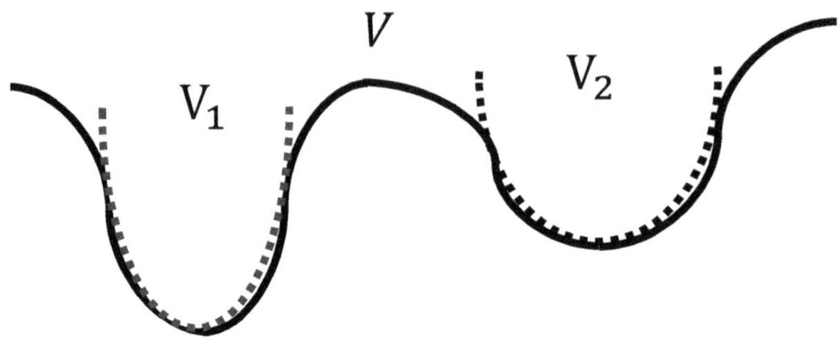

FIGURE 5.7 Schematic in one dimension of a potential V as a function of position x having two local minima. At the two local minima of V, a harmonic potential V_1 is fitted to the left minima and a harmonic potential V_2 is fitted to the right minima.

$$H = H_0 + V = (H_0 + V_1) + (V - V_1) = (H_0 + V_2) + (V - V_2) \tag{5.74}$$

where H_0 is the particle kinetic energy operator, V is a potential with two local minima, V_1 is a harmonic fit to the left potential minimum of V, and V_2 is a harmonic fit to the right potential minimum of V. In addition, let us begin the study of Eq. (5.74) with a consideration of the two auxiliary eigenvalue problems defined by [13–15]

$$(H_0 + V_1)|\varphi_1\rangle = \varepsilon_1 |\varphi_1\rangle \tag{5.75a}$$

and

$$(H_0 + V_2)|\varphi_2\rangle = \varepsilon_2 |\varphi_2\rangle \tag{5.75b}$$

where $|\varphi_1\rangle$ and $|\varphi_2\rangle$ are the vibrational ground states in the respective harmonic wells V_1 and V_2. Specifically, we are concerned with the tunneling of the state $|\varphi_1\rangle$ into the state $|\varphi_2\rangle$ and the tunneling of the state $|\varphi_2\rangle$ into the state $|\varphi_1\rangle$ as mediated by H. A study of this type of tunneling is done by considering the transition matrix elements of H between the two states $|\varphi_1\rangle$ and $|\varphi_2\rangle$.

To do this we consider the Hamiltonian H treated in the basis states $|\varphi_1\rangle$ and $|\varphi_2\rangle$. From Eq. (5.75) the matrix for the tunneling problem associated with Eq. (5.74) in terms of $|\varphi_1\rangle$ and $|\varphi_2\rangle$ is given by the matrix elements [13–15]

$$\langle\varphi_1|\mathrm{H}|\varphi_1\rangle = \varepsilon_1 + \langle\varphi_1|V - V_1|\varphi_1\rangle \approx \varepsilon_1 \tag{5.76a}$$

$$\langle\varphi_1|\mathrm{H}|\varphi_2\rangle = \frac{1}{2}\Delta \tag{5.76b}$$

$$\langle\varphi_2|\mathrm{H}|\varphi_1\rangle = \frac{1}{2}\Delta \tag{5.76c}$$

and

$$\langle\varphi_2|\mathrm{H}|\varphi_2\rangle = \varepsilon_2 + \langle\varphi_2|V - V_2|\varphi_2\rangle \approx \varepsilon_2. \tag{5.76d}$$

In the basis $\{|\varphi_1\rangle, |\varphi_2\rangle\}$ the Hamiltonian then becomes the 2×2 matrix

$$\mathcal{H}_0 = \begin{bmatrix} \varepsilon_1 & \frac{1}{2}\Delta \\ \frac{1}{2}\Delta & \varepsilon_2 \end{bmatrix}. \tag{5.77}$$

If we shift the zero of energy in Eq. (5.77), we can rewrite the Hamiltonian in the form

$$\mathcal{H} = \mathcal{H}_0 - \begin{vmatrix} \frac{1}{2}(\varepsilon_1 + \varepsilon_2) \\ \frac{1}{2}(\varepsilon_1 + \varepsilon_2) \end{vmatrix} = \begin{bmatrix} \frac{1}{2}\varepsilon & \frac{1}{2}\Delta \\ \frac{1}{2}\Delta & -\frac{1}{2}\varepsilon \end{bmatrix} \tag{5.78}$$

where $\varepsilon = \varepsilon_1 - \varepsilon_2$. Consequently, it then follows from Eq. (5.78) that the dynamics of the system considered in the tunneling basis $\{|\varphi_1\rangle, |\varphi_2\rangle\}$ reduces to the study of the eigenvalue problem [13–15]

$$\mathcal{H}|\psi\rangle = E|\psi\rangle. \tag{5.79}$$

Solving the eigenvalue problem in Eq. (5.79), we find two distinct eigenvalues given by

$$E_\pm = \pm\frac{1}{2}\sqrt{\varepsilon^2 + \Delta^2} = \pm E_a \tag{5.80a}$$

along with their corresponding eigenvectors

$$|\psi_+\rangle = a_+|\varphi_1\rangle + b_+|\varphi_2\rangle \tag{5.80b}$$

and

$$|\psi_+\rangle = a_-|\varphi_1\rangle + b_-|\varphi_2\rangle \tag{5.80c}$$

where

$$(a_\pm, b_\pm) = \frac{\left(\Delta, -\varepsilon \pm \sqrt{\varepsilon^2 + \Delta^2}\right)}{\sqrt{\Delta^2 + \left[-\varepsilon \pm \sqrt{\varepsilon^2 + \Delta^2}\right]^2}}. \tag{5.80d}$$

Next consider the occupancy of the two-level states at thermal equilibrium for the system considered as a function of temperature T. From a standard thermodynamics treatment, the average energy of the two-level states system as a function of temperature is given by [13–15]

$$\langle E\rangle = -E_a \tanh(\beta E_a), \tag{5.81}$$

and the contribution to the specific heat of the two-level system is then obtained as

$$C_{2-level} = \frac{\partial}{\partial T}\langle E\rangle = k_B(\beta E_a)^2 \operatorname{sech}^2(\beta E_a) \tag{5.82}$$

where $\beta = \frac{1}{k_B T}$ and k_B is the Boltzmann constant. This is the result for a single two-level state, but presumably the amorphous network has a range of different two-level states distributed throughout the infinite random medium. Due to environmental differences, each of these two-level states will be characterized by different sets of parameters ε, Δ, and E_a.

Due to the lack of long-range order in the network, the two-level states mechanism will display a range of different ε and Δ as set by the local geometry. Consequently, the set of all the two-level states in the amorphous material will exhibit a density of energy levels $D(E_a)$ characterizing their presence in the medium. As a consequence of the distribution $D(E_a)$, the total heat capacity contribution to the system resulting from the two-level states mechanism is given by [13–15]

$$C_{2-level\,total} = \int_0^\infty dE_a\, D(E_a) k_B (\beta E_a)^2 \operatorname{sech}^2(\beta E_a) \tag{5.83}$$

$$= k_B T \int_0^\infty d(\beta E_a) D(E_a) k_B (\beta E_a)^2 \operatorname{sech}^2(\beta E_a).$$

Notice that in the case that $D(E_a)$ is a constant independent of E_a it follows from Eq. (5.83) that

$$C_{2-level\,total} \sim T. \tag{5.84}$$

In this regard, from general arguments in the literature of two-level states, the presence of tunneling geometries in the network contributing to a constant density of two-level energy states is strongly indicated. These arguments are also reinforced by simulation studies on amorphous networks. Consequently, Eqs. (5.83) and (5.84) explain the low-temperature heat capacity of amorphous materials mentioned in Eq. (5.73a).

In the case of the thermal conductivity, the heat current is carried by the phonons in the amorphous materials, and it is the scattering of these phonons by the two-level states mechanisms distributed throughout the amorphous network which contributes an additional component to the thermal resistivity. In this regard, the thermal conductivity is given by [13–15]

$$\kappa = \frac{1}{3} C v_s l \tag{5.85}$$

where C is the phonon heat capacity. The two-level systems now contribute to the scattering of the phonons through the mean free path l. From the increased rate of scattering of low energy phonons at low temperatures arising from the presence of two-level states, it can again be shown by general arguments that [13–16]

$$\kappa \sim T^2 \tag{5.86}$$

accounts for the thermal conductivity in the region $T \leq 1^\circ K$. In this regard, it is found that in the region in which the two-level states mechanism operates and contributes a significant scattering channel, the mean free path increases with decreasing temperature as the two-level states are frozen out of the scattering processes. This effect is the origin of the T^2 dependence of the thermal conductivity. On the other hand, in pure crystalline materials where no such scattering mechanisms are present, the mean free path is temperature independent, being set by the size of the crystal.

To obtain a complete understanding of the thermal conductivity over the entire $0^\circ K \leq T \leq 100^\circ K$ range, one must resort to computer modeling [13–16]. The thermal conductivity properties as outlined earlier are found to be consistent with a variety of computer simulation studies. These all indicate the presence of its three basic features found between $0^{\circ} K \le T \le 100^{\circ} K$. In this regard, in Figure 5.6 some simulation studies of a variety of different amorphous materials are presented [16]. The results for these materials are all consistent with our earlier descriptions of the thermal conductivities of general amorphous materials. Notice, in regard to our earlier discussion of recursive methods, that the phonon contributions to the thermal conductivity in Figure 5.6 were computed using recursive methods.

A major issue remaining in the study of the field of amorphous materials, however, is to obtain a more detailed understanding of the changes in the geometry of the amorphous networks which account for the mechanisms behind the two-level states model. In this regard, the tunneling transitions between different network geometries of an amorphous medium often appear to be quite subtle in nature, so that this offers a major challenge to the study of these classes of materials. This is certainly a topic of future efforts.

As a final topic we look at systems known as quasiperiodic crystals and quasicrystals. These are materials which exhibit atomic structures midway between crystalline materials and amorphous materials. They are not true crystals because, while they exhibit long-range order, they do not display translational symmetry in space. Examples of some interesting types of quasicrystals involve materials exhibiting $5-8-,10-$ or $12-$fold axes of rotational symmetry. This type of rotational symmetry is known to be inconsistent with the translational symmetries of a crystal lattice. The question then is what type of structure can have long-range order in the absence of translational symmetry and can support point group symmetries inconsistent with crystalline geometry.

5.9 TILINGS

A quasiperiodic crystal is a material that has an atomic structure exhibiting long-range order in the absence of translational symmetry [17–24]. True crystals, on the other hand, are materials with long-range order as well as translational symmetry. This is an important distinction because systems with translational symmetry are restricted in the so-called point group symmetries they can display. In this regard, the point group symmetries are the various rotational and reflectional symmetries found along axes, planes, and at points in the crystal lattice. As a result of these restrictions, it is found that there are only 5 different general crystal lattice types in $2-$dimensional space and $24-$different crystal lattice types in $3-$dimensional space [17–24]. These types constitute the Bravais lattices forming the basis of the description of the crystal structure of crystalline materials.

Let us start with a consideration of a $1-$dimensional crystal lattice [17]. (See Figure 5.8 for a schematic representation of the lattice of a $1-$dimensional crystal.) In $1-$dimension, there is only one Bravais lattice type, composed as a periodic array of lattice points separated by a distance a along the crystal axis. The lattice exhibits long-range order because, given the position of a point on the crystal axis, all of the other points of the lattice are determined periodically relative to it. The density of points in the lattice is then defined by [17]

$$\rho(x) = \sum_{n=-\infty}^{\infty} \delta(x - na) \tag{5.87}$$

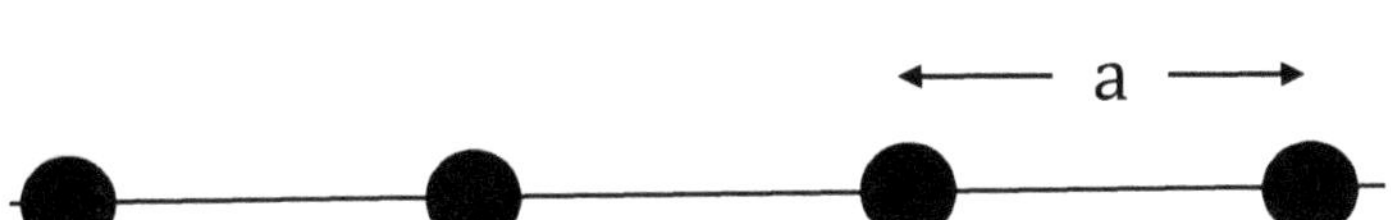

FIGURE 5.8 A schematic of a $1-$dimensional crystal lattice. Shown in the figure are four atoms with the nearest-neighbor separation a, indicated on the figure.

where the sum is over a lattice of $N \to \infty$ points on the x–axis, i.e., the crystal axis. If we Fourier transform the density of lattice points, we find in Fourier transform space that

$$\hat{\rho}(k) = \int_{-\infty}^{\infty} \rho(x) e^{-ikx} dx = \int_{-\infty}^{\infty} \sum_{n=-\infty}^{\infty} \delta(x - na) e^{-ikx} dx \tag{5.88}$$

$$= \sum_{n=-\infty}^{\infty} e^{-ikna} = N \sum_{m=-\infty}^{\infty} \delta_{k, \frac{2\pi}{a} m}.$$

This gives a discrete set of points in k–space forming the so-called reciprocal lattice which complements the direct lattice defined in Eq. (5.87).

The lattice defined in Eq. (5.87) is seen to exhibit translational symmetry because [17]

$$\rho(x + la) = \sum_{n=-\infty}^{\infty} \delta(x + la - na) = \sum_{n=-\infty}^{\infty} \delta(x - na) = \rho(x). \tag{5.89}$$

By the same reasoning, it follows from Eq. (5.88) that

$$\hat{\rho}\left(k + \frac{2\pi}{a} l\right) = \hat{\rho}(k) \tag{5.90}$$

and the reciprocal lattice is also periodic with period $\frac{2\pi}{a} l$ and invariant under the translations in k–space indicated in Eq. (5.90). The translational symmetry in the direct space lattice then carries over to a translational symmetry in the reciprocal space lattice.

Next, let us consider a 1–dimensional set of points on the x–axis which has long-range order but lacks translational symmetry. This set of points would be a form of a 1–dimensional quasicrystal. To do this, consider extending our consideration of Eq. (5.87) to treat a generalization of its form to a density of points on the lattice defined by [17]

$$\rho(x) = \sum_{n=-\infty}^{\infty} \delta(x - na) + \sum_{n=-\infty}^{\infty} \delta(x - n\pi a). \tag{5.91}$$

Here both the first term of the sum

$$\rho_1(x) = \sum_{n=-\infty}^{\infty} \delta(x - na) \tag{5.92a}$$

and the second term of the sum

$$\rho_2(x) = \sum_{n=-\infty}^{\infty} \delta(x - n\pi a) \tag{5.92b}$$

separately generate periodic lattices. However, one periodic array of points is of period a and the other periodic array of points is of period πa. The periods of the two sets of points forming the complete set of points on the x–axis are therefore incommensurate with each other, i.e., their periods differ by a transcendental number. In this regard, the two lattices that are generated separately by each density $\rho_1(x)$ and $\rho_2(x)$ can never be made to match up. As a result, their sum in Eq. (5.91) is not periodic. Consequently, the lattice generated by the sum in Eq. (5.91) is not translationally invariant. It does, however, have long-range order arising separately from each of the sums in Eq. (5.92). This then shows how long-range order can exist in a set of points in the absence of translational symmetry.

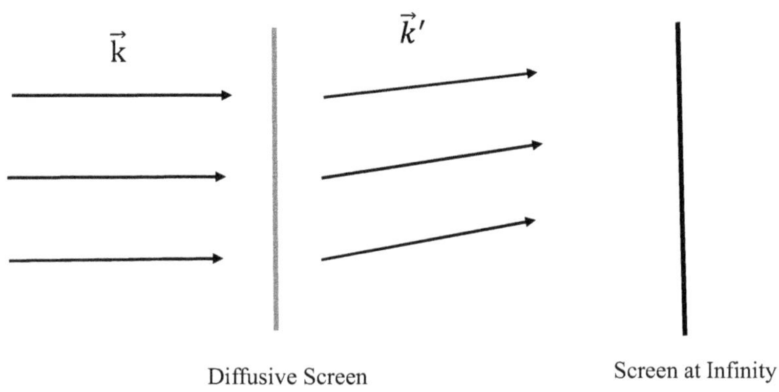

FIGURE 5.9 A schematic of the standard diffraction experiment. Planewave radiation is normal incident from the left on a diffraction screen of slits in the $x-y$ plane. It is diffracted by the screen and formed into a diffraction pattern at the far right imaging screen at infinity. The slits in the diffracting screen are spatially arrayed on the screen in the x –direction and of infinite length in the y –direction.

5.9.1 Scattering from the Incommensurate Lattice

Consider the following scattering problem [17–21]. A planewave of radiation is normal incident on a screen in the $x-y$ plane composed of uniform identical parallel slits which transmit some of the incident radiation. (See Figure 5.9 for a schematic of the problem.) The slits are infinite in length along the y –axis, and the transmitted radiation forms an image on another screen at infinite separation along the z –axis from the diffracting screen. This, then, defines a standard diffraction experiment.

We begin by considering a periodic array of identical slits along the x –axis in the diffracting screen. Assume that there are $2N+1 \to \infty$ slits uniformly placed on the diffracting screen, each with a nearest-neighbor separation a. The transmission scattering from the array of slits is described by a potential in the $x-y$ plane defined by [17–21]

$$V(x) = \sum_{n=-\infty}^{\infty} v\delta(x - na). \tag{5.93}$$

The elastic scattering amplitude from the diffracting screen, for a wave $e^{i\vec{k}|\cdot\vec{r}}$ incident from the left normal to the $x-y$ plane of the diffracting screen and scattering to a transmitted wave $e^{i\vec{k}'\cdot\vec{r}}$ leaving the diffraction screen from the right, is given in the Born approximation by the transition matrix

$$\langle \vec{k}' | V(x) | \vec{k} \rangle = \int_{-\infty}^{\infty} dxdy \sum_{n=-N}^{N} v\delta(x - na) e^{i(\vec{k}' - \vec{k})\cdot x\hat{i}} \tag{5.94}$$

$$= L \sum_{n=-N}^{N} v e^{i\Delta k_x na} = Lv \frac{\sin\left[\frac{1}{2}\Delta k_x (2N+1)a\right]}{\sin\left[\frac{1}{2}\Delta k_x a\right]}$$

where the limit that $N \to \infty$ is taken and $L \to \infty$ is the length of the screen along the y –axis. The result in Eq. (5.94) is the standard Born result for the scattering from a diffraction grating with $2N+1$ slits of nearest-neighbor separation a.

On the other hand, if the grating were ruled with an incommensurate array of slits (i.e., one modeled on the incommensurate system defined in Eq. (5.91)), the resulting potential would be the composition of two incommensurate gratings. The scattering potential for this case would have the form [17]

$$V(x)=\sum_{n=-\infty}^{\infty} v\delta(x-na)+\sum_{n=-\infty}^{\infty} u\delta(x-n\pi a), \tag{5.95}$$

and the Born scattering amplitude, obtained by replacing Eq. (5.93) by Eq. (5.95), would be given by the $N\rightarrow\infty$ limit of

$$\langle \vec{k}^{'} \mid V(x) \mid \vec{k} \rangle = Lv\frac{\sin\left[\frac{1}{2}\Delta k_x(2N+1)a\right]}{\sin\left[\frac{1}{2}\Delta k_x a\right]}+Lu\frac{\sin\left[\frac{1}{2}\Delta k_x(2N+1)\pi a\right]}{\sin\left[\frac{1}{2}\Delta k_x \pi a\right]}. \tag{5.96}$$

Consequently, properties associated with each of the two incommensurate periodicities would be exhibited in the total transmitted scattering pattern. As a result, the scattering pattern would be the sum of the scattering from each of the two incommensurate arrays of different periodicity.

Another type of $1-$dimensional sequence of points that will have an important role in our discussions of quasicrystals involves Fibonacci point distributions. These are point sets defined on an axis so as to exhibit intervals between neighboring points arrayed by a Fibonacci sequence. The Fibonacci distribution generated along the axis has long-range order as it is obtained algorithmically as a deterministic sequence, but it has an absence of translational symmetry along the axis of its definition. It will be useful in extending our discussions of quasicrystals to the treatment of higher dimensional quasicrystals and for a comparison with experimental data. We now consider the properties of the Fibonacci sequence.

5.9.2 Fibonacci Sequential Array in $1-$Dimension

We begin with a discussion of the generation according to a Fibonacci distribution of a sequential set of real numbers denoting points positioned on the $x-$axis. In this distribution, the separation between neighboring points on the $x-$axis is either a short length S or a long length L where [17–24]

$$\frac{L}{S}=\tau=\frac{1+\sqrt{5}}{2}, \tag{5.97}$$

and the occurrence of S and L in the sequence is determined algorithmically based on the Fibonacci integers. (Note: Here the irrational number τ is known as the golden ratio and is of interest in number theory in the context of the study of the Fibonacci integers [22].)

The occurrence in the sequence of the S and L segments is then ordered so that the resulting sequence of Fibonacci points $\{x_n\}$ is given from the recursive relationship [21]

$$x_n=n+\alpha+\frac{1}{\tau}E\left[\frac{n}{\tau}+\beta\right]. \tag{5.98}$$

where n is an integer labelling the position of successive points on the $x-$axis, α and β are arbitrary real numbers, and $E[x]$ is the greatest integer less than x. Given the set $\{x_n\}$ the density of points on the $x-$axis is then defined by

$$\rho(x)=\sum_n \delta(x-x_n), \tag{5.99a}$$

and the Fourier transform of the density of points is obtained from [21]

$$\hat{\rho}(k) = \lim_{N\to\infty}\left[\frac{1}{N}\sum_n \exp(ikx_n)\right] \tag{5.99b}$$

where $N \to \infty$ is the total number of points in the set $\{x_n\}$. As in our earlier discussions, the Fourier transform in Eq. (5.99b) is directly related to the scattering from a Fibonacci array of atoms on the chain or from a Fibonacci array of slits in a diffracting screen.

Let us next consider the problem of the diffractive screen in Figure 5.9 for a Fibonacci array of diffracting slits. Now the scattering potential $V(x)$ characterizing the screen has the form [17, 21]

$$V(x) = \sum_n v\delta(x - x_n) \tag{5.100a}$$

where n runs over the slits in the screen and v characterizes the scattering through the diffraction slits. From Eq. (5.100a) we then define the scattering amplitude from the screen in the Born approximation as

$$\langle \vec{k}^{'} | V(x) | \vec{k} \rangle = v\frac{1}{N}\sum_n \exp\left[i\left(\vec{k}^{'} - \vec{k}\right)\cdot x_n\hat{i}\ \right] \tag{5.100b}$$

$$= v\frac{1}{N}\sum_n \exp\left[i\Delta\vec{k}\cdot x_n\hat{i}\ \right] \propto \sum_n \exp\left[i\Delta\vec{k}\cdot x_n\hat{i}\ \right].$$

Here $\Delta\vec{k} = \vec{k}^{'} - \vec{k}$ is the change in wavevector from the wave incident on the screen, $\vec{k}$, into the diffracted wave $\vec{k}^{'}$, and the incident wave again is taken at normal incidence to the diffracting screen. Following some considerable algebra, it is found that [17, 21]

$$\frac{1}{N}\sum_n \exp\left[i\Delta\vec{k}\cdot x_n\hat{i}\ \right] = \sum_{p,q}\left[\frac{\sin\left(\frac{X}{2}\right)}{\frac{X}{2}}e^{i\gamma}\right]\delta\left(\Delta k_x^{'} - k_{pq}\right), \tag{5.101a}$$

$$X = 2\pi q - \frac{k_{pq}}{\tau}, \tag{5.101b}$$

$$k_{pq} = \frac{2\pi}{1+\frac{1}{\tau^2}}\left[p + \frac{q}{\tau}\right], \tag{5.101c}$$

and

$$\gamma = k_{pq}\alpha - \left[2\pi q - \frac{k_{pq}}{\tau}\right]\beta + \frac{X}{2} \tag{5.101d}$$

where p and q are integers.

The scattering amplitude obtained from Eqs. (5.100) and (5.101) is found to be a dense set of delta-function peaks in k-space with a distribution of amplitudes. This indicates that the system, in the absence of translational invariance, exhibits long-range order. Among the intensity distribution of diffraction lines, it can be shown that the brightest lines of the diffraction pattern are determined by the condition that [17, 21]

$$X = \frac{2\pi\tau}{1+\tau^2}(q\tau - p) \tag{5.102}$$

be small. These bright lines form the most prominent features in the diffraction pattern from the 1−dimensional Fibonacci screen and will enter into our later discussion of the electron and optical diffraction from higher dimensional quasicrystals [17, 21–24].

5.9.3 Fibonacci Sequential Array in Higher Dimensions

Next let us generalize the ideas of incommensurate periodicities to a consideration of higher dimensional systems. This will lead to the definition of higher dimensional quasiperiodic crystals and to discussions of quasicrystal scattering patterns which are inconsistent with those arising from traditional crystalline materials. It also provides an explanation of the existence of quasicrystal systems having long-range order but an absence of translational symmetry. We begin the treatment with a review of the restrictions placed on the point group symmetries in a crystal lattice arising from its translational symmetries. This provides a distinction between crystal and quasicrystal materials. Following this, a treatment of the scattering from 2−dimensional quasicrystals is discussed.

5.9.4 Symmetry Restrictions

Only certain sets of point group symmetries are consistent with the translational symmetry of a crystal lattice. An example of this is seen in the set of allowed rotational symmetries of the crystal lattice. Specifically, the only allowed axes of rotational symmetry consistent with the translational symmetry of a crystal are 1−,2−,3−,4 −, and 6−fold axes of rotational symmetry. A demonstration of this can be made by considering the schematic figures shown in Figure 5.10.

First consider Figure 5.10a. This shows a set of lattice points on a line forming part of a 2− or 3−dimensional crystal lattice so that the crystal has translational symmetry along the line containing the points [25]. The points labeled A, O, and B are three consecutive lattice points on the axis line, and the segments $\overline{AO}$, and $\overline{OB}$ are nearest neighbor separations of length a. The translational symmetry of the lattice along the line is such that the crystal lattice is translated into itself under parallel translations of length na along the line. Here n is an integer.

Assume that passing through O and perpendicular to the page is a rotational symmetry axis. Under a rotation about this axis through the angle θ the crystal lattice is rotated into itself. Figure 5.10b then shows a clockwise rotation of θ about this axis [25]. The rotation is seen to rotate the line so that the lattice point B is moved to a new lattice point B' separated from O by the distance a.

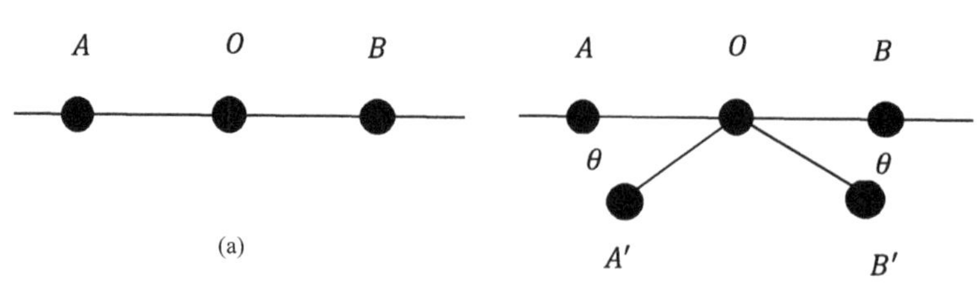

FIGURE 5.10 Schematic figures of: a) three consecutive lattice points labeled A, O, and B on a line of translational symmetry of a crystal lattice. b) the lattice point B' is arrived at by rotating the point B clockwise by θ about the axis through the point O and the lattice point A' is arrived at by rotating A counterclockwise by θ about the axis through the point O.

Similarly, the figure also shows an anticlockwise rotation of the axis by θ which rotates the point A to a new lattice point A' separated from O by the distance a. After these two rotations, the points A' and B' are found to be on a line parallel to that passing through A and B. From trigonometry, the separation of $\overline{A'B'}$ on their line is given by [25]

$$d = 2a\cos\theta. \tag{5.103}$$

However, from the rotational symmetry, A' and B' must be points of the lattice. Consequently, their separation along the line passing through them is ma for some integer m.

As a result, it follows from Eq. (5.103) that

$$d = 2a\cos\theta = ma, \tag{5.104}$$

relating θ to m. From Eqs. (5.103) and (5.104) it is then found that

$$\cos\theta = \frac{m}{2}, \tag{5.105}$$

and from the requirement that $|\cos\theta| \le 1$ we see that the only allowed solutions for θ are

$$\theta = \frac{2\pi}{p} \text{ for } p = 1,2,3,4,6. \tag{5.106}$$

These are the $1-,2-,3-,4-$, and $6-$fold axes of rotational symmetry consistent with the translational properties of the crystal lattice [25].

With these symmetry conditions made clear for systems with long-range order and translational symmetry, we can next discuss the properties of higher dimensional quasicrystals. In quasicrystals there is an absence of translational symmetry so that many types of point group symmetries absent in crystals can be present in quasicrystals.

5.9.5 Quasicrystals in Higher Dimensions: Surface Scattering

Many materials have been found which exhibit long-range order consistent with Bragg diffraction but also exhibit point group symmetries which are not consistent with a crystal lattice [24]. Examples of this behavior arise mostly in alloys of Al, however, some examples are found in alloys which do not contain Al. (See Table 5.1 for a short list of materials exhibiting quasicrystal behavior.) The main features distinguishing these materials as quasicrystals are that they exhibit a Bragg diffraction pattern composed of distinct diffraction spots, and the material can be aligned so as to exhibit a rotational symmetry of the diffractive pattern (and, consequently, of the material) which is outside the $1-,2-,3-,4-$, and $6-$fold axes of rotational symmetry observed in crystal lattices. In this regard, common rotational axes found in the quasicrystals are $5-,8-,10-$ or $12-$fold axes

TABLE 5.1
Some Examples of Quasicrystal Alloys [20]

$Al-based$	$non-Al-based$
$Al-Cu-Fe$	$Ga-Mg-Zn$
$Al-Pd-Mn$	$U-Pd-Si$
$Al-Ni-Co$	$Ti-Mn-Si$

of rotation. These properties are directly determined through the application of X-ray and electron diffraction [24].

Many of the materials condensed into the quasicrystal format are a metastable state of matter. They are not an equilibrium state in the sense that the lowest energy configuration has not been achieved. In this regard, they are usually made by a process in which the material is quickly taken from a liquid state to a metastable quasicrystal solid. Often it is found that the material exhibits a periodic behavior along one direction but is quasiperiodic in the remaining two spatial directions. More recently, however, annealed systems have been studied which also exhibit quasicrystal behavior [17, 18, 24].

An interesting discussion of quasicrystal diffraction patterns has recently involved a comparison of the electron diffraction patterns from quasicrystals formed as atomic arrays with the optical diffractive patterns of manufactured screens exhibiting $5-, 8-, 10-$ or $12-$fold axes of rotational symmetry [24]. These comparisons provide a guide to some of the basic features of patterns exhibiting the diffractive characteristics of quasicrystals and how the ordering of quasicrystals can be understood in terms of their Fourier transform characteristics. In this regard, the emphasis of a certain set of Fourier components in the patterning over another set supports the rotational symmetries exhibited.

In a first type of experiment [24], screens are made by superimposing a number of $1-$dimensional Fibonacci patterns arrayed in different directions on the screen to make a higher dimensional diffractive pattern. In these superpositions, the axes of the different $1-$dimensional Fibonacci patterns occur in different directions on the screen to form a total pattern with the desired rotational symmetry [24]. Consequently, the superimposed screens generate a quasicrystal-like diffractive array for light exhibiting long-range order with rotational symmetries not allowed in a true crystal lattice [24]. In another type of experiment [24], optical screens are made based on atomic models designed to explain the Bragg scattering of electrons from quasicrystal materials. These optical screen patterns are then used to make a diffracting screen to diffract an incident optical field. In both experiments, the geometries of the optical and electron diffraction patterns have been made to agree [24].

As a final point, we note that there is a little distinction between the definition of a quasicrystal and a quasiperiodic crystal. Quasiperiodic crystals exhibit long-range order in the absence of translational symmetry. On the other hand, quasicrystals, in addition, are taken to also exhibit a minimum separation between atoms or slits and to exhibit long-range order along certain so-called star axes of the materials [17–24]. In this sense, $1-$dimensional quasiperiodic crystals are not technically termed as quasicrystals. For a further detailed discussion of these aspects of the subject, the reader is referred to the literature [17–25].

5.9.6 Penrose Tiling and Geometry

Related to the quasicrystal problem is the mathematical problem of tiling surfaces. Specifically, it was shown by Penrose that a two-dimensional surface could be tiled by two different tiles in such a way that a two-dimensional plane is completely covered by non-overlapping tiles [17, 21–23]. This arrangement is termed a Penrose tiling and is found to exhibit long-range order and $5-$fold rotational symmetry. The Penrose tiling has been used to develop many of the atomic models applied to explain the surface scattering from atomic quasicrystal arrays. They also figure into the generation of the $1-$dimensional Fibonacci set of points used in our earlier discussions [17, 18].

As a final note we mention that in the course of the development of the theory of quasicrystals a number of methods have been developed to generate sets of points forming a quasicrystal. One such interesting method is used to generate the $1-$dimensional Fibonacci set of points [17–19, 21 –25]. In this method we start with a square lattice of points defined in the plane such that the sides of the square at the origin are parallel to the $x-$ and $y-$axes of space. The axis of the Fibonacci set of points is then defined as a line with a slope equal to τ in the plane of the square lattice [17

–21]. By perpendicularly projecting onto the line lattice point within a small strip above the line, the projected points are shown to generate the set $\{x_n\}$ of the $1-$dimensional Fibonacci set of points applied in our earlier considerations.

REFERENCES

1. Gedde, U. W. and Hedenqvist, M. S. 2019. *Fundamental Polymer Science* (Springer Nature Switzerland, AG).
2. Treloar, L. R. G. 1975. *The Physics of Rubber Elasticity* (Oxford University Press, Oxford).
3. Schirmacher, W. 2015. *Theory of Liquids and Other Disordered Media* (Springer, Cham).
4. Perepechko, I. I. 1981. *An Introduction to Polymer Physics* (Mir Publishers, Moscow).
5. Klinger, M. 2013. *Glassy Disordered Systems: Glass Formation and Universal Anomalous Low-Energy Properties* (World Scientific Publisher, Singapore).
6. Pauling, L. 1935. The Structure and Entropy of Ice and of Other Crystals with Some Randomness of Atomic Arrangement. *Journal of the American Chemical Society* 57(12): 2680–2684.
7. DiMarzio, E. A. and Stillinger, F. H. 1964. Residual Entropy of Ice. *Journal of Chemical Physics* 40, 1577–1581.
8. Takada, A., Conradt, R., and Richet, P. 2015. Residual Entropy and Structural Disorder in Glass: A Review of History and an Attempt to Resolve Two Apparently Conflicting Views. *Journal of Non-Crystaline Solids* 429, 33–44.
9. Lieb, E. L. 1967. Residual Entropy of Square Ice. *Physical Review* 162, 162–172.
10. Lieb, E. H. 2004. Residual Entropy of Square Ice. In: Nachtergaele, B., Solovej, J. P., Yngvason, J. (eds) *Condensed Matter Physics and Exactly Soluble Models* (Springer, Berlin, Heidelberg).
11. Haydock, R. 1980. The Recursion Solution of the Schrodinger Equation. *Solid State Physics* 35, 216–294 (Academic Press Inc. New York).
12. Woodruff, D. M., Anlage, S. M., and Smith, D. L. 1987. Density-of-States Calculations within the Recursion Method. *Physical Review B* 38, 1725–1734.
13. Kelly, M. J. 1975. Model Amorphous Semiconductor Structures: Tight-Binding s-band Electron Structure. *Journal of Non-Crystalline Solids* 18, 55–64.
14. Phillips, W. A. 1987. Two-Level States in Glass. *Reports on Progress in Physics* 50, 1657–1708.
15. Leggett, A. and Vural, D. C. 2013. Tunneling Two-Level Systems Model of the Low-Temperature Properties of Glasses: Are Smoking-Gun Tests Possible. *The Journal of Physical Chemistry* 8, 12966–12971.
16. Malyshkin, V. and McGurn, A. R. 1996. Computer-Simulation Study of the Thermal Conductivity of Amorphous Insulating Solids. *Physical Review B* 54, 2980–2983.
17. Janot, C. 2012. *Quasicrystals: A Primer* (Oxford University Press, Oxford).
18. Fujiwara, T. and Ishii, Y. 2008. Introduction to Quasicrystals. In: Blackman, J. A., (ed) *Handbook of Metal Physics* (Elsevier B V, Amsterdam).
19. Macia-Barber, E. 2021. *Quasicrystals: Fundamentals and Applications* (CRC Press, Boca Raton).
20. Goldman, A. I. and Widom, M. 1991. Quasicrystal Structure and Properties. *Annual Review of Physical Chemistry* 42, 685–729.
21. Levine, D. and Steinhardt, P. J. 1986. Quasicrystal. I. Definition and Structure. *Physical Review B* 34, 596–616.
22. Jagannathan, A. 2021. The Fibonacci Quasicrystal: Case Study of Hidden Dimensions and Multifractality. *Reviews of Modern Physics* 93, 045001-1 to 045001-38.
23. Avishai, Y. and Berend, D. 1990. Transmission through a One-Dimensional Fibonnacci Sequence of δ-function Potentials. *Physical Review B* 41, 5492–5499.
24. Ferralis, N., Szmodis, A. W., and Diehl, R. D. 2004. Diffraction from One- and Two-dimensional Quasicrystalline Gratings. *American Journal of Physics* 72(9), 1241–1246.
25. Groot, S. 2018. The Fibonacci Quasicrystal. Institute of Theoretical Physics, Utrecht University.

6 Annealed Models, the Replica Method, and Spin Glass

In this chapter we look at some interesting techniques that have been developed to analytically treat disorder in the thermodynamics of many-body insulating magnetic systems [1–12]. As developed, they can be applied to both crystalline and amorphous magnetic materials. The first technique involves the ideas of annealing and annealed models [1–4], and the second involves a mathematical trick referred to as the replica method [5–12]. The techniques developed in both methodologies are based on treating the disorder in magnetic systems using novel analytical approaches. These are developed as a way of simplifying the mathematics of the problems of disorder and providing important new methodologies for formulating an understanding of the properties displayed in disordered media.

Annealed models [1–4] are a method of studying disordered media in which the disorder in the system is allowed to be in thermodynamic equilibrium with the thermodynamic model it is disordering. In this regard, the impurity concentration in the system is regulated by the introduction of a chemical potential into the problem. The resulting model is of interest in its own right or as an indicator of some of the types of behaviors that might be expected in nonannealed or quenched disordered models. In this regard, the annealed models also have found application in understanding the general geometric and connectivity properties of geometric disorder found in random networks and addressed in studies of percolation theory [3, 4].

The other method of interest to us is the replica method [6, 7] which forms the basis for the treatment of many types of studies of the important types of magnetic media known as spin glasses [8–10]. The replica method is based on an identity involving the expression of the logarithm of a variable in terms of a specific limiting form. This is ultimately used to simplify the relationship between the free energy and the partition function of a statistical mechanical system in the presence of quenched disorder. In this regard, quenched disorder requires the determination of the average of the free energy over the disorder rather than the average of the partition function.

A focus of our application of replica techniques is in the recent important study of spin glass. Spin glass is an important new type of magnetic material involving a peculiar form of random spin ordering of the atomic moments in an amorphous magnetic medium. This ordering usually arises due to the presence of various different frustrating magnetic interactions in the system [8–10]. Spin glass is observed as a property of dilute magnetic impurities embedded in the matrix of a nonmagnetic metal and is found at specific concentration intervals of the magnetic impurity atoms. Early examples of spin glass materials [9] were alloy formed of small dilutions of $3d$ transition metals of Fe and Mn in a noble metal matrix, e.g., $AuFe, AuMn, CuMn, AgMn, PtMn$, $PdMn$, and $AuCr$. For simplicity of considerations, we will focus on lattice models which have been proposed to describe features of spin glass materials. In regard to these lattice models, the replica method is a type of mathematical trick used to reduce the complexity of the study of some of the unusual properties of spin glasses. In this sense, both annealed models and replica method applications in this chapter are formulations which endeavor to simplify the treatment of impurities in magnetic media by rewriting the thermodynamic problems in a format which is easier to treat.

We first discuss some of the aspects of annealed models of magnetic materials [1–4]. These are problems of magnetic disorder in which the distribution of disordering impurities in the system is in thermodynamic equilibrium with the system taken as a whole. It is different from the so-called quenched solution of a disorder model, as in the quenched system the disorder is fixed in the system and does not adjust itself to attain a total thermodynamic equilibrium. The annealed model represents a simplification in the study of disordered media, as it is found that solutions to annealed

 DOI: 10.1201/9781032674919-6

models can be obtained by mapping them onto the solutions of pure system models in the absence of disorder. This provides a trick which greatly reduces the complexity of these types of thermodynamic problems because pure systems are easier to treat than disordered systems. Annealed models offer some useful simple systems for approximating the behavior of their nonannealed (quenched) counterparts [1–4].

As an example of applications of the method of annealing, the usefulness of annealed model solutions is also illustrated by considering their application to some interesting problems in the field of percolation theory [3, 4]. Problems in percolation theory involve the study of the properties of lattice connectivity upon bond and site deletions from the lattice. Specifically, in a lattice formed as an array of sites connected together with connecting bonds between the sites, bonds or sites can be randomly removed from the system to change its connectivity. One of the questions of interest in the modified lattices is: what is the probability of a site being a member of an infinite cluster of sites connected by a path of bonds. This is one of a number of problems of connectivity and clustering which is studied as a function of site or bond concentration in the lattice and has important applications in the physics of many types of clustering systems.

In addition, the thermodynamics of annealed magnetic models (i.e., the specific heat, susceptibilities, magnetization, etc.) are also of interest in their own right [1, 2]. For the specifics of these studies and for comparisons with experimental systems, the reader is referred to the literature [1–4].

The replica method [6] is also introduced to our studies as another type of mathematical trick. The trick in this case allows us to obtain the free energy of a disordered magnetic system by studying the powers of its partition function. By using an identity of the logarithm function and assumptions of the analyticity of powers of the partition function, the computation of the free energy of complex disordered systems is simplified. The resulting process allows us to compute the average of the free energy over the disorder in the system, yielding a representation of the thermodynamics of the quenched disorder of the system. As we shall show, this trick has found important applications in the treatment of amorphous magnetism and spin glass [6–10]. We shall find that spin glass is an important focus of our discussions presented in this chapter and provides a new important class of magnetic materials available to technology [8–10].

As a final topic, we also consider some of the general properties of new types of solid state magnetic systems known as spin glass [8–10]. At low temperatures, these materials exhibit transitions to new states of magnetic ordering. The magnetic atoms of the spin glass are ordered in the sense that the magnetic moment of each atom is frozen about a spatial direction. However, the spatial direction of each spin appears to be arbitrarily chosen. Consequently, in the absence of an externally applied magnetic field, the system has a zero net magnetization even though the spins are ordered. The ordering in spin glass, however, is quite different from other orderings that are commonly displayed by magnetic materials and requires the introduction of new measures of its presence and properties. Unlike in other magnetic materials, the state of a spin glass is not an absolute minimum of the free energy but is a metastable state, so that the system may be found in one of a series of possible local free energy minima. In this sense, it is similar to the metastable atomic configurations of amorphous media discussed in Chapter 5.

In this regard, the ideas of magnetic ordering in ferromagnetic and antiferromagnetic materials are simplistic and well understood, but the ordering in spin glasses is of a more complex nature. In ferromagnets and antiferromagnets, chemical bonds exist between atoms favoring either the parallel or antiparallel alignment of the spins on neighboring magnetic atoms in the crystal lattice. In ferromagnetic materials, the atomic spins align parallel to each other so that the material can exhibit a transition to a permanently magnetized state [8–10]. Antiferromagnetic materials, on the other hand, tend to favor an antiparallel alignment between neighboring spins so that they display a permanent antiparallel arrangement of neighboring spins. In both cases, the magnetic ordering is easily visualized, and the equilibrium states are an absolute minimum of the free energy.

Spin glass, however, is a much more complicated arrangement of the ordering of the moments of the magnetic atoms. The complexity again arises from the magnetic interaction between the spins of the atoms, developing from their chemical bonds. Specifically, the nature of the bonding between atoms is found to depend strongly on their spatial separations, varying not only in strength but also in the sign of the energy of interactions between parallel configurations of pairs of atomic moments. In an amorphous arrangement of atoms, then, the interaction between an atom and its neighbors varies in strength and its tendency to favor a ferromagnetic or antiferromagnetic ordering between spin pairs in the system. In this regard, some neighbor pairs will favor ferromagnetic interactions while others will favor antiferromagnetic interactions. In the spin glass state, the competition between local ferromagnetic and antiferromagnetic interactions results in a net zero magnetization in the presence of ordering of the individual spins along arbitrary directions.

This statistical distribution and the interplay between local ferromagnetic and antiferromagnetic ordering are often referred to as frustration [8–10]. In its state of frustration, the ordered spin glass system fails to develop either a ferromagnetic or an antiferromagnetic ordering. These aspects of frustration also account for the existence of many local minima in the spin glass free energy which, in turn, allow for hysteresis effects to appear in some of the properties of the system.

To begin the discussion, we present a general introduction to the formulation of the treatment of annealed models and to the ideas involved in the study of the free energy of spin glass disorder using the replica method. These techniques will be applied in the remaining course of the chapter to the study of specific examples of annealed systems and spin glass materials.

6.1 DISORDER IN MAGNETIC MATERIALS

There are a number of types of disorder that are found in insulating magnetic materials and a number of methods that have been developed to study them. In general, in this regard, insulating magnetic systems are described by Hamiltonians characterizing the interactions of atomic spins (i.e., the angular momenta of magnetic atoms and their associated magnetic moments) associated with the magnetic atoms composing the materials. An example of such a magnetic system is the Heisenberg Hamiltonian, which has the form [1–4]

$$H = -\sum_{\langle i,j \rangle} J_{i,j} \vec{S}_i \cdot \vec{S}_j \tag{6.1}$$

where $\langle i, j \rangle$ indicates a sum over unique pairs i, j of atomic sites, $J_{i,j}$ is the exchange interaction between the pairs i, j of atomic sites, and $\{\vec{S}_l\}$ is the total set of atomic angular momenta (referred to in the study of magnetism as spins) at each atomic site of the material.

Another example of the Hamiltonian of an insulating material that will be of interest in our following discussions is the Ising Hamiltonian [1–4]. This has the form

$$H = -\sum_{\langle i,j \rangle} J_{i,j} \sigma_i \sigma_j \tag{6.2}$$

where the set of atomic angular momenta $\{\vec{S}_l\}$ is replaced by a set of Ising spins $\{\sigma_l\}$ having the property that $\sigma_l = \pm 1$. In this regard the Ising model is a simplification of the Heisenberg model in which the non-commuting spin angular momentum operators are replaced by the commuting Ising spins $\{\sigma_l\}$. The Ising model represents a great reduction of the difficulty of the mathematical treatment of the magnetic system from that involved in the study of the Heisenberg model. In addition, there are materials which are accurately modeled in terms of either the Heisenberg or Ising models of magnetism.

The determination of the thermodynamic properties of these models comes from the evaluation of their partition function defined as

$$Z = Tre^{-\beta H} \tag{6.3}$$

where $\beta = \frac{1}{k_B T}$, k_B is the Boltzmann constant, T is the temperature, and the trace is over the states of $\{\vec{S}_l\}$ in the case of the Heisenberg model or $\{\sigma_l\}$ in the case of the Ising model. The free energy is then defined from Eq. (6.3) as

$$F = -k_B T \ln Z, \tag{6.4}$$

and from the free energy the thermodynamics of the materials are generated by a number of derivative forms.

Disorder can be introduced into magnetic materials by alloying them so that the exchange interactions $J_{i,j}$ spatially vary throughout the alloy, or so that the spins $\{\vec{S}_l\}$ or $\{\sigma_l\}$ vary in the alloy as a result of atomic replacement. If these features of disorder are statistically distributed in the alloyed material, the average free energy $\bar{F}$ can be obtained from Eq. (6.4) by averaging over the statistical configurations of the impurities so that [1–10]

$$\bar{F} = -k_B T \overline{\ln Z} \tag{6.5}$$

where the bar over the expressions represents the statistical average over the distribution of impurities. As in the case of the system in the absence of disorder, the thermodynamics of the materials in the presence of impurity disorder are generated from Eq. (6.5) by the application of a number of derivative forms.

A number of techniques have been developed to treat the evaluation of the form in Eq. (6.5) for a wide variety of systems. These generally involve various perturbation and series expansion approaches, Padé approximants, renormalization group techniques (to be discussed in a later chapter), and computer modeling methods [1–4]. In this chapter we shall look at two different types of methods for the study of impurity problems in insulating magnetic materials. These include the method of annealed models and the replica method. For discussions of the other methods the reader is referred to treatments in the literature or to other chapters in this text.

In the annealed model treatment, the study of Eq. (6.5) is replaced by the study of the annealed free energy. This has the form [1–4]

$$\bar{F}_{An} = -k_B T \ln \bar{Z} \tag{6.6}$$

where the bar again represents an average over the statistical distribution of impurities. Now, in the case of the annealed model, the averaging over the statistical configurations of the impurity disorder is performed on the partition function rather than on the logarithm of the partition function. This reduces the difficulty of the treatment but is not the same as averaging the logarithm of the partition function in Eq. (6.5). In the annealed model, the impurity disorder of the material is allowed to enter into a thermodynamic equilibrium with the thermodynamics of the system. This is not the case with the free energy obtained from Eq. (6.5). However, in some cases, this difference between the two problems can be minimized.

In the replica method, on the other hand, a trick is used to develop an alternative form for dealing with the average in Eq. (6.5). This method is particularly useful in the study of spin glass models and will be applied in a variety of approaches to the study of spin glass [6–10].

In the replica method, we consider the study of the form [6–10]

$$Z^n \tag{6.7}$$

where $n \geq 0$ is an integer and Z is the partition function of the disordered system of our interest. If we analytically continue Eq. (6.7) treating n as a real number, then it follows from an identity defining the logarithm function that [6]

$$\ln Z = \lim_{n \to 0} \frac{Z^n - 1}{n} \tag{6.8a}$$

so that upon averaging over the statistical distribution of impurity configurations in the system it follows that [6–10]

$$\overline{\ln Z} = \lim_{n \to \infty} \frac{\overline{Z^n} - 1}{n}. \tag{6.8b}$$

We shall find in the course of our discussions that it is often easier to compute $\overline{Z^n}$ rather than $\overline{\ln Z}$. Consequently, $\overline{\ln Z}$ determined from Eq. (6.8b) can be usefully applied to the study of the thermodynamics of the disordered impurity system. In this regard, the replicant method is found to be particularly useful in studies of the thermodynamics of spin glass.

We begin the discussions with a study of some basic applications of the annealed model method to magnetic systems.

6.2 ANNEALED MODELS

As an application of the annealed model method, we consider an Ising model defined on the square lattice and subject to a bond random disorder [1–4]. This provides an interesting illustration of annealing techniques as the Ising model in 2−dimensions can be treated exactly by analytical methods both in the absence of disorder and in the presence of annealed disorder. Results for the two models are then expressed in terms of closed-form expressions involving standard functions.

For an Ising model defined on a square lattice with nearest neighbor interactions and in the absence of impurities, the Hamiltonian is written as [1–4]

$$H = -J \sum_{\langle i,j \rangle} \sigma_i \sigma_j \tag{6.9a}$$

where $\sigma_l = \pm 1$, $\langle i,j \rangle$ denotes a sum over nearest-neighbor pairs on the lattice, and $J > 0$ is the ferromagnetic exchange coupling between nearest-neighbor pairs of Ising spins. (Notice that the discussions presented here can also be generalized to treat the antiferromagnetic system in which $J < 0$). From the theory of statistical mechanics, it follows that the partition function of Eq. (6.9a) is given by [1–4]

$$Z_{Ising}(J,T) = Tr_{\{\sigma_l\}} \{\exp - \beta H\} = Tr_{\{\sigma_l\}} \left\{ \exp\left(\beta J \sum_{\langle i,j \rangle} \sigma_i \sigma_j \right) \right\} \tag{6.9b}$$

$$= Tr_{\{\sigma_l\}} \prod_{\langle i,j \rangle} \{\exp(\beta J \sigma_i \sigma_j)\}$$

where the trace is over all the possible configurations of the set of Ising spins $\{\sigma_l\}$, $\beta = \frac{1}{k_B T}$, k_B is the Boltzmann constant, T is the temperature, and $\langle i,j \rangle$ indicates either a sum or product over nearest-neighbor pairs on the lattice. Considering the far right-hand side of Eq. (6.9b), we see that, because the Ising spin operators are classical variables, the sum in the exponent can be rewritten as a product of exponents. The partition function in Eq. (6.9b) is then exactly solvable for the square lattice Ising model, with the solution commonly given in standard texts on statistical mechanics.

An interesting, related problem of magnetic disorder is next defined on the square lattice by modifying the Hamiltonian in Eq. (6.9) to treat the case of bond disorder. To do this, consider the modified form of Eq. (6.9) given by

$$H = -J\sum_{\langle i,j\rangle} c_{i,j}\sigma_i\sigma_j \tag{6.10}$$

where we have introduced into the Hamiltonian operators $c_{i,j}$ representing a random disorder associated with the exchange couplings. Specifically, the operators $c_{i,j}$ are defined to have the statistical properties such that randomly and uniformly throughout the system [1–4]

$$c_{i,j} = 1 \text{ with probability } p \tag{6.11a}$$

and

$$c_{i,j} = 0 \text{ with probability } 1-p. \tag{6.11b}$$

This represents a system of bond disorder in which some bonds have been removed at random uniformly throughout the system. The resulting disordered problem in Eq. (6.10) can be solved exactly within the context of the annealed model of disorder.

To do this consider that the partition function for Eq. (6.10) is then given by the form [1–4]

$$Z_{Annealed}(T) = Tr_{\{\sigma_l, c_{l,k}\}}\left\{\exp\left(-\beta H + \beta\mu\sum_{\langle i,j\rangle} c_{i,j}\right)\right\} \tag{6.12a}$$

$$= Tr_{\{\sigma_l, c_{l,k}\}}\left\{\exp\left(\beta\sum_{\langle i,j\rangle}\left[Jc_{i,j}\sigma_i\sigma_j + \mu c_{i,j}\right]\right)\right\} = Tr_{\{\sigma_l, c_{l,k}\}}\prod_{\langle i,j\rangle}\left\{\exp\left(\beta Jc_{i,j}\sigma_i\sigma_j + \beta\mu c_{i,j}\right)\right\}$$

where $\beta = \dfrac{1}{k_B T}$, k_B is the Boltzmann constant, T is the temperature, and μ is a chemical potential introduced into the problem so as to set the thermodynamic average

$$\overline{c_{i,j}} = p \tag{6.12b}$$

where the bar over the $c_{i,j}$ denotes an average over the possible bond configurations of the system. Notice that in Eq. (6.12) we have introduced the disorder in the system by treating it as part of the statistical mechanics problem, i.e., the impurities are in thermodynamic equilibrium with the spins in the system. In this regard, its presence in the system is governed by a chemical potential setting the average of the bond occupancy operators $c_{i,j}$.

We begin the study of Eq. (6.12a) by first evaluating the trace over the $c_{i,j}$ associated with each of the Ising bonds $-Jc_{i,j}\sigma_i\sigma_j$. For this, we can look separately at each bond term of the product on the far right of Eq. (6.12a) so as to separately consider the terms of the form [1–4]

$$Tr_{\{c_{i,j}\}}\left\{\exp\left(\beta Jc_{i,j}\sigma_i\sigma_j + \beta\mu c_{i,j}\right)\right\} = 1 + \exp\left(\beta J\sigma_i\sigma_j + \beta\mu\right) \tag{6.13}$$

$$= 1 + \left(\cosh\beta J + \sigma_i\sigma_j\sinh\beta J\right)e^{\beta\mu}.$$

Notice here that in the evaluation of Eq. (6.13) we have used the identity that

$$\left(\sigma_i \sigma_j\right)^2 = 1. \tag{6.14}$$

Consequently, $\sigma_i \sigma_j$ is the highest power of $\sigma_i \sigma_j$ that can enter into Eq. (6.13).

Applying Eq. (6.13) in Eq. (6.12a) we find that

$$Z_{Annealed}\left(T\right) = Tr_{\{\sigma_l, c_{l,k}\}} \prod_{\langle i,j \rangle} \left\{ \exp\left(\beta J c_{i,j} \sigma_i \sigma_j + \beta \mu c_{i,j}\right)\right\} \tag{6.15}$$

$$= Tr_{\{\sigma_l\}} \prod_{\langle i,j \rangle} \left\{1 + \left(\cosh \beta J + \sigma_i \sigma_j \sinh \beta J\right) e^{\beta\mu}\right\}.$$

However, from the properties of $\sigma_i \sigma_j$ applied to Eq. (6.15) it follows that

$$1 + \left(\cosh \beta J + \sigma_i \sigma_j \sinh \beta J\right) e^{\beta\mu} = Y e^{W \sigma_i \sigma_j} \tag{6.16}$$

where

$$Y^2 = \left[1 + \left(\cosh \beta J + \sinh \beta J\right) e^{\beta\mu}\right]\left[1 + \left(\cosh \beta J - \sinh \beta J\right) e^{\beta\mu}\right] \tag{6.17}$$

and

$$e^{2W} = \frac{1 + \left(\cosh \beta J + \sinh \beta J\right) e^{\beta\mu}}{1 + \left(\cosh \beta J - \sinh \beta J\right) e^{\beta\mu}}. \tag{6.18}$$

Consequently, from Eq. (6.15) considering a system of $N \to \infty$ Ising spins we find from Eqs. (6.16) through (6.17) that [1–4]

$$Z_{Annealed}\left(T\right) = Tr_{\{\sigma_l\}} \prod_{\langle i,j \rangle} \left\{ Y e^{W \sigma_i \sigma_j} \right\} \tag{6.19}$$

$$= Y^{2N} Z_{Ising}\left(W, T\right)$$

where we use the fact that on a square lattice of $N \to \infty$ Ising spins there are $2N$ bonds. It follows that the annealed problem of the disordered system maps onto the original Ising model problem in Eq. (6.9b).

Next, we show an application of the mapping of the annealed model of a disordered Ising model onto the Ising problem of a system considered in the absence of disorder.

6.2.1 An Example: Percolation Theory

As an example of the application of the annealed model let us consider a basic problem in percolation theory [3, 4]. Specifically, in percolation theory we study the statistical properties of the connectivity of sites in a lattice as the bonds between lattice sites are removed. Such problems of lattice connectivity can be of importance in determining many of the physical properties of materials. Examples are the conductivity in disordered materials in which electron transport is by hopping between atoms in the material or in studies of the stability under stress of atomic lattices upon the removal of atomic bonds in the network [3]. In addition, percolation properties of lattices can be

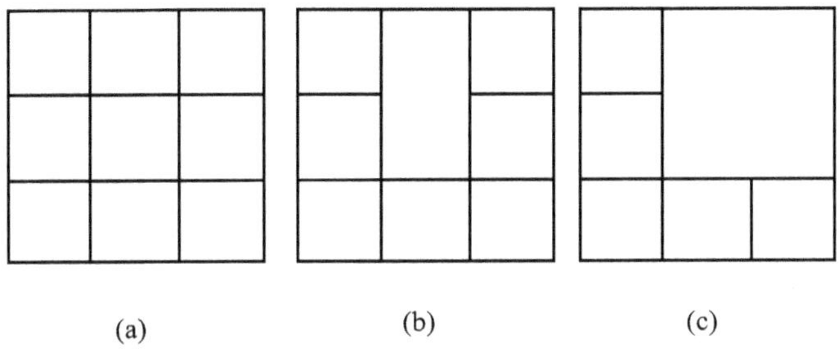

FIGURE 6.1 Schematic figure of: a) a portion of a complete square lattice composed of bonds and sites, b) a portion of a square lattice with a single bond removed (i.e., taken from the upper center of the lattice), and c) a portion of a square lattice with a single site removed (i.e., taken from the upper right-hand side of the lattice). Notice that in removing the site from the lattice, two horizontal bonds and two vertical bonds have been taken from the lattice. Also, in b) a single horizontal bond has been removed, but we could have also illustrated the point by removing a single vertical bond. Both horizontal and vertical bonds figure into the bond concentration.

shown to be directly related to the low-temperature properties of certain types of disordered Ising models. This will be an aid in the following discussion.

In percolation theory [1–4] we consider a lattice of sites (i.e., points of the lattice) which are connected together by nearest-neighbor bonds (i.e., lines connecting nearest-neighbor lattice points). For example, in Figure 6.1a the square lattice is shown, composed of lattice sites (points) and nearest-neighbor bonds (shown as lines) connecting the sites. In percolation theory there are two types of problems of interest. In the first type, bonds are removed at random from the lattice so that there remains a fractional concentration $0 \leq p_{bond} \leq 1$ of bonds uniformly distributed in the lattice. This is illustrated in Figure 6.1b showing the lattice in Figure 6.1a with a single bond removed from the square lattice. (Note that in Figure 6.1b we have removed a horizontal bond, but we could have equally removed a vertical bond.) In the second type, sites are removed at random from the lattice so that there remains a fractional concentration of sites $0 \leq p_{site} \leq 1$ uniformly distributed in the lattice. This is illustrated in Figure 6.1c showing the lattice in Figure (6.1a with a site removed from the square lattice. Notice that when a site is removed from the system, the four bonds connecting it to the rest of the lattice are also removed.

An interesting set of problems studied in percolation theory [1–4] is: First, at what concentration p_{bond} of bonds (p_{site} of sites) uniformly distributed on the lattice does an infinite cluster of connected sites begin to exist in the system? In this regard, an infinite cluster of sites is one composed of an infinite cluster of lattice sites which are connected to one another through a sequence of bonds in the cluster.

As an example, if $p_{bond} = 0$ ($p_{site} = 0$) there cannot be an infinite cluster of connected sites in the lattice. In the $p_{bond} = 0$ case here are no paths between sites in the lattice, and in the $p_{site} = 0$ case here are no sites or bonds in the lattice. On the other hand, if either $p_{bond} = 1$ or $p_{site} = 1$ there is an infinite cluster of connected sites composed of all of the sites in the lattice. Somewhere between these two limits there must be a concentration p^*_{bond} or p^*_{site}, known as the critical concentration, at which an infinite cluster first develops in the lattice.

A second problem of interest is: What is the probability of a site on the lattice belonging to an infinite cluster for $p_{bond} \neq 0$ ($p_{site} \neq 0$)? In this regard, this probability is nonzero for $p_{bond} \geq p^*_{bond}$ ($p_{site} \geq p^*_{site}$) and zero otherwise.

The case of the bond or site percolation is directly related to the zero-temperature magnetization of an array of nearest-neighbor ferromagnetic coupled Ising spins. At zero temperature, the nonzero magnetization of the system arises from an infinite connected cluster of lattice sites. Small clusters have magnetization which averages to zero over all the clusters in the system. Only the infinite

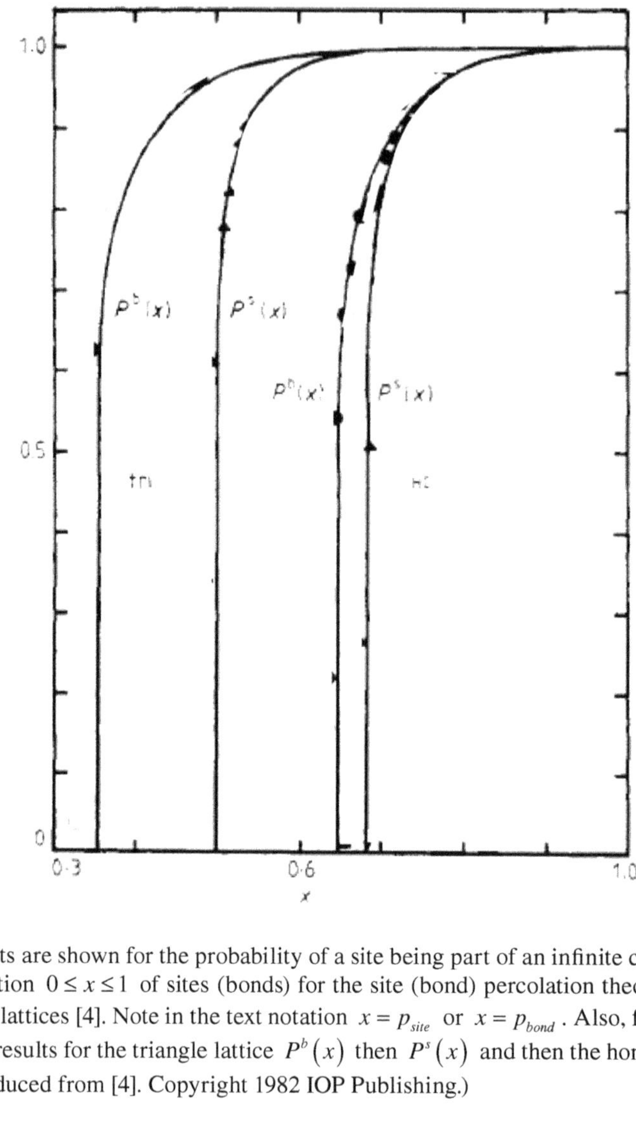

FIGURE 6.2 Results are shown for the probability of a site being part of an infinite cluster $P^s(x)$ ($P^b(x)$) versus the concentration $0 \le x \le 1$ of sites (bonds) for the site (bond) percolation theory problem on the triangle and hexagonal lattices [4]. Note in the text notation $x = p_{site}$ or $x = p_{bond}$. Also, from left to right on the figure are presented results for the triangle lattice $P^b(x)$ then $P^s(x)$ and then the honeycomb lattice $P^b(x)$ then $P^s(x)$. (Reproduced from [4]. Copyright 1982 IOP Publishing.)

cluster gives a nonzero magnetization. Consequently, the critical percolation concentration occurs at the concentration at which the zero-temperature Ising magnetization first becomes nonzero. Similarly, the probability of a site being in an infinite cluster is proportional to the zero-temperature magnetization of the magnetic system defined on the percolation lattice. As a result, the important properties of percolation theory can be determined from the zero-temperature magnetization of the Ising model, and an approximation for the magnetization is given by the annealed model considerations of the Ising model.

A study of the percolation theory results in the context of two-dimensional Ising models has been made for the Ising model defined on the triangular, hexagonal, and Kagome lattice Ising models. On these lattices, the Ising model is exactly solvable in the absence of disorder so that the annealed disordered systems can be readily treated [4].

In Figure 6.2 results are shown for the probability of a site being part of an infinite cluster versus the concentration $0 \le x \le 1$ of sites (bonds) for the site (bond) percolation theory problem on the triangle and hexagonal lattices [4]. (Note that here we use the notation $x = p_{site}$ or $x = p_{bond}$, respectively, for the site or bond problems.) These were obtained from a consideration of the annealed model solutions of the Ising models on these respective lattices and its resulting zero temperature magnetization. The results are in very close agreement with results from numerical simulation studies in nonannealed lattices.

We next turn to the discussion of another type of disorder that can exist in the Ising model. This is a type of disordered system involving frustration effects, realized in some magnetic materials known as spin glasses. These are materials in which the exchange couplings between the spins of the system are randomly distributed over both positive and negative values [6–10]. The study of spin glass is an example of a disordered material which can be approached using the ideas of the replica method. While these ideas have been extended to more general types of spin glass models than those found in the context of the Ising model, the Ising model is the simplest representation of the ideas of spin glass. As such, Ising model spin glasses will be the systems studied here.

6.3 THE REPLICA METHOD APPLIED TO SPIN GLASS

In an Ising model with spin glass behavior, a more complex type of system disorder exists than that treated in our earlier discussions of annealed models [6–10]. The disorder in spin glass involves the random occurrence of interactions tending to favor a mixture of parallel as well as antiparallel couplings between the different interacting pairs of spins. Consequently, some pairs of spins will try to condense into ferromagnetic pairs while others will try to condense into antiferromagnetic pairs [6–10]. At low temperatures, these interactions cause a net ordering of the spins such that the spatial orientation of each spin is frozen into a spatial direction but frozen in such a way as to form an apparently random pattern over the entire material [6]. Consequently, in the absence of an externally applied magnetic field, the ordered spins do not exhibit a net magnetization [6–10]. As an additional important feature, in the frozen state, the random orientation of spins also does not change with time [6]. To illustrate these aspects of the treatment of spin glass, in the following discussions we shall consider the simplest models of spin glass developed in the context of the Ising model. For the study of more complex spin models, the reader is referred to the literature [6–10].

Specifically, the form of the Ising model we shall initially treat is given by the Hamiltonian [6]

$$H = -\sum_{\langle i,j\rangle} J_{i,j}\sigma_i\sigma_j - h\sum_i \sigma_i \tag{6.20}$$

where the indices i and j run over the sites of a crystal lattice, $\langle i, j\rangle$ represents a sum over pairs of nearest-neighbor sites on the lattice, $J_{i,j}$ is a randomly distributed exchange coupling between the nearest-neighbor Ising spins σ_i and σ_j which interact through the exchange energy $-J_{i,j}\sigma_i\sigma_j$, and h is an externally applied field. In our discussions, we shall treat the system as being defined on a crystal lattice though this simplification is not necessary [6–10], and many spin glasses are defined on amorphous networks. In this regard, the focus of our treatment is to give an illustration of a system exhibiting the essential ideas of spin glass behavior with a minimum of complications.

The randomness in the model in Eq. (6.20) enters through the statistical distribution of the nearest-neighbor exchange couplings $\{J_{i,j}\}$. These are distributed between each pair of nearest-neighbor sites throughout the system according to a Gaussian probability distribution $P(J_{i,j})$. The distribution of each nearest-neighbor coupling in the system is then taken to have the form [6–10]

$$P(J_{i,j}) = \sqrt{\frac{1}{2\pi J^2}} \exp\left[-\frac{1}{2}\left(\frac{J_{i,j}}{J}\right)^2\right] \tag{6.21}$$

where J is the variance of the distribution and $\int_{-\infty}^{\infty} dJ_{i,j} P(J_{i,j}) = 1$. Notice that, consistent with our earlier discussion of the distribution of the exchange couplings in spin glass, $P(J_{i,j})$ is equally distributed over positive and negative values of $J_{i,j}$. This accounts for the scattering of ferromagnetic and antiferromagnetic pairs through the system. It also follows that each nearest-neighbor coupling in the model is distributed by the same probability distribution given in Eq. (6.21).

The focus of our discussions is on evaluating the quenched limit of the free energy of the Hamiltonian [6–10] in Eq. (6.20). This is the limit in which the free energy is averaged over the configurations of the exchange coupling given in Eq (6.21). It is not the same as the annealed limit, treated earlier in this chapter, in which the partition function is averaged over the configurations of couplings in the system. In fact, in the context of the annealed model, the systems we discuss here do not exhibit spin glass phases. The quenched solutions, however, do exhibit regions of spin glass behavior and are used to investigate the conditions required for a condensation of the systems into a spin glass state. The spin glass ordering is found in the low-temperature regions of the models discussed and provides representations of spin glass behavior found in experimental materials [6–10]. As a first step in this process, an appropriate indicator for measuring the presence of a spin glass ordering in the system is formulated.

6.3.1 Order Parameter

To measure the degree of order in the spin glass phase we develop a so-called order parameter [6–10] which indicates the degree to which the system has condensed into a spin glass. The order parameter must account for the ordering of each spin in its spatial direction to form a random array of spins which, throughout space, are a relatively time-independent configuration [6]. For the Ising model one such measure was introduced by Edward and Anderson [6], and we will look at this indicator first.

In the Edward-Anderson order parameter [6–10] the configuration of spins in the system at one particular time (denoted by γ) is compared with the configuration of the same spins at another time (denoted by δ) which is infinitely separated in time from the original time (denoted by γ). This comparison is done while keeping the set of couplings $\{J_{i,j}\}$ fixed. Consequently, the set of spins at time γ is denoted $\{\sigma_i^\gamma\}$ while the set of spins at time δ is denoted $\{\sigma_i^\delta\}$. In terms of these two sets of spin snapshots of the system, the Edward-Anderson order parameter is written as [6]

$$q = \frac{1}{N}\sum_i \sigma_i^\gamma \sigma_i^\delta \tag{6.22a}$$

or upon performing a quenched averaging over the possible coupling configurations $\{J_{i,j}\}$ as

$$q = \frac{1}{N}\overline{\left\langle \sum_i \sigma_i^\gamma \sigma_i^\delta \right\rangle} \tag{6.22b}$$

where i runs over all the $N \to \infty$ spins in the system, $\langle\ \rangle$ indicates a thermodynamic average, and the bar notation indicates a configurational average over the set of possible coupling configurations $\{J_{i,j}\}$. In regard to the measure in Eq. (6.22), notice that in an infinite system we expect that the system self-averages [6–10] so that both measures in Eq. (6.22) should agree and, by a similar reasoning, should be independent of the lattice site i.

From the definitions in Eq. (6.22) we see that in the absence of any correlation between the two sets $\{\sigma_i^\gamma\}$ and $\{\sigma_i^\delta\}$ which are separated by an infinite time interval, the order parameter is $q = 0$. In this limit the material is not a spin glass. On the other hand, for $q \neq 0$ in the absence of a net magnetization a correlation is present in the two systems of randomly ordered spins at infinite time separations, and, as a result, the material displays a spin glass phase [6]. Consequently, the indicator q measures the degree of ordering in the apparently random orientation of the spins and whether or not they are frozen into a spin glass state. The idea in our treatment of the free energy, then, is to obtain q from a study of the free energy of the spin glass model and to look for the conditions for spin glass to exist.

The ideas developed in the formulation of the Edwards-Anderson order parameter [6] can now be generalized into the study of an order parameter accessible to replica method techniques [6]. This generalized order parameter will allow us to apply the identity in Eq. (6.8) to our study of the spin glass phase.

6.3.2 Generalization to the Replica Method

In the replica method [6] we develop from the spin glass Hamiltonian in Eq. (6.20) a set of n replica Hamiltonians $\{H^\alpha\}$ where the αth member of this set is denoted by

$$H^\alpha = -\sum_{\langle i,j\rangle} J_{i,j}\sigma_i^\alpha\sigma_j^\alpha - h\sum_i \sigma_i^\alpha. \tag{6.23}$$

Here $\alpha = 1,2,\ldots,n$ labels the n replcas, $\sigma_i^\alpha = \pm 1$ is the spin at the ith lattice site of the αth replica of the Ising model in Eq. (6.20), and the set of exchange couplings $\{J_{i,j}\}$ are the same for each member of the set of identical replica Hamiltonians $\{H^\alpha\}$. Consequently, while the spin configurations in each of the replica Hamiltonians are independent of each other, the exchange couplings are the same in each replica [6–10].

One can then form a replica partition function obtained from the set of n replica Hamiltonians $\{H^\alpha\}$ and defined by [6] (Note: for simplicity in the following we take $h = 0$).

$$Z_{rep} = Z^n = Tr_{\{\sigma_i^\delta\}} \exp\left(-\beta\sum\nolimits_\alpha H^\alpha\right) \tag{6.24}$$

$$= Tr_{\{\sigma_i^\delta\}} \prod_\alpha \prod_{\langle i,j\rangle} \exp\left[\beta J_{i,j}\sigma_i^\alpha\sigma_j^\alpha\right]$$

where $\alpha = 1,2,\ldots,n$, the trace is made over the set of all replica spins $\{\sigma_l^\delta\}$ on each lattice site, and where

$$Z = Tr_{\{\sigma_l\}} \exp(-\beta H) \tag{6.25}$$

is the partition function of the Hamiltonian in Eq. (6.20) taken for $h = 0$. Notice that in Eqs. (6.24) and (6.25) $\beta = \dfrac{1}{k_B T}$, there is also no interaction between the spins of the different Hamiltonians in $\{H^\alpha\}$, and $Z_{rep} = Z^n$ is just the nth power of the partition function given in Eq. (6.25). We shall now consider an application of the identity in Eq. (6.8) (given by $\ln Z = \lim\limits_{n\to 0}\dfrac{Z^n - 1}{n}$) to relate the logarithm of the partition function to the partition function raised to the nth power.

In our next step towards this end, we consider the evaluation of the average of the replica partition function over all its possible nearest-neighbor exchange coupling configurations. From Eq. (6.24) this average is written as [6]

$$\overline{Z_{rep}} = \overline{Z^n} = \int\left[\prod_{\langle i,j\rangle} dJ_{i,j}\right]\left[\prod_{\langle i,j\rangle} P\left(J_{i,j}\right)\right] Z_{rep} \tag{6.26}$$

$$= \int\left[\prod_{\langle i,j\rangle} P\left(J_{i,j}\right) dJ_{i,j}\right] Tr_{\{\sigma_i^\delta\}} \exp\left(-\beta\sum\nolimits_\alpha H^\alpha\right)$$

where $\prod_{\langle i,j\rangle} P(J_{i,j})$ (specifically the notation $\langle i,j\rangle$) indicates a product of all the nearest-neighbor bond probability distributions in the lattice, and the sum on α runs over $\alpha = 1,2,\ldots\ldots,n$ representing the n replicas of the Hamiltonian. Considering the case in which $h = 0$ we then find from Eq. (6.24) that [6]

$$Z_{rep} = Tr_{\{\sigma_i^\delta\}}\left\{\prod_{\langle i,j\rangle}\prod_{\alpha}\exp\left[\beta J_{i,j}\sigma_i^\alpha\sigma_j^\alpha\right]\right\} \tag{6.27}$$

and

$$\begin{aligned}\left[\prod_{\langle i,j\rangle}P(J_{i,j})dJ_{i,j}\right]Z_{rep} &= \left[\prod_{\langle i,j\rangle}P(J_{i,j})dJ_{i,j}\right]Tr_{\{\sigma_i^\delta\}}\left\{\prod_{\langle i,j\rangle}\prod_{\alpha}\exp\left[\beta J_{i,j}\sigma_i^\alpha\sigma_j^\alpha\right]\right\} \\ &= Tr_{\{\sigma_i^\delta\}}\left\{\left[\prod_{\langle i,j\rangle}P(J_{i,j})dJ_{i,j}\right]\prod_{\langle i,j\rangle}\prod_{\alpha}\exp\left[\beta J_{i,j}\sigma_i^\alpha\sigma_j^\alpha\right]\right\} \\ &= Tr_{\{\sigma_i^\delta\}}\left\{\left[\prod_{\langle i,j\rangle}P(J_{i,j})dJ_{i,j}\right]\prod_{\langle i,j\rangle}\exp\left[\beta J_{i,j}\sum_{\alpha}\sigma_i^\alpha\sigma_j^\alpha\right]\right\} \\ &= Tr_{\{\sigma_i^\delta\}}\left\{\prod_{\langle i,j\rangle}f(J_{i,j},dJ_{i,j})\right\}\end{aligned} \tag{6.28a}$$

where

$$f(J_{i,j},dJ_{i,j}) = P(J_{i,j})dJ_{i,j}\exp\left[\beta J_{i,j}\sum_{\alpha}\sigma_i^\alpha\sigma_j^\alpha\right] \tag{6.28b}$$

defines the contribution of each exchange bond $J_{i,j}$ in the product contributing to the partition function.

Consider next the evaluation of the integral $\int_{-\infty}^{\infty} f(J_{i,j},dJ_{i,j})$. This generates the average of the bond factor $\exp\left[\beta J_{i,j}\sum_{\alpha}\sigma_i^\alpha\sigma_j^\alpha\right]$ over all the possible values of the bond exchange coupling $J_{i,j}$. This average is obtained by integrating Eq. (6.28b) over the range $-\infty < J_{i,j} < \infty$ of possible exchange configurations so that [7–10]

$$\begin{aligned}\int_{-\infty}^{\infty} f(J_{i,j},dJ_{i,j}) &= \int_{-\infty}^{\infty} dJ_{i,j}\exp\left[\beta\sum_{\alpha}J_{i,j}\sigma_i^\alpha\sigma_j^\alpha\right]P(J_{i,j}) \\ &= \int_{-\infty}^{\infty} dJ_{i,j}\exp\left[\beta J_{i,j}\sum_{\alpha}\sigma_i^\alpha\sigma_j^\alpha\right]\sqrt{\frac{1}{2\pi J^2}}\exp\left[-\frac{1}{2}\left(\frac{J_{i,j}}{J}\right)^2\right] \\ &= \sqrt{\frac{1}{2\pi J^2}}\int_{-\infty}^{\infty} dJ_{i,j}\exp\left[\beta J_{i,j}\sum_{\alpha}\sigma_i^\alpha\sigma_j^\alpha - \frac{1}{2}\left(\frac{J_{i,j}}{J}\right)^2\right] \\ &= \sqrt{\frac{1}{2\pi J^2}}\int_{-\infty}^{\infty} dJ_{i,j}\exp\left\{-\frac{1}{2J^2}\left[\left(J_{i,j} - J^2\beta\sum_{\alpha}\sigma_i^\alpha\sigma_j^\alpha\right)^2 - \left(J^2\beta\sum_{\alpha}\sigma_i^\alpha\sigma_j^\alpha\right)^2\right]\right\}\end{aligned} \tag{6.29}$$

$$= \left[\sqrt{\frac{1}{2\pi J^2}} \int_{-\infty}^{\infty} dJ_{i,j} \exp\left\{ -\frac{1}{2J^2} \left(J_{i,j} - J^2 \beta \sum_{\alpha} \sigma_i^{\alpha} \sigma_j^{\alpha} \right)^2 \right\} \right] \exp\left[\frac{\beta^2 J^2}{2} \left(\sum_{\alpha} \sigma_i^{\alpha} \sigma_j^{\alpha} \right)^2 \right]$$

$$= C(\beta, J) \exp\left[\frac{\beta^2 J^2}{2} \left(\sum_{\alpha} \sigma_i^{\alpha} \sigma_j^{\alpha} \right)^2 \right].$$

Using the single bond result in Eq. (6.29), the average of the replica partition function, $\overline{Z_{rep}}$, over the entire set of possible exchange bonds $\{J_{ij}\}$ can be obtained. In this way, it follows that [7–10]

$$\overline{Z_{rep}} = \overline{Z^n} = Tr_{\{\sigma_i^{\delta}\}} \left\{ \prod_{\langle i,j \rangle} \left(C(\beta, J) \exp\left[\frac{\beta^2 J^2}{2} \left(\sum_{\alpha} \sigma_i^{\alpha} \sigma_j^{\alpha} \right)^2 \right] \right) \right\} \tag{6.30}$$

$$= D(\beta, J) Tr_{\{\sigma_i^{\delta}\}} \left\{ \prod_{\langle i,j \rangle} \exp\left[\frac{\beta^2 J^2}{2} \left(\sum_{\alpha} \sigma_i^{\alpha} \sigma_j^{\alpha} \right)^2 \right] \right\}$$

where $D(\beta, J)$ comes from the factors of $C(\beta, J)$, and the critical properties of the material arise from the trace on the far right side of Eq. (6.30). Consequently,

$$\overline{Z_{rep}} = \overline{Z^n} \propto Tr_{\{\sigma_i^{\delta}\}} \left\{ \prod_{\langle i,j \rangle} \exp\left[\frac{\beta^2 J^2}{2} \left(\sum_{\alpha=1}^{n} \sigma_i^{\alpha} \sigma_j^{\alpha} \right)^2 \right] \right\} \tag{6.31}$$

$$= Tr_{\{\sigma_i^{\delta}\}} \left\{ \prod_{\langle i,j \rangle} \exp\left[\frac{\beta^2 J^2}{2} \sum_{\alpha,\gamma=1}^{n} \sigma_i^{\alpha} \sigma_j^{\alpha} \sigma_i^{\gamma} \sigma_j^{\gamma} \right] \right\}$$

where here we have explicitly written the sum over α as running over the n replicas of the systems.

The study of the far right form in Eq. (6.31) allows us to determine the conditions needed to observe a low-temperature phase transition in the system to a spin glass ordering in which $q \neq 0$.

(Here the order parameter q is defined as in Eq. (6.22) and involves the results for two different replicas treated in the $n \to 0$ limit.) In addition, notice that the disorder due to the random nature of the exchange couplings is removed from the partition function in Eq. (6.31), and $\overline{Z_{rep}} = \overline{Z^n}$ is essentially the partition function of a nonrandom system dependent on the temperature, on n, and on the variance of the distribution of exchange couplings in the original disordered model. This absence of disorder then acts as a simplification of the original random formulation [7–10].

A variety of techniques have been applied to approximate the solutions of $\overline{Z_{rep}} = \overline{Z^n}$ and study them as a function of n and T. These include both analytical and computer modeling approaches [8–10]. The free energy is ultimately obtained using the identity in Eq. (6.8b), which yields the free energy form [6]

$$\overline{F} = -k_B T \overline{\ln Z} = -k_B T \lim_{n \to 0} \frac{\overline{Z^n} - 1}{n} = -k_B T \lim_{n \to 0} \frac{\overline{Z_{rep}} - 1}{n}. \tag{6.32}$$

In their original paper, Edward and Anderson [6] used a type of variational method to approximate the free energy in Eq. (6.32), determining the properties of the system at the spin glass transition. In this approximation, they showed for a 3-dimensional system that near the critical temperature $T_c = \sqrt{\frac{2J^2}{9k_B^2}}$ for $T < T_c$

$$q(T) = \frac{1}{2}\left[\left(\frac{T_c}{T}\right)^2 - 1\right] \tag{6.33a}$$

and near $T = 0$

$$q(T) = 1 - \sqrt{\frac{8}{3\pi}\frac{T}{T_c}}. \tag{6.33b}$$

Here $q(T)$ is the Edward and Anderson [6] order parameter defined between two unmagnetized configurations of the system which are infinitely separated in time. It indicates the presence of the spin glass phase below a critical temperature T_c found in the system at low temperatures and with a growing ordered spin glass phase at decreasing temperatures. We shall have more to say about the replica order parameter in the following section.

Since their original work, other studies of spin glass have been made. One analytical study which is of great interest in this regard is the Sherrington-Kirkpatrick theory of spin glass. We now look at this approach to the study of the spin glass phase.

6.4 SHERRINGTON-KIRKPATRICK THEORY

Next, we study another model of spin glass known as the Sherrington-Kirkpatrick theory [7]. This is a model which is based on a generalization of the form in Eq. (6.20) and offers an interesting exactly solvable system illustrating many of the features displayed by spin glass [7–10]. It is solvable in the sense that it involves an application of the replica trick introduced in Subsection 6.3.2. In this regard, it also allows for an investigation of some of the difficulties arising in the use of the replica trick method.

Specifically, consider the Ising model given by the form [7–10]

$$H = -\sum_{\langle i,j\rangle} J_{i,j}\sigma_i\sigma_j - h\sum_i \sigma_i \tag{6.34}$$

where now the indices i and j run over all the sites of a crystal lattice, $\langle i,j\rangle$ represents a sum over pairs of sites on the lattice, $J_{i,j}$ is a randomly distributed exchange coupling between the Ising spins σ_i and σ_j on the i and j lattice sites, and h is an externally applied field. In this model, all pairs of lattice spins interact with one another through the set of random couplings $\{J_{i,j}\}$ which are no longer restricted to being nearest-neighbor interactions. Whereas the model in Eq. (6.20) only involved nearest-neighbor spin interactions, in the model in Eq. (6.34) every pair of spins in the system has a spin pair interaction of the form $-J_{i,j}\sigma_i\sigma_j$. As with our earlier considerations of Eq. (6.20), the restrictions to Ising spins and to a crystal lattice are made for the convenience of the presentation.

In the model in Eq. (6.34) the exchange couplings $\{J_{ij}\}$ are statistically distributed such that each coupling $J_{i,j}$ is characterized by the same probability distribution regardless of the separation between the i and j lattice sites. For our discussions we take the exchange coupling in the system to be distributed by the Gaussian probability distribution having the form [7–10]

$$P(J_{i,j}) = \sqrt{\frac{N}{2\pi J^2}}\exp\left[-\frac{N}{2}\left(\frac{J_{i,j}}{J}\right)^2\right] \tag{6.35}$$

where J is the variance of the distribution, $N \to \infty$ is the number of lattice sites in the model, and $\int_{-\infty}^{\infty} dJ_{i,j}P(J_{i,j}) = 1$. Notice that the factor of N is introduced into Eq. (6.35) as in the present model each site i in the lattice is coupled to the other $N \to \infty$ sites of the lattice by nonzero couplings of the form $\{J_{i,j}\}$. For the total energy of the resulting Ising model to scale as N (i.e., for the energy of the system to be an extensive quantity) the $J_{i,j}$ must, consequently, be characterized by an N-dependent distribution of the form in Eq. (6.35). In addition, notice that, as in our previous discussions involving the distribution in Eq. (6.21), the distribution in Eq. (6.35) is equally weighted to couplings over both positive and negative values of $J_{i,j}$. This introduces an element of frustration into the system which ultimately leads to a spin glass transition involving a time-independent randomly ordered configuration of spins.

Applying the earlier ideas of the replica method, based on the spin glass Hamiltonian in Eq. (6.34) we introduce a set of n replica Hamiltonians $\{H^\alpha\}$ where the αth member of this set is denoted by [7–10]

$$H^\alpha = -\sum_{\langle i,j\rangle} J_{i,j}\sigma_i^\alpha\sigma_j^\alpha - h\sum_i \sigma_i^\alpha. \tag{6.36}$$

Here $\alpha = 1,2,\ldots,n$ labels the n replicas, $\sigma_i^\alpha = \pm 1$ is the spin at the ith lattice site of the αth replica of the Ising model in Eq. (6.34), and the set of exchange couplings $\{J_{i,j}\}$ are the same for each member of the set of replica Hamiltonians $\{H^\alpha\}$. Note, however, that now the present couplings $\{J_{i,j}\}$ extend over the entire lattice and each coupling is statistically distributed by Eq. (6.35). As per our earlier considerations of the replica method, the different replicas in the set $\{\alpha\}$ can be considered as states of the system separated by large or infinite time intervals.

From Eqs. (6.34) through (6.36) the replica partition function is defined in terms of the nth power of the partition function of the single system of Eq. (6.34). Consequently, the thermodynamic and configurational average of the nth power of the single system partition function is written as

$$\overline{Z_{rep}} = \overline{Z^n} = \int\left[\prod_{\langle i,j\rangle} dJ_{i,j}P(J_{i,j})\right]Z^n = \int\left[\prod_{\langle i,j\rangle} dJ_{i,j}P(J_{i,j})\right]Z_{rep} \tag{6.37}$$

$$= \int\left[\prod_{\langle i,j\rangle} P(J_{i,j})dJ_{i,j}\right]Tr_{\{\sigma_i^\gamma\}}\exp\left(-\beta\sum_\alpha H^\alpha\right)$$

where the replica partition function $Z_{rep} = Z^n$ is the partition function formed from the n replica Hamiltonians in Eq. (6.36), $\langle i,j\rangle$ indicates a product of all the bond pairs in the lattice, the sum on α runs over the $\alpha = 1,2,\ldots..,n$ replica Hamiltonians in Eq. (6.36), and the trace is taken over all the replica spin configurations on the lattice.

Notice that the configurational average over the $\{J_{i,j}\}$ introduces an interaction between the spins of the different replica systems. In this regard, from Eq. (6.37) it then follows that

$$\overline{Z^n} = \int\left[\prod_{\langle i,j\rangle} dJ_{i,j}P(J_{i,j})\right]Tr\exp\left(\beta\sum_{\langle i,j\rangle} J_{i,j}\sum_{\alpha=1}^{n}\sigma_i^\alpha\sigma_j^\alpha + \beta h\sum_{i=1}^{N}\sum_{\alpha=1}^{n}\sigma_i^\alpha\right) \tag{6.38a}$$

where the trace is again over the set $\{\sigma_i^\delta\}$ of lattice spins and their replica variables. In the case that $h = 0$ and using Eq. (6.35), we find that Eq. (6.38a) is further reduced to [7–10]

$$\overline{Z^n} = Tr\int\prod_{\langle i,j\rangle}\left\{\frac{1}{J}\sqrt{\frac{N}{2\pi}}dJ_{i,j}\exp\left[-\frac{N}{2J^2}J_{i,j}^2 + \beta J_{i,j}\sum_{\alpha=1}^{n}\sigma_i^\alpha\sigma_j^\alpha\right]\right\} \tag{6.38b}$$

$$= Tr\prod_{\langle i,j\rangle}\exp\left(\frac{\beta^2 J^2}{2N}\sum_{\alpha,\gamma=1}^{n}\sigma_i^\alpha\sigma_j^\alpha\sigma_i^\gamma\sigma_j^\gamma\right)$$

$$= Tr\exp\left(\frac{\beta^2 J^2}{2N}\sum_{\langle i,j\rangle}\sum_{\alpha,\gamma=1}^{n}\sigma_i^\alpha\sigma_j^\alpha\sigma_i^\gamma\sigma_j^\gamma\right).$$

Here in the second equality, the integrals over the set of $\{J_{i,j}\}$ have been evaluated over the range $-\infty < J_{i,j} < \infty$.

Next consider in the $N \to \infty$ limit the sum following the second equality of Eq. (6.38b). We find that [7–10]

$$\sum_{\langle i,j\rangle}\sum_{\alpha,\gamma=1}^{n}\sigma_i^\alpha\sigma_j^\alpha\sigma_i^\gamma\sigma_j^\gamma = \sum_{\langle i,j\rangle}\sum_{\alpha,\gamma=1}^{n}\sigma_i^\alpha\sigma_i^\gamma\sigma_j^\alpha\sigma_j^\gamma \tag{6.39}$$

$$= \frac{1}{2}\left[-Nn^2 + \sum_{i,j}\sum_{\alpha,\gamma=1}^{n}\sigma_i^\alpha\sigma_i^\gamma\sigma_j^\alpha\sigma_j^\gamma\right] = -\frac{1}{2}Nn^2 + \frac{1}{2}\sum_{\alpha,\gamma}\left(\sum_i\sigma_i^\alpha\sigma_i^\gamma\right)^2$$

$$= \frac{1}{2}\left[N^2 n - Nn^2\right] + \sum_{\langle\alpha,\gamma\rangle}\left(\sum_i\sigma_i^\alpha\sigma_i^\gamma\right)^2 \cong \frac{1}{2}N^2 n + \sum_{\langle\alpha,\gamma\rangle}\left(\sum_i\sigma_i^\alpha\sigma_i^\gamma\right)^2$$

where $\langle\alpha,\gamma\rangle$ indicates a sum over pairs of different replicas α and γ. Using this in Eq. (6.38b), it then follows that [7–10]

$$\overline{Z^n} = \exp\left(\frac{N\beta^2 J^2 n}{4}\right) Tr\exp\left(\frac{\beta^2 J^2}{2N}\sum_{\langle\alpha,\gamma\rangle}\left(\sum_i\sigma_i^\alpha\sigma_i^\gamma\right)^2\right). \tag{6.40}$$

In the resulting form of the partition function in Eq. (6.40) notice that the trace is over the set $\{\sigma_i^\alpha\}$ of different lattice spins and their replica variables, and that only the replica spins on a single lattice site are coupled together by the nonlinear interaction in the second exponential. In this regard, the problem reduces to a set of isolated spins on each lattice site so that a spin on the lattice only couples to its replicas. The partition function in Eq. (6.40) can now be linearized in the couplings of the replica spins.

The linearization of Eq. (6.40) is made on the basis of an identity involving Gaussian integrals. Specifically, the identity has the general form [7–10]

$$\int_{-\infty}^{\infty} dx e^{-ax^2 + bx} = \sqrt{\frac{\pi}{a}}e^{\frac{b^2}{4a}} \tag{6.41}$$

where on the left side of the equation b occurs linearly in the argument of the exponential, whereas on the right side of the equation the nonlinear factor b^2 is found in the argument of the exponential. Using the identity in Eq. (6.41), we can now write the identity

$$\int_{-\infty}^{\infty} dq_{\alpha\gamma}\exp\left[-\frac{N\beta^2 J^2}{2}q_{\alpha\gamma}^2 + \beta^2 J^2\left(\sum_i\sigma_i^\alpha\sigma_i^\gamma\right)q_{\alpha\gamma}\right] \tag{6.42}$$

$$= \sqrt{\frac{2\pi}{N\beta^2 J^2}}\exp\left[\frac{\beta^2 J^2}{2N}\left(\sum_i\sigma_i^\alpha\sigma_i^\gamma\right)^2\right].$$

Applying Eq. (6.42) in Eq. (6.40), it then follows that [7–10]

$$\overline{Z^n} = \exp\left(\frac{N\beta^2 J^2 n}{4}\right)\times \tag{6.43}$$

$$\times\int\prod_{\langle\alpha,\gamma\rangle}\left[\sqrt{\frac{N\beta^2J^2}{2\pi}}dq_{\alpha\gamma}\exp\left(-\frac{N\beta^2J^2}{2}q_{\alpha\gamma}^2\right)\right]Tr\exp\left(\beta^2J^2\sum\nolimits_{\langle\delta,\varepsilon\rangle}q_{\delta\varepsilon}\sum_i\sigma_i^\delta\sigma_i^\epsilon\right)$$

and the couplings between the replicas spins on the ith site are now linear in $\sigma_i^\alpha\sigma_i^\gamma$. In addition, we note the symmetry that $q_{\alpha\gamma} = q_{\gamma\alpha}$ in our considerations [7–10].

From Eq. (6.32) the average free energy of the spin glass is then obtained from Eq. (6.43) using the replica trick given by [6]

$$\overline{F} = -k_BT\overline{\ln Z} = -k_BT\lim_{n\to0}\frac{\overline{Z^n}-1}{n}. \tag{6.44}$$

In the following we shall study Eqs. (6.32) and (6.44) written in terms of $\overline{Z^n}$ as given in Eq. (6.43). The focus is on determining the behaviors of $\overline{Z^n}$ and $\overline{F}$ as functions of the general analytically continued variable n with particular attention to the $n\to0$ limit.

To evaluate Eq. (6.44) for general n the method of steepest descent [11] is used. (A brief review of this method [11] is found in Appendix A.) In the method of steepest descent $\overline{Z^n}$ is taken to be the extremal value in $\{q_{\alpha\gamma}\}$ of its integrand is taken such that as $n\to0$ we also have $\overline{Z^n}\to1$. Consequently, from Eqs. (6.43) and (6.44) we find that [7–10]

$$\overline{F} = -k_BT\frac{1}{n}\left\{\exp\left(\frac{\beta^2J^2nN}{4}\right)\exp\left(-\frac{N\beta^2J^2}{2}\sum\nolimits_{\langle\alpha,\gamma\rangle}q_{\alpha\gamma}^2\right)Tr\exp\left(\beta^2J^2\sum\nolimits_{\langle\delta,\varepsilon\rangle}q_{\delta\varepsilon}\sum_i\sigma_i^\delta\sigma_i^\epsilon\right)-1\right\} \tag{6.45a}$$

where the trace is over the entire set of spins $\{\sigma_l^\delta\}$ where l runs over the lattice sites and δ runs over all the replicas. In the course of our discussion, the extremal values of $\{q_{\alpha\gamma}\}$ will be determined, but for now, it is assumed that they are a priori known.

To perform a further simplification on Eq. (6.45a), notice that all the sites labeled by different i are disconnected from one another so that we are left with N single site systems. This then allows us to rewrite Eq. (6.45a) into the form [7–10]

$$\overline{F} = -k_BT\frac{1}{n}\left\{\exp\left(\frac{\beta^2J^2nN}{4}\right)\exp\left(-\frac{N\beta^2J^2}{2}\sum\nolimits_{\alpha,\gamma}q_{\alpha\gamma}^2\right)\left[Tr\left(\exp\left[\beta^2J^2\sum_{\delta,\varepsilon}q_{\delta\varepsilon}\sigma^\delta\sigma^\epsilon\right]\right)\right]^N-1\right\} \tag{6.45b}$$

where now σ^α and σ^γ are spins on a generic single site of the lattice, and the trace in Eq. (6.45b) is only over replica spin pairs of σ^α and σ^γ with $\alpha\neq\gamma$. In the limit, $n\to0$ we expect that all three of the exponentials in Eq. (6.45b) should approach unity. Consequently, in the limit as $n\to0$ for $N\gg1$, it follows from Eq. (6.45b) that [7–10]

$$\overline{F}\to -k_BTN\left\{-\frac{\beta^2J^2}{4n}\sum_{\alpha\neq\gamma}q_{\alpha\gamma}^2+\frac{1}{n}\ln\left(Tr\ e^L\right)+\frac{1}{4}\beta^2J^2\right\} \tag{6.45c}$$

where

$$L = \beta^2 J^2 \sum_{\langle \delta,\varepsilon \rangle} q_{\delta\varepsilon} \sigma^\delta \sigma^\varepsilon = \frac{1}{2}\beta^2 J^2 \sum_{\delta \neq \varepsilon} q_{\delta\varepsilon} \sigma^\delta \sigma^\epsilon . \tag{6.45d}$$

Next, we determine the set of $\{q_{\alpha\gamma}\}$ which extremizes the free energy form in Eq. (6.45).

To extremize $\bar{F}$ in Eq. (6.45c) we compute [7–10]

$$\frac{d\bar{F}}{dq_{\alpha\gamma}} = -k_B TN \left\{ -\frac{\beta^2 J^2}{2n} q_{\alpha\gamma} + \frac{\beta^2 J^2}{2n} \frac{Tr \sigma^\alpha \sigma^\gamma e^L}{Tr e^L} \right\} = 0 \tag{6.46a}$$

so that from Eq. (6.46a) it then follows that the extremizing set of $\{q_{\alpha\gamma}\}$ is given by [8–10]

$$q_{\alpha\gamma} = \frac{Tr\left(\sigma^\alpha \sigma^\gamma e^L\right)}{Tr\left(e^L\right)}. \tag{6.46b}$$

In the limit as $n \to 0$, we next show that $q_{\alpha\gamma}$ in Eq. (6.46b) is the same as the order parameter for the spin glass transition defined in Eq. (6.22b) by the form [8–10]

$$q_{\alpha\gamma} = \frac{1}{N} \overline{\left\langle \sum_i \sigma_i^\alpha \sigma_i^\gamma \right\rangle} = \overline{\left\langle \sigma^\alpha \sigma^\gamma \right\rangle} \tag{6.46c}$$

where subscripts for the two replicas α and γ have now been included in the left side of Eq. (6.46c), and we account for the self-averaging of the system in the limit that $N \to \infty$.

In making this comparison we remind the reader that the order parameter in Eq. (6.46c) is defined by the $n \to 0$ limit of [8–10]

$$\overline{\frac{Tr \sigma_i^\alpha \sigma_i^\gamma \exp\left(-\beta \sum_\delta H^{(\delta)}\right)}{Tr \exp\left(-\beta \sum_\delta H^{(\delta)}\right)}} = \overline{\left\langle \sigma_i^\alpha \sigma_i^\gamma \right\rangle} = \overline{\left\langle \sigma^\alpha \sigma^\gamma \right\rangle} \tag{6.47a}$$

where the bar over the formulas indicates an average over the configurations of $\{J_{i,j}\}$. In particular, our focus in the proceeding discussions is now showing for $n \to 0$ that [8–10]

$$q_{\alpha\gamma} = \frac{Tr\left(\sigma^\alpha \sigma^\gamma e^L\right)}{Tr\left(e^L\right)} = \overline{\frac{Tr \sigma_i^\alpha \sigma_i^\gamma \exp\left(-\beta \sum_\delta H^{(\delta)}\right)}{Tr \exp\left(-\beta \sum_\delta H^{(\delta)}\right)}} = \overline{\left\langle \sigma^\alpha \sigma^\gamma \right\rangle} \tag{6.47b}$$

where the important point in our treatment is to prove the validity of the second equality in Eq. (6.47b). Once this is established, it follows that $q_{\alpha\gamma} = \overline{\left\langle \sigma^\alpha \sigma^\gamma \right\rangle}$, and our equality is demonstrated.

First note that as $n \to 0$ the denominator in Eq. (6.47a) is just $Z^n \to 1$ so that [8–10]

$$\overline{\frac{Tr \sigma_i^\alpha \sigma_i^\gamma \exp\left(-\beta \sum_\delta H^\delta\right)}{Tr \exp\left(-\beta \sum_\delta H^\delta\right)}} \to \overline{Tr \sigma_i^\alpha \sigma_i^\gamma \exp\left(-\beta \sum_\delta H^\delta\right)} = \overline{\left\langle \sigma_i^\alpha \sigma_i^\gamma \right\rangle} = \overline{\left\langle \sigma^\alpha \sigma^\gamma \right\rangle}, \tag{6.48}$$

and the resulting expression for the two-spin average $\overline{\left\langle \sigma_i^\alpha \sigma_i^\gamma \right\rangle}$ looks mathematically similar to the average of the replica partition function in Eq. (6.37). Consequently, to evaluate the right side of Eq. (6.48) we proceed as in Eqs. (6.38a) and (6.38b) in our considerations of $\bar{Z}$, writing [7–10]

$$\overline{Tr\sigma_i^\alpha\sigma_i^\gamma \exp\left(-\beta\sum_\delta H^\delta\right)} = \tag{6.49}$$

$$= \int\left[\prod_{\langle i,j\rangle} dJ_{i,j}P\left(J_{i,j}\right)\right] Tr\ \sigma_i^\alpha\sigma_i^\gamma \exp\left(\beta\sum\nolimits_{\langle i,j\rangle} J_{i,j}\sum_{\delta=1}^{n}\sigma_i^\delta\sigma_j^\delta\right)$$

$$= Tr\int\prod_{\langle i,j\rangle}\left\{\frac{1}{J}\sqrt{\frac{N}{2\pi}}dJ_{i,j}\sigma_i^\alpha\sigma_i^\gamma \exp\left[-\frac{N}{2J^2}J_{ij}^2 + \beta J_{i,j}\sum_{\delta=1}^{n}\sigma_i^\delta\sigma_j^\delta\right]\right\}$$

$$= Tr\sigma_i^\alpha\sigma_i^\gamma \exp\left(\frac{\beta^2J^2}{2N}\sum_{\langle i,j\rangle}\sum_{\delta,\varepsilon=1}^{n}\sigma_i^\delta\sigma_j^\delta\sigma_i^\varepsilon\sigma_j^\epsilon\right)$$

where the trace is made again over the set $\left\{\sigma_i^\mu\right\}$. Applying the method used in Eqs. (6.39) through (6.40), Eq. (6.49) can then be written in the form

$$\overline{Tr\sigma_i^\alpha\sigma_i^\gamma \exp\left(-\beta\sum_\delta H^\delta\right)} = \exp\left(\frac{N\beta^2J^2n}{4}\right) Tr\ \sigma_i^\alpha\sigma_i^\gamma \exp\left(\frac{\beta^2J^2}{2N}\sum_{\langle\delta,\varepsilon\rangle}\left(\sum_i\sigma_i^\delta\sigma_i^\epsilon\right)^2\right). \tag{6.50}$$

The $\left(\sum_i\sigma_i^\delta\sigma_i^\varepsilon\right)^2$ nonlinearity in the set of sums in Eq. (6.50) is then removed using the identity in Eq. (6.42). This gives [7–10]

$$\overline{Tr\sigma_i^\alpha\sigma_i^\gamma \exp\left(-\beta\sum_\delta H^\delta\right)} = \exp\left(\frac{N\beta^2J^2n}{4}\right)\times \tag{6.51}$$

$$\times\int\prod_{\langle\delta,\varepsilon\rangle}\left(\sqrt{\frac{N\beta^2J^2}{2\pi}}dq_{\delta\varepsilon}\right)\exp\left(-\frac{N\beta^2J^2}{2}\sum\nolimits_{\langle\delta,\varepsilon\rangle}q_{\delta\varepsilon}^2\right) Tr\ \sigma_i^\alpha\sigma_i^\gamma \exp\left(\beta^2J^2\sum\nolimits_{\langle\delta,\varepsilon\rangle}q_{\delta\varepsilon}\sum_i\sigma_i^\delta\sigma_i^\epsilon\right).$$

From the definition in Eq. (6.45d) and applying the method of steepest descent (see Appendix A), we then find that

$$\overline{Tr\sigma_i^\alpha\sigma_i^\gamma \exp\left(-\beta\sum_\delta H^\delta\right)} = \exp\left(\frac{N\beta^2J^2n}{4}\right)\times \tag{6.52a}$$

$$\times\int\prod_{\langle\delta\varepsilon\rangle}\left(\sqrt{\frac{N\beta^2J^2}{2\pi}}dq_{\delta\varepsilon}\right)\exp\left(-\frac{N\beta^2J^2}{2}\sum\nolimits_{\langle\delta\varepsilon\rangle}q_{\delta\varepsilon}^2\right) Tr\ \sigma_i^\alpha\sigma_i^\gamma e^L$$

$$= Tr\left(\sigma^\alpha\sigma^\gamma e^L\right) = \frac{Tr\left(\sigma^\alpha\sigma^\gamma e^L\right)}{Tr\left(e^L\right)}.$$

Here in the final equality, we use $Tr\left(e^L\right)\to 1$ as $n\to 0$. Going back to Eq. (6.48) we then see from Eq. (6.52a) that as $n\to 0$

$$\frac{Tr\left(\sigma^{\alpha}\sigma^{\gamma}e^{L}\right)}{Tr\left(e^{L}\right)}=\frac{\overline{Tr\,\sigma_i^{\alpha}\sigma_i^{\gamma}\exp\left(-\beta\sum_{\delta}H^{(\delta)}\right)}}{Tr\exp\left(-\beta\sum_{\delta}H^{(\delta)}\right)} \tag{6.52b}$$

which is what we needed to show [7–10]. This demonstrates the equivalence in Eq. (6.47b).

6.4.1 Symmetric Sherrington-Kirkpatrick Solution

Next, we obtain a specific solution for the extremizing set of $\{q_{\alpha\gamma}\}$. Since the replica Hamiltonians are identical to one another, it is natural to look for an extremizing solution for which $q_{\alpha\gamma}=q$ for all $\alpha\neq\gamma$, i.e., $q_{\alpha\gamma}$ does not depend on its subscripts α and γ. This maintains the symmetry between the replicas of the Hamiltonian [7–10] in Eq. (6.36) and generates the so-called symmetric solution. Under this condition, our free energy form in Eq. (6.45b) then becomes

$$\bar{F}=-k_BT\frac{1}{n}\left\{\exp\left(\frac{\beta^2J^2nN}{4}\right)\exp\left(-\frac{N\beta^2J^2n(n-1)}{4}q^2\right)\left[Tr\left(\exp\left[\beta^2J^2q\sum_{\alpha,\gamma}\sigma^{\alpha}\sigma^{\gamma}\right]\right)\right]^N-1\right\} \tag{6.53}$$

where σ^{α} and σ^{γ} are replica spins on the same arbitrarily chosen lattice site.

From Eq. (6.41) we next develop the identity [6–11]

$$\frac{1}{\sqrt{2\pi}}\int dze^{-\frac{z^2}{2}}\exp\left[\beta J\sqrt{q}\sum_{\alpha}\sigma^{\alpha}z\right]=\exp\left[\frac{\beta^2J^2q}{2}\left(\sum_{\alpha}\sigma^{\alpha}\right)^2\right]. \tag{6.54}$$

This allows us to further simplify the trace in Eq. (6.53) so that we find that [7–11]

$$\left[Tr\left(\exp\left[\beta^2J^2q\sum_{\langle\alpha,\gamma\rangle}\sigma^{\alpha}\sigma^{\gamma}\right]\right)\right]^N= \tag{6.55}$$

$$=\left[Tr\left(\exp\left[\frac{\beta^2J^2q}{2}\sum_{\alpha,\gamma}\sigma^{\alpha}\sigma^{\gamma}\right]-n\frac{\beta^2J^2q}{2}\right)\right]^N$$

$$=\left\{Tr\left(e^{-\frac{\beta^2J^2}{2}nq}\frac{1}{\sqrt{2\pi}}\int dze^{-\frac{z^2}{2}}\exp\left[\beta J\sqrt{q}\sum_{\alpha}\sigma^{\alpha}z\right]\right)\right\}^N$$

$$=\left\{e^{-\frac{\beta^2J^2}{2}nq}\frac{1}{\sqrt{2\pi}}\int dze^{-\frac{z^2}{2}}Tr\exp\left[\beta J\sqrt{q}\sum_{\alpha}\sigma^{\alpha}z\right]\right\}^N.$$

Here in the sum $\sum_{\alpha,\gamma}\sigma^{\alpha}\sigma^{\gamma}$ after the first equality both α and γ run over the n replicas, and the factor of $-n\frac{\beta^2J^2q}{2}$ removes the terms with $\alpha=\gamma$ from the sum $\frac{\beta^2J^2q}{2}\sum_{\alpha,\gamma}\sigma^{\alpha}\sigma^{\gamma}$.

Applying Eq. (6.55) in Eq. (6.53), the free energy takes the form [7–10]

$$\bar{F}=-k_BT\frac{1}{n}\times \tag{6.56a}$$

$$\times\left\{\exp\left(\frac{\beta^2J^2nN}{4}\right)\exp\left(-\frac{N\beta^2J^2n(n-1)q^2}{4}\right)\left\{e^{-\frac{\beta^2J^2}{2}nq}\frac{1}{\sqrt{2\pi}}\int dze^{-\frac{z^2}{2}}Tr\exp\left[\beta J\sqrt{q}\sum_\alpha\sigma^\alpha z\right]\right\}^N-1\right\}$$

Similarly, using the properties of logarithms in Eq. (6.53), the free energy is also given by

$$\bar{F}==-k_BT\frac{1}{n}\times\left\{\exp\left(\frac{\beta^2J^2nN}{4}\right)\exp\left(-\frac{N\beta^2J^2n(n-1)q^2}{4}\right)\exp N\ln\left(Tre^L\right)-1\right\}. \quad (6.56b)$$

Consequently, we see that [7–10]

$$\ln\left(Tr\ e^L\right)=-\frac{\beta^2J^2}{2}nq+\ln\left\{\frac{1}{\sqrt{2\pi}}\int dze^{-\frac{z^2}{2}}Tr\exp\left[\beta J\sqrt{q}\sum_\alpha\sigma^\alpha z\right]\right\} \quad (6.57)$$

$$=-\frac{\beta^2J^2}{2}nq+\ln\left\{\frac{1}{\sqrt{2\pi}}\int dze^{-\frac{z^2}{2}}\left[2\cosh\left(\beta J\sqrt{q}z\right)\right]^n\right\}$$

$$=-\frac{\beta^2J^2}{2}nq+\ln\left\{\frac{1}{\sqrt{2\pi}}\int dze^{-\frac{z^2}{2}}\exp\left\{n\ln\left[2\cosh\left(\beta J\sqrt{q}z\right)\right]\right\}\right\}.$$

Here after the second equality the trace over the replica spins $\{\sigma^\alpha\}$ has been taken, and after the third equality the properties of logarithms have again been applied. We shall next focus on simplifying Eqs. (6.56) and (6.57) in the limit that $n\to 0$.

In the limit that $n\to 0$, from Eq. (6.57) it follows that [7–10]

$$\frac{1}{n}\ln\left(Tr\ e^L\right)\to-\frac{\beta^2J^2}{2}q+\frac{1}{n}\ln\left\{\frac{1}{\sqrt{2\pi}}\int dze^{-\frac{z^2}{2}}\left\{1+n\ln\left[2\cosh\left(\beta J\sqrt{q}z\right)\right]\right\}\right\} \quad (6.58a)$$

$$=-\frac{\beta^2J^2}{2}q+\frac{1}{n}\ln\left\{1+\frac{1}{\sqrt{2\pi}}\int dze^{-\frac{z^2}{2}}n\ln\left[2\cosh\left(\beta J\sqrt{q}z\right)\right]\right\}$$

$$\to-\frac{\beta^2J^2}{2}q+\frac{1}{\sqrt{2\pi}}\int dze^{-\frac{z^2}{2}}\ln\left[2\cosh\left(\beta J\sqrt{q}z\right)\right]$$

where we have used the limiting form $\ln(1+x)\to x$ for $|x|\ll 1$. Applying Eq. (6.58a) in Eq. (6.56) it is then found that [7–10]

$$\bar{F}=-N\frac{\beta J^2}{4}(1-q)^2-N\frac{k_BT}{\sqrt{2\pi}}\int dze^{-\frac{z^2}{2}}\ln\left[2\cosh\left(\beta J\sqrt{q}z\right)\right], \quad (6.58b)$$

so that extremizing $\bar{F}$ with respect to q gives

$$\frac{\partial\bar{F}}{\partial q}=\frac{\beta J^2}{2}(1-q)-\frac{1}{\sqrt{2\pi}}\frac{J}{2\sqrt{q}}\int dze^{-\frac{z^2}{2}}z\tanh\left(\beta J\sqrt{q}z\right)=0 \quad (6.59)$$

Consequently, the extremum of q is given by the solutions of [7–10]

$$q = 1 - \frac{1}{\sqrt{2\pi}} \int dz e^{-\frac{z^2}{2}} \left[\operatorname{sech}\left(\beta J \sqrt{q} z \right) \right]^2 = \frac{1}{\sqrt{2\pi}} \int dz e^{-\frac{z^2}{2}} \left[\tanh\left(\beta J \sqrt{q} z \right) \right]^2. \tag{6.60}$$

This is a nonlinear equation which must be solved self-consistently for the order parameter q as a function of temperature.

As we shall presently see, an investigation of the solutions of Eq. (6.60) indicates a region of nonzero solutions for q when $\frac{k_B T}{J} < 1$. This indicates the possibility of a spin glass phase in this region. On the other hand, it is found that $q = 0$ for $\frac{k_B T}{J} > 1$. In this region the system exists in a paramagnetic phase [7–10]. We next consider the transition between these two regions in more detail.

6.4.2 The Glass Transition

In the limit that $q \to 0$ the free energy in Eq. (6.58) can be Taylor expanded in q. This allows us to study the nature of the phase transition from a state in which $q = 0$ to one in which $q \neq 0$. This encompasses the region of the transition between paramagnetic and spin glass phases [7–10], i.e., states without a net magnetization.

To second order in q, the average free energy as the transition is approached is given by

$$\bar{F} = -N \frac{\beta J^2}{4} - N \ln 2 - N \frac{\beta J^2}{4} \left(1 - \beta^2 J^2 \right) q^2 + O\left(q^3 \right), \tag{6.61}$$

so that extremizing this form for $\bar{F}$ in q it follows from Eq. (6.61) that [7–10]

$$q = \beta^2 J^2 q + O\left(q^2 \right). \tag{6.62}$$

For $|q| \ll 1$ we then find from Eq. (6.62) that a nonzero solution of q, indicating the phase transition from the paramagnetic state, should first occur under the condition that [7–10]

$$\beta^2 J^2 = 1 \text{ or } \frac{k_B T}{J} = 1. \tag{6.63}$$

This provides the critical temperature $T_c = \frac{J}{k_B}$ for the transition to a nonzero q.

However, to ensure that the transition in the region $\frac{k_B T}{J} < 1$ is to a $q \neq 0$ spin glass state we also must show that in this region [7–10]

$$\overline{\left\langle \sigma_i^\alpha \right\rangle} = \overline{\frac{Tr \sigma_i^\alpha \exp\left(-\beta \sum_\delta H^{(\delta)} \right)}{Tr \exp\left(-\beta \sum_\delta H^{(\delta)} \right)}} = 0. \tag{6.64}$$

This requires that there is no net magnetization in the system for $q \neq 0$. If this were not the case, the transition in the absence of an externally applied field would exhibit a net magnetization (i.e., $\overline{\left\langle \sigma_i^\alpha \right\rangle} \neq 0$). This would indicate a type of ferromagnetic ordering in the $q \neq 0$ phase rather than a spin glass state.

In this regard, we argue as in our treatment of Eq. (6.48) that as $n \to 0$, $Z^n \to 1$ so that [7–10] from Eq. (6.64)

$$\overline{\left\langle \sigma_i^\alpha \right\rangle} \to \overline{Tr \sigma_i^\alpha \exp\left(-\beta \sum_\delta H^\delta \right)}. \tag{6.65}$$

For the case in which $P(J_{i,j})$ is symmetrically distributed over both positive and negative $J_{i,j}$, the system should not exhibit a predisposition to favor any net average spin direction. Consequently, it then follows for $q \neq 0$ that $\overline{\langle \sigma_l^\alpha \rangle} \to 0$ as $n \to 0$, and in this case the solution from Eq. (6.60) exhibits a spin glass transition. (Note: A more detailed analytical treatment of this result will be provided in the next section. There we will consider a generalization of the model to one supporting transitions to ferromagnetic orderings as well as to the spin glass transition.)

Next let us briefly look at a case in which the system can exhibit a net magnetization as well as the spin glass phase. This involves the discussion of an asymmetric probability density $P(J_{i,j})$, allowing us to analytically treat $\overline{\langle \sigma_l^\alpha \rangle}$ as per our discussions in Eq. (6.65).

6.5 GENERALIZED FORM OF THE SHERRINGTON-KIRKPATRICK MODEL

In the original treatment of their model, Sherrington and Kirkpatrick [7] studied a more general statistical distribution of exchange couplings than that considered in Eq. (6.35). Specifically, the distribution they treated was of the form [7–10]

$$P(J_{i,j}) = \sqrt{\frac{N}{2\pi J^2}} \exp\left[-\frac{N}{2J^2} \left(J_{i,j} - \frac{J_0}{N} \right)^2 \right], \tag{6.66}$$

which reduces to that in Eq. (6.35) in the limit in which $J_0 = 0$. Notice that, whereas the $J_0 = 0$ distribution in Eq. (6.66) is symmetric over positive and negative values in $J_{i,j}$, the distribution of $J_{i,j}$ in Eq. (6.66) for $J_0 \neq 0$ is asymmetrically distributed over the positive and negative values of $J_{i,j}$. This asymmetry allows the system, under certain conditions, to exhibit a net magnetization, indicative of a kind of ferromagnetic ordering. The generalized model based on Eq. (6.66) then allows for an additional discussion of the transition from the spin glass to a ferromagnetic state as a function of $J_0 > 0$. We shall next briefly summarize the derivation of the theory for the generalization based on Eq. (6.66) and describe the conditions on $J_0 > 0$ under which the system transitions from a spin glass to a ferromagnetic behavior. The derivation closely parallels that given earlier for the $J_0 = 0$ model so we shall only briefly outline the differences between the two derivations in the following discussions.

Applying the distribution in Eq. (6.66) in Eqs. (6.37) and (6.38a), we find that the average of the replica partition function over the distributions of $\{J_{i,j}\}$ is now given by [7–10]

$$\begin{aligned}
\overline{Z^n} &= Tr \int \prod_{\langle i,j \rangle} \left\{ \frac{1}{J} \sqrt{\frac{N}{2\pi}} dJ_{ij} \exp\left[-\frac{N}{2J^2} \left(J_{i,j} - \frac{J_0}{N} \right)^2 + \beta J_{ij} \sum_{\alpha=1}^{n} \sigma_i^\alpha \sigma_j^\alpha \right] \right\} \\
&= Tr \prod_{\langle i,j \rangle} \exp\left(\frac{\beta^2 J^2}{2N} \sum_{\alpha,\gamma=1}^{n} \sigma_i^\alpha \sigma_j^\alpha \sigma_i^\gamma \sigma_j^\gamma + \frac{\beta J_0}{N} \sum_{\alpha=1}^{n} \sigma_i^\alpha \sigma_j^\alpha \right) \\
&= Tr \exp\left(\sum_{\langle i,j \rangle} \left[\frac{\beta^2 J^2}{2N} \sum_{\alpha,\gamma=1}^{n} \sigma_i^\alpha \sigma_j^\alpha \sigma_i^\gamma \sigma_j^\gamma + \frac{\beta J_0}{N} \sum_{\alpha=1}^{n} \sigma_i^\alpha \sigma_j^\alpha \right] \right)
\end{aligned} \tag{6.67}$$

in place of the $J_0 = 0$ result in Eq. (6.38b). Here the form following the second equality results from the integration over the $\{J_{ij}\}$ and is seen to now introduce an addition J_0 dependent term in the exponentials. Making use of the spin identity in Eq. (6.39), Eq. (6.67) is then rewritten as [7–10]

$$\overline{Z^n} = \exp\left(\frac{N\beta^2 J^2 n}{4}\right) Tr \exp\left(\frac{\beta^2 J^2}{2N}\sum_{\langle\alpha,\gamma\rangle}\left(\sum_i \sigma_i^\alpha \sigma_i^\gamma\right)^2 + \frac{\beta J_0}{2N}\sum_\alpha\left(\sum_i \sigma_i^\alpha\right)^2\right). \tag{6.68}$$

Notice that in the limit of $J_0 = 0$ this reduces to the previous result for $\overline{Z^n}$ given in Eq. (6.40).

Next, we apply the Gaussian identity in Eq. (6.42) along with a similar identity involving a new set of variables $\{m_\alpha\}$ labeled by the replica indices $\{\alpha\}$. Specifically, from Eq. (6.41) we have the identity [7–11]

$$\int_{-\infty}^{\infty} dm_\alpha \exp\left[-\frac{N\beta J_0}{2} m_\alpha^2 + \beta J_0\left(\sum_i \sigma_i^\alpha\right) m_\alpha\right] \tag{6.69}$$

$$= \sqrt{\frac{2\pi}{N\beta J_0}} \exp\left[\frac{\beta J_0}{2N}\left(\sum_i \sigma_i^\alpha\right)^2\right],$$

and applying Eqs. (6.42) and (6.69) to Eq. (6.68) we find that

$$\overline{Z^n} = \exp\left(\frac{N\beta^2 J^2 n}{4}\right) \int \prod_{\langle\alpha,\gamma\rangle}\left(\sqrt{\frac{N\beta^2 J^2}{2\pi}} dq_{\alpha\gamma}\right) \int \prod_\alpha \left(\sqrt{\frac{N\beta J_0}{2\pi}} dm_\alpha\right) \times \tag{6.70}$$

$$\exp\left(-\frac{N\beta^2 J^2}{2}\sum_{\langle\alpha,\gamma\rangle} q_{\alpha\gamma}^2\right) \exp\left(-\frac{N\beta J_0}{2}\sum_\alpha m_\alpha^2\right) \times$$

$$Tr \exp\left(\beta^2 J^2 \sum_{\langle\delta,\varepsilon\rangle} q_{\delta\varepsilon} \sum_i \sigma_i^\delta \sigma_i^\epsilon + \beta J_0 \sum_\alpha\left(\sum_i \sigma_i^\alpha\right) m_\alpha\right)$$

Here in Eq. (6.70) the trace is again made over the set $\{\sigma_l^\delta\}$ of lattice spins and their replicas. The result in Eq. (6.70) now generalizes Eq. (6.43) so that in the current expression for $\overline{Z^n}$ with $J_0 \neq 0$ there are two Gaussian integrals, one written in terms of $\{q_{\alpha\gamma}\}$ and another in terms of $\{m_\alpha\}$.

Proceeding as in Eqs. (6.43) through (6.45b), the replica trick is again used to generate from Eq. (6.70) the configuration average of the free energy. For this application, the method of steepest descent is also applied to write the average free energy in terms of the integrand in Eq. (6.70) evaluated at $\{q_{\alpha\gamma}\}$ and $\{m_\alpha\}$ chosen as the extrema of the integrand. (See Appendix A for a discussion of the method of steepest descent.) In this manner, we find [6–10]

$$\overline{F} = -k_B T \lim_{n\to 0} \frac{1}{n}\left[\overline{Z^n} - 1\right] \tag{6.71a}$$

where

$$\overline{Z^n} = \exp\left(\frac{\beta^2 J^2 nN}{4}\right) \exp\left(-\frac{N\beta^2 J^2}{2}\sum_{\langle\alpha,\gamma\rangle} q_{\alpha\gamma}^2\right) \exp\left(-\frac{N\beta J_0}{2}\sum_\alpha m_\alpha^2\right) \tag{6.71b}$$

$$\left[Tr \exp\left(\beta^2 J^2 \sum_{\langle\alpha,\gamma\rangle} q_{\alpha\gamma}\sigma^\alpha\sigma^\gamma + \beta J_0 \sum_\alpha \sigma^\alpha m_\alpha\right)\right]^N.$$

(Notice that the free energy form in Eq. (6.71) reduces to that of Eq. (6.45b) in the $J_0 = 0$ limit.) In addition, as in our earlier considerations, $\sigma^\alpha \sigma^\gamma$ indicates the product of two different replica spins taken on a typical site of the lattice (Note, the similar notation for the single spin σ^α.), and the trace is over these replica spins on the same lattice site.

Following our earlier discussions of the $J_0 = 0$ system we next take the limit of the free energy in Eq. (6.71) as $n \to 0$, and apply some logarithm identities in Eq. (6.71b). It follows from these considerations that [7–10]

$$\bar{F} \to -k_B TN\left\{-\frac{\beta^2 J^2}{4n}\sum_{\alpha\neq\gamma} q_{\alpha\gamma}^2 - \frac{\beta J_0}{2n}\sum_\alpha m_\alpha^2 + \frac{1}{n}\ln\left(Tr\ e^L\right) + \frac{1}{4}\beta^2 J^2\right\} \tag{6.72a}$$

where

$$L = \beta^2 J^2 \sum_{\langle\alpha,\gamma\rangle} q_{\alpha\gamma}\sigma^\alpha\sigma^\gamma + \beta J_0 \sum_\alpha \sigma^\alpha m_\alpha \tag{6.72b}$$

$$= \frac{1}{2}\beta^2 J^2 \sum_{\alpha\neq\gamma} q_{\alpha\gamma}\sigma^\alpha\sigma^\gamma + \beta J_0 \sum_\alpha \sigma^\alpha m_\alpha .$$

The results in Eq. (6.72) are again found to reduce to those in Eqs. (6.45c) and (6.45d) in the limit as $J_0 = 0$.

At this point, the values of $\{q_{\alpha\gamma}\}$ and $\{m_\alpha\}$ extremizing the free energy $\bar{F}$ are determined in terms of simple analytic forms from Eq. (6.72). These values of $\{q_{\alpha\gamma}\}$ and $\{m_\alpha\}$ then represent the extrema for application in the steepest descent treatment of Eq. (6.71). The process used to determine the extrema is the same as that applied earlier in Eqs. (6.46) through (6.52) for the $J_0 = 0$ case and is now briefly reviewed.

The extremizing values $\{q_{\alpha\gamma}\}$ of $\bar{F}$ in Eq. (6.72) are obtained from the conditions [7–10]

$$\frac{\partial \bar{F}}{\partial q_{\alpha\gamma}} = 0. \tag{6.73a}$$

As in our earlier considerations (See Eqs. (6.46) through (6.52).) these yield [7–10]

$$q_{\alpha\gamma} = \frac{Tr\left(\sigma^\alpha\sigma^\gamma e^L\right)}{Tr\left(e^L\right)} \tag{6.73b}$$

where for $n \to 0$ it follows that $Tr\left(e^L\right) \to 1$. Consequently, we find that [7–10]

$$q_{\alpha\gamma} \to Tr\left(\sigma^\alpha\sigma^\gamma e^L\right) \to \overline{\sigma^\alpha\sigma^\gamma} \tag{6.73c}$$

applying the method in Eq. (6.52), and again the $q's$ are related to the spin glass order parameter.

Similarly, the extremizing values $\{m_\alpha\}$ of $\bar{F}$ are obtained from

$$\frac{\partial \bar{F}}{\partial m_\alpha} = 0 \tag{6.74a}$$

yielding [7–10]

$$m_\alpha = \frac{Tr\left(\sigma^\alpha e^L\right)}{Tr\left(e^L\right)} \tag{6.74b}$$

where for $n \to 0$ it follows that $Tr\left(e^{L}\right) \to 1$. We then find that [7–10]

$$m_{\alpha} \to Tr\left(\sigma^{\alpha} e^{L}\right) \to \overline{\left\langle \sigma_{i}^{\alpha} \right\rangle} = \overline{\left\langle \sigma^{\alpha} \right\rangle} \tag{6.74c}$$

following the discussions of Eqs. (6.64) and (6.65), and again the $m's$ are related to the magnetization.

As in our discussions of the $J_0 = 0$ solutions, we next look for symmetric solutions of the replica systems. Specifically, these assume that $q_{\alpha\gamma} = q$, and now in addition that $m_{\alpha} = m$. This is the assumption that the order parameters are independent of the replica subscripts.

6.5.1 The Symmetric Solutions

As in the earlier treatment of the symmetric probability distribution function for $\{J_{i,j}\}$, we next assume a solution [7–10] with $q_{\alpha\gamma} = q$ and $m_{\alpha} = m$. Following the methodology used for $J_0 = 0$ in the discussions in Eqs. (6.53) through Eq. (6.60), we obtain for $J_0 \neq 0$ the free energy from Eq. (6.72) considered in the limit that $q_{\alpha\gamma} = q$ and $m_{\alpha} = m$. This is shown to be given by [7–10]

$$\bar{F} \to -k_B TN \left\{ -\frac{\beta^2 J^2}{4n} n(n-1) q^2 - \frac{\beta J_0}{2n} nm^2 + \frac{1}{n} \ln\left(Tr\ e^{L}\right) + \frac{1}{4} \beta^2 J^2 \right\} \tag{6.75a}$$

where q and m are the extrema of the free energy in Eq. (6.75a). Treating the $n \to 0$ limit of the trace term in Eq. (6.75a), it then follows that

$$\ln\left(Tr\ e^{L}\right) = \ln\left\{ 1 + n \int \frac{dz\ e^{-\frac{z^2}{2}}}{\sqrt{2\pi}} \ln\left[2\cosh\left(\beta J \sqrt{q} z + \beta J_0 m\right)\right] - \frac{n}{2} \beta^2 J^2 q \right\}. \tag{6.75b}$$

From Eq. (6.75) as $n \to 0$ we then obtain the average free energy [7–10]

$$\bar{F} = -N \frac{\beta J^2}{4} (1-q)^2 + N \frac{1}{2} J_0 m^2 - N \frac{k_B T}{\sqrt{2\pi}} \int dz e^{-\frac{z^2}{2}} \ln\left[2\cosh\left(\beta J \sqrt{q} z + J_0 m\right)\right], \tag{6.76}$$

which for $J_0 = 0$ reduces to the result in Eq. (6.58b).

In addition, extremizing Eq. (6.76) in q we find [7–10]

$$q = 1 - \frac{1}{\sqrt{2\pi}} \int dz e^{-\frac{z^2}{2}} \left[\operatorname{sech}\left(\beta J \sqrt{q} z + \beta J_0 m\right)\right]^2 \tag{6.77}$$

$$= \frac{1}{\sqrt{2\pi}} \int dz e^{-\frac{z^2}{2}} \left[\tanh\left(\beta J \sqrt{q} z + \beta J_0 m\right)\right]^2$$

which is the self-consistent equation to be solved for q. (Notice that for $J_0 = 0$ Eq. (6.77) reduces to our result for $J_0 = 0$ in Eq. (6.60).) A similar consideration can also be made for m, and following this we find for m the self-consistent equation [7–10]

$$m = \frac{1}{\sqrt{2\pi}} \int dz e^{-\frac{z^2}{2}} \tanh\left(\beta J \sqrt{q} z + \beta J_0 m\right). \tag{6.78}$$

Notice in Eq. (6.78) that for $J_0 = 0$ the solution gives $m = 0$ which is consistent with our earlier discussions of the $J_0 = 0$ model.

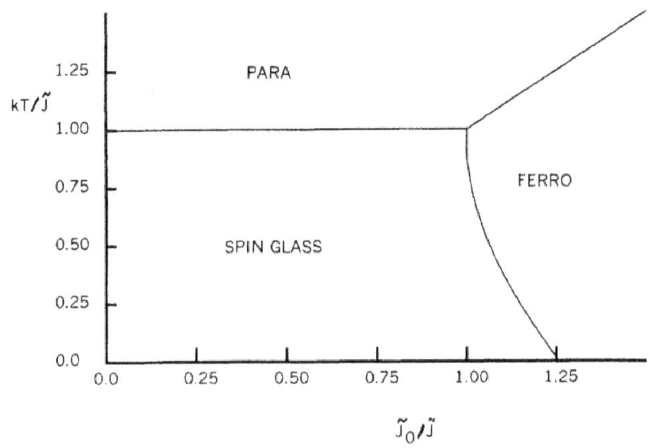

FIGURE 6.3 The phase diagram of spin-glass [7] presented in the space of $\frac{kT}{\tilde{J}} = \frac{k_B T}{J}$ versus $\frac{\tilde{J}_0}{\tilde{J}} = \frac{J_0}{J}$. The three regions of the $q_{\alpha\gamma} = q$, $m_\alpha = m$ assumption are indicated as paramagnetic, ferromagnetic, and spin glass behaviors. In a more advanced treatment, relaxing the $q_{\alpha\gamma} = q$ and $m_\alpha = m$ assumptions, the spin glass region between the $\frac{J_0}{J} = 1$ line and approximately centered on the present left boundary between the spin glass and ferromagnetic regions is found to be occupied by a mixed state consisting of spin glass and ferromagnetic domains [7]. (Reprinted figure with permission from [7]. Copyright 1975, American Physical Society.)

The three equations in Eqs. (6.76) through (6.78) now constitute the system of equations yielding the solutions of the $J_0 \neq 0$ model in the context of the replica trick. These equations can be solved numerically to obtain the phase diagram for the disordered spin system in terms of J, J_0, and T. The resulting solutions exhibit regions of paramagnetic, ferromagnetic, and spin glass behaviors. The new spin glass phase, in particular, is distinguished by a lack of magnetization in the presence of a nonzero order parameter $q(T)$. In the course of our discussions, we shall also find by studying the form of the free energy at the phase boundaries that the transitions between the different phase regions ultimately involve both first and second order transitions between the various phases.

In Figure 6.3 the phase diagram of the spin glass is presented in a plot of $\frac{k_B T}{J}$ versus $\frac{J_0}{J}$ for a numerical solution of Eqs. (6.76) through (6.78). This is the original plot given in the work of Sherrington and Kirkpatrick [7]. From the plot we see a region of spin glass is found at low temperatures $\frac{k_B T}{J} \leq 1$ and for $J_0 < J$. Outside the area of spin glass are regions at elevated temperatures $\frac{k_B T}{J} > 1$ which exhibit paramagnetic behavior. In addition, in the far right of the figure, the system is also found to exhibit a transition to a region of ferromagnetic order at low temperatures.

In the original work of Sherrington and Kirkpatrick [7] three phases are present, composed as the paramagnetic, ferromagnetic, and spin glass phases. The new transition from the spin glass to the paramagnetic phase is found, from the free energy form in Eq. (6.61), to be a second order phase transition, exhibiting a continuous transition at $\frac{k_B T}{J} = 1$ of $q \to 0$ from the spin glass to the paramagnetic state. In addition, a transition to a ferromagnetic phase is found at low temperature on the right-hand side of the phase diagram in Figure 6.3. Later, upon relaxing the requirements that $q_{\alpha\gamma} = q$ and $m_\alpha = m$ some interesting changes will be seen to be presently observed at the spin glass interface with the ferromagnetic region. These lead to a new type of behavior as the system goes from a spin glass to a ferromagnetic region [8–13].

Specifically, in our next discussions, when the $q_{\alpha\gamma} = q$ and $m_\alpha = m$ assumptions for the forms of the solution are relaxed, a new region is identified in the phase diagram for $J_0 \geq J$ at low temperatures $\frac{k_B T}{J} < 1$. The new region is composed of a region of mixed ferromagnetic and spin glass phases. The mixed phase is located in part of the present spin glass and ferromagnetic regions in Figure 6.3, occurring between $J_0 \geq J$ and approximately centered on the left boundary of the ferromagnetic region in Figure 6.3. The existence of such a mixed state in the transition from pure spin glass to a pure ferromagnetic state is indicative of a first-order transition between the boundaries of the regions of pure spin glass phase and the pure ferromagnetic phase [8–13].

We next note that a problem has been found to exist with the Sherrington-Kirkpatrick theory of our current considerations. This is observed in its $J_0 = 0$ form and concerns the limit of the theory in the spin glass phase as $T \to 0$. Specifically, in this limit a region of negative entropy arises in the present form of the theory. This problem is eventually resolved in a modification of the treatment of the solutions involving $\{q_{\alpha\gamma}\}$ and $\{m_\alpha\}$. We now consider these aspects of the model and the resolution of the negative entropy problem.

6.6 THE NEGATIVE ENTROPY PROBLEM

A problem with the $J_0 = 0$ solution of the Sherrington-Kirkpatrick model [7, 8] is found in the spin glass phase at very low temperatures under the assumption, introduced at Eq. (6.53), that $q_{\alpha\gamma} = q$. Specifically, under these conditions the thermodynamics of the solution obtained for q enters a region of negative entropy, and this is a violation of the third law of thermodynamics. The resolution of this problem ultimately involves relaxing the conditions that $q_{\alpha\gamma} = q$ between all the different replicas of the Hamiltonian. We begin the present treatment by demonstrating the existence of the region of negative entropy.

To demonstrate the difficulty, we begin by considering the $T \to 0$ limit (i.e., the limit that $\beta \to \infty$) of the self-consistent equation for q given in Eq. (6.60). In this temperature limit it follows from Eq. (6.60) that to first order in T the order parameter has the form [8]

$$q(T) \to 1 - \sqrt{\frac{2}{\pi}}\frac{k_B T}{J}. \tag{6.79a}$$

Similarly, from a consideration of the free energy in Eq. (6.58) and applying the result in Eq. (6.79a), it also follows [8] that as $T \to 0$

$$\overline{F} \to N\left(-\frac{1}{\sqrt{2\pi}}J + \frac{k_B T}{2\pi}\right). \tag{6.79b}$$

The entropy in terms of the free energy in this limit is then given by applying standard thermodynamic relationships to obtain [8]

$$S = -\frac{\partial \overline{F}}{\partial T} = -N\frac{k_B}{2\pi} \approx -Nk_B 0.16 < 0 \tag{6.79c}$$

which is anomalously negative.

The negative entropy result in Eq. (6.79c) represents a fundamental problem [7–10] which is overcome by treating a more general form of $\{q_{\alpha\gamma}\}$ than that obtained assuming $q_{\alpha\gamma} = q$. To understand what is going on in terms of the free energy itself, we start with the free energy form in Eq. (6.72) and make a stability analysis of the model solution. Note that for this analysis the generalized model in which J_0 can be nonzero is treated, and as per our earlier considerations the extremized free energy is tested under the assumptions that $q_{\alpha\gamma} = q$ and $m_\alpha = m$. In this regard, the considerations in Eqs. (6.77) through (6.79) were earlier made for the $J_0 = 0$ case of this model, and our present discussions are of a more general nature.

We begin by rewriting Eq. (6.72) in terms of new variables defined by [8–13]

$$y_{\alpha\gamma} = J\beta q_{\alpha\gamma} \text{ and } x_\alpha = \sqrt{J_0\beta}\, m_\alpha. \tag{6.80}$$

In terms of these variables, Eq. (6.72) then becomes

$$\bar{F} \to -k_B TN\left\{-\frac{1}{4n}\sum_{\alpha\neq\gamma} y_{\alpha\gamma}^2 - \frac{1}{2n}\sum_\alpha x_\alpha^2 + \frac{1}{n}\ln\left(Tr\, e^L\right) + \frac{1}{4}\beta^2 J^2\right\} \tag{6.81a}$$

where

$$L = \beta J \sum_{\langle\alpha,\gamma\rangle} y_{\alpha\gamma}\sigma^\alpha\sigma^\gamma + \sqrt{\beta J_0}\sum_\alpha \sigma^\alpha x_\alpha \tag{6.81b}$$

$$= \frac{1}{2}\beta J \sum_{\alpha\neq\gamma} y_{\alpha\gamma}\sigma^\alpha\sigma^\gamma + \sqrt{\beta J_0}\sum_\alpha \sigma^\alpha x_\alpha.$$

To determine the nature of the free energy extrema at the $q_{\alpha\gamma} = q$ (i.e., $y_{\alpha\gamma} = J\beta q = y$), $m_\alpha = m$ (i.e., $x_\alpha = \sqrt{J\beta}m = x$) solution obtained in Eqs. (6.76) through (6.78), we expand the free energy in Eq. (6.81) about x and y. To make this expansion, consider the variables [8–13]

$$y_{\alpha\gamma} = y + \delta_{\alpha\gamma} \text{ and } x_\alpha = x + \epsilon_\alpha \tag{6.82a}$$

where

$$y = J\beta q \text{ and } x = \sqrt{J_0\beta}\, x \tag{6.82b}$$

and $\{\epsilon_\alpha, \delta_{\alpha\gamma}\}$ are a set of infinitesimal deviations from x and y. In this way, by studying the change about the proposed minimum of the free energy at x and y, it is found that x and y obtained earlier do not minimize the solutions within the spin glass regions. Specifically, in this region of the phase diagram in Figure 6.3, the free energy at x and y exhibits a saddle point rather than a minimum.

Notice, however, that the x and y solutions within the paramagnetic and ferromagnetic regions of Figure 6.3 do represent minima of the free energy. It is only after the spin glass critical points in Figure 6.3 have been passed and the region labeled as spin glass has been entered that the x and y no longer represent minima of the free energy. Consequently, in the region of spin glass in Figure 6.3 a different form of solution must be applied to study the free energy form. The new form [8–13] of the solution must minimize the free energy of the system by assuming a less restrictive form of $\{q_{\alpha\gamma}\}$ and $\{m_\alpha\}$.

The consideration of the corrections to the forms in the set of $\{q_{\alpha\gamma}\}$ and $\{m_\alpha\}$ to accurately represent the spin glass phase is now treated. These corrected forms ultimately lead to a minimization of the free energy in Eq. (6.72) within the spin glass phase in Figure 6.3.

6.7 A BETTER TREATMENT OF THE FREE ENERGY

A first attempt at finding the correct forms of $\{q_{\alpha\gamma}\}$ and $\{m_\alpha\}$ for minimizing the free energy [8–13] in Eq. (6.72) involved introducing more parameters into $\{q_{\alpha\gamma}\}$ than just treating it by the single variables $q_{\alpha\gamma} = q$ for all α and γ and $m_\alpha = m$ for all α. These additional parameters are chosen to enter the free energy form in such a manner that they provide additional variables which are available to extremize the free energy form. In this context, an early suggested form was one in which $m_\alpha = m$ was chosen to be a constant and $\{q_{\alpha\gamma}\}$ was chosen to be composed of block submatrices. Specifically, for a set $\{q_{\alpha\gamma}\}$ forming an $n \times n$ matrix, a matrix decomposition of it was made in the form of $n_1 \times n_1$ block submatrices where $\frac{n}{n_1}$ is an integer [8–13]. Similarly, the process can ultimately be continued so that the $n_1 \times n_1$ block matrices are subsequently decomposed in terms of $n_2 \times n_2$ block submatrices where $\frac{n_1}{n_2}$ is an integer, etc. In this way, more and more parameters are introduced into the free energy which are available for its minimization [8–13].

In this regard, an example of such a $\{q_{\alpha\gamma}\}$ matrix is given by the 6×6 matrix composed of four 3×3 matrices and given by [8–10, 12]

$$\{q_{\alpha\gamma}\} = \begin{vmatrix} 0 & q_1 & q_1 & q_0 & q_0 & q_0 \\ q_1 & 0 & q_1 & q_0 & q_0 & q_0 \\ q_1 & q_1 & 0 & q_0 & q_0 & q_0 \\ q_0 & q_0 & q_0 & 0 & q_1 & q_1 \\ q_0 & q_0 & q_0 & q_1 & 0 & q_1 \\ q_0 & q_0 & q_0 & q_1 & q_1 & 0 \end{vmatrix}. \tag{6.83}$$

We see from Eq. (6.83) that the off-diagonal block matrices only involve q_0 while the on-diagonal block matrices involve only 0 and q_1. The form in Eq. (6.83) is also easily generalized to treat $4 \times 4, 8 \times 8$, 10×10, etc. matrices, all exhibiting a similar type of decomposition in terms of $2 \times 2, 4 \times 4$, 5×5, etc. matrices [8–10, 12]. Other parameterizations can also be formulated in this way in terms of the block matrix format and discussions of these are found in the literature [8–13].

For an illustration, we next consider the evaluation of Eq. (6.72) for the case in which $m_\alpha = m$ and $\{q_{\alpha\gamma}\}$ are matrices of the form in Eq. (6.83). This gives the free energy per site obtained (in an ultimate consideration of the analytically continued $n \to 0$ limit) by the methodology used to obtain Eq. (6.76) from Eq. (6.72a). In this way we find that [8–10, 12]

$$f = \frac{\overline{F}}{N} \to -k_B T \left\{ -\frac{\beta^2 J^2}{4n} \sum_{\alpha \neq \gamma} q_{\alpha\gamma}^2 - \frac{\beta J_0}{2n} \sum_\alpha m_\alpha^2 + \frac{1}{n} \ln\left(Tr\ e^L\right) + \frac{1}{4}\beta^2 J^2 \right\} \tag{6.84a}$$

and as $n \to 0$

$$f \to \frac{\beta J^2}{4} \left\{ (n_1 - 1) q_1^2 - n_1 q_0^2 + 2q_1 - 1 \right\} - \frac{\ln 2}{\beta} + \frac{1}{2} J_0 m^2 \tag{6.84b}$$

$$-\frac{1}{\beta n_1} \int \frac{e^{-\frac{u^2}{2}}}{\sqrt{2\pi}} \ln \int \frac{e^{-\frac{v^2}{2}}}{\sqrt{2\pi}} \left[\cosh \beta \left(J\sqrt{q_0}\, u + J\sqrt{q_1 - q_0}\, v + J_0 m \right) \right]^{n_1} du dv$$

where $n \times n$ is the dimension of the $\{q_{\alpha\gamma}\}$ matrix in Eq. (6.83), $n_1 \times n_1$ is the dimension of the block matrices composing Eq. (6.83), and the final limit is taken as the analytically continued $n \to 0$ limit. Here on the far right in Eq. (6.84b) we take the limit that $n \to 0$ while regarding n_1, q_0, q_1, and m as new analytically continued parameters chosen to minimize f. (Note that this is similar in spirit to our consideration in Eq. (6.76) in which q was retained as an extremizing parameter of the free energy in the $n \to 0$ limit. Also, notice that on the far right of Eq. (6.84b), the dimension n_1 is relaxed and allowed to become a new parameter for the minimization of the free energy.)

Proceeding similar to the method used to obtain Eqs. (6.77) and (6.78), we extremize [8–10, 12] Eq. (6.84b) in the analytically continued variables n_1, q_0, q_1, and m. From this we obtain four equations to be solved self-consistently for m, q_0, q_1, and n_1. These are given for m, q_0, q_1 by [8, 12]

$$m = \frac{1}{\sqrt{2\pi}} \int du e^{-\frac{u^2}{2}} \frac{\int dv e^{-\frac{v^2}{2}} \left(\cosh\left[A(u,v)\right]\right)^{n_1} \tanh A(u,v)}{\int dv e^{-\frac{v^2}{2}} \left(\cosh\left[A(u,v)\right]\right)^{n_1}} \tag{6.85a}$$

obtained from $\frac{\partial f}{\partial m} = 0$,

$$q_0 = \frac{1}{\sqrt{2\pi}} \int du e^{-\frac{u^2}{2}} \left\{ \frac{\int dv e^{-\frac{v^2}{2}} \left(\cosh\left[A(u,v)\right]\right)^{n_1} \tanh A(u,v)}{\int dv e^{-\frac{v^2}{2}} \left(\cosh\left[A(u,v)\right]\right)^{n_1}} \right\}^2 \tag{6.85b}$$

obtained from $\frac{\partial f}{\partial q_0} = 0$, and

$$q_1 = \frac{1}{\sqrt{2\pi}} \int du e^{-\frac{u^2}{2}} \frac{\int dv e^{-\frac{v^2}{2}} \left(\cosh\left[A(u,v)\right]\right)^{n_1} \left(\tanh A(u,v)\right)^2}{\int dv e^{-\frac{v^2}{2}} \left(\cosh\left[A(u,v)\right]\right)^{n_1}} \tag{6.85c}$$

obtained from $\frac{\partial f}{\partial q_1} = 0$, where

$$A(u,v) = \beta\left(J\sqrt{q_0}u + J\sqrt{q_1 - q_0}v + J_0 m\right). \tag{6.85d}$$

A self-consistent equation for n_1 is similarly obtained from $\frac{\partial f}{\partial n_1} = 0$ and involves a complicated form which can be found in the literature [8, 12]. Notice that for $q_0 = q_1 = q$ these equations reduce to our earlier treatment of the symmetric formulation.

In the new formulation based on Eqs. (6.83) through (6.85), the spin glass region is located in the same area of the plot in Figure 6.3. In addition, considering the $J_0 = 0$ solutions of Eqs. (6.84) and (6.85) in the $T = 0$ spin glass phase, the spin glass entropy [8, 12] is found to be $S \approx -Nk_B 0.01$. This compares favorably to the result $S \approx -Nk_B 0.16$ of the original $q_{\alpha\gamma} = q$ formulation given in Eq. (6.79).

While in the new treatment the entropy is still negative, it is less negative than that in Eq. (6.79). This is indicative that the resolution of the treatment of the spin glass phase lies in more complex forms of the spin glass order parameter. Specifically, these forms are ultimately based on block matrices similar to those treated here.

6.7.1 Towards the Best Solution of the Sherrington-Kirkpatrick Model

In this section we conclude by giving an idea of how the ultimate form of the expression of the Parisi free energy [8, 12, 13] for the spin glass phase is obtained [8–10]. In this regard, the ideas upon which the formulation is based involve the introduction of additional parameters into the free energy by considering expressions for $\{q_{\alpha\gamma}\}$ involving the elements of larger and larger $n\times n$ block form matrices. These $n\times n$ matrices are chosen to decompose into block forms similar to the type described earlier in the context of the treatment of the particular matrix discussed in Eqs. (6.83) through (6.85). In this way, the Parisi solution in its final form is given in the continuum limit of the system as the matrices become infinite in size. We begin our discussions by considering the introduction of additional parameters into $\{q_{\alpha\gamma}\}$ and follow this to a treatment of the continuum limit solution for the free energy.

The proposed solution [8, 12, 13] generated in Eqs. (6.84) through (6.85) involved taking the forms $m_\alpha = m$ and $q_{\alpha\gamma} = q_0$ or q_1 which are generated upon the introduction of block submatrices of dimension $n_1 \times n_1$. These parameters were obtained from the division of a $n\times n = 6\times 6$ matrix into a block form matrix involving four $n_1 \times n_1 = 3\times 3$ block submatrices where $\frac{n}{n_1} = 2$ is an integer. This generalization allowed us to study the free energy in terms of the four new analytically continued parameters $\{q_0, q_1, m, n_1\}$ rather than the two $\{q, m\}$ parameters used initially in the Sherrington-Kirkpatrick theory [7].

In this way, for example, from Eq. (6.84a) we see that the term in the free energy [8] given by $\frac{\beta J^2}{4n}\sum_{\alpha\neq\gamma} q_{\alpha\gamma}^2$ transforms in terms of the new parameters $\{q_0, q_1, m, n_1\}$ as

$$\frac{\beta J^2}{4n}\sum_{\alpha\neq\gamma} q_{\alpha\gamma}^2 \to \frac{\beta J^2}{4}\left\{(n_1 - 1)q_1^2 - n_1 q_0^2 + 2q_1 - 1\right\} \tag{6.86a}$$

$$= \frac{\beta J^2}{4n}\left\{n^2 q_0^2 + \left(q_1^2 - q_0^2\right)\frac{n}{n_1} n_1^2 - n q_1^2\right\}.$$

For this particular division of the $n\times n = 6\times 6$ matrix only four $n_1 \times n_1 = 3\times 3$ submatrices were used. Later, of course, in the discussions of this system the parameter n was analytically continued to the reals and the $n \to 0$ limit of the analytic continuation was taken. During this process the resulting free energy was also minimized using the new real variables $\{q_0, q_1, m, n_1\}$.

Continuing this process, we represent $\{q_{\alpha\gamma}\}$ in the form of higher dimensional matrices. For this representation, the $n\times n$ matrix $\{q_{\alpha\gamma}\}$ is expressed in terms of a hierarchy of block matrix subdivisions while introducing more parameters into the free energy form [8–13]. (The reader is referred to the literature [8–10] for the details of the structure of the block matrices used in this process.) Specifically, we next consider the case $m_\alpha = m$ for block submatrices of side dimensions $n_1, n_2, \ldots, n_{K+1}$ and composed of parameters $q_{\alpha\gamma} = q_0, q_1, q_2, \ldots, q_K$. This then introduces more parameters for the minimization of the free energy form in the $n \to 0$ limit.

In this way, for example, it is found from Eq. (6.84a) that [8] in the analytically continued limit $n \to 0$ we have for general K

$$\frac{\beta J^2}{4n}\sum_{\alpha\neq\gamma} q_{\alpha\gamma}^2 \to \frac{\beta J^2}{4n}\left\{q_0^2 n^2 + \left(q_1^2 - q_0^2\right)\frac{n}{n_1} n_1^2 + \left(q_2^2 - q_1^2\right)\frac{n}{n_1}\frac{n_1}{n_2} n_2^2 + \ldots - q_K^2 n\right\} \tag{6.86b}$$

$$= \frac{\beta J^2}{4} n \sum_{i=0}^{K} \left(n_i - n_{i+1}\right) q_i^2 .$$

where $n_0 = n$ and $n_{K+1} = 1$. Notice that for $K = 1$ the result in Eqs. (6.84) and (6.86b) for $\frac{\beta J^2}{4n}\sum_{\alpha\neq\gamma} q_{\alpha\gamma}^2$ is obtained so that the expression is consistent with the treatment in Eqs. (6.84) and (6.86a).

Taking the continuum limit we can next convert the sum on the right-hand side of Eq. (6.86b) into an integral as the $\{q_{\alpha\gamma}\}$ matrix size $n \times n \to \infty \times \infty$. To do this we identify $n_{i+1} - n_j \to dx$ and $q_i \to q(x)$ so that for the $n \to 0$ limit of the analytically continued parameter n we find

$$\frac{\beta J^2}{4n}\sum_{\alpha\neq\gamma} q_{\alpha\gamma}^2 \to -\frac{\beta J^2}{4}\int_0^1 dx q^2(x). \tag{6.86c}$$

This gives an integral representation of the sum on the far left of Eq. (6.86). Now by similar means the other expressions entering into the free energy can be all written into the form of integrals rather than as sums. Using Eq. (6.86c) along with these additional results, a total free energy form is obtained in the continuum limit.

Proceeding in a similar way with the remaining terms of the free energy in Eq. (6.84), a reformulation of the problem in terms of a free energy involving a continuum representation is obtained. In this way, it is found that the free energy per site in the limit as $n \to 0$ is given by [8]

$$f = -\frac{\beta J^2}{4}\left(1 + \int_0^1 dx q^2(x) - 2q(1)\right) - \int dx \frac{e^{-\frac{x^2}{2}}}{\sqrt{2\pi}} f_0\left(0, \sqrt{q(0)}x\right) \tag{6.87}$$

where $f_0(x,h)$ is a solution of

$$\frac{\partial}{\partial x} f_0(x,h) = -\frac{J^2}{2}\frac{dq}{dx}\left[\frac{\partial^2}{\partial h^2} f_0(x,h) + x\left(\frac{\partial}{\partial h} f_0(x,h)\right)^2\right] \tag{6.88}$$

with initial condition $f_0(1,h) = \ln 2\cosh \beta h$.

The format in Eqs. (6.87) and (6.88) ultimately gives a resolution of the earlier difficulties arising in the treatment of the free energy with a limited number of parameters $\{q_{\alpha\gamma}\}$ and $\{m_\alpha\}$. For further details of the mathematical treatment of the formulation of the free energy in the continuum limit given in Eqs. (6.87) and (6.88), the reader is referred to the literature [8–13].

APPENDIX A

Appendix A: Method of Steepest Descent [11]

We present a review of the method of steepest descent with a focus on real functions and for simplicity treat functions of a single real variable. Specifically, consider the evaluation of an integral of the form

$$I(s) = \int g(z) e^{sf(z)} dz \tag{6.89}$$

where $s > 0$ and $f(z)$ is a function with a sharp maximum centered at z_0 so that $f''(z_0) < 0$. The object of the method of steepest descent is to obtain an analytic approximation for the value of the integral $I(s)$ based on these assumptions.

We begin by Taylor series expanding $f(z)$ and $g(z)$ about the maximum z_0. We find to second order in the series that

$$f(z) = f(z_0) - \frac{1}{2}\left|f''(z_0)\right|(z - z_0)^2 \tag{6.90a}$$

and

$$g(z)=g(z_0)+g'(z_0)(z-z_0)+\frac{1}{2}g''(z_0)(z-z_0)^2 \tag{6.90b}$$

Substituting the results in Eq. (6.90) into Eq. (6.89) we approximate

$$I(s)\approx g(z_0)e^{sf(z_0)}\int e^{-\frac{1}{2}s|f''(z_0)|(z-z_0)^2}dz \tag{6.91}$$

where the integral in Eq. (6.91) can be reduced to the Gaussian form where

$$\int e^{-\frac{1}{2}az^2}dz=\sqrt{\frac{2\pi}{a}}. \tag{6.92}$$

Applying Eq. (6.92) in Eq. (6.91), it follows that

$$I(s)\approx g(z_0)e^{sf(z_0)}\sqrt{\frac{2\pi}{s|f''(z_0)|}}. \tag{6.93}$$

REFERENCES

1. Syozi, I. 1972. *Transformation of Ising Models in Phase Transitions and Critical Phenomena Vol 1*, ed. C. Domb and M. S. Green (Academic, New York), 109–135.
2. Syozi, I. and Miyazima, S. 1966. A Statistical Model for the Dilute Ferromagnet. *Progress of Theoretical Physics* 36, 1083–1094.
3. Essam, J. W. 1980. Percolation Theory. *Reports on Progress in Physics* 43, 833–912.
4. McGurn, A. R. 1982. Percolation Theory and the Annealed Model of Random Ferromagnetism. *Journal of Physical C* 15, 111–128.
5. Binder, K. and Young, A. P. 1986. Spin Glasses: Experimental Facts, Theoretical Concepts, and Open Questions. *Reviews of Modern Physics* 59, 801–963.
6. Edwards, S. F. and Anderson, P. W. 1975. Theory of Spin Glasses. *Journal of Physics F: Metal Physics* 5, 965–974.
7. Sherrington, D. and Kirkpatrick, S. 1975. Solvable Model of a Spin-Glass. *Physical Review Letters* 35, 1792–1795.
8. Nishimorri, H. 2001. *Statistical Physics of Spin Glasses and Information Processing An Introduction* (Clarendon Press, Oxford).
9. Kawamura, H. and Taniguchi, T. 2015. Spin Glasses. In: BuschowK. H. J. (ed) *Handbook of Magnetic Materials* vol. 24 (Elsevier, Amsterdam), 1–137.
10. Bolthausen, E. and Bovier, A. 2007. *Spin Glasses* (Springer, Zurich).
11. Arfken, G. B. and Weber, H. J. 2005. *Mathematical Methods for Physicists*, 6th Edition (Elsevier, Amsterdam).
12. Parisi, G. 1979. Towards a Mean Field Theory for Spin Glasses. *Physics Letter A* 73, 203–205.
13. Parisi, G. 1980. A Sequence of Approximated Solutions to the S-K Model for Spin Glass. *Journal of Physics A: Mathematical and General* 13, L115–L121.

7 Diffusion, Localization, and Turing Patterns

In this chapter we look at some interesting techniques that have been developed to analytically study the diffusion and localization properties of particles moving in disordered media [1–7]. Both of these types of processes involve particles scattering from a randomly disordered medium, with the scattering from the medium being a dominant determinant of the nature of the particle motion. In diffusion processes [1, 2] a concentration gradient in the particle density distribution in space acts, through the mechanism of osmotic pressure, to cause the particle distribution to change in space and time. This is seen in systems of both classical and quantum mechanical particles and is a mechanism of particle motion which is important in both chemical reaction processes and in the functioning of the flow of electrons and holes in a variety of semiconductor devices. The equations describing such processes are of the standard type known as diffusion equations [1, 2].

Localization processes [6–14], on the other hand, arise from the phase propagation of waves, and the constructive and destructive interference of the wave-like excitations as they randomly scatter during their propagation through a disordered medium. In this regard, as a consequence of the self-interference of the scattered waves, sequences of scattering events in random media are found which tend to inhibit the propagation of wave-like excitations. In the limit of significant random scattering, localization processes can dominate so that the wave-like excitations become localized or bound in the medium to a region of space from which they cannot leave. This limit of confining excitations to a bounded region of the random medium is known as the limit of strong localization. It is different from weak localization, which is just a tendency to inhibit propagation through the random medium by offering a phase-coherent resistance to such motion [6–13]. Both strong and weak localization are multiple scattering processes, each being a topic under the general study of Anderson localization.

Our discussions will begin with a treatment of diffusion. We start these by deriving and developing an understanding of the Boltzmann transport equation [1] as it attempts to generate the general dynamics of the statistical distribution of particles. This distribution characterizes a density of particles as they are distributed in space, momenta, and time. The Boltzmann equation does this by providing a formulation of the equations of motion of the distribution function of the particles as treated either in classical or quantum mechanics. The resulting theory is then applied by us in the development of Fick's law and ultimately of the diffusion equation for the many-body system [1–5]. As an example of the Boltzmann formulation, an application is also made to develop an understanding of diffusive transport of particles characterized by the Maxwell-Boltzmann and Fermi-Dirac distributions [1].

Following this, as a preliminary to the study of localization, a treatment of the Kubo linear response theory is given [6–9]. This formulation leads to an expression for the electrical conductivity in terms of two-particle Green's functions, ultimately offering a general expression for the conductivity of a material [6, 7]. As an example of the Green's function theory, the general formulation is first reduced to Ohm's law of conductivity (i.e., the classical Drude theory of conductivity) in the context of a many-body Green's function approach. The higher order corrections to the Ohm's law limit are then also studied. These involve quantum corrections to the Drude theory, which is a theory based on classical mechanics, and also provides a first principles calculation of the relaxation time for the Ohm's law conductivity. The higher order corrections are shown to lead to the Anderson localization of the electrons and the Anderson metal–insulator transition [6, 7].

DOI: 10.1201/9781032674919-7

Ultimately, the two-particle Green's function approach is modified to the consideration of optical scattering problems [7, 12, 13]. Specifically, a treatment of the scattering of light from randomly rough metal surfaces is given in terms of Green's functions and the weak localization of surface electromagnetic waves on the scattering interface. Weak localization is observed in the diffuse scattering of light from the surface, found as an enhancement in the diffuse scattering in the back-scattering direction [12, 13]. The processes involved in the localization of light are shown to be intimately related to those responsible for localization in the electron localization problem. A comparison is made, however, of how the two-particle Green's function enters differently in the electron conductivity and in the light scattering problems.

In the course of the treatment of localization, the Ioffe-Regal criterion [8] of localization is introduced as well as an explanation of the origin of localization in terms of double passage effects and phase coherence [2–5, 7, 12, 13]. The scaling theory of localization due to Abrahams, et al. [14] is also explained as an example of scaling renormalization. This leads to the important conclusion that localization always exists in $1-$ and $2-$dimensional systems in the presence of arbitrarily weak disorder but that bands of localized and nonlocalized states are found in disordered $3-$dimensional materials.

As a final topic of this chapter, we renew the discussions of diffusion by considering an interesting set of problems involving the morphology of patterns that can be developed in naturally occurring systems subject to diffusive processes [15–20]. Specifically, it is found in certain types of physical [16–18], chemical [16–18], and biological systems [16, 17, 20] exhibiting reactive equilibrium between their constituents that, in the additional presence of a diffusive dynamics, spatial patterns can develop. These patterns are manifest in the form of a spatial distribution of the system's constituents, and the pattern generated by the system is a type of system self-organization resulting from the interaction of the reactive equilibrium with the process of diffusion [15–17]. Patterns generated in this way are known as Turing patterns [17]. They are responsible for the explanation of a variety of spatial orderings which occur spontaneously in nature. We shall find that Turing patterns [17] are generated from instability in the system and the conditions required for this generation involve the study of stability theory applied to the equations of motion of the system.

7.1 DIFFUSION

In statistical mechanics, the transport properties of systems of particles are often studied within the context of the Boltzmann transport equation. In this approach, the dynamics of the particles in the system are characterized by a distribution function of the form [1–5]

$$f(\vec{r},\vec{p},t) \tag{7.1}$$

where $f(\vec{r},\vec{p},t)$ represents at time t the density of particles at the position $\vec{r}$ per spatial volume and at the momentum $\vec{p}$ per volume in momentum-space. Consequently, in this representation, for a system of $N\to\infty$ particles in a $d-$dimensional spatial volume $V\to\infty$ it follows that

$$n_N(\vec{r},\vec{p},t) = f(\vec{r},\vec{p},t)\,d^d r d^d p \tag{7.2}$$

is the number of particles at time t located in the volume element $d^d r$ at $\vec{r}$ and in the momentum volume element $d^d p$ at $\vec{p}$. Consequently, for a system in which the number of particles is conserved, it follows that integrating the distribution function over all possible space and momentum coordinates gives

$$\int_{-\infty}^{\infty} d^d r \int_{-\infty}^{\infty} d^d p f(\vec{r},\vec{p},t) = N. \tag{7.3}$$

Similarly, the thermodynamic properties of the system such as the average energy, specific heat, and momentum as functions of time are then ultimately determined in terms of various integrals involving the particle distribution function $f(\vec{r},\vec{p},t)$ and its dynamical evolution in space and time [1].

Let us consider a system of noninteracting particles such as is commonly used to model conductivity in the electron gas theory (hole gas theory) of conduction electrons (conduction holes) in metals and semiconductors or in some applications in classical plasma physics. In this model, the identically charged particles are initially treated as a gas of noninteracting particles obeying fermi or Maxwell-Boltzmann statistics [1]. This representation of the system comes about as the different particle-particle interactions in the fluid system tend to cancel one another out so that the system behaves to leading order as a gas of noninteracting particles. The particle-particle scattering, however, is found to enter in higher order corrections to the independent particle model, and under certain conditions, these higher order terms to the system dynamics lead to important corrections to the independent particle model. Some of these corrections will be encountered and treated in our later discussions. The simplified treatment of the noninteracting particle model, however, is found in many systems to give a general semi-quantitative treatment of many of the thermodynamic properties of electrons or holes in conducting materials.

7.1.1 Gas Model

In the gas model of a system of identical charges, the distribution of particles at time t evolves from its form at time $t-dt$ so that [1–5]

$$f(\vec{r},\vec{p},t)=f\left(\vec{r}-\vec{v}(\vec{p})dt,\vec{p}-\vec{F}dt,t-dt\right) \tag{7.4a}$$

where $\vec{F}$ is a constant force arising from an externally applied field (e.g., an electromagnetic field and/or gravitational field) and we have introduced the simplifying notation $\vec{v}(\vec{p})=\dfrac{\vec{p}}{m}$ where m is the mass of the particles in the gas. Here we see that a particle at time t located at $\vec{r}$ and having momentum $\vec{p}$ came from its previous time $t-dt$ location at $\vec{r}-v(\vec{p})dt$ where it had momentum $\vec{p}-\vec{F}dt$. The expression in Eq. (7.4a) assumes that the gas of noninteracting particles does not interact with the medium it propagates through and, consequently, each particle moves as it would in its single particle trajectory. Later, the effects of the propagation of the gas through a dissipative medium are accounted for by introducing a phenomenological dissipative term to Eq. (7.4a).

Let us begin our study of the gas model by considering the right-hand side of Eq. (7.4a). As $dt\to 0$, a Taylor expansion is made to write [1]

$$f(\vec{r},\vec{p},t)=f\left(\vec{r}-\vec{v}(\vec{p})dt,\vec{p}-\vec{F}dt,t-dt\right) \tag{7.4b}$$

$$=f(\vec{r},\vec{p},t)-dt\left[\vec{v}(\vec{p})\cdot\nabla_r+\vec{F}\cdot\nabla_p+\frac{\partial}{\partial t}\right]f(\vec{r},\vec{p},t)$$

where $\nabla_r=\left(\dfrac{\partial}{\partial x},\dfrac{\partial}{\partial y},\dfrac{\partial}{\partial z}\right)$ is the spatial gradient and $\nabla_p=\left(\dfrac{\partial}{\partial p_x},\dfrac{\partial}{\partial p_y},\dfrac{\partial}{\partial p_z}\right)$ is the momentum-space gradient. From Eq. (7.4b) it then follows in the absence of a dissipative medium that

$$\left[\vec{v}(\vec{p})\cdot\nabla_r+\vec{F}\cdot\nabla_p+\frac{\partial}{\partial t}\right]f(\vec{r},\vec{p},t)=0 \tag{7.4c}$$

governs the motion of the particle distribution function. Consequently, each particle is found to move through space subject only to its initial conditions and the external force $\vec{F}$.

With the inclusion of dissipative losses, originating in the scattering properties of a conductive medium the gas is moving through, we generalize Eq. (7.4c) to the form [1]

$$\left[\vec{v}(\vec{p})\cdot\nabla_r+\vec{F}\cdot\nabla_p+\frac{\partial}{\partial t}\right]f(\vec{r},\vec{p},t)=\frac{\partial}{\partial t}f(\vec{r},\vec{p},t)\bigg|_{collision}. \tag{7.5}$$

Here $\frac{\partial}{\partial t}f(\vec{r},\vec{p},t)\Big|_{collision}$ is usually introduced as a phenomenological term accounting for the variety of elastic and inelastic scattering processes in a dissipative or resistive medium. It is also a topic of kinetic transport theory, and the reader is referred to the literature for a more detailed discussion [1–4] of the origins of $\frac{\partial}{\partial t}f(\vec{r},\vec{p},t)\Big|_{collision}$. Here we will treat it as a phenomenological loss term.

In our following discussions of Eq. (7.5) and its solutions for $f(\vec{r},\vec{p},t)$, expressions are obtained for the properties arising from conductive and diffusive transport processes in the gas-dissipative-medium system. These will include the effects of both independent particle motion and dissipative collisions with the conductive medium through which the particles propagate.

To arrive at an equation accounting for the diffusive transport of the particle density in the medium we begin by integrating both sides of Eq. (7.5) over all of momentum space [1–4]. This will allow us to generate the standard form of the diffusion equation for the spatial particle density as a function of space and time variables. (Notice in the following we treat the case of motion in a 3–dimensional isotropic medium. The treatment of the case of other dimensional systems and with additional restrictions is a straightforward generalization of these discussions which we leave to the reader.) In this way we find that [1]

$$\int d^3p\left[\vec{v}(\vec{p})\cdot\nabla_r+\vec{F}\cdot\nabla_p+\frac{\partial}{\partial t}\right]f(\vec{r},\vec{p},t)=\int d^3p\frac{\partial}{\partial t}f(\vec{r},\vec{p},t)\bigg|_{collision}. \tag{7.6}$$

This is now evaluated term by term to obtain an equation only involving space and time variables.

Considering the last term in the bracket on the left of Eq. (7.6) we have [1]

$$\int d^3p\frac{\partial}{\partial t}f(\vec{r},\vec{p},t)=\frac{\partial}{\partial t}\int d^3pf(\vec{r},\vec{p},t)=\frac{\partial}{\partial t}n(\vec{r},t) \tag{7.7}$$

where $n(\vec{r},t)=\int d^3pf(\vec{r},\vec{p},t)$ is the particle density at $\vec{r},t$, i.e., the number of particles at $\vec{r},t$ in the volume d^3r is given by $n(\vec{r},t)d^3r$. Similarly, from the second term in the brackets on the left, it follows that [1]

$$\int d^3p\vec{F}\cdot\nabla_pf(\vec{r},\vec{p},t)=\vec{F}\cdot\int d^3p\nabla_pf(\vec{r},\vec{p},t)=0. \tag{7.8}$$

Here use is made of the boundary conditions that $f(\vec{r},\vec{p}\to\pm\infty,t)\to 0$ and the results for the integrals of the individual components given by [1]

$$\int_{-\infty}^{\infty}dp_i\frac{\partial}{\partial p_i}f(\vec{r},\vec{p},t)=f(\vec{r},\vec{p}\to\infty,t)-f(\vec{r},\vec{p}\to-\infty,t)\to 0 \tag{7.9}$$

for $i=x,y,z$. The remaining first term on the left side requires the most involved evaluation, which we now consider.

For the remaining integral in the brackets on the left side of Eq. (7.6) it then follows that [1]

$$\int d^3p\vec{v}(\vec{p})\cdot\nabla_rf(\vec{r},\vec{p},t)=\nabla_r\cdot\int d^3p\vec{v}(\vec{p})f(\vec{r},\vec{p},t), \tag{7.10}$$

where we remind the reader of the notational simplification $\vec{v}(\vec{p}) = \frac{\vec{p}}{m}$. Notice also from Eq. (7.4a) considered in the weak scattering limit that we can approximate

$$f(\vec{r},\vec{p},t) = f_0\left(\vec{r} - \vec{v}(\vec{p})\tau, \vec{p} - \vec{F}\tau, t - \tau\right) \tag{7.11}$$

$$= \left\{1 - \tau\left[\vec{v}(\vec{p})\cdot\nabla_r + \vec{F}\cdot\nabla_p + \frac{\partial}{\partial t}\right]\right\} f_0(\vec{r},\vec{p},t)$$

$$= \left\{1 - \tau\left[\vec{v}(\vec{p})\cdot\nabla_r + \vec{F}\cdot\nabla_p\right]\right\} f_0(\vec{r},\vec{p},t)$$

where $f_0(\vec{r},\vec{p},t)$ is the equilibrium distribution function for the system with $\vec{F} = 0$, τ is the relaxation time of the system to return to its equilibrium state, and we notice that for the equilibrium distribution $\frac{\partial}{\partial t} f_0(\vec{r},\vec{p},t) = 0$. Using Eq. (7.11) in Eq. (7.10) then gives [1]

$$\int d^3p\vec{v}(\vec{p})\cdot\nabla_r f(\vec{r},\vec{p},t) = \nabla_r \cdot \int d^3p\vec{v}(\vec{p})\left\{1 - \tau\left[\vec{v}(\vec{p})\cdot\nabla_r + \vec{F}\cdot\nabla_p\right]\right\} f_0(\vec{r},\vec{p},t). \tag{7.12a}$$

Our focus is now to evaluate each of the integrals occurring on the right side of Eq. (7.12a).

From Eq. (7.12a) we note that based on the spatial isotropy of $f_0(\vec{r},\vec{p},t)$ the integral

$$\int d\vec{p}\vec{v}(\vec{p}) f_0(\vec{r},\vec{p},t) = 0. \tag{7.12b}$$

Considering the second term on the right side in Eq. (7.12a) we find that [1]

$$-\nabla_r \cdot \int d^3p\tau\overrightarrow{v}(\vec{p})\vec{v}(\vec{p})\cdot\nabla_r f_0(\vec{r},\vec{p},t) \tag{7.12c}$$

$$= -\frac{\partial}{\partial x}\int d^3p\tau v_x^2 \frac{\partial}{\partial x} f_0(\vec{r},\vec{p},t) - \frac{\partial}{\partial y}\int d^3p\tau v_y^2 \frac{\partial}{\partial y} f_0(\vec{r},\vec{p},t) - \frac{\partial}{\partial z}\int d^3p\tau v_z^2 \frac{\partial}{\partial z} f_0(\vec{r},\vec{p},t)$$

$$= -\frac{\partial}{\partial x}\int d^3p\tau v_x^2 \frac{\partial}{\partial n} f_0(\vec{r},\vec{p},t)\frac{\partial n}{\partial x} - \frac{\partial}{\partial y}\int d^3p\tau v_x^2 \frac{\partial}{\partial n} f_0(\vec{r},\vec{p},t)\frac{\partial n}{\partial y} - \frac{\partial}{\partial z}\int d^3p\tau v_z^2 \frac{\partial}{\partial n} f_0(\vec{r},\vec{p},t)\frac{\partial n}{\partial z}$$

where due to the spatial isotropy of $f_0(\vec{r},\vec{p},t)$ only terms involving v_i^2 for $i = x, y, z$ survive the integration and terms such as $v_x v_y$, $v_y v_z$, etc. drop out because they are odd functions. The result in Eq. (7.12c) can then be written in the form

$$-\frac{\partial}{\partial x}D_x\frac{\partial n}{\partial x} - \frac{\partial}{\partial y}D_y\frac{\partial n}{\partial y} - \frac{\partial}{\partial z}D_z\frac{\partial n}{\partial z} \tag{7.13a}$$

where

$$D_i = \int d^3p\tau v_i^2 \frac{\partial}{\partial n} f_0(\vec{r},\vec{p},t) \text{ for } i = x, y, z \tag{7.13b}$$

are the diffusion coefficients of the system. Consequently, for an isotropic medium, it then follows that $D_i = D$ for $i = x, y, z$ so that Eq. (7.13a) reduces to the form [1]

$$-\nabla \cdot \left[D\nabla_r n \right]. \tag{7.13c}$$

Finally, considering the third term on the right side in Eq. (7.12a), it follows that

$$-\nabla_r \cdot \int d^3 p\vec{v}(\vec{p})\tau\vec{F}\cdot\nabla_p f_0(\vec{r},\vec{p},t) = -\frac{\partial}{\partial x}F_x\int d^3 pv_x\tau\frac{\partial}{\partial p_x}f_0(\vec{r},\vec{p},t) \tag{7.14}$$

$$-\frac{\partial}{\partial y}F_y\int d^3 pv_y\tau\frac{\partial}{\partial p_y}f_0(\vec{r},\vec{p},t) - \frac{\partial}{\partial z}F_z\int d^3 pv_z\tau\frac{\partial}{\partial p_z}f_0(\vec{r},\vec{p},t)$$

where due to the isotropy of $f_0(\vec{r},\vec{p},t)$ only terms involving F_i, v_i and p_i for $i = x, y, z$ survive the integration. The result in Eq. (7.14) can then be written as [1]

$$\frac{\partial}{\partial x}\eta_x F_x + \frac{\partial}{\partial y}\eta_y F_y + \frac{\partial}{\partial z}\eta_z F_z = \nabla \cdot \left[\ddot{\eta}\cdot\vec{F}n \right] \tag{7.15a}$$

where

$$\ddot{\eta} = \begin{vmatrix} \eta_x & 0 & 0 \\ 0 & \eta_y & 0 \\ 0 & 0 & \eta_z \end{vmatrix} \tag{7.15b}$$

with

$$\eta_i = -\frac{1}{n}\int d^3 p\tau v_i \frac{\partial}{\partial p_i} f_0(\vec{r},\vec{p},t) \text{ for } i = x, y, z \tag{7.15c}$$

being the particle mobility coefficients. It follows from the symmetry of the system in an isotropic medium that $\eta_x = \eta_y = \eta_z = \eta$.

Assembling the results in Eqs. (7.6) through (7.15), it follows that, in the approximation that $\int d^3 p \frac{\partial}{\partial t} f(\vec{r},\vec{p},t)\Big|_{collision} \approx 0$ (i.e., the relaxation collision processes are small), we find that [1]

$$\frac{\partial n(\vec{r},t)}{\partial t} = -\nabla \cdot \left[-D\nabla_r n(\vec{r},t) + \ddot{\eta}\cdot\vec{F}n(\vec{r},t) \right] \tag{7.16}$$

describes the diffusion of the system in space and time. The relationship in Eq. (7.16) is seen to be the standard diffusion equation for a gas propagating in a scattering medium, subject to an external force, and for the case of a system having isotropic diffusion coefficients.

If we compare Eq. (7.16) with the continuity equation for the conservation of the particles of a gas moving through space, an identification can be made for the particle current density in the medium. In this regard, the continuity equation (i.e., the statement that the gas particles move through space without being created or destroyed so that their number remains unchanged) has the form [1]

$$\frac{\partial n(\vec{r},t)}{\partial t} = -\nabla \cdot \vec{j}(\vec{r},t) \tag{7.17a}$$

where $n(\vec{r},t)$ is the particle density in space, and the particle current density is given by

$$\vec{j}(\vec{r},t) = n(\vec{r},t)\vec{v}(\vec{r},t) \tag{7.17b}$$

where $\vec{v}(\vec{r},t)$ is the particle velocity at position $\vec{r}$ and time t. Comparing Eqs. (7.16) and (7.17), we find that it is natural to associate the particle current density with the form

$$\vec{j}(\vec{r},t) = -D\nabla_r n(\vec{r},t) + \ddot{\eta}\cdot\vec{F}n(\vec{r},t). \tag{7.18}$$

This defines the current density arising from both diffusion and the externally applied field.

For the case in which $\vec{F} = 0$ it follows that the current density in Eq. (7.18) is given by [1]

$$\vec{j}(\vec{r},t) = -D\nabla_r n(\vec{r},t), \tag{7.19}$$

known as Fick's law for the diffusion current density. This describes the dynamics of particles moving under the osmotic pressure arising from their concentration gradient. On the other hand, for the case in which $-D\ \nabla_r n(\vec{r},t) = 0$ in Eq. (7.18) we find that [1]

$$\vec{j}(\vec{r},t) = n(\vec{r},t)\ddot{\eta}\cdot\vec{F} \tag{7.20a}$$

so that

$$\ddot{\sigma}_p = n\ddot{\eta} \tag{7.20b}$$

is the particle conductivity tensor. (Note: For a charged gas with particle density n and particle charge q in an external electric field $\vec{E}$ the electrical conductivity tensor is $\ddot{\sigma} = qn\ddot{\mu}$ where $\ddot{\mu} = q\ddot{\eta}$). Here the motion arises solely from the effects of an external field.

The combined form of the current density in Eq. (7.18), then, treats the general cases in which both diffusive processes and processes arising from fields are at play. This, for example, is the case commonly found in semiconductor junctions in which there is a gradient in the particle hole concentrations as well as a variety of forces from junction fields [1–4].

We next study the solutions of Eq. (7.16) for the cases of an electron gas subject to a Maxwell-Boltzmann and a fermi distribution [1]. These provide models of conduction electrons in semiconductor and metallic materials [1–4].

7.1.2 Maxwell-Boltzmann Distribution

As an example of the diffusion equation in Eq. (7.16), consider the determination of the diffusion and mobility coefficients for a plasma subject to the Maxwell-Boltzmann distribution [1]. This provides an example of the treatment of diffusion in semiconductors and other more general dilute plasmas composed of low concentrations of identical charges. In these types of systems fermi statistics are not important so that the charged particles in a uniform gas are distributed by the Maxwell-Boltzmann distribution given by

$$f_0(\vec{r},\vec{p},t) = \frac{n}{(2\pi mk_BT)^{\frac{3}{2}}}\exp\left(-\frac{p^2}{2mk_BT}\right). \tag{7.21a}$$

Here m is the mass of the identical particles with a particle dispersion relation $\epsilon = \frac{p^2}{2m}$ and momenta $\vec{p}$ in the gas,

$$n = \int d^3 p f_0(\vec{r}, \vec{p}, t) \tag{7.21b}$$

is the density of particles per volume, and T is the temperature.

From Eqs. (7.13) and (7.21) it follows that the diffusion coefficients for particles characterized by isotropic homogeneous transport are given for $i = x, y, z$ by [1]

$$\begin{aligned} D_i &= \tau \int d^3 p v_i^2 \frac{\partial}{\partial n} f_0(\vec{r}, \vec{p}, t) = \frac{\tau}{3} \frac{1}{(2\pi m k_B T)^{\frac{3}{2}}} \int d^3 p v^2 \exp\left(-\frac{p^2}{2mk_B T}\right) \\ &= \frac{1}{3} v^2 \tau = D \end{aligned} \tag{7.22}$$

where use is made of the isotropy of the gas to write $v_i^2 = \frac{1}{3} v^2$, it is assumed that τ is energy and momentum independent, and D is the isotropic diffusion coefficient. Similarly, from a consideration of Eqs. (7.15) and (7.21) it follows that the particle mobility in the system is

$$\begin{aligned} \eta_i &= -\tau \frac{1}{n} \int d^3 p v_i \frac{\partial}{\partial p_i} f_0(\vec{r}, \vec{p}, t) = -\tau \frac{1}{(2\pi m k_B T)^{\frac{3}{2}}} \int d^3 p v_i \frac{\partial}{\partial p_i} \exp\left(-\frac{p^2}{2mk_B T}\right) \\ &= \eta. \end{aligned} \tag{7.23a}$$

Again, due to the isotropy of the distribution function, η is isotropic in space, representing the mobility of the particle transport.

In the particular case that the particles each carry charge q and are subject to an external electric force $\vec{F} = q\vec{E}$, the particle charge mobility is defined in terms of η by

$$\mu = q\eta. \tag{7.23b}$$

Considering the results in Eqs. (7.22) and (7.23a), it follows that the charge mobility μ is related to the particle diffusion coefficient by [1]

$$\mu = \frac{q}{k_B T} D. \tag{7.23c}$$

In terms of this notation, the associated electrical conductivity tensor of the isotropic system is $\sigma = nq\mu$ or in the full tensor format by $\ddot{\sigma} = qn\ddot{\mu}$.

The relationship $\frac{q}{k_B T} D = \mu$ for the isotropic system is known as the Einstein equation [1]. It connects the charge mobility to the particle diffusion coefficient in dilute plasmas described by Maxwell-Boltzmann statistics. We shall next see that the relationship in Eq. (7.23c) takes a quite different form in a degenerate fermi gas.

7.1.3 Fermi Distribution

As a second example of the diffusion equation in Eq. (7.16), consider the diffusion and mobility coefficients for the case of plasmas subject to the $T = 0$ fermi distribution [1]. Again, the isotropic gas is composed of particles of mass m with a particle dispersion relation $\epsilon = \frac{p^2}{2m}$ and momenta $\vec{p}$ in the gas. This provides an example of the treatment of diffusion in highly degenerate plasmas such as metals. For these systems, the charged particles in a uniform gas are distributed by the $T = 0$ fermi distribution defined by [1]

$$f_0(\vec{r},\vec{p},t) = \frac{2}{(2\pi\hbar)^3} \text{ for } |\vec{p}| \le p_f \tag{7.24}$$

$$= 0 \text{ otherwise}$$

where $p_f = \hbar\left(\frac{3\pi^2 N}{V}\right)^{\frac{1}{3}}$ is the fermi momentum.

From Eqs. (7.13) and (7.24) it follows that the diffusion coefficients for particles characterized by isotropic homogeneous transport are given for $i = x, y, z$ by [1]

$$D_i = \frac{\partial}{\partial n}\int d^3p\tau v_i^2 f_0(\vec{r},\vec{p},t) = \frac{1}{3}\frac{\partial}{\partial n}\int_0^{\epsilon_f} \tau v^2 \frac{1}{(2\pi\hbar)^3} 4\pi(2m)^{\frac{3}{2}}\sqrt{\epsilon}d\,\epsilon \tag{7.25}$$

$$= \frac{2\sqrt{2}\sqrt{m}}{3\pi^2\hbar^3}\frac{\partial}{\partial n}\int_0^{\epsilon_f} \tau v^2\,\epsilon\,\sqrt{\epsilon}d\,\epsilon = D$$

where from the isotropy of the gas $v_i^2 = \frac{1}{3}v^2$, the mass of the particles is m, $v^2 = \frac{2\epsilon}{m}$, the fermi energy is $\epsilon_f = \frac{p_f^2}{2m}$, and D is the isotropic diffusion coefficient [1]. In fermi gas systems, it is common to choose an energy-dependent relationship for the relaxation time τ in the general form of a power law in the energy. Consequently, we shall assume a general power law form given by [1]

$$\tau = a\,\epsilon^r. \tag{7.26}$$

With this form, it follows from Eqs. (7.25) and (7.26) that [1]

$$\boldsymbol{D} = \frac{4\sqrt{2}\sqrt{m}}{3\pi^2\hbar^3}\frac{\partial}{\partial n}\int_0^{\epsilon_f} \tau v^2\,\epsilon\,\sqrt{\epsilon}d\,\epsilon = \frac{4\sqrt{2}\sqrt{m}}{3\pi^2\hbar^3}a\,\epsilon_f^{r+\frac{3}{2}}\frac{\partial\epsilon_f}{\partial n} \tag{7.27}$$

$$= \frac{4\sqrt{2}\sqrt{m}}{3\pi^2\hbar^3}a\,\epsilon_f^{r+\frac{3}{2}}\frac{2}{3}\frac{\epsilon_f}{n}$$

where $\frac{\partial\epsilon_f}{\partial n} = \frac{2}{3}\frac{\epsilon_f}{n}$.

Similarly, from Eqs. (7.15) and (7.21) it follows that the particle mobility in the system is [1]

$$\eta_i = -\frac{1}{n}\int d^3p\tau v_i\frac{\partial}{\partial p_i}f_0(\vec{r},\vec{p},t) = \frac{1}{n}\int f_0(\vec{r},\vec{p},t)\frac{\partial}{\partial p_i}(v_i\tau)d^3p \tag{7.28a}$$

$$= \frac{1}{n} \int f_0(\vec{r}, \vec{p}, t) \left[\frac{\tau}{m} + \frac{v^2}{3} \frac{\partial \tau}{\partial \epsilon} \right] d^3 p = \eta.$$

Applying the form in Eq. (7.26) then gives the charge mobility μ as

$$\mu = q\eta = \frac{q}{n} \int f_0(\vec{r}, \vec{p}, t) \left[\frac{\tau}{m} + \frac{v^2}{3} \frac{\partial \tau}{\partial \epsilon} \right] d^3 p = \frac{q}{n} \frac{4\sqrt{2}\sqrt{m}}{3\pi^2 \hbar^3} a \, \epsilon_f^{r+\frac{3}{2}}. \tag{7.28b}$$

In a comparison of Eqs. (7.27) and (7.28b) and applying the mobility relationship $\mu = q\eta$, it is seen that [1]

$$\mu = \frac{3}{2} \frac{qD}{\epsilon_f}. \tag{7.29}$$

This relationship is different from that of the Maxwell-Boltzmann given in Eq. (7.23) where the difference arises specifically from the difference in the nature of the particle statistics.

We shall now see that the ideas of diffusion, in its limiting form of the dynamics of a system, enter into the study of another set of phenomena known as Anderson localization. Specifically, we treat systems of excitations which lose their ability for diffusive motion and become bound or localized to a bounded region of space. This is a phenomenon encountered, for example, in some forms of the metal–insulator transition observed in certain conducting materials and arises from the wave nature of quantum mechanical particles [6–14].

7.2 ANDERSON LOCALIZATION

In classical mechanics whether or not a particle is confined to moving in a bounded region of space is a simple matter [6–14]. Given the potential energy in which the particle moves, the region of motion of the particle is straightaway determined by the kinetic energy of the particle. In the mechanics of wave phenomena, however, the motion of a wave-particle or excitation in a region of potential energy is determined not only by its kinetic energy but by the wavelengths of the wave information carried. Specifically, as the wave is scattered by the random disorder of a conducting medium, it has a tendency to phase interfere with itself during its propagation through the disorder. The constructive and destructive interference of the various scattered components of the wave makes the condition for the determination of the confinement of the excitation to a bounded region of space much more complicated than that for the classical wave. This is a property of the wave nature of the excitation and occurs in both classical and quantum mechanical systems.

One of the earliest attempts at understanding the difference in the propagation of waves and particles through random media was the formulation of the Ioffe-Regel criterion [8] for the onset of localization. This provides a rough idea as to when so-called localization effects should become important in transport systems.

7.2.1 Ioffe-Regel Condition and Double Passage Effects

In the simple Drude model [1–4] for conductivity we consider each particle in the system as a classical particle moving independently of the other particles in the system while interacting with a viscous conducting medium and a constant externally applied force [2–5, 6–8]. The viscous force of the conducting medium provides a drag on the particle motion similar to that of the air resistance on a particle falling through the atmosphere. Consequently, each particle in the conducting system eventually transports through the medium with a constant terminal velocity, and this determines the character of the transport in the conducting medium.

The particle motion in the conductor in the presence of a uniform position-independent external force is then characterized by a simple single particle equation of $1-$dimensional motion in a viscous medium. This has the general form

$$m\left(\ddot{x}+\frac{\dot{x}}{\tau}\right)=F \tag{7.30}$$

where m is the particle mass, τ is a relaxation time arising from the viscosity of the medium, and F is the constant $1-$dimensional external force pushing the particle through along the $x-$axis. In the case of steady state motion, we have $\ddot{x}=0$ so that the constant drift velocity of the particle is given by

$$\dot{x}=\frac{F}{m}\tau, \tag{7.31}$$

being proportional to the relaxation time. Consequently, under the condition that $\frac{F}{m}\neq 0$ the drift velocity is nonzero unless $\tau=0$, so that when $\tau\neq 0$ the particles are always conducted through the medium. On the other hand, when $\tau=0$ the terminal velocity is zero, indicating a complete restraint of the classical particle at a single point in space, e.g., they are forcefully confined at a point in space. As a result, the particles fail to transport through the system. This relaxation process, then, is the picture of transport for classical particles.

When one treats quantum systems or classical waves, however, the motion of the excitations is in addition subject to the wave information carried by them. This additional feature led to speculation by Ioffe and Regel [8] that such wave-particles exhibit more general localization properties than those found for classical particles. We next consider these new features of localization for wave-like excitations.

Ioffe and Regel [8] proposed a simple criterion for the condition of the breakdown of the single particle picture of transport when the particles are wave-like, carrying additional phase information in their transport through the conducting medium. This is found in the case of quantum particles and in classical treatments of elastic and electromagnetic waves. The breakdown in the picture of particle transport found in such systems arises due to the phase interference occurring in them. These types of effects become important when wave excitations are scattered by random disordered conducting media.

Specifically, for wave excitations to propagate through the scattering medium and remain recognizable as waves, we expect the distance a particle travels between successive scattering in the medium must contain many wavelengths. Otherwise, the lengths between scattering events become too short and the particle scattering too intense for the excitations to retain their characterization as waves. In this sense, we would expect the wave nature of the particle to break down when

$$l=\dot{x}\tau=\frac{F}{m}\tau^2\leq\lambda \text{ or } kl=\frac{2\pi}{\lambda}l\leq 1 \tag{7.32a}$$

where l approximates the mean free path between scattering events and $k=\frac{2\pi}{\lambda}$ is the wavenumber characterizing the excitation. This is a statement that the mean free path between scatterings is smaller than the wavelength of the excitation. Under these conditions, we expect the scattering to be intense enough to stop the excitation from propagating through the medium. On the other hand, as

$$kl=\frac{2\pi}{\lambda}l\gg 1 \tag{7.32b}$$

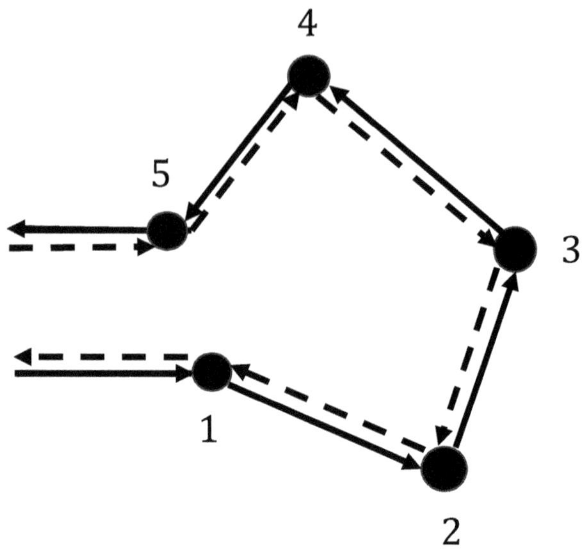

FIGURE 7.1 The double passage effect in the scattering from a random medium. Two rays are incident on a random medium of point scatterers. In the figure one wave scatters from the random medium by the process 1,2,3,4,5 (solid) and a second scatters from the random medium in the time-reversed process 5,4,3,2,1 (dashed). The two rays exiting from the medium add phase coherently to enhance the scattering from the medium in the backward direction. This is due to the equal lengths of the two time-reversed paths.

the scattering of the wave is weak so that the mean free path is very large, and the motion of the wave should only be weakly affected by the disorder.

Another way of seeing the importance of phase-coherent scattering of wave-like excitations in random media is to understand the so-called double passage effect. This is illustrated in Figure 7.1 in which we follow the passage through a random scattering medium of two rays carrying wave information [7]. The medium is composed of point scatterers randomly positioned in space. The two waves enter the random medium traveling parallel to one another, scatter from the random media in a time-reversed sequence of events and emerge from the random medium again moving parallel to one another. The incoming and outgoing waves are then moving antiparallel to one another. Consequently, one wave scatters from the medium in the scattering sequence 1,2,3,4,5 and the other wave scatters from the same medium in the reverse scattering sequence 5,4,3,2,1. The paths are time-reversed trajectories in the medium. As a result, the lengths of each path are the same and for a wave propagating along the two paths the waves emerging from the random medium add phase coherently. This shows up as an enhanced backscattering from the medium [7]. This enhancement then arises from a similar mechanism to that found in Bragg diffraction from crystals. In the case of the random medium all such time-reverse scattering processes with antiparallel inputs and outputs contribute to the enhancement.

We see in the double passage effect that the disordered medium prefers to backscatter waves that are incident upon the medium [7]. This provides a resistance to motion in the random medium. If the backscattering is large enough, the wave will no longer propagate through the random medium and the effect is known as strong localization. In this case the wave is confined in a bounded region of the medium. On the other hand, if the backscattering is small enough, the wave encounters an increase in resistance to propagation through the random medium and the effect is known as weak localization. In this case the wave experiences an increased resistance to propagation through the medium.

We next look at a detailed mathematical treatment of strong and weak localization effects. This involves first reformulating the conduction problem into a transport theory involving Green's functions. In this regard, we begin with a consideration of the Green's function approach to conductivity.

7.2.2 Green's Functions

Our objective is to study the problem of the $T = 0$ conductivity of a uniform isotropic electron gas as it propagates through a randomly disordered medium. As the electrons move through the conducting medium, they are scattered by the random disorder of the medium, which is characterized as being statistically uniform over all space. The electron gas is composed as a system of noninteracting quantum mechanical electrons of effective masses m with a particle energy dispersion relation $\epsilon\left(\vec{k}\right)=\frac{\hbar^2k^2}{2m}$ and momentum $\vec{p}=\hbar\vec{k}$, i.e., the electron-electron forces average out so that, as an approximation, the electrons do not interact with one another but propagate through the medium as independent particles. Consequently, at $T = 0$ the electrons occupy all the low energy single particle energy levels in the system up to a maximum energy level known as the fermi energy $\epsilon=\epsilon_f$. This is done in such a way that each energy level is occupied with a spin up and a spin down electron [6].

In the absence of electron-electron interactions, however, the electrons do interact with both the random scattering medium through which they transport and with an externally applied uniform electric field. The medium is randomly disordered throughout all space, with the disorder being statistically uniform in space. This supplies an impedance to the electron motion. As a second electron interaction, the applied field is taken to be of the form [6]

$$E\left(t\right)=E_0e^{-i\omega t}+E_0^*e_0^{i\omega t} \tag{7.33}$$

where E_0 is the amplitude of the electric field, and ω is the field frequency. The expression in Eq. (7.33) is a 1−dimensional field which pushes the electrons along the field axis. Consequently, in response, a time-dependent current density is generated along the field, given by

$$j\left(t\right)=\sigma\left(\omega\right)E_0e^{-i\omega t}+\sigma^*\left(\omega\right)E_0^*e_0^{i\omega t} \tag{7.34}$$

$$=\sigma\left(\omega\right)E_0e^{-i\omega t}+\sigma\left(-\omega\right)E_0^*e_0^{i\omega t}$$

where $\sigma\left(\omega\right)$ is the frequency dependent conductivity with the property that for a real current $\sigma^*\left(\omega\right)=\sigma\left(-\omega\right)$. Our focus in the discussions of this system is on a study of the properties of the conductivity of the electron gas.

In the considerations we are interested in the properties of the real part of the conductivity because it is responsible for the dissipation of the electric current transported through the conducting medium. On the other hand, the imaginary part of the conductivity acts primarily to set the relative phase response of the current to the field. The idea is to determine the conditions under which the real part of the conductivity vanishes so that the electrons are not transported through the medium. In the limit that $\omega\rightarrow 0$ this is the condition for electron localization. To this end, we begin with a study of the power dissipated to the conducting medium by the field.

Consider then the average power P dissipated by the applied field in the electron gas and conducting medium. From electrodynamics this is given by a standard result for the power lost to the system when it supports a current. Specifically, in terms of the applied electric field and the current density it generates, the average power supplied by the field to the system is obtained from the time-averaged expression [6]

$$P = V\overline{E(t)j(t)} = V\overline{\left(E_0 e^{-i\omega t} + E_0^* e_0^{i\omega t}\right)\left(\sigma(\omega)E_0 e^{-i\omega t} + \sigma^*(\omega)E_0^* e_0^{i\omega t}\right)} \tag{7.35}$$

$$= V\left[\sigma^*(\omega)|E_0|^2 + \sigma(\omega)|E_0|^2\right] = 2V|E_0|^2\sigma_R(\omega)$$

where $\sigma_R(\omega)$ is the real part of the conductivity, V is the volume of the random medium, and the bar over the expression indicates the time average over one period of the motion. From Eq. (7.35) it then follows that at zero conductivity (i.e., for $\sigma_R(\omega) = 0$) the current through the system and the power dissipation are both zero.

Another way of generating an expression for P is by microscopically considering the scattering transitions of the electron gas as it interacts with the random disordered conducting medium and the applied field. These considerations treat the detailed microscopic processes of power dissipation from the fields in the scattering of the individual electrons of the gas. In this approach, as the gas scatters from a single particle quantum state $|\alpha$ to a single particle quantum state $|\beta$ the energy change of the gas is given by [6]

$$\epsilon_{\beta\alpha} = \hbar\omega_{\beta\alpha} = \epsilon_\beta - \epsilon_\alpha \tag{7.36}$$

where ϵ_α is the energy of the single particle in the state $|\alpha\rangle$ and ϵ_β is the energy of the single particle in the state $|\beta\rangle$. (Note that here $\{|\gamma\rangle\}$ is a complete orthonormal set of single particle states of the gas.) It then follows that if $\epsilon_{\beta\alpha} > 0$ the gas receives energy from the field and if $\epsilon_{\beta\alpha} < 0$ the gas gives energy to the field. By accounting for such energy exchanges between the gas and the field and computing the rate of energy exchange for these processes, the power dissipated by the fields in the conducting system is obtained.

In this picture, the power transferred from the applied field to the conducting system is obtained from [6]

$$P = \sum_{\alpha,\beta} \epsilon_{\beta\alpha} r_{\alpha\beta} \tag{7.37}$$

where $r_{\alpha\beta}$ is the rate of transition from the single particle state $|\alpha\rangle$ to the single particle state $|\beta\rangle$, and the transition rates $r_{\alpha\beta}$ in Eq. (7.37) are calculated within the context of the fermi golden rule. The rates $r_{\alpha\beta}$ in Eq. (7.37) are, consequently, expressed in the form [6]

$$r_{\alpha\beta} = f_\alpha(1 - f_\beta)W_{\alpha\beta}. \tag{7.38}$$

were f_α is the probability that the gas is in the single particle state $|\alpha\rangle$, $(1 - f_\beta)$ is the probability that the single particle state $|\beta\rangle$ is unoccupied so that the gas can transition into it, and $W_{\alpha\beta}$ is the ideal rate of transition assuming transitions can always be made from the state $|\alpha\rangle$ to the final state $|\beta\rangle$ during its interaction with the external field.

The transition rate $W_{\alpha\beta}$ in Eq. (7.38) is computed in the context of first-order time-dependent perturbation theory [6]. This allows us to determine the transition rate as an expression quadratic in the applied field, ultimately leading to a linear conductivity which is independent of the applied field. Following these considerations, then, yields the transport coefficients of the system in the so-called region of linear response.

For this chapter, we will only be interested in the linear response of the system to the applied field and will ignore nonlinear effects. Higher order terms of the time-dependent perturbation theory contain corrections to the conductivity involving higher powers of the field; however, these are known experimentally to be small. Such higher order terms are the subject of nonlinear optics, which is important for some technological considerations but are not a consideration in our treatment. The reason for our limitations to a linear response theory is that electrodynamic phenomena are generally reasonably well explained by linear response theories [6].

The interaction of a uniform external field, applied along the x −axis, with the electrons in the gas is given by

$$H_I^g(x,t) = \sum_i ex_i E(t) \tag{7.39a}$$

where x_i is the coordinate along the x −axis of the ith conduction electron; $-e$ is the charge on an electron where $e > 0$ is the unit of electrical charge; and i runs over all the conduction electrons. When the gas is characterized as an antisymmetric many-particle wavefunction formed from the individual single particle states, Eq. (7.39a) accounts for the detailed interaction of each of the electrons in the gas with the field in the determination of the transition matrix. However, for our considerations of the electron gas of independent particles, a simplification can be made in the study of the interaction of the field in Eq. (7.39a) with the particles of the gas. This is done by focusing on the interaction of one particular electron state in the gas with the field, computing its transition rate, and using the fact that the gas is formed from independent identical particles to obtain the gas transition rate as a sum involving the single particle transition rates weighted by the Fermi-Dirac occupancy of the single particle states.

Specifically, let us focus on one of the electrons in the gas with coordinate x on the x −axis, i.e., $x \in \{x_j\}$. This single electron has an interaction with the applied field described by the form [6]

$$H_I(x,t) = exE(t) = ex\left[E_0 e^{-i\omega t} + E_0^* e_0^{i\omega t}\right]. \tag{7.39b}$$

From this interaction term let us next determine the single-particle transition rate of the electron mediated by $H_I(x,t)$. The application of Eq. (7.39b) is seen to lead to a simplification of the determination of the power, obtained from Eq. (7.37), arising from the nature of the electrons as identical particles. In this consideration, it follows from first order time-dependent perturbation theory that the rate for a transition of the gas from the initial single-particle state $|\alpha\rangle$ with energy ϵ_α to the final single-particle state $|\beta\rangle$ with energy ϵ_β is given from Eq. (7.39b) by the standard form [6]

$$W_{\alpha\beta} = \frac{2\pi}{\hbar}\left|\langle\alpha|exE_0|\beta\rangle\right|^2\left[\delta(\hbar\omega - \epsilon_{\beta\alpha}) + \delta(\hbar\omega + \epsilon_{\beta\alpha})\right] \tag{7.40}$$

$$= \frac{2\pi e^2 |E_0|^2}{\hbar}\left|\langle\alpha|x|\beta\rangle\right|^2\left[\delta(\hbar\omega - \epsilon_{\beta\alpha}) + \delta(\hbar\omega + \epsilon_{\beta\alpha})\right].$$

This transition rate does not account for the occupancies of single-particle states $|\alpha\rangle$ and $|\beta\rangle$ in the gas, which must be consistent with Fermi-Dirac statistics and the Pauli exclusion principle.

Accounting for the Fermi-Dirac occupancies of the particles in the gas, we then find from Eqs. (7.37) through (7.40) that the power generated by the interaction is [6]

$$P = \frac{1}{2}\sum_{\alpha,\beta} \varepsilon_{\beta\alpha}\frac{2\pi}{\hbar}e^2|E_0|^2\left|\langle\alpha|x|\beta\rangle\right|^2\left[\delta(\hbar\omega - \epsilon_{\beta\alpha}) + \delta(\hbar\omega + \epsilon_{\beta\alpha})\right] f_\alpha(1 - f_\beta) \tag{7.41}$$

$$+ \frac{1}{2}\sum_{\alpha,\beta} \varepsilon_{\alpha\beta}\frac{2\pi}{\hbar}e^2|E_0|^2\left|\langle\alpha|x|\beta\rangle\right|^2\left[\delta(\hbar\omega - \epsilon_{\alpha\beta}) + \delta(\hbar\omega + \epsilon_{\alpha\beta})\right] f_\beta(1 - f_\alpha)$$

$$= \frac{\pi}{\hbar} e^2 |E_0|^2 \sum_{\alpha,\beta} |\langle \alpha | x | \beta \rangle|^2 \left[\varepsilon_{\beta\alpha} (f_\alpha - f_\beta) \delta(\hbar\omega - \epsilon_{\beta\alpha}) + \varepsilon_{\alpha\beta} (f_\beta - f_\alpha) \delta(\hbar\omega - \epsilon_{\alpha\beta}) \right]$$

$$= \frac{2\pi}{\hbar} e^2 |E_0|^2 \sum_{\alpha,\beta} \hbar\omega_{\beta\alpha} |\langle \alpha | x | \beta \rangle|^2 (f_\alpha - f_\beta) \delta(\hbar\omega - \epsilon_{\beta\alpha})$$

where note that the sums are over all possible orthonormal single-particle states of the gas. The result in Eq. (7.41) then expresses the power given by the field to the system in terms of the microscopic configurational changes of the gas-medium system induced by Eq. (7.39b).

From a comparison of Eqs. (7.35) and (7.41), it consequently follows that [6, 7]

$$\sigma_R(\omega) = \frac{\pi e^2}{V} \sum_{\alpha,\beta} \omega_{\beta\alpha} |\langle \alpha | x | \beta \rangle|^2 (f_\alpha - f_\beta) \delta(\hbar\omega - \epsilon_{\beta\alpha}) \tag{7.42}$$

which can be further simplified by some dynamical considerations. Specifically, from the Heisenberg equations of motion

$$p_x = m \frac{dx}{dt} = \frac{i}{\hbar} m [Hx - xH] \tag{7.43a}$$

we find that

$$\langle \alpha | p_x | \beta \rangle = im\omega_{\alpha\beta} \langle \alpha | x | \beta \rangle \tag{7.43b}$$

where m is the mass of the electron and $\omega_{\alpha\beta} = -\omega_{\beta\alpha} > 0$. From Eqs. (7.43) and (7.42) it then follows that the real part of the conductivity is rewritten into the form [6]

$$\sigma_R(\omega) = \frac{\pi e^2}{m^2 V} \sum_{\alpha,\beta} |\langle \alpha | p_x | \beta \rangle|^2 \frac{f_\alpha - f_\beta}{\omega_{\beta\alpha}} \delta(\hbar\omega - \hbar\omega_{\beta\alpha}). \tag{7.44}$$

This expression relates the real part of the conductivity to the elements of the momentum matrix $\langle \alpha | p_x | \beta \rangle$ between the single-particle eigenstates of the electron gas.

Once the real part of the conductivity is determined, by arguments involving the causality of the system, the imaginary part of the conductivity can be obtained. These arguments employ standard relations in electrodynamics referred to as the Kramers-Kronig relationships [6], and the reader is referred to the literature for a discussion of these [6]. Specifically, it is found that the imaginary part of the conductivity $\sigma_I(\omega)$ is related to its real part by

$$\sigma_I(\omega) = -\frac{1}{\pi} P \int_{-\infty}^{\infty} d\omega' \frac{\sigma_R(\omega')}{\omega' - \omega} \tag{7.45a}$$

and the real part of the conductivity $\sigma_R(\omega)$ is related to the imaginary part by

$$\sigma_R(\omega) = \frac{1}{\pi} P \int_{-\infty}^{\infty} d\omega' \frac{\sigma_I(\omega')}{\omega' - \omega}. \tag{7.45b}$$

In Eq. (7.45) the principal part of the integrals (indicated by the P) is taken.

We now return to the expression for the real part of the conductivity in Eq. (7.44) and rewrite it in terms of the Green's functions of the electron gas [6]. In this regard, let us define the $T = 0$ Green's function $G_0^{(\mp)}(\epsilon)$ for an electron gas with a fermi energy ϵ_f by the operator forms

$$G_0^{(-)}(\epsilon) = \lim_{\delta \to 0^+} \frac{1}{[\epsilon - i\delta - H]} \text{ for } \epsilon < \epsilon_f \tag{7.46a}$$

and

$$G_0^{(+)}(\epsilon) = \lim_{\delta \to 0^+} \frac{1}{[\epsilon + i\delta - H]} \text{ for } \epsilon > \epsilon_f \tag{7.46b}$$

where ϵ is the Green's function energy and H is the Hamiltonian operator of the electron gas. Note that these are the standard single-particle Green's functions we have encountered earlier in the text. Separating the Green's functions in Eq. (7.46) into their real and imaginary parts, it follows that [6]

$$G_0^{(-)}(\epsilon) = \lim_{\delta \to 0^+} \left\{ \frac{\epsilon - H}{(\epsilon - H)^2 + \delta^2} + \frac{i\delta}{(\epsilon - H)^2 + \delta^2} \right\} \text{ for } \epsilon < \epsilon_f \tag{7.47a}$$

$$= \lim_{\delta \to 0^+} \frac{\epsilon - H}{(\epsilon - H)^2 + \delta^2} + i\pi\delta(\epsilon - H)$$

and

$$G_0^{(+)}(\epsilon) = \lim_{\delta \to 0^+} \left\{ \frac{\epsilon - H}{(\epsilon - H)^2 + \delta^2} - \frac{i\delta}{(\epsilon - H)^2 + \delta^2} \right\} \text{ for } \epsilon > \epsilon_f \tag{7.47b}$$

$$= \lim_{\delta \to 0^+} \frac{\epsilon - H}{(\epsilon - H)^2 + \delta^2} - i\pi\delta(\epsilon - H).$$

As we shall presently see, the expressions in Eq. (7.47), in particular, provide useful representations of the energy delta function in terms of the imaginary parts of the Green's functions. Later these will allow us to write the conductivity of the gas in terms of the Green's functions of the gas.

To introduce the Green's functions into our expressions for the conductivity, we start at $T = 0$ with a consideration of the identity [6]

$$\frac{f_\alpha - f_\beta}{\omega_{\beta\alpha}} \delta(\hbar\omega - \epsilon_{\beta\alpha}) = \int d\,\epsilon\, \delta(\epsilon - \epsilon_\beta + \hbar\omega)\delta(\epsilon - \epsilon_\alpha) \frac{f(\epsilon) - f(\epsilon + \hbar\omega)}{\omega} \tag{7.48}$$

$$= \frac{f_\alpha - f_\beta}{\omega_{\beta\alpha}} \delta(\hbar\omega - \hbar\omega_{\beta\alpha})$$

where the fermi distribution function at $T = 0$ is

$$f(\epsilon) = 1 \text{ for } \epsilon < \epsilon_f \tag{7.49a}$$

$$= 0 \text{ otherwise}$$

and in this notation

$$f_\alpha = f(\epsilon_\alpha) \text{ and } f_\beta = f(\epsilon_\beta). \tag{7.49b}$$

The result in Eq. (7.48) can be applied in Eq. (7.44) to rewrite Eq. (7.44) into the form [6]

$$\sigma_R(\omega) = \frac{\pi e^2}{m^2 V}\sum_{\alpha,\beta}\left|\langle\alpha|p_x|\beta\rangle\right|^2 \int d\epsilon\, \delta(\epsilon-\epsilon_\beta+\hbar\omega)\delta(\epsilon-\epsilon_\alpha)\frac{f(\epsilon)-f(\epsilon+\hbar\omega)}{\omega}. \tag{7.50}$$

$$= \frac{\pi e^2}{m^2 V}\int d\epsilon\, \frac{f(\epsilon)-f(\epsilon+\hbar\omega)}{\omega}\sum_{\alpha,\beta}\langle\alpha|p_x|\beta\rangle\langle\beta|p_x|\alpha\rangle\delta(\epsilon-\epsilon_\beta+\hbar\omega)\delta(\epsilon-\epsilon_\alpha)$$

$$= \frac{\pi e^2}{m^2 V}\int d\epsilon\, \frac{f(\epsilon)-f(\epsilon+\hbar\omega)}{\omega}\sum_{\alpha,\beta}\langle\alpha|p_x\delta(\epsilon-\epsilon_\beta+\hbar\omega)|\beta\rangle\langle\beta|p_x\delta(\epsilon-\epsilon_\alpha)|\alpha\rangle$$

$$= \frac{\pi e^2}{m^2 V}\int d\epsilon\, \frac{f(\epsilon)-f(\epsilon+\hbar\omega)}{\omega}\sum_{\alpha,\beta}\langle\alpha\,|\,p_x\delta(\epsilon+\hbar\omega-H)\,|\,\beta\rangle\langle\beta\,|\,p_x\delta(\epsilon-H)\,|\,\alpha\rangle.$$

The delta functions in the last equality of Eq. (7.50) are next rewritten using the forms in Eq. (7.47) for the Green's functions.

This rewrite gives [6]

$$\sigma_R(\omega) = \frac{e^2\hbar}{\pi m^2 V}\int d\epsilon\, \frac{f(\epsilon)-f(\epsilon+\hbar\omega)}{\hbar\omega}\sum_{\alpha,\beta}\langle\alpha\,|\,p_x ImG_0^{(+)}(\epsilon+\hbar\omega)\,|\,\beta\rangle\langle\beta\,|\,p_x ImG_0^{(+)}(\epsilon)\,|\,\alpha\rangle \tag{7.51a}$$

$$= \frac{e^2\hbar}{\pi m^2 V}\int d\epsilon\, \frac{f(\epsilon)-f(\epsilon+\hbar\omega)}{\hbar\omega}Tr\left\{p_x ImG_0^{(+)}(\epsilon+\hbar\omega)\,p_x ImG_0^{(+)}(\epsilon)\right\}$$

for the real part of the conductivity in terms of the Green's functions of the electron gas. The resulting expression in Eq. (7.51a) can also be shown to hold in the case of general scattering of the electrons from random disorder developed in the conducting medium, i.e., in cases in which the imaginary parts of the Green's function are no longer delta functions but are sharply peaked functions of energy. In this general case the $G_0^{(+)}$ Green's functions are replaced by $G^{(+)}$ Green's functions corrected to include scattering effects. In the case of this renormalization, we find that [6]

$$\sigma_R(\omega) = \frac{e^2\hbar}{\pi m^2 V}\int d\epsilon\, \frac{f(\epsilon)-f(\epsilon+\hbar\omega)}{\hbar\omega}\sum_{\alpha,\beta}\langle\alpha\,|\,p_x ImG^{(+)}(\epsilon+\hbar\omega)\,|\,\beta\rangle\langle\beta\,|\,p_x ImG^{(+)}(\epsilon)\,|\,\alpha\rangle \tag{7.51b}$$

$$= \frac{e^2\hbar}{\pi m^2 V}\int d\epsilon\, \frac{f(\epsilon)-f(\epsilon+\hbar\omega)}{\hbar\omega}Tr\left\{p_x ImG^{(+)}(\epsilon+\hbar\omega)\,p_x ImG^{(+)}(\epsilon)\right\}$$

describes the conductivity in terms of the Green's functions renormalized by impurity scattering. For a demonstration of the general form in Eq. (7.51b) the reader is referred to the literature [7]. In the following, however, we shall use the generalized result without further discussion.

At $T = 0$ in the limit that $\omega \to 0$, there are a number of limiting forms for the terms entering Eq. (7.51) that provide a simplification. These limits are as follows [6]:

Considering the temperature factors, it follows from a Taylor expansion that

$$\frac{f(\epsilon)-f(\epsilon+\hbar\omega)}{\hbar\omega} \to \frac{f(\epsilon)-\left[f(\epsilon)+\hbar\omega\dfrac{df(\epsilon)}{d\epsilon}\right]}{\hbar\omega} = -\frac{df(\epsilon)}{d\epsilon} = \delta(\epsilon-\epsilon_f) \tag{7.52a}$$

where use is made of the properties of the Fermi-Dirac distribution $f(\epsilon)=\frac{1}{e^{\frac{(\epsilon-\mu)}{k_B T}}+1}$ with the chemical potential obeying the limiting relationship $\lim_{T\to 0}\mu \to \epsilon_f$. Notice that in the $T=0$ limit the low energy single particle states are occupied up to the fermi surface at the energy ϵ_f.

In the presence of weak scattering disorder, the Green's function in Eqs. (7.46b) and (7.47b) develops a renormalization of its particle energies, adjusting their real and imaginary values. As a result, the Green's function in Eq. (7.51b) averaged over the random disorder becomes of the general form

$$\left\langle G^{(+)}(\epsilon)\right\rangle = \frac{1}{\left[\epsilon-\Sigma^{(+)}-H\right]} = \frac{\epsilon-H-\Sigma_R^{(+)}}{\left(\epsilon-H-\Sigma_R^{(+)}\right)^2+\left(\Sigma_I^{(+)}\right)^2} - i\frac{\Sigma_I^{(+)}}{\left(\epsilon-H-\Sigma_R^{(+)}\right)^2+\left(\Sigma_I^{(+)}\right)^2} \tag{7.52b}$$

where $\Sigma^{(+)}=\Sigma_R^{(+)}-i\Sigma_I^{(+)}$ is the complex renormalization of the energy arising from scattering with the disordered medium, and $\langle\rangle$ is an average over the disorder. This assumes that the disorder is described by a set of random configurations characterized by a set of statistical properties and that the average properties of the medium are uniform over space. (Note: The origins of the form in Eq. (7.52b) will be discussed later. For now, however, it is sufficient to see that for weak scattering Eq. (7.52b) reduces to the original delta-function formulation of Eq. (7.47). Specifically, we find that the limit $\lim_{\Sigma_I^{(+)}\to 0^+}\frac{\Sigma_I^{(+)}}{\left(\epsilon-H-\Sigma_R^{(+)}\right)^2+\left(\Sigma_I^{(+)}\right)^2}\to \pi\delta\left(\epsilon-H-\Sigma_R^{(+)}\right)$ is realized).

Next, we configurationally average the $T=0$ sum in Eq. (7.51b) over all the possible statistical configurations of the weak disorder in the system so that the average of the sum in Eq. (7.51b) becomes

$$\sum_{\alpha,\beta}\langle\langle\alpha\,|\,p_x ImG^{(+)}(\epsilon+\hbar\omega)\,|\,\beta\rangle\langle\beta\,|\,p_x ImG^{(+)}(\epsilon)\,|\,\alpha\rangle\rangle \tag{7.52c}$$

where the outer $\langle\rangle$ indicates an average over all possible configurations of the randomly disordered medium. Due to the highly peaked nature of the imaginary parts of the Green's functions in the weak scattering limit, we can further approximate

$$\sum_{\alpha,\beta}\langle\langle\alpha\,|\,p_x ImG^{(+)}(\epsilon+\hbar\omega)\,|\,\beta\rangle\langle\beta\,|\,p_x ImG^{(+)}(\epsilon)\,|\,\alpha\rangle\rangle \tag{7.52d}$$

$$\approx \sum_{\alpha,\beta}\langle\langle\alpha\,|\,p_x ImG^{(+)}(\epsilon+\hbar\omega)\,|\,\beta\rangle\;\langle\beta\,|\,p_x ImG^{(+)}(\epsilon)\,|\,\alpha\rangle\rangle$$

i.e., we assume that the average of the product of Green's functions is approximated by the product of the average Green's functions. (Note again in Eqs. (7.52c) and (7.52d) that $\{|\gamma\rangle\}$ is a complete basis of single-particle electron gas states.)

Considering Eq. (7.52d) in a basis of the gas states composed of single-particle energy and momentum planewave states, it follows that

$$\sum_{\alpha,\beta}\langle\langle\alpha\,|\,p_x ImG^{(+)}(\epsilon+\hbar\omega)\,|\,\beta\rangle\;\langle\beta\,|\,p_x ImG^{(+)}(\epsilon)\,|\,\alpha\rangle\rangle \tag{7.52e}$$

$$= \sum_{\vec{k},\vec{k}'} \hbar k_x \hbar k_x' Im\tilde{G}^{(+)}(\epsilon + \hbar\omega, k)\delta_{\vec{k},\vec{k}'} Im\tilde{G}^{(+)}(\epsilon, k')\delta_{\vec{k}',\vec{k}}$$

$$= \sum_{\vec{k}} \left|\langle\vec{k} \mid \frac{p_x}{m} \mid \vec{k}\rangle\right|^2 \left|Im\tilde{G}^{(+)}(\epsilon, k)\right|^2$$

where we have used $\langle\langle\vec{k} \mid G^{(+)}(\epsilon,k) \mid \vec{k}'\rangle\rangle = \tilde{G}^{(+)}(\epsilon,k)\delta_{\vec{k},\vec{k}'}$ in the representation of single particle planewave states. Specifically, $|\vec{k}\rangle$ and $|\vec{k}'\rangle$ are single particle states of momenta $\hbar\vec{k}$ and $\hbar\vec{k}'$ and at $T = 0$ the sum in the last equality is over the energy states at and below ϵ_f.

In Eq. (7.52e) we also note that

$$\langle\vec{k} \mid \frac{p_x}{m} \mid \vec{k}\rangle = v_x(\vec{k}) = \frac{\hbar k_x}{m} \tag{7.53}$$

where $\{|\vec{k}\rangle\}$ are the planewave states of the gas particles for which $\epsilon(\vec{k}) = \frac{\hbar^2 k^2}{2m}$ and $p_x = \hbar k_x$.

Applying the results in Eq. (7.51) through Eq. (7.53), it then follows that for weak disorder [6]

$$\sigma_R(\omega) = 2\frac{e^2\hbar}{\pi V}\int d\,\epsilon\left(-\frac{df(\epsilon)}{d\epsilon}\right)\sum_{\vec{k}} v_x(\vec{k})^2 \left|Im\tilde{G}^{(+)}(\epsilon, \vec{k})\right|^2 \tag{7.54}$$

$$= 2\frac{e^2\hbar}{\pi V}\sum_{\vec{k}} v_x(\vec{k})^2 \left|Im\tilde{G}^{(+)}(\epsilon(k), \vec{k})\right|^2$$

where the sum over $\vec{k}$ is now at the fermi surface, and the factor of 2 accounts for the spin degeneracy of the gas. This form will next be shown to yield Ohm's law in the weak scattering limit.

The resulting forms in Eqs. (7.51b) and (7.54) express the real part of the conductivity in terms of the Green's functions of the gas. Before applying the relationship in Eq. (7.51b) to a study of the localization properties of the gas, we first demonstrate the Ohm's law limit of Eq. (7.54) for the conductivity of the electron gas.

7.2.3 Ohm's Law

The Ohm's law result for the electrical conductivity can be obtained in the format of Eq. (7.54) considered in the weak scattering limit [6]. In this regard, we next consider the weak scattering limit of Eq. (7.54) leading to the linear, field independent, DC conductivity. The result is commonly generated in introductory physics classes, based on a classical mechanics treatment. In this context, it yields the so-called Drude theory.

The Green's functions in Eqs. (7.52b) and (7.54) were determined for the system in the limit of very small or no scattering. As we have mentioned in our discussions of Eq. (7.52b), with the inclusion of electron scattering by the randomly disordered medium, the energies of the particles are renormalized. In this way, the particles are found to develop a shift in their real energies as well as to develop a lifetime associated with the imaginary part of their complex energy correction. This means that the imaginary contributions to the denominators $i\delta \to i0^+$ in Eq. (7.46) develop into a small nonzero value, and a small shift is also added to the real energy of the excitations.

To account for this scattering correction let us then replace both $G_0^{(-)}(\epsilon)$ and $G_0^{(+)}(\epsilon)$ in Eq. (7.46) by the general operator forms [6]

$$\langle G^{(-)}(\epsilon)\rangle = \frac{1}{\left[\epsilon - \Sigma^{(-)} - H\right]} \text{ for } \epsilon < \epsilon_f \tag{7.55a}$$

and

$$\langle G^{(+)}(\epsilon)\rangle = \frac{1}{\left[\epsilon - \Sigma^{(+)} - H\right]} \text{ for } \epsilon > \epsilon_f \tag{7.55b}$$

where $\Sigma^{(\mp)} = \Sigma_R^{(\mp)} \pm i\Sigma_I^{(\mp)}$ arises as a self-energy correction from the weak impurity scattering and $\langle\rangle$ indicates a configurational average over the disorder in the scattering medium. Here, $\Sigma_R^{(\mp)}(\epsilon)$ represents the small shifts in the energy levels of the gas particles and $\Sigma_I^{(\mp)}(\epsilon)$ represents the lifetimes developed in the energy levels of the gas particles. Both of these shifts, arising from scattering by the random disorder, will be explained later in this chapter. For now, let us assume that Eq. (7.55) correctly accounts for the scattering.

For the independent gas particles, characterized by renormalized Green's functions including scattering with the random medium, it follows from Eq. (7.55) that [6]

$$\langle\langle \vec{k} \mid G^{(+)}(\epsilon,\vec{k}) \mid \vec{k}\rangle\rangle = \langle \vec{k} \mid \frac{1}{\left[\epsilon - \Sigma^{(+)} - \epsilon(k)\right]} \mid \vec{k}\rangle = \frac{\epsilon - \Sigma_R^{(+)} - \epsilon(k) - i\Sigma_I^{(+)}}{\left(\epsilon - \Sigma_R^{(+)} - \epsilon(k)\right)^2 + \left(\Sigma_I^{(+)}\right)^2}. \tag{7.56}$$

Applying Eq. (7.56) in Eq. (7.54) we then find for $\vec{k}$ at the spherical fermi surface of the gas that

$$\sigma_R(\omega \to 0) = 2\frac{e^2\hbar}{\pi V}\sum_{\vec{k}} v_x^2(\vec{k})\left|Im\tilde{G}^{(+)}(\epsilon_f,\vec{k})\right|^2 \tag{7.57a}$$

$$= 2\frac{e^2\hbar}{\pi V}\sum_{\vec{k}} \frac{1}{3}v^2(\vec{k})\frac{\left(\Sigma_I^{(+)}\right)^2}{\left\{\left(\epsilon_f - \Sigma_R^{(+)} - \epsilon(k)\right)^2 + \left(\Sigma_I^{(+)}\right)^2\right\}^2}$$

$$= \frac{2}{3}\frac{e^2\hbar}{\pi V}v_f^2\sum_{\vec{k}} \frac{\left(\Sigma_I^{(+)}\right)^2}{\left\{\left(\epsilon_f - \Sigma_R^{(+)} - \epsilon(k)\right)^2 + \left(\Sigma_I^{(+)}\right)^2\right\}^2}.$$

The sum over $\vec{k}$ in Eq. (7.57a) can, consequently, be rewritten as an integral $\sum_{\vec{k}} \to \frac{V}{(2\pi)^3}\int d^3k$ where V is the volume of the gas. The result of this conversion, subsequently, is in turn converted to an integral over energy. In this process, use is made of the dispersion relation $\epsilon(k) = \frac{\hbar^2k^2}{2m}$ and the differential relationship $dk = \frac{d\epsilon}{\hbar v_f}$.

In the energy format it is found from Eq. (7.57a) that the integrand $\frac{\left(\Sigma_I^{(+)}\right)^2}{\left\{\left(\epsilon_f - \Sigma_R^{(+)} - \epsilon(k)\right)^2 + \left(\Sigma_I^{(+)}\right)^2\right\}^2}$ for $\left|\Sigma_R^{(+)}\right|, \left|\Sigma_I^{(+)}\right| \ll 1$ is highly peaked about the fermi surface energy so the bulk of the integral in Eq. (7.57a) comes from contributions near the fermi surface. Consequently, it follows from Eq. (7.57a) that [6]

$$\sigma_R(\omega \to 0) = \frac{2}{3}\frac{e^2\hbar}{\pi} v_f \frac{1}{(2\pi)^3 \hbar} 4\pi k_f^2 \int d\,\epsilon \frac{\left(\Sigma_I^{(+)}\right)^2}{\left\{\left(\epsilon_f - \Sigma_R^{(+)} - \epsilon(k)\right)^2 + \left(\Sigma_I^{(+)}\right)^2\right\}^2} \tag{7.57b}$$

$$= \frac{e^2}{m}\left[\frac{2}{(2\pi)^3}\frac{4\pi}{3}k_f^3\right]\frac{\hbar}{2\Sigma_I} = \frac{e^2\tau n}{m},$$

were $n = \frac{2}{(2\pi)^3}\frac{4\pi}{3}k_f^3$ is the carrier density, $v(\vec{k}_f) = \frac{p_f}{m} = \frac{\hbar k_f}{m}$, and we have taken the mean free time between successive scattering of an electron to be $\tau = \frac{\hbar}{2\Sigma_I}$ evaluated at the fermi surface. (Note that in the evaluation of Eq. (7.57b), we have also applied the two integral identities

$$\int \frac{dx}{\left(x^2+c^2\right)^n} = \frac{1}{2c^2(n-1)}\left[\frac{x}{\left(x^2+c^2\right)^{n-1}} + (2n-3)\int \frac{dx}{\left(x^2+c^2\right)^{n-1}}\right] \text{ for } n > 2 \tag{7.58a}$$

and

$$\int \frac{dx}{\left(x^2+c^2\right)^2} = \frac{1}{2c^2}\int \frac{dx}{x^2+c^2} = \frac{1}{2c^3}\arctan\frac{x}{c} \tag{7.58b}$$

during the evaluation).

The result in Eq. (7.57b) gives the standard form for the Drude conductivity of the electron gas model. In this regard, by expressing the result in Eq. (7.57b) in terms of the density of states in energy of the electron gas at the fermi surface $\rho(\epsilon_f) = \frac{3}{2}\frac{n}{\epsilon_f}$ and using the relationship from Eq. (7.22) for the Einstein diffusion coefficient of a particle at the fermi surface $D = \frac{1}{3}v_f^2\tau$, the Ohmic conductivity is often equivalently rewritten into the form [6]

$$\sigma_R(\omega \to 0) = \frac{e^2\tau n}{m} = \frac{2\epsilon_f \rho(\epsilon_f)}{3}\frac{e^2\tau}{m} = \frac{p_f^2}{3m}\frac{e^2\tau}{m}\rho(\epsilon_f) = e^2 D\rho(\epsilon_f). \tag{7.59}$$

This relates the conductivity to the Einstein diffusion coefficient evaluated at the fermi surface.

We see that, in the limit of weak scattering, Eq. (7.57) reduces to the form of Drude's law obtained from a classical mechanical treatment of the model for electrons moving in a viscous medium. As the scattering increases in intensity, however, higher order terms in the Green's function consideration of scattering become important. Involved in these changes, the approximation of replacing the configurational average of a product of Green's functions with the product of the configurational averages of the two Green's functions also loses its validity. This leads to further corrections involving the renormalization of the conductivity and ultimately to the development of localization in the system.

Next a qualitative discussion is given of the origins of the strong localization of the conduction electrons arising from their scattering with the disordered medium. In the course of the discussion, the origins of the self-energy renormalizations in Eqs. (7.52b) and (7.55) will be explained. This treatment is followed by another example of localization in the form of weak localization effects involving the scattering of light from rough metallic surfaces. In optical systems, the treatment of single and two-particle Green's functions in the development of optical scattering is again found to lead to interesting scattering effects associated with localization. The focus of both electron and optical studies is on advancing the Green's function treatment of scattering models and showing the origins of Anderson localization in the study of two-particle Green's functions.

7.2.4 Origins of Anderson Localization: Strong Localization of Electrons

We continue the discussions of Eq. (7.51b) for the consideration of the electron scattering in the conducting media with increasing electron-medium interactions. The focus is on developing a qualitative understanding of the renormalization of the electron Green's functions and ultimately of the origins of strong and weak localization effects in the electron motion. To begin with, a study of the origins of the Green's function self-energy corrections $\Sigma^{(\pm)}(\epsilon) = \Sigma_R^{(\pm)}(\epsilon) \mp i\Sigma_I^{(\pm)}(\epsilon)$ is presented.

Consider the treatment of the Green's function renormalization in the presence of a position-dependent scattering potential operator U representing the interaction of the electrons with the random medium. In the presence of the scattering potential, the Green's functions of the electron gas in Eq. (7.46) become [6]

$$G_U^{(-)}(\epsilon) = \lim_{\delta \to 0^+} \frac{1}{\left[\epsilon - i\delta - H - U\right]} \quad \text{for } \epsilon < \epsilon_f \tag{7.60a}$$

and

$$G_U^{(+)}(\epsilon) = \lim_{\delta \to 0^+} \frac{1}{\left[\epsilon + i\delta - H - U\right]} \quad \text{for } \epsilon > \epsilon_f, \tag{7.60b}$$

where H is the Hamiltonian operator of the electron gas in the absence of interactions of the electrons with the transport medium. The limits in Eq. (7.60) can be rewritten in the form of a scattering sequence expressed by an operator recursion relation. This is given by the form [6]

$$G_U^{(\pm)}(\epsilon) = G_0^{(\pm)}(\epsilon) + G_0^{(\pm)}(\epsilon) U G_U^{(\pm)}(\epsilon) \tag{7.61a}$$

$$= G_0^{(\pm)}(\epsilon) + G_0^{(\pm)}(\epsilon) U G_0^{(\pm)}(\epsilon) + G_0^{(\pm)}(\epsilon) U G_0^{(\pm)}(\epsilon) U G_0^{(\pm)}(\epsilon)$$

$$+ G_0^{(\pm)}(\epsilon) U G_0^{(\pm)}(\epsilon) U G_0^{(\pm)}(\epsilon) U G_0^{(\pm)}(\epsilon) + \ldots,$$

where from Eq. (7.46) we again define

$$G_0^{(-)}(\epsilon) = \lim_{\delta \to 0^+} \frac{1}{\left[\epsilon - i\delta - H\right]} \quad \text{for } \epsilon < \epsilon_f \tag{7.61b}$$

and

$$G_0^{(+)}(\epsilon) = \lim_{\delta \to 0^+} \frac{1}{\left[\epsilon + i\delta - H\right]} \quad \text{for } \epsilon > \epsilon_f. \tag{7.61c}$$

The sequence in Eq. (7.61a) is known as the Dyson equation for the single particle Green's functions and is a standard equation of Green's function theory. It represents the renormalization of the Green's function arising from the multiple scattering of the particles from the potential U introduced into the system.

Averaging $G_U^{(\pm)}(\epsilon)$ over a distribution of configurations of the random disordered medium, it follows from Eq. (7.61) that

$$\langle G_U^{(\pm)}(\epsilon)\rangle = G_0^{(\pm)}(\epsilon) + G_0^{(\pm)}(\epsilon)\langle U\rangle G_0^{(\pm)}(\epsilon) + G_0^{(\pm)}(\epsilon)\langle U G_0^{(\pm)}(\epsilon) U\rangle G_0^{(\pm)}(\epsilon) + \ldots \tag{7.62}$$

where $\langle\rangle$ indicates an average of the disorder in U over a statistical set of configurations representing the randomness in the system. In this respect, the average properties displayed by the disorder are considered to have well-characterized statistical properties and to be statistically the same throughout the conducting medium [6].

As a simplification of the problem, consider a system defined such that $\langle U\rangle = 0$. In this case, for weak scattering, the statistical averages simplify so that [6]

$$\langle G_U^{(\pm)}(\epsilon)\rangle = G_0^{(\pm)}(\epsilon) + G_0^{(\pm)}(\epsilon)\langle UG_0^{(\pm)}(\epsilon)U\rangle G_0^{(\pm)}(\epsilon) + \ldots \tag{7.63}$$

$$\approx G_0^{(\pm)}(\epsilon) + G_0^{(\pm)}(\epsilon)\langle UG_0^{(\pm)}(\epsilon)U\rangle\langle G_U^{(\pm)}(\epsilon)\rangle.$$

From Eq. (7.63) it then follows, for $\delta \to 0^+$ and in the limit of a weak renormalization of the Green's functions by the disorder, that

$$\left[\epsilon - H - \Sigma^{(\pm)}\right]\langle G_U^{(\pm)}(\epsilon)\rangle \approx 1 \tag{7.64a}$$

where

$$\Sigma^{(\pm)} \approx \langle UG_0^{(\pm)}(\epsilon)U\rangle \tag{7.64b}$$

represents the so-called self-energy correction generated from the interaction with the conducting medium.

The result in Eq. (7.64) is then rewritten in terms of a series having the form [6]

$$\langle G_U^{(\pm)}(\epsilon)\rangle = G_0^{(\pm)}(\epsilon) + G_0^{(\pm)}(\epsilon)\Sigma^{(\pm)}\langle G_U^{(\pm)}(\epsilon)\rangle \tag{7.65}$$

$$= G_0^{(\pm)}(\epsilon) + G_0^{(\pm)}(\epsilon)\langle UG_0^{(\pm)}(\epsilon)U\rangle\langle G_U^{(\pm)}(\epsilon)\rangle.$$

This is another Dyson equation, now involving the self-energy corrections $\Sigma^{(\pm)}$. By using Feynman diagram techniques, we next show that the processes in Eq. (7.65) can be simply understood in terms of a graphical illustration of the sequence of scattering events.

The graphical representation of Eq. (7.65) is shown in Figure 7.2a. In this diagrammatic representation of the Green's function, $\langle G_U^{(\pm)}(\epsilon)\rangle$ is denoted by a bold solid line, $G_0^{(\pm)}(\epsilon)$ is denoted by a light solid line, and $\Sigma^{(\pm)} = \langle UG_0^{(\pm)}(\epsilon)U\rangle$ is shown as a filled circle. From the diagram, $\langle G_U^{(\pm)}(\epsilon)\rangle$ is found to evolve as a time development involving successive Green's function propagations followed by self-energy scatterings. Specifically, in the sequence a propagation by $G_0^{(\pm)}(\epsilon)$ is followed by a scattering event involving $\langle UG_0^{(\pm)}(\epsilon)U\rangle$, and the scattering is followed by a propagation by $G_0^{(\pm)}(\epsilon)$, etc. This process continues as the electron travels across the entire random medium. Consequently, a series of scattering events is generated as a sum of diagrams in which the particle scatters $0,1,2,3,\ldots$ times as it passes through the medium.

It is important to note here that the self-energy in Eqs. (7.64b) and (7.65) involves mainly elastic scattering processes, but that other inelastic processes can also enter into the corrections to the Green's function self-energy. At low temperature, however, these processes tend to be a smaller contribution than those of the elastic scattering [7]. Consequently, they have been ignored here. In this regard, the reader is referred to the literature for a consideration of the effects of inelastic scattering [7].

A similar diagrammatic representation can also be given for the time development of the product of Green's functions in Eqs. (7.52d) and (7.52e) which enters the expressions for the conductivity of the gas of electrons interacting with the random scattering potential U given in Eqs. (7.51b) and

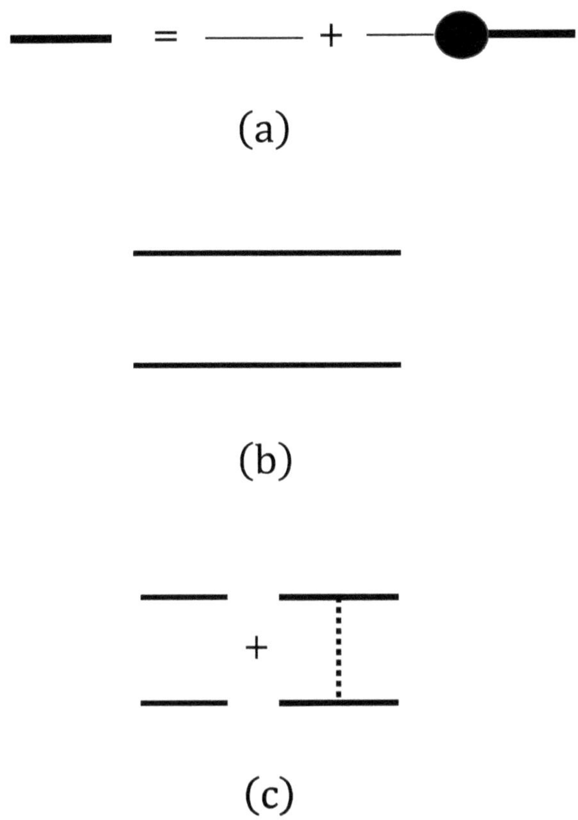

FIGURE 7.2 Diagrammatic representations in terms of Feynman diagrams portraying the time development of the scattering from the random potential U. Representations are given of: a) Eq. (7.65), i.e., $\langle G_U^{(\pm)}(\epsilon)\rangle = G_0^{(\pm)}(\epsilon) + G_0^{(\pm)}(\epsilon)\Sigma^{(\pm)}\langle G_U^{(\pm)}(\epsilon)\rangle$ where $\langle G_U^{(\pm)}(\epsilon)\rangle$ is represented by a bold solid line, $G_0^{(\pm)}(\epsilon)$ is represented by a light solid line, and $\Sigma^{(\pm)}$ by a filled circle, b) the right-hand side of Eq. (7.69b), i.e., $\langle G_U^{(+)} G_U^{(+)*}\rangle$ where the upper line is $\langle G_U^{(+)}\rangle$ and the lower line is $\langle G_U^{(+)*}\rangle$, and c) the right-hand side of Eq. (7.70), i.e., $\langle G_U^{(+)}\rangle\langle G_U^{(+)*}\rangle + \langle G_U^{(+)}\rangle\langle G_U^{(+)*}\rangle\langle UU^*\rangle\langle G_U^{(+)}\rangle\langle G_U^{(+)*}\rangle$ where the upper lines are $\langle G_U^{(+)}\rangle$ and the lower lines are $\langle G_U^{(+)*}\rangle$ and the interaction $\langle UU^*\rangle$ is the vertical dashed line.

(7.60). Specifically, in the calculation of the conductivity of our electron gas we need to determine the sum [6]

$$\sum_{\vec{k},\vec{k}'} \langle\langle \vec{k} \,|\, p_x Im G_U^{(+)}(\epsilon + \hbar\omega) \,|\, \vec{k}'\rangle\, \langle \vec{k}' \,|\, p_x Im G_U^{(+)}(\epsilon) \,|\, \vec{k}\rangle\rangle \tag{7.66a}$$

where again $\langle\rangle$ represents a statistical average over the scattering potential of the system and the basis of the electron gas is given in terms of single-particle energy and momentum states of the individual electrons.

Notice here we have taken $G_U^{(+)}$. to represent the Green's function in the presence of the random disordered potential U, and the sum is in terms of the average of the products of two Green's functions over the random disorder. To facilitate obtaining Eq. (7.66a), however, we can first study the properties of another function known as the two-particle Green's function. Once these are determined, the result in Eq. (7.66a) can be expressed in terms of these two-particle Green's functions. This is useful as the two-particle Green's functions contain much of the physics involved in Eq. (7.66a), displaying it in a simple format. We shall see this again later in our discussions of weak localization in optical scattering. In the context of optical scattering, the differential reflection coefficient is directly related to the optical two-particle Green's function.

The study of the sum in Eq. (7.66a) is simplified by first studying a simple product of two Green's functions. In this regard, let us simplify the considerations by treating the more general form known as the two-particle Green's function. This generalization is written as

$$\langle G_U{}^{(+)}\left(\vec{q}',\vec{k}'\right) G_U{}^{(+)*}\left(\vec{q},\vec{k}\right)\rangle \tag{7.66b}$$

where the Green's functions now depend on the incoming and outgoing scattering electron states as denoted by the wavevectors $\vec{k},\vec{k}'$ and $\vec{q},\vec{q}'$, respectively. (Notice, however, the absence between Eqs. (7.66a) and (7.66b) of the prefactors involving the x–components of the wavevectors in Eq. (7.66b).) The form in Eq. (7.66a) is a part of the more specified expression in Eq. (7.66b) and can be obtained from it through a process involving integration. In addition, note that the wavevector dependence of the two-particle Green's function is found to come from the self-energy correction $\Sigma^{(\pm)} \approx \langle UG_0^{(\pm)}(\epsilon)U\rangle$ which in wavevector space is given by

$$\Sigma^{(\pm)}\left(\vec{q},\vec{k}\right) \approx \frac{1}{V}\int d^3r_1 \int d^3r_2 e^{i\vec{q}\cdot\vec{r}_1}\langle U\left(\vec{r}_1\right) G_0^{(\pm)}\left(\vec{r}_1,\vec{r}_2\right) U\left(\vec{r}_2\right)\rangle e^{i\vec{k}\cdot\vec{r}_2}. \tag{7.66c}$$

The result shown in Eq. (7.66c) is the Fourier transform of the direct space self-energy correction.

Consequently, the important point here is that from Eq. (7.66b) the result for Eq. (7.66a) can ultimately be obtained, and the physics contained in both are essentially the same. The average in Eq. (7.66b) is referred to as a two-particle Green's function because its product of two Green's functions describes the motion of two particles through the system. This is done in a similar way that a single particle Green's function tracks the motion of a single particle through the system as a succession of free particle motions interrupted by scattering events. In the following we shall outline how Eq. (7.66b) is studied and how it ultimately leads to the origins of Anderson localization.

We begin with a study of certain types of processes known as ladder processes or ladder diagram scatterings between the propagation of two particles [6, 7]. These provide a renormalization of the relaxation time characterizing the propagation between successive scattering events in the system, ultimately entering into the form of the Drude conductivity.

To simplify the discussions, however, let us restrict our considerations to statistical disorders known as Gaussian random disorders [12]. These facilitate the development of analytic treatments of disordered systems. Other more general types of disorder can be handled by computer simulation methods, but these are not of interest here. It has been shown, however, that the qualitative features of localization developed in both analytical and simulation methods agree [7]. In this regard, while simulations are helpful in the study of random systems, their numerical nature often does not allow them to illustrate the basic physical mechanisms giving rise to Anderson localization as transparently as do analytical approaches. This is why we focus here on analytical approaches.

In Gaussian random disorder [7–12], we consider the scattering potential $U\left(\vec{r}\right)$ of the independent particles to be a function of position $\vec{r}$ and to be defined so that the configuration average of single scattering events gives zero, i.e.,

$$\langle U\left(\vec{r}\right) = 0\rangle. \tag{7.67}$$

Continuing to products of multiple $U\left(\vec{r}\right)$'s, the statistical average of products of an even number of potentials is written in terms of products of pair averages, i.e., $\langle U\left(\vec{r}\right)U\left(\vec{r}'\right)\rangle$ is a pair average. For example, we write the configurational average of the production of the four potentials $U\left(\vec{r}_1\right)U\left(\vec{r}_2\right)U\left(\vec{r}_3\right)U\left(\vec{r}_4\right)$ as [12]

$$\langle U(\vec{r}_1)U(\vec{r}_2)U(\vec{r}_3)U(\vec{r}_4)\rangle = \langle U(\vec{r}_1)U(\vec{r}_2)\rangle\langle U(\vec{r}_3)U(\vec{r}_4)\rangle + \langle U(\vec{r}_1)U(\vec{r}_3)\rangle\langle U(\vec{r}_2)U(\vec{r}_4)\rangle \quad (7.68a)$$

$$+\langle U(\vec{r}_1)U(\vec{r}_4)\rangle\langle U(\vec{r}_2)U(\vec{r}_3)\rangle$$

where a sum of the products of all the different pairings of $\left\{U(\vec{r}_1),U(\vec{r}_2),U(\vec{r}_3),U(\vec{r}_4)\right\}$ is made. Similarly, all averages of higher products of an even number of potentials are expressed as sums in terms of the possible pairings of the pair products. Unlike the average of an even number of products of $U(\vec{r})$'s, however, the average of all odd numbers of potentials is set to zero. Consequently, as an example

$$\langle U(\vec{r}_1)U(\vec{r}_2)U(\vec{r}_3)\rangle = \langle U(\vec{r}_1)U(\vec{r}_2)\rangle\langle U(\vec{r}_3)\rangle + \langle U(\vec{r}_1)U(\vec{r}_3)\rangle\langle U(\vec{r}_2)\rangle \quad (7.68b)$$

$$+\langle U(\vec{r}_1)\rangle\langle U(\vec{r}_2)U(\vec{r}_3)\rangle = 0$$

Note, in this regard, that in the average over the randomness of the system, the pair average $\langle U(\vec{r})U(\vec{r}')\rangle$ has the property in most systems that it tends to decrease rapidly with increasing separation $|\vec{r}-\vec{r}'|$. This tends to localize the averages of products of multiple $U(\vec{r})$'s in the system about sets of closely located points. In this regard, the function $\langle U(\vec{r})U(\vec{r}')\rangle$ is often referred to as a pair correlation function because it provides an indication of the predictability of the values of the random function at points separated from one another [7, 12].

In the treatment of the weak scattering limit of the two-particle Green's function in Eq. (7.66b), as a first approximation we write Eq. (7.66b) in the form [6, 7]

$$\langle G_U^{(+)}(\vec{q}',\vec{k}')G_U^{(+)*}(\vec{q},\vec{k})\rangle \approx \langle G_U^{(+)}(\vec{q}',\vec{k}')G_U^{(+)*}(\vec{q},\vec{k})\rangle \quad (7.69a)$$

or taken in a more concise position space notation

$$\langle G_U^{(+)}G_U^{(+)*}\rangle \approx \langle G_U^{(+)}\rangle\langle G_U^{(+)*}\rangle. \quad (7.69b)$$

This is the statement that the average of the two Green's functions is approximated by the product of the averages of the two Green's functions and is related to our earlier discussions of Ohm's law. The diagrammatic representation of Eq. (7.69b) is shown in Figure 7.2b where the upper Green's function line corresponds to $\langle G_U^{(+)}\rangle$ and the lower Green's friction line to $\langle G_U^{(+)*}\rangle$. In this representation all the pair correlators of the scattering potentials have been formed only between potentials acting on the same Green's function line so that there is no interaction between the two different Green's functions. Consequently, for both Green's functions in Figure 7.2b we have used the renormalized form of the Green's functions in Eq. (7.65) which is represented diagrammatically in Figure 7.2a.

In a next level of approximation, we include a correlator scattering acting between the two Green's function lines in Figure 7.2b. This leads to a two-particle Green's function process given by [6, 7]

$$\langle G_U^{(+)}G_U^{(+)*}\rangle \approx \langle G_U^{(+)}\rangle\langle G_U^{(+)*}\rangle + \langle G_U^{(+)}\rangle\langle G_U^{(+)*}\rangle\langle UU^*\rangle\langle G_U^{(+)}\rangle\langle G_U^{(+)*}\rangle \quad (7.70)$$

where U acts on the $\langle G_U^{(+)}\rangle$, U^* acts on the $\langle G_U^{(+)*}\rangle$, and the correlation function $\langle UU^*\rangle$ provides an interaction between the two different Green's functions. The processes in Eq. (7.70) are represented diagrammatically in Figure 7.2c where again the Green's function lines are those of the renormalized Green's functions in Eq. (7.65). This is graphically the next level of approximation for the two-particle Green's function from that in Figure 7.2b and represents a correction to Ohm's law conductivity. Notice that because of the spatial localization of $\langle UU^* \rangle$ the two Green's function lines are scattering in the medium at sites closely located to one another [7].

In analogy with the Dyson equation in Eq. (7.65) for the development of the single particle Green's function, we can also extend the approximation in Figure 7.2c for the two-particle Green's function to a consideration of the scattering series [6, 7]

$$\begin{aligned}\langle G_U{}^{(+)} G_U{}^{(+)*} \rangle &\approx \langle G_U{}^{(+)} \rangle \langle G_U{}^{(+)*} \rangle + \langle G_U{}^{(+)} \rangle \langle G_U{}^{(+)*} \rangle \langle UU^* \rangle \langle G_U{}^{(+)} G_U{}^{(+)*} \rangle \\ &= \langle G_U{}^{(+)} \rangle \langle G_U{}^{(+)*} \rangle + \langle G_U{}^{(+)} \rangle \langle G_U{}^{(+)*} \rangle \langle UU^* \rangle \langle G_U{}^{(+)} \rangle \langle G_U{}^{(+)*} \rangle \\ &+ \langle G_U{}^{(+)} \rangle \langle G_U{}^{(+)*} \rangle \langle UU^* \rangle \langle G_U{}^{(+)} \rangle \langle G_U{}^{(+)*} \rangle \langle UU^* \rangle \langle G_U{}^{(+)} \rangle \langle G_U{}^{(+)*} \rangle + \ldots\end{aligned} \tag{7.71}$$

This is known as the Bethe-Salpeter equation [7, 12], and it sums an increasing sequence of scattering processes involving successive $\langle UU^* \rangle$ interactions between the $\langle G_U{}^{(+)} \rangle \langle G_U{}^{(+)*} \rangle$ Green's function products. In Figure 7.3a a diagrammatic representation of these extended processes in Eq. (7.71) is shown. Notice that in the scattering sequence the two Green's functions encounter the $\langle UU^* \rangle$ interactions in the same order. As we shall soon see, in the consideration of arbitrary scattering processes this need not be the case so that the interactions with U and U^* in a given correlator can be made in different orders during the propagation of $\langle G_U{}^{(+)} \rangle$ and $\langle G_U{}^{(+)*} \rangle$ through the medium.

In the case of the scattering in Eq. (7.71), however, the particles propagate through the system scattering with each other through closely spaced sites in the random medium. This reduces the phase interference of the two waves and promotes an enhanced contribution from those of diagrammatic contributions representing completely phase incoherent scatterings [7].

The scattering processes included in Eq. (7.71), because of their diagrammatic appearance, are known as the ladder diagram processes [6, 7]. They portray sequences of repeated scatterings of the two-particle Green's functions from each other via the impurities in the system made in such a way that each Green's function encounters the same impurities in the same order. In this type of scattering, a net phase coherence tends to develop between the Green's functions during the transit through the scattering medium. The sum of the diagrams is such that each successive term of the series represents an increment of the number of scatterings in the system by one additional scattering event as the medium is traversed. In this way, in the first term of the ladders, the two Green's functions do not encounter each other. In the second term, they scatter once from each other via an impurity; the third term contains two scatterings between the Green's functions via two different impurities, etc.

We next consider how the ladder diagrams influence the electrical conductivity of the electron gas as it moves through the scattering medium. Earlier in Eq. (7.57) a discussion of the conductivity leading to Ohm's law was presented in which the only processes considered were those in which there was no scattering between the $\langle G_U{}^{(+)} \rangle$ and $\langle G_U{}^{(+)*} \rangle$ Green's functions, i.e., the two-particle Green's function was approximated as $\langle G_U{}^{(+)} G_U{}^{(+)*} \rangle \approx \langle G_U{}^{(+)} \rangle \langle G_U{}^{(+)*} \rangle$. This is the first term in the ladder diagram series shown in Figure 7.3a. At this level of approximation, the Ohm's law result followed, being expressed as

$$\sigma_R(\omega \to 0) = \frac{e^2 \tau n}{m} \tag{7.72}$$

where from Eq. (7.57a) it was found that the mean free time between weak scatterings is [6, 7]

$$\tau = \frac{2}{3} \frac{\hbar v_f^2}{\pi V} \frac{m}{n} \sum_{\vec{k}} \frac{\left(\Sigma_I^{(+)}\right)^2}{\left\{\left(\epsilon_f - \Sigma_R^{(+)} - \epsilon(k)\right)^2 + \left(\Sigma_I^{(+)}\right)^2\right\}^2}. \tag{7.73}$$

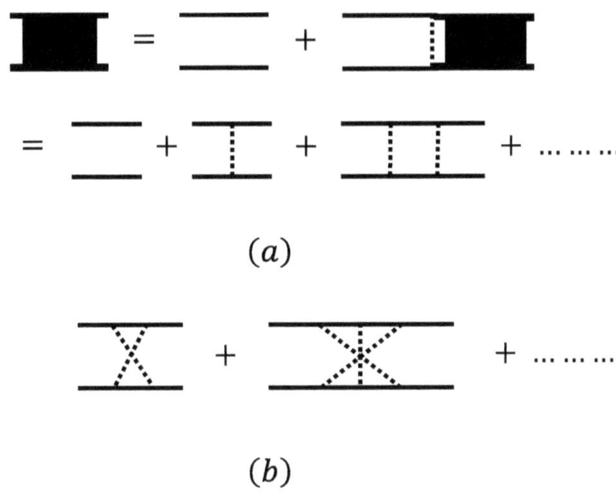

FIGURE 7.3 Schematic figure of: a) the ladder diagram scattering sequence in which the horizontal solid lines represent $\langle G_U^{(+)}\rangle\langle G_U^{(+)*}\rangle$, the vertical dashed lines represent the scattering by $\langle UU^*\rangle$, and the two-particle Green's function $\langle G_U^{(+)}G_U^{(+)*}\rangle$ is represented by the solid rectangle, b) the maximally-crossed diagrammatic contributions to $\langle G_U^{(+)}G_U^{(+)*}\rangle$ in which the horizontal solid lines represent $\langle G_U^{(+)}\rangle\langle G_U^{(+)*}\rangle$ and the vertical dashed lines represent the scattering by $\langle UU^*\rangle$. Notice the reversal in the sequence of $\langle UU^*\rangle$ interactions between the upper and lower Green's function horizontal lines.

Here the sum involves $\vec{k}$ at the fermi surface and Eq. (7.73) relates the mean-free time between collisions to the self-energy renormalization. We shall now see that the remaining ladder diagrams lead to an adjustment in the value of the mean free time.

From Eq. (7.40) we found that the transition rate for scattering from a quantum state $|\alpha\rangle$ to a quantum state $|\beta\rangle$ is given by [6, 7]

$$W_{\alpha\beta} = \frac{2\pi}{\hbar}\left|\langle\alpha \mid exE_0 \mid \beta\rangle\right|^2\left[\delta\left(\hbar\omega - \epsilon_{\beta\alpha}\right) + \delta\left(\hbar\omega + \epsilon_{\beta\alpha}\right)\right]. \tag{7.74}$$

As a result, the rate of transition for an electron in the state $|\vec{k}\rangle$ to scatter into the electron state $|\vec{k}'\rangle$ at the fermi surface is given by

$$\frac{1}{\tau\left(\vec{k}\right)} = \sum_{\vec{k}'} W_{\vec{k}\to\vec{k}'}\left(1 - f_{\vec{k}'}\right) \tag{7.75}$$

where $\left(1 - f_{\vec{k}'}\right)$ is the probability that $\vec{k}'$ is unoccupied. Notice that, from the symmetry of the gas, $\tau\left(\vec{k}\right) = \tau$ is the same for all $\vec{k}$ at the fermi surface. Consequently, τ in Eq. (7.75) is the same as that found in the Ohm's law formula, i.e., $\sigma_R\left(\omega \to 0\right) = \frac{e^2\tau n}{m}$ for the approximation [6, 7] $\langle G_U^{(+)}G_U^{(+)*}\rangle \approx \langle G_U^{(+)}\rangle\langle G_U^{(+)*}\rangle$.

Next, consider including in Eq. (7.75) the additional ladder diagram processes in Figure 7.3a involving interactions between the two Green's function lines. (Note that these contributions of course must be processed into the generation of the conductivity from Eq. (7.52c) and the configuration average of Eq. (7.51b).) As a result of this additional processing, it is found that the mean-free time in Eq. (7.75) is renormalized to the form [6, 7]

$$\frac{1}{\tau(\vec{k})} = \sum_{\vec{k}'} W_{\vec{k}\to\vec{k}'} \left(1 - f_{\vec{k}'}\right)\left(1 - \frac{\vec{k}\cdot\vec{k}'}{kk'}\right). \tag{7.76}$$

This result is commonly obtained in both Boltzmann equation and Green's function treatments of the transport, where we see that the interactions between the Green's function lines introduce an angular dependence into the integral for the transition rates. Note, in this regard, that in the absence of anisotropic scattering, it is found that the $\frac{\vec{k}\cdot\vec{k}'}{kk'}$ contribution drops out of Eq. (7.76).

Consequently, with the inclusion of the ladder diagrams containing interaction rungs in Ohm's law result $\sigma_R(\omega \to 0) = \frac{e^2\tau n}{m}$, the mean-free time in Eq. (7.75) is replaced by that from Eq. (7.76). We then find that in the weak scattering limit the effects of the ladder diagram contributions are a renormalization of the mean-free time between collisions. This is a quantitative rather than a qualitative change in the theory. We next consider a different set of diagrams which introduce qualitative as well as quantitative changes in the conductivity. These are the so-called maximally crossed diagrams which introduce phase-coherent processes into the considerations.

In Figure 7.3b is shown the so-called maximally crossed diagrams [6, 7]. These are also part of the multiple scattering effects determining the Green's function conductivity. From the diagrams, we see that the difference between the ladder and maximally crossed diagrams is that the upper and lower Green's function lines of the maximally crossed diagrams sequentially encounter the impurities in a reversed ordering during their scattering processes. This is different from the ladder diagram processes in which the sequential order of the scattering sequence is the same for each of the upper and lower Green's function lines. A consequence of this difference is that the maximally crossed processes retain a phase coherence between the two Green's function processes which is similar to that discussed in the double passage scattering [6, 7] considered in Figure 7.1.

As in the double passage effect, the maximally crossed diagrams contribute to backscattering enhancements in the system, i.e., the electrons experience an enhanced resistance to their forward propagation through the disorder medium. This reduction in the transport arises from phase coherence scattering and is known as weak localization. It is observed in conducting systems and can be reduced in the presence of an externally applied magnetic field. In this regard, the application of an external magnetic field introduces a violation of time reversal symmetry into the system [7], changing the phase coherence along previously time-reversed paths in the scattering medium. The violation of time reversal symmetry then tends to modify the phase coherence of the maximally crossed scattering processes. This in turn shows up as changes in the conductivity. Later, we shall see similar effects exhibited in the phase-coherent scattering found in optical systems.

In 3−dimensional materials, as the scattering interactions with the disordered medium increase, the weak localization effect can transform to strong localization [6–8]. Strong localization occurs when the backscattering is large enough to confine the electrons to a bounded region of space [6–11]. The diagrammatics accounting for strong localization are based on the maximally crossed diagrams but in a very complicated way [10]. Maximally crossed diagrams enter the process as a renormalizing factor to other diagrams, all of which yield large backscattering contributions. Consequently, the system proceeds to strong localization as a type of phase transition and is found to be responsible for metal–insulator transitions observed as a function of applied pressure in some materials. (The reader is referred to the literature for a more complete treatment of these topics [6–11].) At the point of strong localization all the electrons in the system are localized and at $T = 0$ the medium behaves as an insulator. At finite temperature, however, forms of hopping conductivity can often be observed in which the electrons of the system hop between localized states [6–8]. This is one form of the metal–insulator transition known as the Anderson transition.

In terms of the metal–insulator transition, notice that we have ignored the electron-electron interactions in our model of electron localization [6–8]. This is an important point as at low electron densities the electrons can also become localized solely from their electron-electron interactions, i.e., they crystallize into a nonconducting lattice. This is known as the Mott transition [9] and accounts for metal–insulator transitions in various materials and in some electron devices [8, 10]. In this regard, when localization is found in electron systems there is an amount of ambiguity as to its precise origins, i.e., whether it is from an Anderson or a Mott transition or a combination of both [6–8, 10]. Consequently, it is interesting to look for localization effects in systems in which the excitations do not interact with each other. Such systems are found in optics because photons for most optical applications have no photon-photon interactions [10]. Localization in optical systems then arises solely from Anderson localization. Note that localization in optical systems will now be a present topic of this chapter.

We next turn to an example of localization in the context of an optical system. This provides another illustration of the association of transport properties with two-particle Green's functions and of the importance of ladder and maximally crossed processes in characterizing transport behaviors. In optics, the transport properties are found to be more directly related to the optical two-particle Green's functions than we have found to be the case in the conductivity of electron systems.

7.3 WEAK LOCALIZATION EFFECTS IN THE SCATTERING OF LIGHT FROM RANDOM ROUGH SURFACES

As an additional illustration of the application of ladder and maximally crossed diagrams in the investigation of transport processes, let us treat within the context of classical electrodynamics the scattering of light from a randomly rough metal or semiconductor interface [6–8]. For a basic consideration of such a scattering process, we shall assume that the rough surface is planar on average, exhibits statistically uniform disorder over the entire surface, and that the scattering is elastic, i.e., no frequency shift occurs during the propagation of the light. In this geometry, we consider scattering in which an electromagnetic plane wave is incident from vacuum onto the weakly rough conducting interface, which then scatters the light into a specularly reflected and a diffusely scattered component. The interesting features in this process are observed in the diffusely scattered component and arise from an excited electromagnetic surface wave propagated along the interface. Specifically, the diffuse scattering is shown to exhibit characteristics affected by the weak localization properties of the surface waves on the interface [12].

The interest in our considerations is in conducting surfaces because light incident on such surfaces not only is reflected from the interface but also couples to surface electromagnetic waves. The latter are electromagnetic modes bound to and traveling on the vacuum-conductor interface. In this regard, the surface waves propagate parallel to the mean plane of the surface and have electromagnetic fields which decay rapidly away from the field maxima located at the interface. At some point, these surface waves are scattered by the interface disorder into the vacuum where they propagate away from the surface as diffusely scattered radiation. Light then transports energy not only in the specularly reflected waves but in the propagation of surface electromagnetic waves along the rough interface. Consequently, surface electromagnetic waves transport electromagnetic energy along the interface and can experience weak localization effects on surfaces with weak roughness. Such weak localization effects are observed in anomalous features of the diffusely scattered light [12].

For the geometry of the problem, we consider the scattering of light from a one-dimensional randomly rough surface which is planar on statistical average. (See Figure 7.4 for the scattering geometry.) Above the surface is vacuum and below it is a conducting medium with a complex, frequency-dependent, dielectric constant of the form $\varepsilon(\omega) = \varepsilon_R(\omega) + i\varepsilon_I(\omega)$ where $\varepsilon_R(\omega) < 0$ for surface electromagnetic waves to be supported on the interface. The mean surface is the $x_1 - x_2$ plane and the x_3 –axis is normal to the mean surface. Along the x_1 –axis the surface is randomly rough with a surface profile characterized by a profile function of the form

$$x_3 = \xi(x_1), \tag{7.77}$$

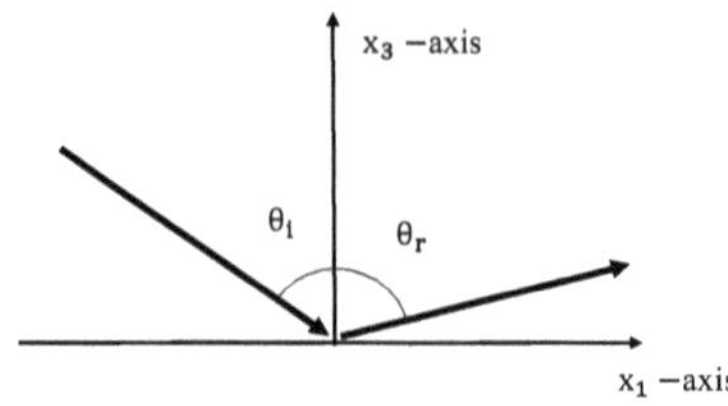

FIGURE 7.4 Schematic of the scattering from the weakly rough surface with surface profile function $x_3 = \xi(x_1)$. The mean surface is in the $x_1 - x_2$ plane so that above the rough surface is vacuum and below the surface is a conducting medium. Light propagates in the $x_1 - x_3$ plane in the vacuum and is incident on the surface with an incident angle θ_i measured in the anticlockwise direction from the normal to the mean interface. The light is reflected from the surface into the vacuum with an angle of reflection θ_r measured in the clockwise direction from the normal to the mean interface.

where the x_3 - coordinate of the surface is expressed in terms of the position on the x_1 - axis. Along the x_2 - axis the surface has complete translational symmetry so that the disorder is only along the x_1 - axis [12].

Specifically, in the context of our model, the surface profile functions are from a set of profiles characterized by Gaussian random statistics [12]. If $\langle\rangle$ indicates a statistical average of the surface profiles, we then find [12]

$$\langle \xi(x_1) \rangle = 0 \tag{7.78a}$$

and the surface profile correlator is defined by

$$\langle \xi(x_1)\xi(x_1^{'}) \rangle = \sigma^2 \exp\left(-\frac{\left| x_1 - x_1^{'} \right|^2}{a^2} \right). \tag{7.78b}$$

Here σ is a measure of the statistical height of the surface profiles above the $x_1 - x_2$ plane and a is a measure of the rate of decay of the profile function correlator along the x_1 −axis. Again, in Gaussian random disorder all the averages of products of surface profile functions are expressed in terms of Eq. (7.78). Consequently, as in our discussion in Eqs. (7.67) and (7.68) the averages of higher products of the profile functions are written in terms of $\langle \xi(x_1) \rangle = 0$ and the pair correlators $\langle \xi(x_1)\xi(x_1^{'}) \rangle$.

An incident electromagnetic wave to the interface propagates in the $x_1 - x_3$ plane above the weakly rough surface with a magnetic field of the form [12]

$$\vec{H}(\vec{x},t) = \left(0, H_2(x_1, x_3 \mid \omega), 0\right) e^{-i\omega t} \tag{7.79}$$

where

$$H_2(x_1, x_3 \mid \omega) = e^{ik_0 x_1 - i\alpha_0(k_0\omega)x_3} + \int \frac{dq}{2\pi} R(q \mid k_0) e^{iqx_1 + i\alpha_0(q\omega)x_3} \tag{7.80a}$$

with the first term on the right in Eq. (7.80a) being the incident wave and the second term the scattered waves. (Note that k_0 is the component of the incident wavevector along the mean surface and q is the component of the scattered wavevector along the mean surface.) In Eq. (7.80a)

$$\alpha_0(q\omega) = \sqrt{\left(\frac{\omega}{c}\right)^2 - q^2}, \; q^2 < \left(\frac{\omega}{c}\right)^2 \tag{7.80b}$$

$$= i\sqrt{q^2 - \left(\frac{\omega}{c}\right)^2}, \; q^2 > \left(\frac{\omega}{c}\right)^2$$

where $\alpha_0(k_0\omega)$ is real for the incident wave. The polarized wave in Eq. (7.80a) is known as a p-polarized wave and differs from a wave with the electric field polarized along the x_2-axis which is known as an s-polarized wave. Here we study p-polarized waves as they can couple through the surface roughness to surface waves whereas s-polarized waves do not. In the following, the focus is then on the determination of the coefficient $R(q \mid k_0)$ in Eq. (7.80a) which is obtained through an application of the electromagnetic boundary conditions at the rough scattering interface.

Upon matching the electromagnetic boundary conditions at the vacuum-conductor interface, it is found that [12]

$$R(q \mid k_0) = 2\pi\delta(q - k_0)R_0(k_0) - 2iG_0(q)T(q \mid k_0)G_0(k_0)\alpha_0(k_0\omega), \tag{7.81a}$$

where for $\alpha(q\omega) = \sqrt{\varepsilon(\omega)\left(\frac{\omega}{c}\right)^2 - q^2}$ with $Re\alpha(q\omega) > 0$ and $Im\alpha(q\omega) > 0$. In Eq. (7.81a)

$$R_0(k_0) = \frac{\varepsilon(\omega)\alpha_0(k_0\omega) - \alpha(k_0\omega)}{\varepsilon(\omega)\alpha_0(k_0\omega) + \alpha(k_0\omega)} \tag{7.81b}$$

is the Fresnel coefficient characterizing the specular reflection of p-polarized light from the average planar interface,

$$G_0(p) = \frac{i\varepsilon(\omega)}{\varepsilon(\omega)\alpha_0(p\omega) + \alpha(p\omega)} \tag{7.81c}$$

is the Green's function propagator of a surface electromagnetic wave bound to the average planar interface in the $x_1 - x_2$ plane and propagating with wavevector p parallel to the planar interface,

$$T(p \mid k) = V(p \mid k) + \int \frac{dq}{2\pi} V(p \mid q) G_0(p) T(q \mid k) \tag{7.81d}$$

is the T-matrix describing the multiple scattering of the surface wave along the rough interface,

$$V(p \mid k) \approx \frac{\varepsilon(\omega) - 1}{\varepsilon(\omega)^2}\hat{\xi}(p - k)\left[\varepsilon(\omega)pk - \alpha(p\omega)\alpha(k\omega)\right] \tag{7.81e}$$

is the scattering potential of the surface wave as it propagates along the rough interface, and

$$\hat{\xi}(p) = \int dx_1 \xi(x_1) e^{-ipx_1} \tag{7.81f}$$

is the Fourier transform of the surface profile function. Notice that $T(p \mid k)$ in Eq. (7.81d) is a T-matrix containing all the surface wave scattering arising from the random surface roughness profile function $\hat{\xi}(p - k)$. We shall presently see that the T-matrix is a renormalizing factor used to convert $G_0(p)$ at the average planar interface into a renormalized Green's function describing the propagation of surface electromagnetic waves on the disordered interface. In this regard, the scattering from the random roughness is contained in the second term in the sum on the right side of Eq. (7.81a).

The expression for $R(q \mid k_0)$ is seen from Eqs. (7.79) and (7.80) to give a formal solution for the radiation fields above the conducting medium where $k_0 = \frac{\omega}{c}\sin\theta_i$ for the incident angle θ_i and $q = \frac{\omega}{c}\sin\theta_r$ for scattered angle θ_r. Notice here that from Figure 7.4 θ_i and θ_r are measured relative to the normal above the mean scattering surface. In this regard, by computing the Poynting vectors of the incident and scattered fields above the interface, the incident and reflected field intensities I_{inc} and I_{ref} are obtained. From the ratio of the reflected and incident intensities, the differential reflection coefficient $\frac{\partial R_f}{\partial \theta_r}$ for the scattering of radiation by the surface then follows. Applying these considerations, we find that [12]

$$\frac{I_{ref}}{I_{inc}} = \int d\theta_r \frac{\partial R_f}{\partial \theta_r} \tag{7.82}$$

where θ_r is the angle of the diffusely reflected light measured from the normal to the surface in the scattering region above the interface.

Specifically, Eq. (7.82) $\frac{\partial R_f}{\partial \theta_r}$ represents the relative amount of scattered intensity of light reflected from the surface at the angle θ_r as measured at infinity. In this way, it is found for radiation incident on the surface at an angle of incidence θ_i that the differential reflection coefficient for light diffusely reflected at the reflected angle θ_r is given by [12]

$$\frac{\partial R_f}{\partial \theta_r} = \frac{\omega}{2\pi c}\frac{\cos^2\theta_r}{\cos\theta_i}\frac{\langle |R(q_0 \mid k_0)|^2 \rangle}{L} \tag{7.83}$$

where $q_0 = \frac{\omega}{c}\sin\theta_r$, $k_0 = \frac{\omega}{c}\sin\theta_i$, and $L \to \infty$ is the length of the surface along the x_1 −axis.

To develop an understanding of the physical processes going on in the diffuse scattering, it is helpful to further pursue the study of the Green's functions describing the motion of the surface waves propagating along the rough vacuum-metal interface [12]. We shall ultimately find that the scattering in Eq. (7.83) is expressed in terms of these Green's functions so that their properties control the diffuse scattering and not some integral forms based on them. To this end, from Eqs. (7.81c), (7.81d), and (7.81e) the Green's function of a surface wave as renormalized by the scattering on the disordered surface is expressed in terms of a scattering sequence described by a standard Dyson equation. This is given by the recursive equation [12]

$$\begin{aligned} G(p \mid k) &= 2\pi\delta(p-k)G_0(k) + G_0(p)\int \frac{dq}{2\pi} V(p \mid q)G(q \mid k) \\ &= 2\pi\delta(p-k)G_0(k) + G_0(p)V(p \mid k)G_0(k) \\ &\quad + G_0(p)\int \frac{dq}{2\pi} V(p \mid q)G_0(q)V(q \mid k)G_0(k) + \ldots \\ &= 2\pi\delta(p-k)G_0(k) + G_0(p)T(p \mid k)G_0(k). \end{aligned} \tag{7.84a}$$

The sequence represents a series of propagations and scatterings of the surface waves as they move along the interface.

As we have seen in our earlier discussions in Eqs. (7.63) through (7.65), the Dyson equation offers a description of the renormalization of the Green's function operators due to multiple scattering. Upon configurational averaging Eq. (7.84a) over the set of surface profiles, we then find that the average of the renormalized Green's function over profile configurations is [12]

$$\langle G(p\,|\,k)\rangle \approx 2\pi\delta(p-k)G_0(k)+G_0(p)\Sigma(p)\langle G(p\,|\,k)\rangle \tag{7.84b}$$

where

$$2\pi\delta(p-k)\Sigma(p)=\langle\int\frac{dq}{2\pi}V(p\,|\,q)G_0(q)V(q\,|\,k)\rangle \tag{7.84c}$$

represents the so-called self-energy correction $\Sigma(p)$ generated from the interaction with the surface profile function.

Similar to our earlier considerations of the electron propagation in the disordered conducting medium, the process of surface wave propagation characterized by the single-particle surface wave Green's function is represented diagrammatically in Figure 7.5a. (Compare this to the diagrams in Figure 7.2a for the renormalization of the electron Green's functions in the treatment of electrical conductivity.) As noted earlier, the renormalization of the surface wave Green's function is found to

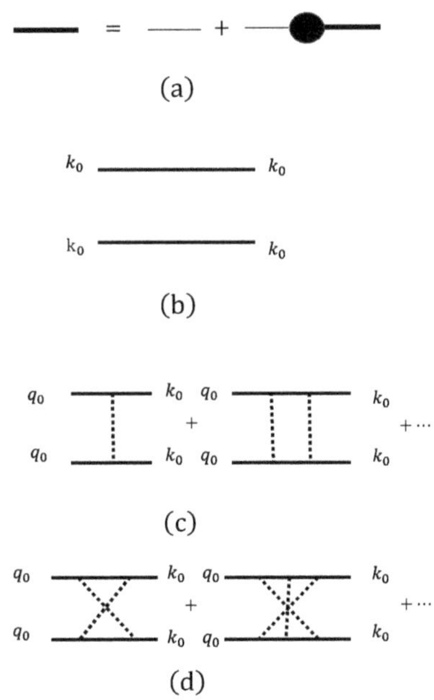

FIGURE 7.5 Diagrammatic representation of: a) The Dyson equation for the renormalized Green's function averaged over the surface roughness (heavy horizontal solid line). Here the flat surface Green's function is the thin horizontal solid line and the self-energy Σ from the scattering potentials is the solid circle. b) The first term in the ladder diagram sum, composed of two noninteracting heavy horizontal solid lines, representing the averaged renormalized Green's functions. This diagram contributes to the specular scattering. c) The remaining ladder diagrams composed of the averaged renormalized Green's functions interacting with each other by surface scattering. Here the surface scattering interactions between the Green's functions are shown by dashed vertical lines. d) The series of maximally-crossed diagrams composed of averaged renormalized Green's functions interacting with each other by surface scattering indicated by the dashed lines. Note that in accord with our notation for the two-particle Green's function $\langle|G(q_0\,|\,k_0)|^2\rangle$ in Eq. (7.85), we have labeled the q_0 and k_0 (reflected and incident) sides of the Green's function lines.

involve a sum of a series of terms each involving increasing numbers of scattering processes. These contribute to a self-energy renormalization of the Green's functions.

The differential reflection coefficient in Eq. (7.83) can ultimately be expressed in terms of the renormalized surface wave Green's function in Eq. (7.84). This is done by noting that the operator form $G_0(p)T(p\,|\,k)G_0(k)$ occurs in both Eq. (7.81a) for $R(q\,|\,k_0)$ and in Eq. (7.84) for $G(p\,|\,k)$. Applying this, we find from Eqs. (7.84) and (7.83) that with $q_0 \neq k_0$ the differential reflection coefficient for the diffuse scattering can be rewritten in terms of the Green's function form [12]

$$\frac{\partial R_f}{\partial \theta_r} = \frac{2}{\pi}\frac{\omega^3}{c^3}\cos^2\theta_r \cos\theta_i \frac{\langle |G(q_0\,|\,k_0)|^2 \rangle}{L}. \tag{7.85}$$

(Note: The specular reflection is proportional to $2\pi\delta(q_0 - k_0)$ and is not a consideration here).

The result of our discussions in Eq. (7.85) expresses the diffuse scattering from the rough surface in terms of the average over the surface disorder of the two-particle Green's function

$$\langle |G(q_0\,|\,k_0)|^2 \rangle \tag{7.86}$$

where $\langle\rangle$ denotes an average over the rough surface profiles.

As in the case of electron conductivity, the transport processes of the optical surface waves along the disordered interface are written in terms of two-particle Green's functions. This, in fact, is a general property of many transport calculations in a variety of different dynamical systems. For example, two-particle Green's function treatments in addition exist for the thermal conductivity and the scattering from magnetic, electron, phonon systems, etc. On the other hand, we also have found in an earlier chapter that the single-particle Green's functions yield the densities of states of the excitations in a variety of many-body systems, and this again is a feature found in the study of many types of single-particle excitation. Consequently, it can be said as a general rule that single-particle Green's functions relate to the excitation spectra and lifetimes of the excitations in a many-body system, while two-particle Green's functions deal with the transport properties of the system.

Next, let us consider the structure of the two-particle Green's function as it gives rise to weak localization effects in the diffuse scattering. As was the case in the considerations of electrical conductivity, the processes of interest in the study of diffuse scattering are those of the ladder diagrams and those of the maximally crossed diagrams, but now their relationship to the transport is much simpler than in the case of electrons. These are shown, respectively, in Figures 7.5b, 7.5c, and 7.5d. The ladder diagrams are composed as the sum of the diagram in Figure 7.5b of two noninteracting Green's function lines which contribute to the specular reflection from the mean surface and the sum of sequentially interacting Green's function lines shown in Figure 7.5c. (Compare Fig.7.5b with Figure 7.2b in the electron problem and Figure 7.5c with Figure 7.3a in the electron problem.) The latter contributes a diffuse scattering at general angles. The maximally crossed diagrams, on the other hand, are shown in Figure 7.5d and contain the weak localization effects. (Compare Figure 7.5d with Figure 7.3b in the electron problem.) The diffuse scattering from the surface involves contributions from the diagrams in Figures 7.5c and 7.5d, while the specular scattering from the mean surface is from the diagram in Figure 7.5b as well as a contribution from the $q_0 = k_0$ terms in Figures 7.5c and 7.5d.

In Figure 7.5 the renormalized Green's functions are shown as the heavy horizontal lines while the dashed lines connecting the two horizontal Green's functions are pair correlators of the scattering potential in Eq. (7.81e), i.e.,

$$\langle V^* V \rangle \propto \langle \xi^* \xi \rangle \tag{7.87}$$

which are found to be proportional to the statistical correlators $\langle \xi^* \xi \rangle$ of the random surface profile functions [12]. As we have seen, the process of the renormalization of the single particle surface wave Green's functions is shown in Figure 7.5a where the renormalization is similar to that followed earlier in the electron conduction problem treated in Eqs. (7.63) and (7.65). The processes essentially are based on the application of the single particle Dyson equation expressed in terms of the self-energy correction $\Sigma \propto \langle VV \rangle \propto \langle \xi\xi \rangle$. These renormalized single-particle Green's functions are important in formulating the two-particle Green's functions. In this regard, we next study the two-particle Green's function which enters through Eq. (7.85) for the determination of the differential reflection coefficient.

Now we focus our considerations on the two-particle Green's functions and their renormalization involving the ladder contributions shown in Figures 7.5b and 7.5c. As with the treatment of electron conductivity, this is determined by the Bethe-Salpeter equation, yielding a sequence of ladder scattering processes. Note, as an additional point, that in Figure 7.5b and 7.5c, based on the notation $\langle |G(q_0 \mid k_0)|^2 \rangle$ for the two-particle Green's function occurring in Eq. (7.85), we have labeled the q_0 and the k_0 momenta sides of the ladder diagrams. This is for the convenience of the reader.

The ladder contributions are then followed by the addition to them of the renormalizations from the maximally crossed diagrams. These processes are shown in Figure 7.5d, where again, based on the notation $\langle |G(q_0 \mid k_0)|^2 \rangle$ for the two-particle Green's function, we have labeled the q_0 and k_0 momenta sides of the diagrams. As per our earlier discussions of the electron system, the weak localization effects in optical scattering are found to arise due to the phase coherence of the maximally crossed diagrams. For weakly rough surfaces, they contribute a sharp backscattering peak to the differential reflection coefficients.

In the context of the two sets of diagrams, considerations of the diagrams in Figure 7.5 show that the differential reflection coefficient is a sum of contributions of the form [12]

$$\frac{\partial R_f}{\partial \theta_r} = \left(\frac{\partial R_f}{\partial \theta_r}\right)_{Lad} + \left(\frac{\partial R_f}{\partial \theta_r}\right)_{Max}. \tag{7.88}$$

Here $\left(\frac{\partial R_f}{\partial \theta_r}\right)_{Lad}$ is a contribution from the ladder diagrams (shown in Figure 7.5c) giving a contribution to the diffuse scattering at general angles above the surface. These processes are to leading order $\langle \xi^* \xi \rangle$ in the surface roughness correlator. The second term on the right $\left(\frac{\partial R_f}{\partial \theta_r}\right)_{Max}$ is the contribution of the maximally crossed diagrams (shown in Figure 7.5d), containing the weak localization contribution. These first enter as scattering processes in leading order $\langle \xi^* \xi^2 \rangle$ and give contributions of the form [12]

$$\left(\frac{\partial R_f}{\partial \theta_r}\right)_{Max} \propto \frac{\Delta_{tot}}{(q_0 + k_0)^2 + 4\Delta_{tot}^2} \tag{7.89a}$$

where $q_0 = \frac{\omega}{c}\sin\theta_r$, $k_0 = \frac{\omega}{c}\sin\theta_i$, and

$$\Delta_{tot} \approx \frac{1}{l} \tag{7.89b}$$

where l is the mean-free path of the surface electromagnetic waves on the interface. Here the interest is in long l so that Eq. (7.89a) is sharply peaked about $q_0 = -k_0$. This type of scattering is in the direction opposite to that of the incident wave, i.e., it is a backscattering into the incident beam. In Figure 7.6, an example of scattering from such a system is given.

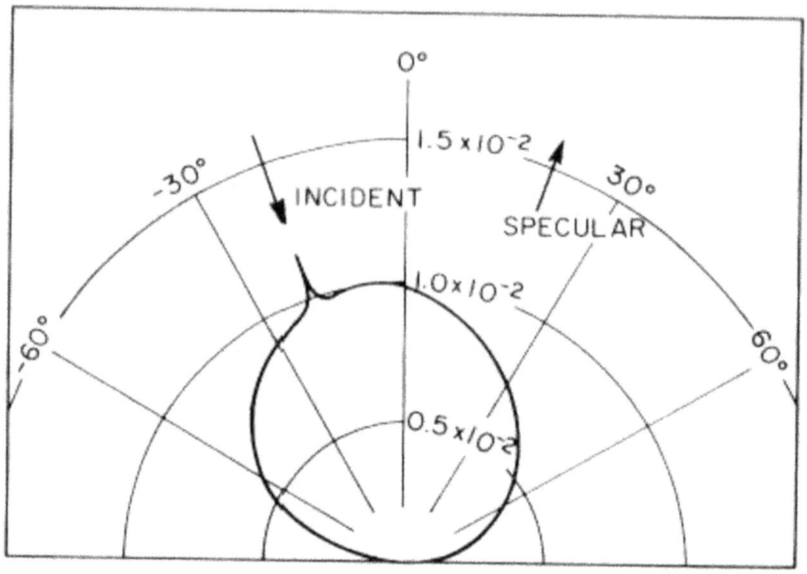

FIGURE 7.6 A polar plot of the differential reflection coefficient versus the angle of reflection for light diffusely scattered from a weakly rough vacuum-silver surface [12]. Note the enhanced backscattering peak in the direction opposite the incident plane wave. (Reprint with permission from Ref. [12], Copyright 1990, Elsevier.)

In Figure 7.6 theoretical results for the differential reflection coefficient are presented as a function of diffuse scattering angle for a 1−dimensionally rough silver surface [12]. The feature of our interest is the backscattering which is shown as a small peak in the back reflected radiation. Note that the width of the peaks in the figures is proportional to $\Delta_{tot} \approx \frac{1}{l}$. These results have since been demonstrated by computer simulation studies and by experimental results [12].

Note that in optics the enhanced backscattering process not only occurs in the surface scattering from rough surfaces supporting surface electromagnetic waves [13]. It also occurs in the volume scattering of particulates suspended in 3-dimensional media and in disordered 2−dimensional and layered media. The reader is, again, referred to the literature [7, 13] for a discussion of these. In this regard, in the presence of sufficiently random disorder [7] light can ultimately be confined to a bounded region of space within 3− dimensional media with volume disorder. This would be the region of strong localization.

We next turn to a discussion of the question of what the properties of localization are as related to the dimensionality of the system in which it is investigated. This is given in an important result involving considerations based on the ideas of scaling theory and renormalization. While the discussions will be applied to transport in electronic systems, the results can be generalized to treat transport processes in general physical systems in the presence of random disorder [7].

7.4 STRONG LOCALIZATION AND ITS DEPENDENCE ON DIMENSIONALITY

In this section, considerations of the dimensionality of electronic systems are introduced into the study of their $T = 0$ resistance and conductance in the context of localization and the metal–insulator transition [14]. These considerations ultimately are based in and lead to discussions of the nature of the electron wavefunctions in 1−, 2−, and 3-dimensional disordered media. The focus is on the study of certain scaling relationships relating the conductance of an object to its lengths and the conductivity of the material of its composition, i.e., how do the transport properties change as the size or length scale of the conducting medium changes. This will allow us to make general statements about the nature of the localization of the conduction electrons in their transporting medium and their changes at a metal–insulator transition.

In this regard, the disordered material in which the electrons propagate will be taken to appear to be homogeneous on length scales much greater than the electron mean free paths but, nevertheless, randomly disordered on much smaller scales. Specifically, if we consider the conductance of a homogeneous cube of material of side L, our focus is on determining how the conductance between opposite faces of the cube behaves as a function of $L \to \infty$. Similarly, in the $2-$dimensional considerations of a homogeneous film of material, the focus is on how the conductance between opposite edges of a square of the film behaves as a function of the edge length L in the limit as $L \to \infty$. Finally, in $1-$dimensional considerations of a homogeneous line or wire of material, discussions center on the conductance between the opposite ends of a line or wire as a function of the length L of the line as $L \to \infty$. We shall find that these considerations lead to a variety of interesting general conclusions regarding the Anderson localization properties of the conduction electrons in the materials composing such structures [14].

In the case that the resistivity of the material is a constant ρ_d (i.e., independent of L) in each of our $d-$dimensional considerations, these questions are easily answered. In a $3-$dimensional system, from elementary physics we know that the resistance of a material of length L and cross-sectional area $A = L^2$, measured between the ends of L, is given by

$$R = \rho_3 \frac{L}{A} = \rho_3 \frac{1}{L} \propto \frac{1}{L}. \tag{7.90a}$$

This indicates that the resistance scales as $\frac{1}{L}$ in the length L characterizing the size of the cube. Similarly, in a $2-$dimensional system, the resistance of a square of material of edge length L, measured between opposite edges is given by

$$R = \rho_2 \frac{L}{L} = \rho_2, \tag{7.90b}$$

and in a $1-$dimensional system, the resistance of a material of length L, measured between the ends of L is given by

$$R = \rho_1 L \propto L. \tag{7.90c}$$

Consequently, in general, we find that for a $d-$dimensional cube of side L that its resistance $R(L)$ is related to L by

$$R(L) = \rho_d L^{2-d} \propto L^{2-d} \tag{7.91}$$

provided ρ_d is a constant independent of L.

Since the conductance of a system is the reciprocal of the resistance, it then follows from Eq. (7.91) that for the conductance $G(L)$ we find the scaling relationship

$$G(L) = \sigma_d L^{d-2} \propto L^{d-2} \tag{7.92}$$

where $\sigma_d = \frac{1}{\rho_d}$ is the conductivity of the media. Note that a combination of fundamental constants which has the units of conductance is given by $\frac{e^2}{\hbar}$, consequently, this combination can be used to define from Eq. (7.92) a dimensionless conductance $g(L)$ as [14]

$$g(L) = \frac{G(L)}{\frac{e^2}{\hbar}} = \frac{\hbar G(L)}{e^2}. \tag{7.93}$$

This dimensionless function facilitates some of the discussions to follow. Similarly, the resistance can be written in the dimensionless form $r(L) = \frac{1}{g(L)}$. We next relax the condition that $\sigma_d = \frac{1}{\rho_d}$ is a constant and allow it to depend on the statistical structure of the system randomness.

By a consideration of various localization models Abrahams, et al. [14] came to an important conclusion regarding the dependence of $g(L)$ on L, i.e., how $g(L)$ scales with L in a system with a statistically characterized randomness. Specifically, when considered as a function of $L \gg l$, where l is the mean free path of the electrons, they proposed that as the size of the system is changed, the dimensionless conductance depends only on L. Continuing from these observations, they further suggested that, in the continuum limit of L, the scaling of the conductivity is characterized by the differential form [14]

$$\frac{d \ln g(L)}{d \ln L} = \beta_0(g(L)) = \beta(\ln g(L)) \tag{7.94}$$

where $\beta_0(x)$ and $\beta(\ln x)$ are monotonically increasing functions of x. The form in Eq. (7.94) expresses the scaling dependence of $g(L)$ on the size of the system L, where the functions $\beta_0(x)$ and $\beta(\ln x)$ take different forms depending on the dimension of the system. We shall next show that Eq. (7.94) is indeed valid for the system in the limits of very weak and very strong disorder of the conducting medium.

First consider the limit of weak scattering for the conductance $g(L)$ form in Eq. (7.93), i.e., this is the high conductivity limit in which $\sigma_d \to \infty$. Substituting Eq. (7.93) in this limit into Eq. (7.94), it then follows that [14]

$$\frac{d \ln g(L)}{d \ln L} = \frac{d \ln \frac{\hbar}{e^2} \sigma_d L^{d-2}}{d \ln L} = \frac{d \ln L^{d-2}}{d \ln L} = (d-2)\frac{d \ln L}{d \ln L} = (d-2) \tag{7.95a}$$

$$= \beta(\ln g(L))$$

so that $\beta(\ln g(L)) = (d-2)$ is a constant independent of $\ln g(L)$. This confirms our scaling relationship in Eq. (7.94) for the particular case of weak scattering. In this regard, note that as $\sigma_d \to \infty$ it also follows that [14]

$$\ln g(L) = \ln \frac{\hbar}{e^2} \sigma_d L^{d-2} = \ln \frac{\hbar}{e^2} \sigma_d + (d-2) \to \infty. \tag{7.95b}$$

Consequently, it is concluded that for $\sigma_d \to \infty$ the system approaches a limit given by

$$(\ln g(L), \beta(\ln g(L))) \to (\infty, d-2). \tag{7.95c}$$

In the other extreme limit of low conductivity, however, localization begins to enter the system in the strong scattering limit in which $g(L) \to 0$ or $\ln g(L) \to -\infty$. This case is considered next.

For the other extreme, notice that in the strong scattering limit localization begins to enter the system and $g(L) \to 0$ with increasing L. For this case the system begins to transition to localized electron states arising from the highly phase-coherent scattering discussed earlier in Subsection 7.2.1 on the Ioffe-Regel condition and double passage effects. We, consequently, expect in this limit a form for the conductance given for $L \gg \xi$ by [14]

$$g(L) \approx g_a e^{-\frac{L}{\xi}}, \tag{7.96}$$

i.e., there is an exponential decay to zero with increasing L. Here ξ is the so-called localization length, characterized as the radius of the region in which localized electron wavefunctions are confined. The exponential form in Eq. (7.96) is expected because the conduction in such an array of localized states would arise from tunneling processes between the localizing electron modes. The tunneling processes set in as an electron hops between wavefunction tails of the electronic states.

Notice that applying Eq. (7.96) in Eq. (7.94), it then follows that [14]

$$\frac{d\ln g(L)}{d\ln L} = \frac{d\ln g_a e^{-\frac{L}{\xi}}}{d\ln L} = -\frac{1}{\xi}\frac{dL}{d\ln L} = -\frac{1}{\xi}L = \ln\frac{g(L)}{g_a} = \beta(\ln g(L)) \tag{7.97a}$$

so that $\beta(\ln g(L)) = \ln\frac{g(L)}{g_a}$. Consequently, the scaling relation in Eq. (7.94) holds in the $g(L) \to \infty$ limit in Eq. (7.95) and also in the $g(L) \to 0$ limit in Eq. (7.97a). In addition, from Eq. (7.97a) we also see that

$$\left(\ln g(L), \beta(\ln g(L))\right) \to (-\infty, -\infty). \tag{7.97b}$$

It is then a further assumption of scaling theory that Eq. (7.94) is, indeed, valid for all $g(L)$. Note, in this regard, that in the next level of correction to Eq. (7.97) it is shown [14] that in the $g(L) \to 0$ limit

$$\beta(\ln g(L)) = \ln\frac{g(L)}{g_a}\left[1 + \alpha g(L)\right] \tag{7.98}$$

where $\alpha > 0$. Similarly, the next level of correction to Eq. (7.95) in the $g(L) \to \infty$ limit is

$$\beta(\ln g(L)) = \left(d - 2 - \frac{a}{g(L)}\right) \tag{7.99}$$

where $a > 0$. These forms are again consistent with the scaling relationship in Eq. (7.94). In addition to these limiting forms, computer simulation work on a variety of models has also confirmed the scaling form in Eq. (7.94) over the general range of $\ln g(L)$ values [3–7].

In the original paper of Abrahams, et al. [14] the result in Figure 7.7 was presented. The figure shows a plot of $\beta(\ln g(L))$ versus $\ln g(L)$ formed as a conjecture [14]. On the plot are given the limits discussed earlier along with a speculative set of lines connecting these limits. The lines represent monotonically increasing functions connecting the limiting forms, consistent with the belief that $\beta(\ln g(L))$ should continuously increase with increasing $\ln g(L)$. In this regard, it is assumed that no unusual singularities occur in the system [14]. The figure is to a large extent an idealization;

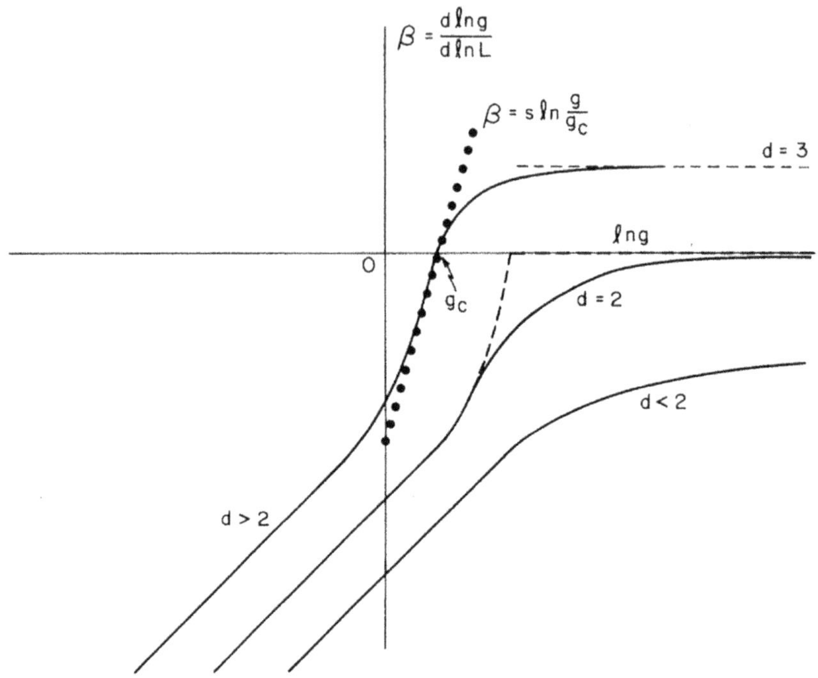

FIGURE 7.7 A plot of $\beta(\ln g(L))$ versus $\ln g(L)$ formed as a conjecture from the paper of Abrahams, et al. [14]. (Reprinted figure with permission from [14]. (Copyright 1979, American Physical Society.)

nevertheless, a number of properties of localization and its dependence on the system dimensionality can be inferred from the general features of the figure.

First consider the line of the 3−dimensional system. (This is the upper curve in the figure.) The point on the $\ln g(L)$ axis at which $\beta(\ln g(L)) = 0$ is a fixed point of the mapping, i.e., under this condition

$$\beta(\ln g(L)) = \frac{d\ln g(L)}{d\ln L} = 0 \tag{7.100a}$$

and $\ln g(L)$ is an unchanging constant. If we go above the $\ln g(L)$−axis, however,

$$\beta(\ln g(L)) = \frac{d\ln g(L)}{d\ln L} > 0 \tag{7.100b}$$

so that $g(L)$ increases with increasing L, but if we go below the $\ln g(L)$−axis

$$\beta(\ln g(L)) = \frac{d\ln g(L)}{d\ln L} < 0 \tag{7.100c}$$

and $g(L)$ decreases with increasing L. Consequently, the fixed point is unstable, and in the region above the fixed point, with an increasing L the conductance is driven to the weak scattering (high conducting) limit. In this region, the system does not experience localization. On the other hand, in the region below the fixed point we find that with increasing L the conductance is driven to the large scattering (low conducting) limit, ultimately resulting in localized modes [14].

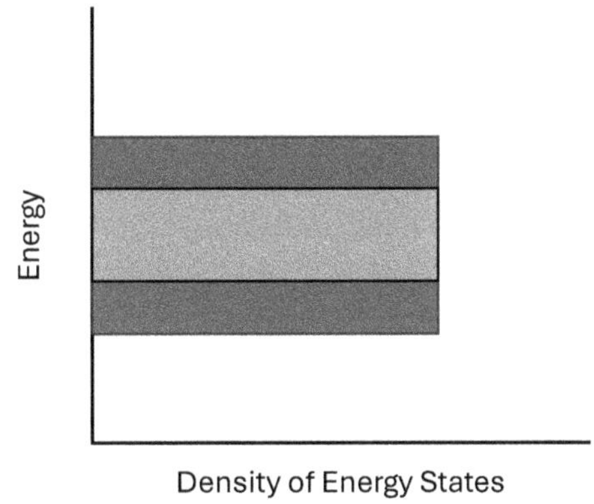

FIGURE 7.8 A schematic of a single band of electron energy states shown as the Energy versus the Density of States. The black regions are localized nonconducting states and the gray region is a band of extended conducting electron states. As the disorder increases, the size of the black region increases while the gray decreases.

On the other hand, the consideration of the conductance of the 2 – and 1 – dimensional systems is most interesting. For these systems we see that [14]

$$\beta\left(\ln g\left(L\right)\right)=\frac{d\ln g\left(L\right)}{d\ln L}<0 \tag{7.101}$$

always. Consequently, as the system becomes infinite $g(L)$ always decreases (i.e., $\ln g(L)$ becomes larger negative) with increasing L. This means in these dimensions that infinite systems must exhibit localization in the presence of arbitrarily weak disorder. Note, in this regard, that though a finite segment of 1 – dimensional copper wire is observed to conduct electrons, the nonzero conductance is only because the localization length ξ of the electron modes in the wire is much larger than the sample size characterized by L, i.e., $\xi \gg L$. We conclude that while at zero temperature 3 – dimensional systems can be composed of both localized and propagating electronic modes, in 2 – and 1 – dimensional systems at zero temperature the electron modes are always localized.

Notice in addition, that in the 3 – dimensional electron gas model the localization properties of the electrons are found to depend on the energy of the electron eigenmodes [7, 9]. With increasing disorder in the conducting medium, the electron energy states at the band edges first become localized and with increasing disorder localized electron states progressively approach the center of the energy bands [9]. So-called mobility edges are established marking the energy boundaries between localized and extended electron modes in the conduction bands [7–9]. See Figure 7.8 for a schematic illustration of this. For a further discussion of the energy band characteristics of localization the reader is referred to the literature [7–9].

As a final topic of consideration, we next look at another phenomenon based on the ideas of diffusive transport. In particular, we investigate some interesting features of systems which combine chemically reacting constituents having concentration gradients in the system and the associated diffusive transport of these constituents. Such systems can be shown under certain conditions to develop equilibrium states organized into spatial patterns.

7.5 TURING PATTERNS

In this section we look at how spatial patterns can be developed in physical, chemical, and biological systems composed of different reacting population components which are also subject to the presence of spatial diffusion [15–20]. The population components of the system tend to develop a steady state through reactions characterized by transition rates. These determine the ability of the different species in the system to transform into each other and to establish a state involving population configurations. On the other hand, diffusion in the system governs the spatial mobilities of the different populations and tends to the establishment of a dynamics in space of the distribution of the different reacting components [15–17]. An example of such a system involving chemical reactions is one in which two or more reacting chemicals can also diffuse through the system, each component with its own characteristic diffusion coefficient. Another is a set of two or more biological populations affecting each other's population numbers while each of the populations exhibits a diffusive spatial mobility. Under the proper conditions each of these models sets the interaction between the transition rates and diffusivity to generate a variety of spatial patterns involving the reacting components [15–17]. These patterns may be static or dynamic in nature. The resulting patterns are known as Turing patterns [17] and are common to a variety of important pattern generations observed in nature. As an introduction to the topic of Turing patterns, in the following we present one of the simpler models generating a $1-$dimensional Turing pattern [15].

Specifically, consider a simple model [15] composed of two different species or types (denote by u and v) which have diffusive properties in the medium in which they exist as well as exhibiting transitions into one another through some types of reactive processes, i.e., these transitions may involve physically, chemically, or biologically mediated events. As a further simplification, we consider a $1-$dimensional system so that the two species and their dynamics are described by the position and time dependent concentrations $u(x,t)$ and $v(x,t)$ where x is the position on the $x-$axis. We begin by treating the system for the case in which there is no interaction between the u and v species, and the dynamics in the resulting model involves only simple diffusion of its two components. Following this, interactive terms involving transitions between the u and v species are added to the model. We shall show that these added interactions can lead to the development of pattern formation in the concentrations of u and v as functions of position x and time t.

For the case in which $u(x,t)$ and $v(x,t)$ do not interact, the model has a simple solution for each component. In their transporting medium, the species u and v spatially diffuse only along the $x-$axis due to their $1-$dimensional concentration gradients. Consequently, the system is described by the set of two independent diffusion equations [15]

$$\frac{\partial u}{\partial t} = D_u \frac{\partial^2 u}{\partial x^2} \tag{7.102a}$$

and

$$\frac{\partial v}{\partial t} = D_v \frac{\partial^2 v}{\partial x^2}. \tag{7.102b}$$

These have solutions which propagate independently of one another in a straightforward diffusive motion.

Next let us assume the components $u(x,t)$ and $v(x,t)$ react with and can transition into one another during the course of their diffusion processes. This interaction results in a change in the diffusive equations in Eq. (7.102) as the concentrations of u and v now depend on each other so that the equations of motion become a coupled set of equations. Consequently, in the presence of a reactive interaction converting some of $u(x,t)$ into $v(x,t)$ and some of $v(x,t)$ into $u(x,t)$ the diffusion equations in Eq. (7.102) take the form [15]

$$\frac{\partial u}{\partial t} = D_u \frac{\partial^2 u}{\partial x^2} + f(u,v) \tag{7.103a}$$

and

$$\frac{\partial v}{\partial t} = D_v \frac{\partial^2 v}{\partial x^2} + g(u,v). \tag{7.103b}$$

Here the partial differentials on the right of the equations characterize the diffusion of $v(x,t)$ and $u(x,t)$ while the terms $f(u,v)$ and $g(u,v)$ characterize the processes in which $v(x,t)$ and $u(x,t)$ transition amongst each other.

In the following discussions of Eq. (7.103), the focus is on finding the fixed points of the system (i.e., the points at which $\frac{\partial u}{\partial t}=\frac{\partial v}{\partial t}=0$) for the case in which u and v are position independent. Once these are determined, the next objective is to determine the stability of these fix points when a spatial perturbation is introduced into the system. This analysis will allow us to show that under the proper conditions a modulated pattern in x can develop in the system [15].

In a first consideration, we assume that the system in Eq. (7.103) has no spatial variation, i.e., $\frac{\partial^2 u}{\partial x^2}=\frac{\partial^2 v}{\partial x^2}=0$, $\frac{\partial u}{\partial x}=0$, $\frac{\partial v}{\partial x}=0$. Consequently, the system reduces to the set [15]

$$\frac{\partial u}{\partial t}=f(u,v) \tag{7.104a}$$

and

$$\frac{\partial v}{\partial t}=g(u,v). \tag{7.104b}$$

The fixed-points of the coupled set in Eq. (7.104) are then given by the solutions of

$$f(\bar{u},\bar{v})=0 \tag{7.105a}$$

and

$$g(\bar{u},\bar{v})=0 \tag{7.105b}$$

where the fixed-point solutions are denoted as $(\bar{u},\bar{v})$. To introduce the possibility of a spatial pattern, however, we must allow for a spatial variation in the concentrations so that the diffusion operators in Eq. (7.103) must be reintroduced into the discussions.

To investigate the stability of the fixed-points of Eq. (7.105) when a spatial modulation is introduced at the fixed-points, let us look for a solution of Eq. (7.103) of the form [15]

$$u(x,t)=\bar{u}+a\exp(\lambda t+iqx) \tag{7.106a}$$

and

$$v(x,t)=\bar{v}+b\exp(\lambda t+iqx) \tag{7.106a}$$

where $a,b,\bar{u},\bar{v}$ are constants with $|a|,|b|\ll 1$,

$$f(\bar{u},\bar{v})=0, \tag{7.107a}$$

and

$$g(\bar{u},\bar{v})=0. \tag{7.107b}$$

This introduces a space and time perturbation into the fix-point solutions which will allow us to determine the stability of the fixed-points under a perturbation having the form of a Fourier spatial component. Specifically, does such a perturbation at the fixed point cause the system to move toward or away from the fixed-point?

We see from Eq. (7.106) that $u(x,t)$ and $v(x,t)$ represent small perturbations about the values $\bar{u},\bar{v}$ giving $f(\bar{u},\bar{v})=g(\bar{u},\bar{v})=0$. To study the effects of the perturbation we develop from Eq. (7.103) a linearized theory about the fixed-points of the system in Eq. (7.103). Substituting Eq. (7.106) into $f(u,v)$ and $g(u,v)$ for $|a|,|b|\ll 1$ we can then Taylor expand about $\bar{u},\bar{v}$ to find that [15]

$$f(u(x,y),v(x,y)) \approx \left(\frac{\partial f}{\partial u}\right)_{\bar{u},\bar{v}} a\exp(\lambda t+iqx)+\left(\frac{\partial f}{\partial v}\right)_{\bar{u},\bar{v}} b\exp(\lambda t+iqx) \tag{7.108a}$$

$$= f_u a\exp(\lambda t+iqx)+f_v b\exp(\lambda t+iqx)$$

and

$$g(u(x,y),v(x,y)) \approx \left(\frac{\partial g}{\partial u}\right)_{\bar{u},\bar{v}} a\exp(\lambda t+iqx)+\left(\frac{\partial g}{\partial v}\right)_{\bar{u},\bar{v}} b\exp(\lambda t+iqx) \tag{7.108b}$$

$$= g_u a\exp(\lambda t+iqx)+g_v b\exp(\lambda t+iqx)$$

where we define $f_u=\left(\frac{\partial f}{\partial u}\right)_{\bar{u},\bar{v}}$, $f_v=\left(\frac{\partial f}{\partial v}\right)_{\bar{u},\bar{v}}$, $g_u=\left(\frac{\partial g}{\partial u}\right)_{\bar{u},\bar{v}}$, and $g_v=\left(\frac{\partial g}{\partial v}\right)_{\bar{u},\bar{v}}$.

Substituting Eq. (7.106) into Eq. (7.103) and using the linearization in Eq. (7.108), it then follows that [15]

$$a\lambda=-D_u q^2 a+f_u a+f_v b \tag{7.109a}$$

and

$$b\lambda=-D_v q^2 b+g_u a+g_v b \tag{7.109b}$$

determine the motions of the perturbed system. These equations represent a set of equations for the linearized perturbational solutions about the fixed-points and are solved for λ as a function of q as well as the coefficients a and b. Notice that the diffusion components of the equation have introduced factors of $D_u q^2$ and $D_v q^2$ into the system of equations. This will allow us to study the stability properties about the fixed-point $(\bar{u},\bar{v})$ of pattern perturbations of the form in Eq. (7.106) for different q.

Written as a matrix equation, Eq. (7.109) becomes [15]

$$\begin{vmatrix}\lambda+D_u q^2-f_u & -f_v\\ -g_u & \lambda+D_v q^2-g_v\end{vmatrix}\begin{vmatrix}a\\ b\end{vmatrix}=0. \tag{7.110}$$

The condition for a solution of Eq. (7.110) is then given by the solvability condition

$$Det\begin{vmatrix}\lambda+D_u q^2-f_u & -f_v\\ -g_u & \lambda+D_v q^2-g_v\end{vmatrix}=0 \tag{7.111}$$

$$=\lambda^2+\left(D_u q^2+D_v q^2-f_u-g_v\right)\lambda+D_u D_v q^4-\left(D_u g_v+D_v f_u\right)q^2+f_u g_v-g_u f_v=0.$$

Defining

$$A(q) = D_u q^2 + D_v q^2 - f_u - g_v \tag{7.112a}$$

and

$$B(q) = D_u D_v q^4 - (D_u g_v + D_v f_u) q^2 + f_u g_v - g_u f_v \tag{7.112b}$$

the solution of Eq. (7.111) for λ in terms of q is then given by [15]

$$\lambda_\pm = \frac{-A(q) \pm \sqrt{[A(q)]^2 - 4B(q)}}{2} \tag{7.112c}$$

corresponding to each of the two solutions of Eq. (7.110).

The solutions of $\lambda_\pm(q)$ as a function of q determine the stability of the linearized forms in Eq. (7.106) given by the corresponding sets of $(a,b) = (a_+, b_+)$ or (a_-, b_-) where

$$\frac{a_\pm}{b_\pm} = \frac{f_v}{\lambda_\pm + D_u q^2 - f_u}. \tag{7.113}$$

From Eq. (7.106) it then follows that the linearized solutions are expressed as [15]

$$u_\pm(x,t) = \bar{u} + a_\pm \exp(\lambda_\pm t + iqx) \tag{7.114a}$$

and

$$v_\pm(x,t) = \bar{v} + b_\pm \exp(\lambda_\pm t + iqx). \tag{7.114a}$$

Consequently, the stability about the $(\bar{u}, \bar{v})$ fixed-point solution as a function of time is determined by the nature of the exponents $\lambda_\pm(q)$.

Specifically, for the fixed-point $(\bar{u}, \bar{v})$ to be stable, we must have $\lambda_\pm(q) < 0$ so that the perturbation decays in time to the fixed-point $(\bar{u}, \bar{v})$. Otherwise, when $\lambda_+(q) > 0$ and/or $\lambda_-(q) > 0$ the system moves away from the fixed-point and towards an instability with wavenumber q. This instability represents the onset of a pattern formed in part with this wavenumber. We next investigate some of the detailed properties of these instabilities.

First notice that in the case $q = 0$ the perturbation in Eq. (7.106) involves a uniform state. The condition for the stability of the system under a perturbation of this wavenumber is determined by the requirement on the exponents in Eqs. (7.106) and (7.114) that [15]

$$\lambda_\pm(q=0) = \frac{f_u + g_v \pm \sqrt{(f_u + g_v)^2 - 4(f_u g_v - g_u f_v)}}{2} < 0. \tag{7.115a}$$

From this we see that it is required that

$$f_u + g_v < 0 \text{ and } f_u g_v - g_u f_v > 0, \tag{7.115b}$$

so that a stable uniform system at a minimum must satisfy the conditions in Eq. (7.115b).

Next consider the case of general q. The stability of the uniform, position independent state in this case requires the condition

$$\lambda_{\pm} = \frac{-A(q) \pm \sqrt{\left[A(q)\right]^2 - 4B(q)}}{2} < 0 . \tag{7.116}$$

In this regard, we note from Eq. (7.112a) and Eq. (7.115b) that due to $D_u, D_u > 0$

$$A(q) = D_u q^2 + D_v q^2 - f_u - g_v > 0. \tag{7.117}$$

Consequently, it follows that if $B(q) < 0$ the λ_+ root in Eq. (7.116) is positive, indicating an instability with a divergence from the uniform state $(\bar{u}, \bar{v})$. On the other hand, when $B(q) > 0$ the real part of the λ_+ root in Eq. (7.116) is negative, indicating the stability of the uniform state $(\bar{u}, \bar{v})$. In this regard, the determining factor in our considerations of Eq. (7.116) is the sign of $B(q)$ which must be positive for $(\bar{u}, \bar{v})$ to be stable.

From Eq. (7.112b) we then note that $B(q) = D_u D_v q^4 - (D_u g_v + D_v f_u) q^2 + f_u g_v - g_u f_v$ is an upward opening parabola in q^2 with a minimum determined by the condition [15]

$$\frac{dB(q)}{dq^2} = 2D_u D_v q^2 - (D_u g_v + D_v f_u) = 0. \tag{7.118}$$

From this it follows that the minimum of the parabola is located at

$$q_{min}^2 = \frac{D_v f_u + D_u g_v}{2D_u D_v} \tag{7.119}$$

where the value of the parabola at the minimum is [15]

$$B(q_{min}) = -\frac{(D_u g_v + D_v f_u)^2}{4D_u D_v} + f_u g_v - f_v g_u . \tag{7.120}$$

Consequently, to have a negative minimum value of $B(q_{min})$ so that the $(\bar{u}, \bar{v})$ point of the system is unstable, we need

$$\frac{(D_u g_v + D_v f_u)^2}{4D_u D_v} > f_u g_v - f_v g_u > 0 \tag{7.121}$$

for the system to be unstable under a q_{min} perturbation. (Note here that the condition $f_u g_v - g_u f_v > 0$ comes from our previously determined conditions in Eq. (7.115b) regarding the stability of $(\bar{u}, \bar{v})$ in the case of $q = 0$). The condition in Eq. (7.121) would make the entire q perturbed system unstable and lead to a pattern formation under the conditions [15]

$$f_u + g_v < 0 \text{ and } f_u g_v - g_u f_v > 0. \tag{7.122}$$

To summarize, for the system to exhibit an instability leading to a pattern formation under a q perturbation to the system, we require from Eq. (7.115b) that

$$f_u + g_v < 0, \tag{7.123a}$$

$$f_u g_v - g_u f_v > 0, \tag{7.123b}$$

and from Eq. (7.121) that

$$D_v f_u + D_u g_v > 2\sqrt{D_u D_v \left(f_u g_v - f_v g_u\right)} > 0 \tag{7.123c}$$

or alternatively written as [15]

$$\frac{D_v}{D_u} f_u + g_v > 2\sqrt{\frac{D_v}{D_u}\left(f_u g_v - f_v g_u\right)} > 0. \tag{7.123d}$$

Under these conditions, the system will develop a patterning. (Notice in this regard that if $f_u > 0$, it is required from Eq. (7.123a) that $g_v < -f_u < 0$ and then by choosing $\frac{D_v}{D_u}$ sufficiently large and positive the inequalities in Eq. (7.123d) can be met. Similarly, if $g_v > 0$, it is required that $f_u < -g_v < 0$ and then by choosing $\frac{D_v}{D_u}$ sufficiently small and positive the inequalities in Eq. (7.123d) can be met.)

The existence of Turing patterns was first proposed in 1952 [17]. Since then, a variety of demonstrations of Turing patterns have been given in physical, chemical, and biological systems. For a discussion of these and more, the reader is referred to the literature [15–20].

REFERENCES

1. Vasilyev, A. 1983. *An Introduction to Statistical Physics* (Mir Publishers, Moscow).
2. Reed-Hill, R. E. 1973. *Physical Metallurgy Principles*, 2nd Edition (D. Van Nostrand Company, New York).
3. Marder, M. P. 2000. *Condensed Matter Physics* (John Wiley & Sons, Inc., New York).
4. Ashcroft, N. W. and Mermin, N. D. 1976. *Solid State Physics* (Harcourt Brace College Publishers, Fort Worth).
5. Animalu, A. O. E. 1977. *Intermediate Quantum Theory of Crystaline Solids* (Prentice-Hall, New Jersey).
6. Economou, E. N. 2006. *Green's Functions in Quantum Physics,* 3rd Edition (Springer-Verlag, Berlin).
7. Akkermans, E. and Montambaux, G. 2007. *Mesoscopic Physics of Electrons and Phonons* (Cambridge University Press, Cambridge).
8. Rammer, J. 2007. *Quantum Field Theory of Non-equilibrium States* (Cambridge University Press, Cambridge).
9. Mott, N. F. and Davis, E. A. 2012. *Electronic Processes in Non-Crystaline Materials* (Oxford University Press, Oxford).
10. Vollhardt, D. and Wolfle, P. 1980. Diagrammatic, Self-Consistent Treatment of the Anderson Localization Problem in $d \leq 2$ Dimensions. *Physical Review B* 22, 4666–4676.
11. Wolfle, P. and Vollhardt, P. 2010. *50 Years of Anderson Localization* (World Scientific, Singapore), 43–71.
12. McGurn, A. R. 1990. Enhances Retroreflection Effects in the Reflection of Light from Randomly Rough Surfaces. *Surface Science Reports* 10(8), 357–410.
13. Sheng, P. 1990. *Scattering and Localization of Classical Waves in Random Media* (World Scientific, Singapore).
14. Abrahams, E., Anderson, P. W., Licciardello, D. C., and Ramakrishnan, T. V. 1979. Scaling Theory of Localization: Absence of Quantum Diffusion in Two Dimensions. *Physical Review Letters* 42, 673–676.
15. Kinashita, S. 2013. *Pattern Formation and Oscillatory Phenomena* (Elsevier Inc., Amsterdam).
16. Cross, M. C. and Hohenberg, P. C. 1993. Pattern Formation Outside of Equilibrium. *Reviews of Modern Physics* 65(3), 854–1086.
17. Turing, A. M. 1952. The Chemical Basis of Morphgenesis. *Philosophical Transactions of the Royal Society* 237, 37–72.
18. Castets, V., Dulos, E., Boissonade, J., and De Kepper, P. 1990. Experimental Evidence of a Sustained Turing-Type Nonequilibrium Chemical Pattern. *Physical Review Letters* 64, 2953–2956.

19. Koch, A. J. and Meinhardt, H. 1994. Biological Patter Formation: From Basic Mechanisms to Complex Structures. *Reviews of Modern Physics* 66, 1481–1507.
20. Kondo, S. and Asai, R. 1995. A Reaction-Diffusion Wave on the Skin of the Marine Angelfish Pomacanthus. *Nature* 376, 765–768.

8 Chaos

An Introduction to Nonlinear Dynamics

In this chapter we shall study the dynamics of some very general types of physical systems that exhibit time dependences described by interesting new forms of difference equations and differential equations [1–10]. These include, for example, population studies in biological systems, studies of the concentrations of chemical species in chemical reactions, or treatments of the nonlinear motion of particles along a one-dimensional axis [1–10]. A common feature of these systems is that they are described by nonlinear equations exhibiting a new important type of dynamics known as chaos. The new chaotic dynamics only exist in nonlinear systems and are absent in linear dynamic models.

In a nonlinear world, things are often more complex than they at first seem [1–8]. The illustration of this is found in the study of chaos, which we now consider [1–5]. Here we shall introduce the topics of the logistic mapping [1–8], simple so-called unimodal mappings related to it [1–4], and some elementary dynamical systems of nonlinear differential equations [1–7]. These are all very basic and simply posed mathematical problems in nonlinear dynamics of general biological, chemical, and mechanical systems. Nevertheless, the simplicity of their nonlinear forms belies the complex motions which they often generate [1–8]. They represent some of the fundamental studies upon which the theory of chaotic systems has been founded.

Specifically, all of these systems are found in some limits to exhibit a complex form of dynamics known as chaotic dynamics or deterministic chaos. This is an aperiodic motion which is highly sensitive to the initial conditions of the system [1–6]. Slightly different initial states of the system, consequently, lead to highly divergent types of dynamical behaviors. These dynamical processes are difficult to quantify and lead to a certain degree of loss of predictability in the motions of the system.

The logistic mapping (See Eq. (8.1) in our following discussions.) is a simple difference-equation mapping [1–6] which was initially developed to simulate population as a function of time. In this regard, it studies the population dynamics in a series of equal time steps in which the population at one time step is determined from that at its previous time step. It is also a nonlinear equation as the population interacts with itself or some outside limiting conditions which cause its dynamics to depend on the second power of the population size [1–4].

Two types of motion are found to dominate the dynamics of the logistic mapping, and the region in which these motions occur is set by the parameters of the mapping and not by changes in the basic form of the mapping. These are classes of periodic, cyclic population dynamics [1–8] or of aperiodic, chaotic dynamics in the population [1–8]. Periodic cycles in a population density are often found in predator-prey studies or in some types of chemical reactions. Here the rates of change in populations or chemical species are mutually dependent on one another so that as one population becomes large, the rate of increase of the other population increases. Chaos, on the other hand, occurs when the nonlinear interactions are sufficiently strong that the population dynamics lose their sense of predictability and the dynamics appear erratic and aperiodic [1–6].

In our studies we shall find that the dynamics of the logistic mapping are determined to a great degree by its fixed-points [1–6] and also on its continuous regions of attraction to chaotic motion [1–6]. These are often referred to as attractors [1–4] because the system, as it evolves in time, often settles down to periodic motions between the fixed-points [1–4] or into a continuum region

DOI: 10.1201/9781032674919-8

of attraction of the system into chaotic trajectories [1–4]. We shall begin with discussions of the periodic fixed-points and see how these develop in a parameter space of the logistic mapping into regions characterized by periodic motions. A discussion is given of the so-called period-doubling approach to the aperiodic chaotic state [1–4] of the system motion as the parameters of the logistic mapping are changed. Chaotic motion is found to occur in a particular region of the parameter space of the logistic mapping in which there is an absence of periodic motions but in which the system wanders through a bounded continuous region of population space [1–6]. For the region of chaos, a study is made of the measures developed in order to characterize the aperiodic, chaotic, motion of the system [1–4].

A treatment involving Lyapunov exponents [1–6] as a means for developing a quantification of the measure of chaos in the systems is introduced. It is developed to measure the separation as a function of time of trajectories having slightly different boundary conditions. Specifically, it will be shown that in periodic, non-chaotic states of the dynamics the Lyapunov exponent is negative [1–6] so that initially close trajectories remain closely separated in time. On the other hand, in chaotic states the Lyapunov exponent of the dynamics is positive [1–6] so that initially close trajectories spatially diverge in time. In this way, the sign of the Lyapunov exponent becomes an indicator of the presence of chaos in a system.

Studying the parameter space of the logistic mapping we shall find it breaks up into regions of periodic motion and regions of chaotic behavior. In this regard, for the treatment of chaotic motions it will be seen that, in parameter regions of mostly chaotic motion, the system can quite unexpectedly break into small windows of parameter space exhibiting periodic dynamics [1–6].

The behaviors found in the dynamics of the logistic mapping will be shown to extend to other difference-equation mappings in a general class of so-called unimodal mappings [1]. Examples of these are the sine mapping and the so-called tent mapping [1]. Consequently, the general properties of the logistic dynamics are found in treatments of other types of systems characterized as population studies. These shared properties may occur both in a qualitative and a quantitative sense and give rise to the ideas of universality [1–10]. The general characteristics are also observed in a variety of experiments made on biological populations and on chemical reactions [1].

Finally, some discussion is given of the chaotic dynamics described by simple systems of non-linear partial differential equations [1–7]. The chaotic properties developed in difference equation and in differential equation systems are contrasted and explained. For example, the definitions and conditions for the existence of chaos and the consequent differences in its occurrence in both these systems are investigated [1, 8].

8.1 THE LOGISTIC MAPPING

The one-dimensional mapping known as the logistic mapping illustrates many of the important principles of chaos in a dynamical system [1–7]. The mapping is a recurrence relation which generates the motion of a system as a series of time steps, determining the state of the system at a particular time step in terms of its state in the previous time step. It does this by mapping the region of the x−axis defined by $0 \le x \le 1$ back into itself so that the motion of the system only involves the $0 \le x \le 1$ interval. (Notice here that depending on the system being considered x may be a variable representing biological population density, chemical concentration, position of a particle in one dimension, etc.) In this regard, the system evolves as a series of equal time steps labeled by consecutive integers $n = 0,1,2,\ldots\ldots$ where x_n is the position of the system at the nth step of the dynamical sequence and x_0 is its initial position. The resulting dynamics of this process are typical of the sequential evolution commonly generated in computer simulation studies of dynamical systems.

Under the logistic mapping a point x in the interval $0 \le x \le 1$ exhibits one of three different behaviors [1–4]. Specifically, x can be mapped back onto itself, x can be mapped into a point $x^+ > x$ on its right, or x can be mapped into a point $x^- < x$ on its left. When x maps onto itself it is known as a fixed-point of the mapping as its value does not change under the mapping. Otherwise, the value is either increased or decreased from its initial value. In this way, the result from an iteration of the mapping exhibits a possible number of isolated fixed-points along the x−axis, with the remaining generated points either moving towards or away from these fixed-points. Consequently, an understanding of the dynamics generated under the mapping is greatly facilitated by the determination of its fixed-points [1–7].

The specific form of the logistic mapping is given as a one-dimensional nonlinear recurrence relationship generating the $(n+1)^{th}$ time step of a system from that of its nth time step. It is given by the form [1–7]

$$x_{n+1} = \mu x_n \left(1 - x_n\right) \text{ for } n = 0,1,2,3,\ldots\ldots. \tag{8.1}$$

where $x_n, x_{n+1} \in \left[0,1\right]$ are, respectively, the values of the system variables at the $n^{th}$and $(n+1)^{th}$ time steps, and $0 < \mu < 4$ is a constant in time setting the nature of the dynamical process. In this realization, the values x_n, x_{n+1} can represent population densities in biological studies, chemical species in studies of reactions, spatial variables in the one-dimensional dynamics of a particle, etc.

The mapping in Eq. (8.1) sends $x_n \in \left[0,1\right]$ into a new $x_{n+1} \in \left[0,1\right]$ in the interval of a step as either a fixed-point or a new point which is to the right or left of its original x_n position on the x−axis [1 –4]. In this sense, the fixed-points of the mapping are very important in representing the types of motions found in the system, and we shall see that the dynamics of the mapping either centers on motions confined to the fixed-points of the mapping or is confined to bounded regions of the x−axis known as a region of attraction [1–7]. Let us begin the study of Eq. (8.1) then by looking at some of the fixed-points of the mapping.

8.1.1 Fixed-Points for $0 < \mu < 1$

To begin [1], consider the mapping defined by Eq. (8.1) for the dynamical parameter μ in the region $0 < \mu < 1$. In this region as x_n ranges over the interval $0 \le x_n \le 1$ it is seen that both $0 < \mu < 1$ and $0 \le \left(1 - x_n\right) \le 1$. From Eq. (8.1) a consequence of this is that

$$x_{n+1} = \mu x_n \left(1 - x_n\right) = \mu \left(1 - x_n\right) x_n < x_n \tag{8.2}$$

so that the value of the $(n+1)^{th}$ step x_{n+1} is always decreased from that of the nth step, x_n. As one increases the number of iterations, it then follows that [1]

$$\lim_{n \to \infty} x_{n+1} \to x^* = 0 \tag{8.3}$$

and under the stated conditions the mapping must have a fixed-point x^* at $x^* = 0$. This means that no matter what initial value $x_0 \in \left[0,1\right]$ is chosen for our sequence of steps, after an infinite number of iterations we ultimately arrive at the fixed-point $x^* = 0$.

In summary, we then see that in the region $0 < \mu < 1$ all initial points x_0 are eventually forced by the mapping to the $x^* = 0$ fixed-point of the mapping [1].

8.1.2 Fixed-Points for $1 < \mu < 3$

The next set of fixed-points of interest [1] come from considering Eq. (8.1) for a single iteration in which μ is restricted to the region $1 < \mu < 3$. Specifically, we search for the fixed-points x^* of the mapping in Eq. (8.1) given by

$$x_{n+1} = x^* = \mu x_n \left(1 - x_n\right) = \mu x^* \left(1 - x^*\right) \tag{8.4a}$$

or

$$x^* = \mu x^*\left(1 - x^*\right). \tag{8.4b}$$

Again, it is found that $x^* = 0$ satisfies the equation, but this fixed-point now leads to an uninteresting result that a zero value of x_n always leads to a zero value of x_{n+1}. (Note that, of course, $x_0 = 1$ also maps to $x_1 = 0$. This dynamics is similarly of little interest to us.)

The other fixed-point solution [1] of Eq. (8.4b) is nonzero and given by the form

$$x^* = 1 - \frac{1}{\mu}. \tag{8.5}$$

Since it is required that $x^* \in [0,1]$ it follows that Eq. (8.5) has no valid nonzero fixed-points in the previously treated interval $0 < \mu < 1$, but for $\mu > 1$ a valid nonzero fixed-point of the mapping exists. Now we focus on the nonzero fixed-point for $\mu > 1$ and consider the properties of the fixed-point. Specifically, a focus is on the stability of the fixed-point solution, i.e., as the mapping is iterated do points x_n located in the neighborhood centered about the nonzero fixed-points x^* move towards or away from the fixed-point.

In this regard, consider the mapping in Eq. (8.1) rewritten into the form

$$x_{n+1} = \mu x_n\left(1 - x_n\right) = f_\mu\left(x_n\right). \tag{8.6}$$

Here we define the nonlinear function [1]

$$f_\mu(x) = \mu x(1 - x). \tag{8.7}$$

We now use Eq. (8.6) to focus on the behavior of points located in the vicinity of the fixed-point x^* defined by

$$x^* = f_\mu\left(x^*\right). \tag{8.8}$$

To do this, let us rewrite Eq. (8.6) into the form

$$x_{n+1} = f_\mu\left(x_n\right) = f_\mu\left(x^* + \left(x_n - x^*\right)\right). \tag{8.9}$$

Next, assume that we are close to the fixed-point so that $x_n - x^*$ is small and a Taylor expansion in $x_n - x^*$ can be made. Expanding to first order in $x_n - x^*$ then gives

$$x_{n+1} = f_\mu\left(x^* + \left(x_n - x^*\right)\right) = f_\mu\left(x^*\right) + \left(x_n - x^*\right)\left.\frac{df_\mu(x)}{dx}\right|_{x=x^*} \tag{8.10}$$

$$= x^* + \left(x_n - x^*\right)\left.\frac{df_\mu(x)}{dx}\right|_{x=x^*}.$$

It, consequently, follows from Eq. (8.10) that

$$\frac{x_{n+1} - x^*}{x_n - x^*} = \left.\frac{df_\mu(x)}{dx}\right|_{x=x^*}. \tag{8.11}$$

Notice in Eq. (8.11) that for

$$\left|\frac{df_\mu(x)}{dx}\right|_{x=x^*} < 1 \tag{8.12}$$

we find

$$\left|x_{n+1} - x^*\right| < \left|x_n - x^*\right|, \tag{8.13}$$

and the sequence of points x_n is seen to approach the fixed-point x^* with increasing n. In this case the fixed-point is said to be an attractor as the fixed-point is approached by the sequence [1].

On the other hand, for

$$\left|\frac{df_\mu(x)}{dx}\right|_{x=x^*} > 1, \tag{8.14}$$

the sequence of points x_n moves away from the fixed-point x^* with increasing n. In this case the fixed-point is said to be a repeller as the sequence moves away from the fixed-point. (See Figure 8.1 for a schematic representing the properties of an attractor and a repeller.)

Consider Eqs. (8.6) through (8.8) for the fixed-point of the logistic mapping in Eq. (8.5) evaluated within the region $1 < \mu < 3$. Evaluating $f_\mu'(x^*) = \left.\frac{df_\mu(x)}{dx}\right|_{x=x^*}$ at the fixed-point $x^* = 1 - \frac{1}{\mu}$ gives [1]

$$\left|f_\mu'(x^*)\right| = |2 - \mu|. \tag{8.15}$$

For the fixed point to be an attractor, we must have $\left|f_\mu'(x^*)\right| = |2-\mu| < 1$ and this occurs only in the region $1 < \mu < 3$. Outside this interval of μ the fixed points in Eq. (8.5) are unstable repellers of the sequence of mapping generated points in their neighborhood.

In general, it is found in the region $0 < \mu < 3$ that all the nonzero initial points x_0 are eventually forced by the mapping to its $x^* = 1 - \frac{1}{\mu}$ fixed-point. At $\mu = 3$, however, a transition is made to a new type of behavior which is next considered.

8.1.3 Fixed-Points for $3 < \mu < 1 + \sqrt{6} \approx 3.44949$

With increasing μ new sets of fixed-points are encountered [1–3]. These fixed-points arise from multiple iterations of the logistic mapping, and, as we shall see, they are associated with increasingly complex periodic behaviors of the nonlinear system. For example, we next consider the fixed-points associated with a double iteration of the logistic mapping. This composition of two successive mappings generates stable fixed-points of the two-iteration mapping [1–3] for parameters μ within the region $3 < \mu < 1 + \sqrt{6} \approx 3.44949$. Viewed in terms of the single-step or iteration mappings, the

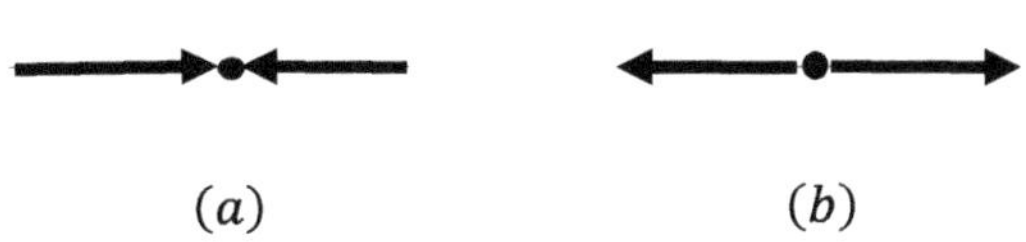

FIGURE 8.1 Schematic of: (a) a stable fixed-point and (b) an unstable fixed-point. The arrows indicate the direction of motion caused by the logistic mapping, and the dots are the fixed-points of the mapping.

fixed-points of two-iteration mapping represent a periodic motion involving the application of two single-step mappings.

The double mapping is composed as an application of Eq. (8.6) twice and is given by [1–3]

$$x_{n+2} = f_\mu\left(f_\mu\left(x_n\right)\right) = f_\mu\left(\mu x_n\left(1-x_n\right)\right) = \mu\mu x_n\left(1-x_n\right)\left[1-\mu x_n\left(1-x_n\right)\right], \quad (8.16a)$$

expressing x_{n+2} in terms of x_n. The fixed-points of this mapping involve $x_{n+2}, x_n \to x_2^*$ where x_2^* is the fixed-point approached by both x_{n+2} and x_n as $n \to \infty$. Similarly, the associated two-iteration mapping given by the composition

$$x_{n+3} = f_\mu\left(f_\mu\left(x_{n+1}\right)\right) = f_\mu\left(\mu x_{n+1}\left(1-x_{n+1}\right)\right) = \quad (8.16b)$$

$$= \mu\mu x_{n+1}\left(1-x_{n+1}\right)\left[1-\mu x_{n+1}\left(1-x_{n+1}\right)\right],$$

expresses x_{n+3} in terms of x_{n+1}. The fixed-points of this mapping involve $x_{n+3}, x_{n+1} \to x_2^{*'}$ where $x_2^{*'}$ is a second fixed-point of the system which is approached by both x_{n+3} and x_{n+1} as $n \to \infty$. The two fixed points x_2^* and $x_2^{*'}$where $x_2^* \neq x_2^{*'}$ then characterize the periodic motion involving two successive single steps of the system of sequential points generated by the single-step mapping [1–3]

$$x_{n+1} = \mu x_n\left(1-x_n\right) = f_\mu\left(x_n\right). \quad (8.17)$$

These are applied successively in the generations of single steps in the system so that between the fixed-points are the two single steps given by $x_2^{*'} = f_\mu\left(x_2^*\right)$ followed by $x_2^* = f_\mu\left(x_2^{*'}\right)$.

We begin by finding the fixed-points of the mappings in Eqs. (8.16a) and (8.16b). These are obtained as the solutions of the fixed-point equation [1–4]

$$x_2^* = \mu^2 x_2^*\left(1-x_2^*\right)\left[1-\mu x_2^*\left(1-x_2^*\right)\right]. \quad (8.18)$$

From Eq. (8.18) two different fixed-point solution of x_2^* are found, given by the form [1–3]

$$x_{2\pm}^* = \frac{1}{2\mu}\left[\mu+1\pm\sqrt{(\mu+1)(\mu-3)}\right]. \quad (8.19)$$

The two fixed-points $x_2^* \neq x_2^{*'}$ in Eq. (8.16) are seen to each take one of the forms $x_{2\pm}^*$ dependent on whether they involve the sequence x_{n+2} and x_n for x_2^* or the sequence x_{n+3} and x_{n+1} for $x_2^{*'}$. We now show that stable period-2 solutions from Eq. (8.16) exist only for μ within the range $3 < \mu < 1+\sqrt{6}$.

In Eq. (8.19) it is seen from the square root term that the two roots $x_{2\pm}^*$ are real only when $\mu \geq 3$. This, consequently, fixes the lower limit of μ in order that Eq. (8.19) represents valid solutions of the period-2 fixed-points. The upper limit of the interval is next determined by the stability condition of the fixed points $x_{2\pm}^*$. Specifically, to determine the stability of the fixed-points of the doubly composed mapping in Eq. (8.16) we need [1]

$$\left|\frac{d}{dx} f_\mu\left(f_\mu\left(x\right)\right)\Big|_{x=x_{2\pm}^*}\right| < 1. \quad (8.20)$$

The condition in Eq. (8.19) follows from a discussion of stability similar to that given in Eq. (8.9) through (8.14) for the mapping $x_{n+1} = f_\mu\left(x_n\right)$ but now the treatment is extended to the doubly composed mapping $x_{n+2} = f_\mu\left(f_\mu\left(x_n\right)\right)$ or $x_{n+3} = f_\mu\left(f_\mu\left(x_{n+1}\right)\right)$ of Eq. (8.16).

From Eqs. (8.16) and (8.20) it then follows that

$$\frac{d}{dx}f_\mu(f_\mu(x)) = \left[\frac{d}{df_\mu(x)}f_\mu(f_\mu(x))\right]\frac{d}{dx}f_\mu(x). \tag{8.21}$$

Evaluating this at either of the two fixed-points $x^*_{2\pm}$ it follows that

$$\frac{d}{dx}f_\mu\left(f_\mu\left(x^*_{2\pm}\right)\right) = \left[\frac{d}{df_\mu\left(x^*_{2\pm}\right)}f_\mu\left(f_\mu\left(x^*_{2\pm}\right)\right)\right]\frac{d}{dx}f_\mu\left(x^*_{2\pm}\right) \tag{8.22}$$

$$= \left[\frac{d}{dx}f_\mu\left(x^*_{2\mp}\right)\right]\frac{d}{dx}f_\mu\left(x^*_{2\pm}\right)$$

$$= \mu\left(1-2x^*_{2\mp}\right)\mu\left(1-2x^*_{2\pm}\right)$$

$$= 4 + 2\mu - \mu^2.$$

where here we have used the relationship that $f_\mu\left(f_\mu\left(x^*_{2\pm}\right)\right) = f_\mu\left(x^*_{2\mp}\right)$ Consequently, for a stable solution at the fixed-point it is required that [1–3]

$$\left|4 + 2\mu - \mu^2\right| < 1 \tag{8.23}$$

or written in another way

$$3 < \mu < 1 + \sqrt{6} \approx 3.44949. \tag{8.24}$$

We see that the two period-2 fixed-points in Eq. (8.19) are only stable fixed-points within the region given in Eq. (8.24).

As an interesting point, it should be noted that $x^*_2 = 0$ is also a fixed-point of Eq. (8.18), and this again arises as the mapping always maps zero onto zero. This is not an interesting result of the mapping, nor is $x^*_2 = 1$ which again only maps to zero.

In general, it is found in the region $3 < \mu < 1 + \sqrt{6} \approx 3.44949$ that all the nonzero initial points x_0 are eventually forces by the composite mappings in Eq. (8.16) to the $x^*_{2\pm}$ fixed-point.

8.1.4 Fixed-Points for $1 + \sqrt{6} \approx 3.44949 < \mu < 3.569946$

In this section, we look at the development of increasingly higher periodic motions [1–3] generated in the system with increasing values of μ. These higher periodic motions continue to arise in the dynamics until ultimately the system enters a state of aperiodic chaotic motion at $\mu = 3.569946$ [1]. This scenario is understood by studying increasingly higher compositions of the single-step mapping treated in Eqs. (8.6) through (8.8). In this regard, remember that a composition of two single-step mappings led in Eq. (8.16) to a mapping involving a period-2 cyclic behavior between two fixed-points of the composite mapping. In the treatment of that case, it was found that single-step motions traveled between the fixed-point of the composite two-step mapping, resulting in period-2 cycles in the system dynamics. We shall see a similar process giving rise in the system to higher order even period cycles and finally to an aperiodic state [1–3]. The fixed-points developed during the course of these considerations are known as the attractors of the mapping as for a given μ all initial points $x_0 \neq 0,1$ eventually move to the fixed-points of the system.

We begin the study with a composition of four single-step mappings to generate mappings involving a set of period-4 fixed-points. Note that we shall come back later to study the composition of three single-step mappings which for the moment are removed from our considerations. For now, the focus is only on compositions of even numbers of single-step mappings.

Consider the composition of four single-step mappings of the form given in Eqs. (8.6) through (8.8). Specifically, the composite recurrence relation is formed as [1]

$$x_{n+4} = f_\mu\left(f_\mu\left(f_\mu\left(f_\mu\left(x_n\right)\right)\right)\right) \tag{8.25}$$

where $f_\mu\left(x\right) = \mu x\left(1-x\right)$. This involves taking four single steps at once with the logistic mapping in Eq. (8.6). Disregarding the uninteresting fixed-point a $x = 0$, there are four nonzero fixed-points obtained from this composite mapping which exist and are stable in the interval $1+\sqrt{6} \approx 3.44949 < \mu < 3.54409$. The four stable fixed-points generated by the mapping are found to evolve continuously from the two stable fixed-points in Eq. (8.19) which are stable over the region $3 < \mu < 1+\sqrt{6} \approx 3.44949$. In this regard, at $\mu = 1+\sqrt{6}$ there is a continuous transition [1–3] from period-2 to period-4 behaviors, and at this transition $x_{2-}^* = 0.43996$ and $x_{2+}^* = 0.84994$.

To illustrate the period doubling process in Figure 8.2a a plot of the fixed-points or attractors of the mapping (denoted x) is given versus μ for the region $0 < \mu < 3.5$. In this plot, there are no chaotic trajectories, and the system always settles to a constant or periodic behavior. For the region $0 < \mu < 1$ all initial points in the system are forced to zero, while periodic behavior first enters the system in the period-2 region, i.e., $3 < \mu < 1+\sqrt{6} \approx 3.44949$. After these regions, there is a period-4 region in the interval $1+\sqrt{6} \approx 3.44949 < \mu < 3.54409$ followed by a period-8 set of fixed-points [1–5].

As in the case of the period-2 mapping in Eq. (8.16), the region of stability of the period-4 mapping is set by the condition that at all the fixed-points of the mapping $\left\{x_4^*\right\}$ in Eq. (8.25)

$$\left|\frac{d}{dx} f_\mu\left(f_\mu\left(f_\mu\left(f_\mu\left(x\right)\right)\right)\right)\Big|_{x=x_4^*}\right| < 1. \tag{8.26}$$

Denoting the four nonzero fixed-points as $\left\{x_{4,1}^*, x_{4,2}^*, x_{4,3}^*, x_{4,4}^*\right\}$, we find as in Eq. (8.21) that

$$\frac{d}{dx} f_\mu\left(f_\mu\left(f_\mu\left(f_\mu\left(x\right)\right)\right)\right)\Big|_{x=x_4^*} = \prod_{\imath=1}^{4} \frac{d}{dx} f_\mu\left(x_{4,i}^*\right). \tag{8.27}$$

This condition ultimately sets the upper limit of the interval of stability, $1+\sqrt{6} \approx 3.44949 < \mu < 3.54409$, of the period-4 mapping. At the upper limit of this interval, a transition is made to a region of stability for a period-8 mapping. In the progression to the chaotic state, it is found that with increasing μ the fixed points of the system progressively double in period at each transition until the period becomes infinite. This is often referred to as the period doubling route to chaos, where chaos ultimately is an aperiodic state.

With increasing μ the period-8 mapping is found to be stable over the interval, $3.54409 < \mu < 3.56441$ and at the upper edge of this region a continuous transformation is made to a period-16 mapping [1 –3]. The period-16 mapping is stable over the region $3.56441 < \mu < 3.56876$ after which it makes a continuous transition to a periodic-32 mapping at $\mu = 3.56876$. In this process, the periods of successive cycles are doubled, and the interval in μ of their stability is successively shortened until an infinite periodic cycle is generated at $\mu = 3.569946$. As we shall see later, the generation of the infinite cycle signals the onset of chaos in the system.

In general, it is found in all these regions of periodic fixed-points that all the nonzero initial points x_0 in the region are eventually forces by the composite mappings to a fixed-point of the region. For

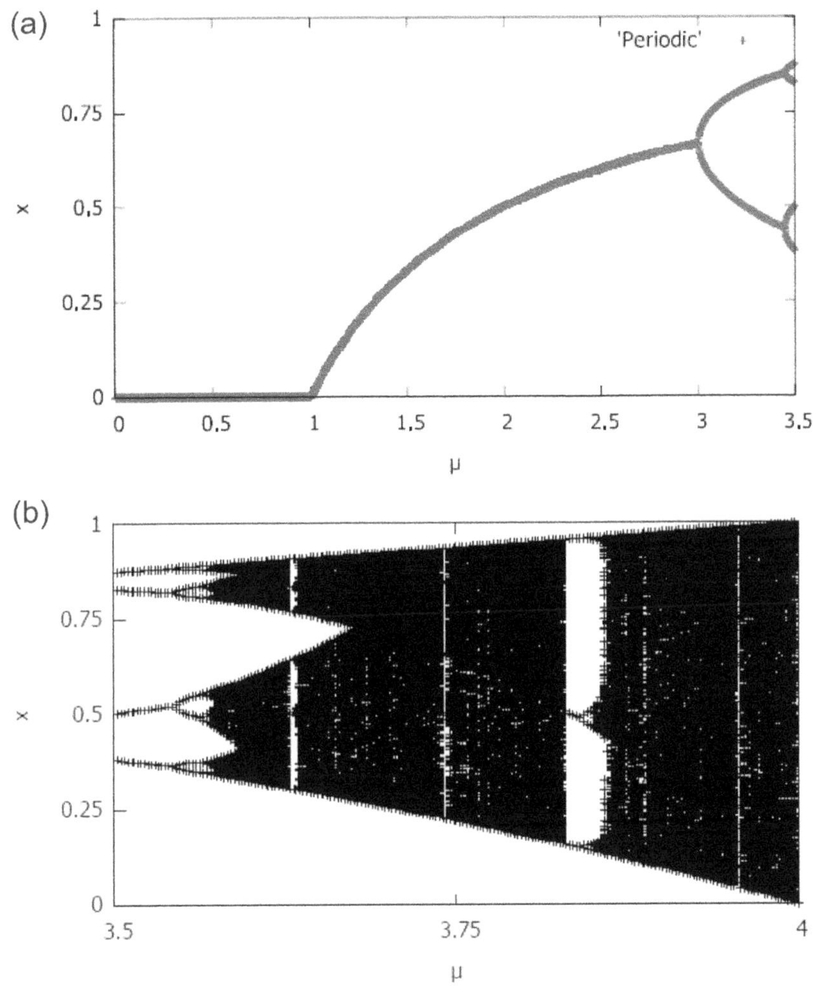

FIGURE 8.2 Plot of the fixed-points of the logistic mapping indicating their x positions versus μ. In a) are given some periodic fixed-points in the region $0 < \mu < 3.5$, and in b) are the fixed-points in the region $3.5 < \mu < 4.0$. Notice that the chaotic behavior of the mapping is contained within the blackened regions of b) and mixed in these are so-called windows of periodic behavior indicated by the white windows mixed in the region of black. Notice that in both a) and b) the logistic mapping eventually sends an initial point $0 < x_0 < 1$ to either the fixed-points or a chaotic or strange attractor depending on the value of μ of the mapping.

example, if x_0 is a general, nonzero point in the $3.56441 < \mu < 3.56876$ region of period-16 fixed-points, the composite mapping generating these fixed points will send x_0 to one of its 16 fixed-points.

8.1.5 Fixed-Points for $1+\sqrt{6} \approx 3.44949 < \mu < 3.5699456$

It is interesting to note that, during the period doubling process, a number of useful relationships can be developed for characterizing the approach to the aperiodic state. In the discussion of this, we begin by introducing some useful simplifying notations for the μ associated with the various states of periodic dynamics [1–3]. Specifically, let μ_p, where p in a positive integer or zero, denote the smallest value of μ at which stable fixed-points of period-2^p first occur. In this notation Table 8.1 lists μ_p for some of the period doubling processes [1–4]. It is found that the period successively doubles with increasing μ and p, and the convergence is rapidly made to the aperiodic limit $\mu_\infty = 3.5699456$.

TABLE 8.1
Values μ_p of μ at Which Period-n Cycles (Here $n = 2^p$) First Appear with Increasing μ during the Period-Doubling Process to μ_∞

p	μ_p
1	3
2	3.4494
3	3.54409
4	3.56441
5	3.56876
∞	3.569946

From Table 8.1 it is seen that the separations $\mu_{p+1} - \mu_p$ in the numerically generated data also decrease rapidly with increasing p. This rapid decrease is quantitatively described from the numerical data by fitting the data with the geometric analytic form

$$\mu_\infty - \mu_p = \frac{c}{\delta^p} \tag{8.28a}$$

where $\delta = 4.6692$ and $c = 2.637$. Following some algebra, it then is found from Eq. (8.28a) that [1]

$$\delta = \lim_{p\to\infty} \frac{\mu_p - \mu_{p-1}}{\mu_{p+1} - \mu_p} = 4.6692 \tag{8.28b}$$

where δ is known as the Feigenbaum constant.

The Feigenbaum constant gives an idea of the lengths of the successive regions in μ of cyclic stability. In particular, at the beginning of the sequence of cycles, it is found that the initial value of this ratio is [1]

$$\frac{\mu_2 - \mu_1}{\mu_3 - \mu_2} = 4.75. \tag{8.28c}$$

Notice that this already is not far from the limiting form in Eq. (8.28b).

The sequence of ratios in Eq. (8.28b) is important as they characterize the rapid development of the system to its limiting state of chaos. Later we shall discuss the additional origins of the Feigenbaum constant in more detail and also introduce a number of other interesting relationships between the fixed points in the approach to chaos. These arise from a consideration of a scaling theory treatment of the system which is developed in studying the properties characterizing the approach to the chaotic state. For now, we turn to a discussion of the quantitative characterization of the basic nature of the chaotic state.

To illustrate the complexity of the behavior in the chaotic state, in Figure 8.2b we plot the x of the fixed points and regions of aperiodic motion into which the trajectories settle versus μ in the region $3.5 < \mu < 4.0$. This is the region of the logistic mapping containing the chaotic motion which, specifically, enters the system at $\mu = 3.569946$ and exists in the system throughout most of the region $3.569946 < \mu \le 4.0$. In this region for the most part, there is aperiodic behavior in which, following some initial transient motion, the system eventually settles down to a well-defined x-region of aperiodic motions [1–3]. The bounded region of these aperiodic trajectories is often referred to as a strange or chaotic attractor. Aperiodicity, however, is not the most important aspect of the motion in chaotic states. Rather, it is the unpredictability of the system dynamics in terms of the initial conditions of the motion. This is next addressed.

In the following, we first develop a definition of what is meant by a chaotic state of the system dynamics. This requires introducing the concept of Lyapunov exponents and an understanding of their applications in treating the divergent properties of chaotic motions. Following this, we return to the study of periodic motions of the system as it approaches chaos and to the development of certain scaling relations that are found to govern the fixed-points of the system in its period doubling to the chaotic state. Finally, we consider periodic motions that have odd periods. Where do these occur and how do they enter the scheme of things?

8.1.6 Chaos and the Lyapunov Exponents

The property which most characterizes a chaotic state is its sensitivity to initial conditions. Specifically, two trajectories differing slightly in initial conditions are found in chaotic systems to exhibit dynamical evolution involving rapidly divergent spatial separations as a function of time. This is not the case in a typical non-chaotic dynamic which usually exhibits a high degree of continuity in its dynamics with changing initial conditions. The focus in the study of chaotic dynamics is then on the characterization of the rate of spatial divergence of two trajectories having infinitesimally different initial conditions. Typically, chaotic trajectories diverge exponentially from one another, and the rate of this divergence is set by a Lyapunov exponent [1–3]. We begin by defining the Lyapunov exponent as a characterization of both chaotic and non-chaotic motions.

As we shall see, the nature of the divergence in chaotic systems is characterized by the Lyapunov exponent. It does this by quantifying the spatial separation of the trajectories in terms of an exponential form involving a time argument, and by considering only trajectories with infinitesimal spatial differences in their initial conditions. In this sense, the Lyapunov exponent measures the change of the argument of the exponent with increasing time and, as a result, the divergence in the system. Consequently, the Lyapunov exponent is large and positive for rapidly divergent trajectories in chaotic systems, whereas in the absence of chaos (i.e., in systems without divergent trajectories), it is zero or less than zero. This provides a straightforward characterization of dynamic systems and a determination of the presence of chaos.

Specifically, if we are given a mapping of the form

$$x_{n+1} = f_\mu\left(x_n\right), \tag{8.29a}$$

then the sequence of time steps $\left\{x_0, x_1, \ldots\ldots, x_n\right\}$ is generated from n iterations or time steps of Eq. (8.29a). In this way, the x_n step is obtained from the initial position x_0 by the composition

$$x_n = f_\mu\left(f_\mu\left(\ldots f_\mu\left(x_0\right)\right)\right) = f_\mu^{(n)}\left(x_0\right) \tag{8.29b}$$

where $f_\mu^{(n)}(\;)$ indicates the composition or successive application of $f_\mu(\;)$ a number of times n. Considering two trajectories of the system, one with initial condition x_0 and the other with initial condition $x_0 + \epsilon$ where ϵ is an infinitesimal, then we define the spatial separation [1–3]

$$L_n = \left|f_\mu^{(n)}\left(x_0 + \epsilon\right) - f_\mu^{(n)}\left(x_0\right)\right| \approx \left|\frac{d}{dx} f_\mu^{(n)}\left(x_0\right)\epsilon\right| \tag{8.30}$$

as the difference between the two trajectories after n iterations. Notice that here we have expanded $f_\mu^{(n)}\left(x_0 + \epsilon\right)$ to first order in ϵ.

In general, it is found after a large number of iterations n (but still a number in which the separation of the trajectories is relatively small in the system) that [1–3]

$$L_n = \epsilon\, e^{\lambda n}. \tag{8.31}$$

This is the exponential dependence (mentioned earlier) on the time variable n for separations which are small compared to the total extent of the system. It is an assumed form which is later verified in numerical and scaling theory studies [1–5] on dynamical systems such as those mentioned in the context of Eq. (8.28). (These verifications are also topics of later discussions of scaling theory presented in this chapter.)

The exponential form in Eq. (8.31) ultimately provides a definition of the Lyapunov exponent. To formulate a reliable definition quantifying the rate of separation of the two trajectories as they evolve, n we focus on the exponent λ. From Eqs. (8.30) and (8.31) it follows that for $\epsilon \to 0$ and $n \gg 1$

$$\lambda = \frac{1}{n}\ln\left(\frac{\left|f_{\mu}^{(n)}\left(x_0+\epsilon\right)-f_{\mu}^{(n)}\left(x_0\right)\right|}{\epsilon}\right) \tag{8.32}$$

$$= \frac{1}{n}\ln\left(\left|\frac{d}{dx}f_{\mu}^{(n)}\left(x_0\right)\right|\right)$$

so that the exponent λ is given by the derivative of the composite mapping [1–5]. Consequently from Eq. (8.31), we see that if $\lambda > 0$ the trajectories in Eq. (8.31) are divergent from one another, this indicates a chaotic state. On the other hand, if $\lambda < 0$ the trajectories are convergent, the system motion is non-chaotic. In the case that $\lambda = 0$ the trajectories neither converge nor diverge, they are again non-chaotic. A better understanding of the form in Eq. (8.32) comes from evaluating the derivative.

Considering Eq. (8.32) in greater detail and proceeding as in Eqs. (8.21) and (8.27) we have

$$\frac{d}{dx}f_{\mu}^{(n)}\left(x_0\right) = \prod_{i=0}^{n-1} f_{\mu}^{'}\left(x_i\right), \tag{8.33}$$

where the product runs over the set of steps $\{x_0, x_1, \ldots\ldots, x_n\}$ of the mapping in Eq. (8.29a) which is composed to generate the composite mapping in Eq. (8.29b). Applying the result in Eq. (8.33) in Eq. (8.32) gives [1]

$$\lambda = \frac{1}{n}\ln\left|\prod_{i=0}^{n-1} f_{\mu}^{'}\left(x_i\right)\right| = \frac{1}{n}\sum_{i=0}^{n-1}\ln\left|f_{\mu}^{'}\left(x_i\right)\right|. \tag{8.34}$$

The value of λ generated in Eq. (8.34) is seen to depend on the set $\{x_0, x_1, \ldots\ldots, x_n\}$ and this appears to be a problem if the system has not settled into its periodic or chaotic limits of motion. However, there is a quick solution to this.

A way of getting around the uncertainty in the set $\{x_0, x_1, \ldots\ldots, x_n\}$ is to consider the limit of Eq. (8.34) as $n \to \infty$. In a periodic system, the infinite application of the mapping in Eq. (8.29a) will eventually settle down to a generation of an infinite sequence of repetitions of the finite number of periodic fixed points of the mapping in Eq. (8.29a). As a result, Eq. (8.34) will depend only on the fixed points of the periodic mapping. On the other hand, in a chaotic system, the aperiodic nature of the dynamics requires an infinite $n \to \infty$ set of $\{x_0, x_1, \ldots\ldots, x_n\}$ to evaluate the exponent. Consequently, in order to properly define the Lyapunov exponent independent of x_0 we take as the formal definition of the Lyapunov exponent, which we denote as λ to be of the form [1]

$$\lambda = \lim_{n\to\infty}\frac{1}{n}\sum_{i=0}^{n-1}\ln\left|f_{\mu}^{'}\left(x_i\right)\right|. \tag{8.35}$$

The definition in Eq. (8.35) is found to be appropriate for the Lyapunov exponents of one-dimensional mappings.

An important point in reference to these considerations is that for periodic fixed-points $\{x_i\}$ of the mapping, $|f_\mu'(x_i)| < 1$ or equivalently $\ln|f_\mu'(x_i)| < 0$ are the conditions which set the stability of the periodic fixed-points. This was discussed earlier in the context of Eq. (8.12). On the other hand, in the case of $|f_\mu'(x_i)| > 1$ or equivalently $\ln|f_\mu'(x_i)| \geq 0$ we have seen in Eq. (8.14) that the mapping is unstable. This is an indication that the system is aperiodic and chaotic.

As an example, consider the period-2 mapping in Eq. (8.16). It has the two fixed-points [1]

$$x_{2\pm}^* = \frac{1}{2\mu}\left[\mu + 1 \pm \sqrt{(\mu+1)(\mu-3)}\right]. \tag{8.36}$$

From Eq. (8.35) the Lyapunov exponent of the system is then given by

$$\lambda_2 = \frac{1}{2}\sum_{i=+}^{-} \ln|f_\mu'(x_{2i})| \tag{8.37}$$

$$= \frac{1}{2}\left\{\ln\mu|1-2x_{2+}^*| + \ln\mu|1-2x_{2-}^*|\right\} = \frac{1}{2}\left\{\ln\mu^2|(1-2x_{2+}^*)(1-2x_{2-}^*)|\right\}$$

where we sum over both of the fixed-point roots, and we find that $\lambda_2 < 0$ in the region $3 < \mu < 1+\sqrt{6}$. On the other hand, for the mapping in Eq. (8.8) involving a single-step yielding a single fixed-point we have [1–3]

$$\lambda_1 = \ln|f_\mu'(x^*)| = \ln|2-\mu|. \tag{8.38}$$

Again in the region $1 < \mu < 3$ it follows that $\lambda_1 < 0$.

8.1.7 Scaling Theory: The Region of Period Doubling to Chaos

In the following, we treat some interesting geometric relationships that exist in the Figure 8.2 plot of the fixed-points of the logistic mapping and its composite mappings in the region of period doubling to chaos. The logistic mapping for a single step or iteration has the general form

$$x_{n+1} = f_\mu(x_n) = f_\mu^{(1)}(x_n) \tag{8.39a}$$

where $f_\mu(x) = \mu x(1-x) = f_\mu^{(1)}(x)$, and for a composition of n single steps [1–3]

$$x_n = f_\mu^{(n)}(x_0) = f_\mu\left(f_\mu\left(f_\mu\left(\ldots f_\mu(x_0)\right)\right)\right) \tag{8.39b}$$

is a composite function generating the nth time step in the system from an initial value x_0. In this subsection we shall see that the geometry of the fixed-points obtained from Eq. (8.39), as plotted in Figure 8.2a, leads to a set of scaling relations involving the fixed-point coordinates in $(x,\mu)-$space. These relationships allow us to characterize the approach of the system to a chaotic state as μ increases from zero to the chaotic transition at $\mu = \mu_\infty = 3.569946$. Notice, during this limiting process, the n of the composite function $f_\mu^{(n)}(\;)$ in Eq. (8.39b) takes the successive period doubling values $n \in \{1,2,4,8,16,32,\ldots,\infty\}$, or written in a simpler format $n = 2^p$ for $p = 0,1,2,\ldots$. This represents the period doubling behavior of the system as its fixed-points approach an aperiodic state of the system in the limit that $n = 2^p \to \infty$.

To begin, consider the fixed-points of the general $n = 2^p$ – step mapping in Eq. (8.39b) for our limiting process. The derivative of the mapping evaluated at one of its stable fixed-points x_0 then satisfies the stability condition discussed earlier in Eq. (8.20), (8.26), and in Eqs. (8.35) through (8.38). Specifically, for the stability of the fixed-point x_0 of $f_\mu^{(n)}(x_0)$ it is required that

$$\left|\frac{d}{dx} f_\mu^{(n)}(x_0)\right| = \left|\prod_{i=0}^{n-1} f_\mu^{'}(x_i)\right| < 1. \tag{8.40a}$$

This is a very general condition for stability which all stable fixed-points must satisfy.

In the following discussions, we start by focusing on a particular subset of the stable fixed-points $\{x_i\}$ from Eq. (8.40a). These are determined by a more restrictive condition than that in Eq. (8.40a). Specifically, we consider fixed-points that satisfy the condition [1–3]

$$\left|\prod_{i=0}^{n-1} f_\mu^{'}(x_i)\right| = 0. \tag{8.40b}$$

The points in this subset are known as super-stable fixed-points of the system, and they will be a focus of a number of important scaling considerations.

Notice that the condition in Eq. (8.40b) can be satisfied by the fixed-point x_0 if the derivative of the logistic mapping satisfies

$$f_\mu^{'}(x_0) = 0. \tag{8.40c}$$

In this regard, it is important to realize that Eq. (8.40c) is also the condition determining the maximum of the function $f_\mu(x) = \mu x(1-x)$ of the single-step transformation in Eq. (8.39a). Consequently, the super-stable fixed-point x_0 of the mapping in Eq. (8.40a) satisfying Eq. (8.40c) also generates the maximum value of $f_\mu(x) = \mu x(1-x)$. Under these conditions let us denote this particular super-stable fixed-point as $x_0 = x_{max}^*$ where the subscript *max* reminds us that x_{max} also generates the maximum of the function $f_\mu(x) = \mu x(1-x)$.

The requirement from Eq. (8.40c) that x_{max}^* determines the maximum of $f_\mu(x)$ implies that

$$f_\mu^{'}(x_{max}^*) = \mu(1 - 2x_{max}^*) = 0, \tag{8.40d}$$

so that in fact $x_{max}^* = \frac{1}{2}$ for the logistic mapping. In addition, the requirement that it is also a super-stable fixed-point of the mapping $f_\mu^{(n)}(x_0)$ means that it must satisfy the fixed-point relation

$$x_{max}^* = \frac{1}{2} = f_\mu^{(n)}\left(x_0 = x_{max}^* = \frac{1}{2}\right) = f_\mu^{(n)}\left(\frac{1}{2}\right). \tag{8.41a}$$

Notice that written more succinctly Eq. (8.41a) is just [1]

$$\frac{1}{2} = f_\mu^{(n)}\left(\frac{1}{2}\right), \tag{8.41b}$$

and is seen to relate μ to n. This is used to determine the value Ω_n of μ at which the nth composition of the logistic mapping $f_{\mu=\Omega_n}^{(n)}(x_{max}^*)$ exhibits the super-stable fixed-point $x_{max}^* = \frac{1}{2}$.

The relationship in Eq. (8.41b) expresses the values of Ω_n supporting periodic super-stable fixed-point solutions at $x = x_{max}^* = \frac{1}{2}$ for each of the n composited mappings used in generating the fix-point plot in Figure 8.2a. Geometrically these values of Ω_n are determined graphically by the intersection of the line $x = \frac{1}{2}$ with the plot of stable periodic fixed-points in Figure 8.2a. The resulting set of $\{\Omega_n\}$ for the $x^*_{max} = \frac{1}{2}$ super-stable fixed-points are next shown to enter into the development of a number of important scaling relations characterizing the geometry of the pattern of fixed-points of the logistic mapping.

We next focus on making measurements in our system relative to the super-stable fixed-points $(\mu, x) = \left(\Omega_n, \frac{1}{2}\right)$ for the consideration of the period-doubling process as the aperiodic limit is approached. This involves making a change of some notation and ultimately a change of variables in our general mappings in Eq. (8.39). In this regard, for these considerations the integers n of interest in the period-doubling process are now written in the form $n = 2^p$ where $p = 0,1,\ldots$. In addition, we further define the constants $R_p = \Omega_n = \Omega_{2^p}$ denoting the values of μ at which the super-stable fixed-point processes for $n = 2^p$ are found. With this notation, we return to a consideration of the original composite mapping in Eq. (8.39b).

In these new considerations, the variable x_0 is now not a super-stable fixed-point but a general point in the near neighborhood of the super-stable fixed-point. Specifically, $x_0 \approx x^*_{max} = \frac{1}{2}$ and the fixed-point in the x-coordinate is at $x^*_{max} = \frac{1}{2}$. In this new notation, then Eq. (8.39b) for $\mu = R_p$ is rewritten in the form [1–3]

$$x_{2^p} = f_{R_p}^{(2^p)}(x_0). \tag{8.42}$$

This represents the generation of the x_{2^p} step from x_0 for $n = 2^p$ and $\mu = R_p$.

Next, let us make a change of variables in Eq. (8.42) using the transformation $y_j = x_j - \frac{1}{2}$ where $j = 0,1,\ldots,2^p$. As a result of the transformation, it follows from Eq. (8.42) that [1–3]

$$y_{2^p} = f_{R_p}^{(2^p)}\left(y_0 + \frac{1}{2}\right) - \frac{1}{2} \equiv \tilde{f}_{R_p}^{(2^p)}(y_0) \tag{8.43}$$

where the new form $\tilde{f}_{R_p}^{(2^p)}(y_0)$ absorbs a factor of $-\frac{1}{2}$. We see that before the transformation in Eq. (8.43) the fixed-point of the mapping in Eq (8.42) was at $(\mu, x) = \left(\Omega_n, \frac{1}{2}\right) = \left(R_p, \frac{1}{2}\right)$ for $n = 2^p$. Following the transformation, the same fixed-point of the resulting mapping in Eq. (8.43) is at $(\mu, y) = (\Omega_n, 0) = (R_p, 0)$. In this regard, the transformation has only shifted the origin of coordinates.

Based on numerical studies of the logistic mapping [1–5], it can be shown that the mappings in Eq. (8.43) for different values of p are related to one another in the near neighborhoods of their super-stable fixed-points. Specifically, it is found in the neighborhood of their super-stable fixed-points that the $p = 0$ mapping is related to the mapping for general p by the relationship [1]

$$\tilde{f}_{R_0}^{(1)}(y_0) = \alpha^p \tilde{f}_{R_p}^{(2^p)}\left(\frac{y_0}{\alpha^p}\right) \tag{8.44}$$

where $p = 0,1,2,\ldots.$ and $\alpha = -2.5029$.

The form in Eq. (8.44) is a scaling relationship which indicates how the geometry of the mapping in the vicinity of the super-stable fixed-points at $(R_0, 0)$ and $(R_p, 0)$ map onto each other. It is a relationship involving homogeneous functions of a type which occur commonly in describing the critical properties of thermodynamic functions in the neighborhood of their critical points at second order phase transitions [1–5].

In addition, numerically it can also be shown that in the limit as the aperiodic state is approached [1–3, 6–12]

$$\lim_{p\to\infty} \alpha^p \tilde{f}_{R_p}^{(2^p)}\left(\frac{y_0}{\alpha^p}\right) = g_0(y_0). \tag{8.45}$$

Here $g_0(y)$ in a finite nonzero so-called universal function characterizing the limiting behavior of the system. Notice, also, that the subscript zero on $g_0(y)$ reminds us of the R_0 on the left side of Eq. (8.44) from which Eq. (8.45) is obtained.

Additional universal functions which prove valuable in the following discussions of scaling relations can also be numerically generated and will now be discussed. These arise from a further consideration of scaling relations similar to that in Eq. (8.44).

Specifically, the scaling relation in Eq. (8.44) can be shown to generalize to the form [1–3, 6–12]

$$\tilde{f}_{R_q}^{(1)}(y_0) = \alpha^p \tilde{f}_{R_{p+q}}^{(2^p)}\left(\frac{y_0}{\alpha^p}\right) \tag{8.46}$$

for q a positive integer where $p = 0,1,2,\ldots..$ and $\alpha = -2.5029$. Notice here that R_0 on the left side of Eq. (8.44) is replaced by R_q and R_p on its right side is replaced by R_{p+q}. It then follows from the relationship in Eq. (8.46) that a universal function is shown to exist. This is defined by the limit [6–12]

$$g_q(y_o) = \lim_{p\to\infty} \alpha^p \tilde{f}_{R_{p+q}}^{(2^p)}\left(\frac{y_0}{\alpha^p}\right) \tag{8.47}$$

which is finite and nonzero. The subscript q on $g_q(y_o)$ references the R_q on the left side of Eq. (8.46).

An addition relationship of the type in Eq. (8.46) involves R_q in the limit that $q \to \infty$ In particular, it can be shown that between the mappings $\tilde{f}_{R_\infty}^{(1)}(y_0)$ and $\tilde{f}_{R_\infty}^{(2)}(y_0)$ there is the scaling relationship of the form [1–3, 6–12]

$$\tilde{f}_{R_\infty}^{(1)}(y_0) = \alpha \tilde{f}_{R_\infty}^{(2)}\left(\frac{y_0}{\alpha}\right). \tag{8.48}$$

This leads to a general universal function $g(y_o)$ given by the functional equation

$$g(y_o) = \alpha g^2\left(\frac{y_0}{\alpha}\right). \tag{8.49}$$

The universal function $g(y_o)$, as we shall see, is of particular interest in developing the properties of the logistic mapping. Specifically, we now obtain a solution for $g(y_o)$.

The functional equation in Eq. (8.49) has a useful symmetry relationship in that, if $g(y)$ is a solution of Eq. (8.49), then so is $\tilde{g}(\xi)$ defined by [1–3, 6–12]

$$\tilde{g}(\xi) = cg\left(\frac{\xi}{c}\right). \tag{8.50}$$

In this regard, substituting Eq. (8.50) into the form $\tilde{g}(\xi) = \alpha\tilde{g}^2\left(\frac{\xi}{\alpha}\right)$ of Eq. (8.49), it then follows that

$$cg\left(\frac{\xi}{c}\right) = \alpha cg\left(\frac{1}{c}cg\left(\frac{1}{c}\frac{\xi}{\alpha}\right)\right) = \alpha cg\left(g\left(\frac{1}{c}\frac{\xi}{\alpha}\right)\right) \tag{8.51a}$$

or

$$g\left(\frac{\xi}{c}\right)=\alpha g\left(g\left(\frac{1}{c}\frac{\xi}{\alpha}\right)\right). \tag{8.51b}$$

A consequence of this is that c in Eq. (8.50) can always be chosen so that at $\xi=0$ the function $\tilde{g}(\xi=0)=1$. Since $\tilde{g}(\xi)$ is a maximum at $\xi=0$ it also follows that $\tilde{g}'(\xi=0)=0$. These two conditions then serve as boundary conditions for the solution of Eq. (8.49), i.e., of $\tilde{g}(\xi)=\alpha\tilde{g}^2\left(\frac{\xi}{\alpha}\right)$.

Applying the results in Eqs. (8.50) through (8.51), an approximate solution of Eq. (8.47) for the form of $g(y_0)$ is obtained. Specifically, for $\alpha=-2.5029$ it is found that [1–15]

$$g(y_0)=1-1.5276y_0^2+0.1048y_0^4+.. \tag{8.52}$$

The function in Eq. (8.52) then approximates the particular scaling form in Eqs. (8.48) and (8.49).

8.1.8 Limiting Forms Involving μ_p and R_p

Earlier in Eq. (8.28a) the Feigenbaum constant δ was defined in terms of the values of μ at which the periodic dynamics changes during the period-doubling process to an aperiodic state. Specifically, in Eq. (8.28) it is found that the Feigenbaum constant characterizes the limiting form of the of ratios [1–12]

$$\delta=\lim_{p\to\infty}\frac{\mu_p-\mu_{p-1}}{\mu_{p+1}-\mu_p}=4.6692. \tag{8.53}$$

Now, in the context of our discussion of scaling theory results, it is interesting to note that a similar relationship holds between the μ values of R_p (Note: we remind the reader that in our system $\mu_p<R_p<\mu_{p+1}$.) such that [1–12]

$$\delta=\lim_{p\to\infty}\frac{R_p-R_{p-1}}{R_{p+1}-R_p}=4.6692. \tag{8.54}$$

The sameness of these ratios can be associated with certain properties of self-similarity found in the geometry of the logistic mapping of fixed-points for its region of period-doubling behavior [1–3]. This has to do with features involving the fractal nature of the mapping geometry in Figure 8.2. The topic of fractal geometry will be covered in the next chapter where more will be said about this aspect of the logistic mapping [1–12].

8.1.9 Scaling Theory in d_p

As an additional point, we note another relationship, similar to those in Eqs. (8.53) and (8.54), exists in the system. This occurs between the x-space variables at different values of R_p. In this regard, consider the mapping in Eq. (8.42) of the form

$$x_{2^p}=f_{R_p}^{(2^p)}(x_0). \tag{8.55}$$

If at $\mu=R_p$ we define the spatial separation along the x-axis given by [1–4, 6–12]

$$d_p=f_{R_p}^{(2^{p-1})}\left(x_0=\frac{1}{2}\right)-\frac{1}{2} \tag{8.56}$$

where d_p is the separation of the nearest neighbor of the fixed point $x_m = \frac{1}{2}$ from $x_m = \frac{1}{2}$ itself, then it is found that

$$\lim_{p\to\infty} \frac{d_p}{d_{p+1}} = \alpha = -2.50280. \tag{8.57}$$

The relationship in Eq. (8.57) is seen to provide yet another limiting form characterizing the dynamics of the logistic mapping.

Now we will look at the features characterizing the chaotic dynamics of the system. These are of a much more complex and less predictable nature than those in the periodic state.

8.1.10 Chaos

We next change the focus from considerations of the periodic dynamics of the logistic mapping to considerations of its chaotic dynamics [1–4]. The chaotic motion of the system is found in Figure 8.2b within the region $\mu > \mu_\infty = 3.569946$. Here, with the exception of a few windows of μ displaying periodic motion, the system exhibits aperiodic motion with Lyapunov exponents (see Eq. (8.35)) greater than zero. In these regions, therefore, small changes in the initial conditions in the system greatly affect the motion [1–4]. Consequently, the chaotic states are indicated in the figure by blackened regions, arising due to the aperiodic dynamics of the logistic mapping [1–10]. These blackened regions contain aperiodic trajectories which, except after possible initial transient behaviors outside the blackened area, settle down to move only over the blackened region. In this regard, the region is known as a strange or chaotic attractor of the motion.

In these dynamics, the mapping generates (for a fixed μ) a series of never-repeating steps tending to fill all of a bounded continuous region of the x−variables (the strange attractor) in which the system moves [1–4]. Infrequently in the blackened regions, windows of μ with periodic motion are also found. These occur as small, whitened regions (little segments of μ) in which the presence of a few dark lines indicates that the motion of the system is periodic. In these windows, the Lyapunov exponents are negative, indicating the absence of chaos and the presence of periodic motion.

8.1.11 Windows of Periodic Motion in the Chaos

In the region, $\mu < \mu_\infty = 3.569946$ we found that, as μ increased from zero, the system dynamics of the logistic mapping successively passed through regions of period-2^p dynamics in the order of increasing $p = 0, 1, 2, \ldots$. These, however, are not the only periodic behaviors found in the mapping. Specifically, in the region $\mu > \mu_\infty = 3.569946$, while the system is mostly chaotic, it is found that there are small regions of periodic motion. These are the whitened windows seen in Figure 8.2b which open briefly between the blackened regions of chaos. The period doubling effect does not carry over to these windows, which contain both even and odd periodic behaviors.

It is found in Figure 8.2b that for increasing $\mu > \mu_\infty = 3.569946$ small periodic windows occur in the system [1] with periods $6, 5, 3, 2\times 3, 5, 6, 4, 6, 5$, and 6. (Note: here 2×3 denotes a period-doubling of the earlier listed period-3 mode.) Of these windows of periodicity, the largest window is the period-3 window [1–3] and the next larger windows are of period-5 and period-6. (See Table 8.2 where values of μ at the left edges of these windows are listed [1–3, 8].) We focus on the period-3 window in our next discussions [1–3].

8.1.12 Period-3 Window

In this subsection we study the location of the period-3 cycles [1–4] in the (μ, x)− plot of fixed-points in Figure 8.2. For the values of small μ presented in Figure 8.2a, only the fixed-points of even

TABLE 8.2
Values of μ at Lower Edge of n −Cycle Windows in the Region of Chaos (Here are the Edges of the Dominant Windows in Figure 8.2b)

Period	μ
6	3.630388
5	3.741121
3	3.828427

period cycles were found. Consequently, we expect the odd period cycles, not involved in the period-double transition to chaos, to occur at the larger μ given in Figure 8.2b. In fact, in the following discussions we shall show that period-3 cycles occur in the large window in Figure 8.2b beginning [1] near $\mu \approx 3.83$. This is done by obtaining an exact expression for the position of the left edge of the period-3 window [1–3].

To study the properties of the period-3 cycle, consider the recompositing of the period-3 fixed-point equation $x = f^{(3)}(x)$ from the forms [13]

$$y = \mu x(1-x) = f_\mu(x), \tag{8.58a}$$

$$z = \mu y(1-y) = f_\mu(y) = f_\mu^{(2)}(x), \tag{8.58b}$$

and

$$x = \mu z(1-z) = f_\mu(z) = f_\mu^{(2)}(y) = f_\mu^{(3)}(x). \tag{8.58c}$$

The stability of the fixed-point solutions requires that $\left|\dfrac{df^{(3)}(x)}{dx}\right| < 1$ so in a window of stability we expect that $\left|\dfrac{df^{(3)}(x)}{dx}\right| < 1$ Consequently, from Eq. (8.58) it follows that at the passage into the region of stability (left edge of the window in Figure 8.2b) we need [13]

$$\frac{df^{(3)}(x)}{dx} = \frac{df^{(3)}(x)}{df^{(2)}(x)}\frac{df^{(2)}(x)}{df(x)}\frac{df(x)}{dx} = \frac{df(z)}{dz}\frac{df(y)}{dy}\frac{df(x)}{dx} = \mu^3(1-2z)(1-2y)(1-2x) = 1, \tag{8.59}$$

and this determines the μ −coordinate at which we enter or leave the window of period-3 stability.

The focus is now on determining the value of μ at the left edge of the period-3 window in Figure 8.2b. To do this, we begin by making a change of variables in Eqs. (8.58) and (8.59) to the new coordinates [13]

$$p = x - \frac{1}{2}, q = y - \frac{1}{2}, t = z - \frac{1}{2}. \tag{8.60}$$

In Eqs. (8.58a), (8.58b), and (8.58c) this, respectively, gives

$$\frac{\mu^2}{4} - \frac{\mu}{2} = \mu^2 p^2 + \mu q, \tag{8.61a}$$

$$\frac{\mu^2}{4} - \frac{\mu}{2} = \mu^2 q^2 + \mu t, \tag{8.61b}$$

and

$$\frac{\mu^2}{4} - \frac{\mu}{2} = \mu^2 t^2 + \mu p. \tag{8.61c}$$

In addition, in Eq. (8.59) the same change of variables gives

$$8\mu t \mu q \mu p = -1. \tag{8.62}$$

If we then define $A = \mu p$, $B = \mu q$, and $C = \mu t$, it follows from Eq. (8.61) that

$$\frac{\mu^2}{4} - \frac{\mu}{2} = A^2 + B = B^2 + C = C^2 + A, \tag{8.63}$$

and from Eq. (8.62) that

$$8ABC = -1. \tag{8.64}$$

The new variables A, B, and C allow for a simplification of the algebra, and the resultant Eqs. (8.63) and (8.64) are now ultimately solved from Eq. (8.60) for x, y, and z. This allows us to find the value of μ at the left edge of the window of period-3 behavior.

To begin the solution, notice that Eqs. (8.63) and (8.64) are invariant under simultaneously changing A to B, B to C, and C to A. As a result of this invariance, let us try to generate the solutions for μ, x, y, and z in terms of factors which are invariant under these symmetries. Consequently, in solving Eqs. (8.63) and (8.64) the solutions will be developed in terms of the invariant factors [13]

$$a = A + B + C, \tag{8.65a}$$

$$b = AB + BC + CA, \tag{8.65b}$$

and

$$c = ABC. \tag{8.65c}$$

From the factors in Eq. (8.65), it is also helpful to note some useful identities.

$$A^2 + B^2 + C^2 = a^2 - 2b, \tag{8.66a}$$

$$A^3 + B^3 + C^3 = a^3 - 3ab + 3c, \tag{8.66b}$$

and

$$A^2B^2 + B^2C^2 + A^2C^2 = b^2 - 2ac. \tag{8.66c}$$

These are, of course, also invariant under simultaneously changing A to B, B to C, and C to A. The sets of identities in Eqs. (8.65) and (8.66) are next used in rewriting and solving Eqs. (8.63) and (8.64).

Now we begin by adding the three forms of $\frac{\mu^2}{4}-\frac{\mu}{2}$ in Eq. (8.63) together to give

$$3\left(\frac{\mu^2}{4}-\frac{\mu}{2}\right)=A^2+B+B^2+C+C^2+A=a^2-2b+a \tag{8.67a}$$

so that we arrive at the important relationship

$$\left(\frac{\mu^2}{4}-\frac{\mu}{2}\right)=\frac{1}{3}\left(a^2-2b+a\right). \tag{8.67b}$$

Similarly, from Eq. (8.63), it also follows that

$$\left(A+B+C\right)\left(\frac{\mu^2}{4}-\frac{\mu}{2}\right)=A\left(A^2+B\right)+B\left(B^2+C\right)+C\left(C^2+A\right) \tag{8.68}$$

$$=a^3-3ab+3c+b$$

and combining Eq. (8.68) with Eq. (8.67b) it then follows that

$$2a^3-7ab-a^2+9c+3b=0. \tag{8.69}$$

Notice that Eq. (8.69) is the first equation in terms of the invariant factors a,b, and c.

From Eq. (8.63) it also follows that

$$\left(A+B+C\right)\left(\frac{\mu^2}{4}-\frac{\mu}{2}\right)=C\left(A^2+B\right)+A\left(B^2+C\right)+B\left(C^2+A\right) \tag{8.70a}$$

and applying Eqs. (8.67b) and (8.66b) we find that

$$CA^2+AB^2+BA^2=A^3+B^3+C^2. \tag{8.70b}$$

Similarly, from Eqs. (8.63) and (8.70b) we finally find that

$$3\left(\frac{\mu^2}{4}-\frac{\mu}{2}\right)^2=\left(A^2+B\right)\left(B^2+C\right)+\left(B^2+C\right)\left(C^2+A\right)+\left(C^2+A\right)\left(A^2+B\right), \tag{8.71}$$

and using Eqs. (8.66) and (8.67b) a second equation in a,b, and c is found. This is given by

$$a^4-4a^3+14ab+a^2+b^2+6ac-4a^2b-3b-18c=0. \tag{8.72}$$

Applying Eqs. (8.64) and (8.65c) for c in Eq. (8.69), it follows that b is of the form [13]

$$b=\frac{16a^3-8a^2-9}{56a-24}. \tag{8.73}$$

Using c from Eqs. (8.64) and (8.65c) and b from Eq. (8.73) in Eq. (8.72) we find

$$24(2a-1)(2a+3)\left(4a^2-6a+9\right)^2=0 \tag{8.74}$$

where the root of interest to us is

$$a=\frac{1}{2}. \tag{8.75}$$

From this it follows from Eq. (8.73) that

$$b=-\frac{9}{4} \tag{8.76}$$

and from Eq. (8.67b) it is then found that

$$\left(\frac{\mu^2}{4}-\frac{\mu}{2}\right)=\frac{7}{4}. \tag{8.77}$$

Solving Eq. (8.77) for μ we, consequently, have [13]

$$\mu=1+2\sqrt{2}\approx 3.828427. \tag{8.78}$$

This gives the left edge of the cycle-3 window in Figure 8.2b.

It is interesting to take a closer look at the transition with increasing μ from the chaotic state into the window of period-3 cyclic states. The transition with increasing μ from chaotic states into the left edge at $\mu=1+2\sqrt{2}\approx 3.828427$ of the period-3 window is made by a process known as intermittency. This occurs in regions of μ close to the transition from the chaotic to the periodic state. As the $\mu=1+2\sqrt{2}\approx 3.828427$ transition is approached, the chaotic states of the system are found to begin to exhibit, mixed into them, time segments of period-3 behavior. In this way, at fixed $\mu<1+2\sqrt{2}\approx 3.828427$ the dynamics is composed of alternating sequences of chaotic and period-3 behaviors going back and forth between one another. Finally, at $\mu=1+2\sqrt{2}\approx 3.828427$ the period-3 behavior overwhelms the system dynamics, and period-3 motion becomes the only dynamics observed. This path between periodic and chaotic behaviors is often referred to as the intermittent pathway to chaos.

8.2 UNIVERSALITY

An interesting aspect of the logistic mapping is that the properties it exhibits are mimicked by other so-called one-dimensional unimodal mappings. These are smooth mappings which are concave downward, displaying a single maximum [1–4]. Examples [1] of such additional unimodal mappings are the sine mapping given by $x_{n+1}=\mu\sin\pi x_n$ and the tent map given by $x_{n+1}=\mu\min\left(x_n,1-x_n\right)$. As in the case of the logistic mapping, these types of mappings are defined over $x\in\left[0,1\right]$ and display behaviors which are qualitatively similar to those of the logistic mapping. In some formats, the new unimodal mappings have characteristics that are the same as those of the logistic mapping [1–4].

These features (i.e., the qualitative and quantitative similarities of the unimodal mappings) are termed universality [1–4]. The ideas of universality are a holdover from the study of the properties of thermodynamic systems which undergo phase transitions. In such thermodynamic systems, the critical properties at a phase transition can depend on only certain broad features of the system Hamiltonian. In a sense, however, the transitions in the dynamics of unimodal mappings are a type of phase transition of the dynamical system from one type of state of motion to another. In this regard, it is also found for the logistic mapping that in the neighborhood of the transition at μ_∞ from periodic to chaotic motion, the behavior of the Lyapunov exponent in the region $\mu - \mu_\infty > 0$ takes the critical form [5]

$$\lambda(\mu) = \lambda_0 (\mu - \mu_\infty)^{\frac{\ln 2}{\ln \delta}} \tag{8.79}$$

where δ is the Feigenbaum constant and λ_0 is a constant. Note that this expression is reminiscent of the critical behavior of a thermodynamic order parameter near the critical temperature. In this context, then, the critical exponent of the Lyapunov exponent is $\frac{\ln 2}{\ln \delta} \approx 0.445$.

Qualitatively, the fixed-point plots of the unimodal mappings are similar in general appearance. They all exhibit, as μ increases from zero, a period-doubling process leading to the onset of chaos at an aperiodic value μ_∞. During the periodic doubling processes, the windows occur with increasing periods 2^p for $p = 0,1,2,3,\ldots.$ in the same order, and the interval in μ of each periodicity-windows decreases [1–8] with increasing p. After the onset of chaos, with increasing μ, windows of periodic motion are also encountered embedded in the system in the same sequence of occurrence in each model as in the logistic mapping [1–10]. The largest window of periodicity embedded in the region of chaos is a period-3 window and the next larger windows are for period-5 and period-6 motions [1–4]. Similar to the logistic mapping, all of the unimodal mappings eventually send a given initial position x_0 to a fixed-point of periodic motion from one of its composite mappings or to a chaotic orbit in a chaotic attractor.

In addition, quantitatively, the limits in Eqs. (8.53) and (8.57) for the logistic mapping are also obeyed in the unimodal mappings [1–8]. Consequently, the Feigenbaum constant is a universal property characterizing the unimodal mappings.

Both qualitative and quantitative universality is also observed experimentally in dynamic systems represented by one-dimensional unimodal mappings [1]. This has been illustrated by a number of examples of mechanical, biological, and chemical systems which display nonlinear dynamics [1]. The reader is referred to the literature for detailed discussions of these [1–8].

We next turn to a brief introduction to a three-dimensional dynamical system exhibiting components of chaotic motion, where the motion is described by a set of nonlinear partial differential equations. For these considerations, we look at one of the earliest considered models of chaotic dynamics [1–4].

8.3 CHAOS IN HIGHER DIMENSIONS: DIFFERENTIAL EQUATIONS

Next, we look at three-dimensional systems and some aspects of chaos found in them. These are systems that, because of their increased dimensionality, have a wider variety of features affecting their time development than the simple fixed-points found in lower dimension systems. In addition to fixed-points, in higher dimensional systems a variety of other so-called bifurcation mechanisms or changes in the topological properties of the equations from one type of general dynamical behavior to another are present. The focus of this brief introduction, in particular, is the treatment of higher dimensional systems in terms of their description by sets of nonlinear partial differential equations. Specifically, we shall see that, under proper conditions, nonlinear partial differential equations also display chaotic behaviors similar to those found in the logistic mapping. The focus is on the qualitative discussion of the nature of some of the chaotic motions encountered in these new types of formulations. For a more complete treatment of the ideas of bifurcation and chaos in higher dimensions, the reader is referred to the literature [1].

Of particular interest to us is the Lorenz equation [1–4] which is an early example of a system found to exhibit chaotic dynamics. It was originally developed as a fluid mechanics model of weather and convective motion but has also been applied to a variety of other nonlinear physical processes, including those found in chemical systems and in lasering processes [1]. Our interest in the following is in the unusual dynamics of its chaotic solution trajectories displayed in the three-dimensional space of coordinates (x, y, z).

The Lorenz equation has the form of a system of three nonlinear differential equations in the variables (x, y, z) studied as functions of time. It is given by the set of coupled equations [1]

$$\dot{x} = \sigma(y - x), \tag{8.80a}$$

$$\dot{y} = rx - y - xz. \tag{8.80b}$$

and

$$\dot{z} = xy - bz, \tag{8.80c}$$

where σ is a constant called the Prandtl number, r is a constant known as the Rayleigh number, b is a parameter used to set the dynamics of the system, and where the solutions are studied for $\sigma, r, b > 0$. The nonlinearity in Eq. (8.80) arises, specifically, from the terms xz in Eq. (8.80b) and xy in Eq. (8.80c), and we shall see that these terms are significant in their effect on the chaotic dynamics of the system [1]. In addition, as a spatial symmetry property, it is found that the system in Eq. (8.80) is also invariant under changing (x, y, z) to $(-x, -y, z)$.

As in the case of the logistic mapping, the dynamics of the system in Eq. (8.80) can be characterized in terms of its fixed-points [1]. These are obtained from a consideration of the solutions of the set

$$0 = \sigma(y - x), \tag{8.81a}$$

$$0 = rx - y - xz, \tag{8.81b}$$

$$0 = xy - bz, \tag{8.81c}$$

and the stability properties of the fixed-points are then obtained by studying the solutions of Eq. (8.80) in their neighborhoods. We shall now give a brief example of how this is done for a simple consideration of the system.

8.3.1 Fixed-Points for $0 < r < 1$

To begin the study of the fixed-points we see that, near the origin, the terms xz and xy become small so that the system in Eq. (8.80) can be linearized [1]. Dropping xz and xy in Eq. (8.80), we then find the linearized set [1]

$$\dot{x} = \sigma(y - x), \tag{8.82a}$$

$$\dot{y} = rx - y, \tag{8.82b}$$

and

$$\dot{z} = -bz. \tag{8.82c}$$

As a result, it follows from Eq. (8.82c) that the z-motion becomes a decay to the $x - y$ plane and the motion within the $x - y$ plane is decoupled from the z-motion. In particular, also note that upon dropping the xz and xy in the fixed-point equations in Eq. (8.81), the fixed-points conditions of Eq. (8.82) are obtained. In this regard, it is found that $(x, y, z) = (0,0,0)$ is a fixed-point of the system of Eq. (8.82). Next, the stability properties at the $(x, y, z) = (0,0,0)$ fixed-point must be determined by considering the solutions of the linearized equations of motion in Eq. (8.82) at the origin.

Substituting into Eq. (8.82) the assumed forms $x = \delta x_0 e^{at}$, $y = \delta y_0 e^{at}$, and $z = \delta z_0 e^{ct}$ representing solutions in the vicinity of the fixed-point at the origin [1], we find a set of algebraic eigenvalue equations for a and c. These are given by the set of equations

$$\begin{vmatrix} -\sigma - a & \sigma \\ r & -1-a \end{vmatrix} \begin{vmatrix} x \\ y \end{vmatrix} = 0 \tag{8.83a}$$

and

$$\left[-b-c\right]z = 0. \tag{8.83b}$$

From the solution of this set, the constants a and c characterizing the motion in the neighborhood of the fixed point are given by

$$a = \frac{1}{2}\left[-(1+\sigma) \pm \sqrt{(1+\sigma)^2 - 4\sigma(1-r)}\right] \tag{8.84a}$$

under the restriction that $0 < r < 1$, and by

$$c = -b. \tag{8.84b}$$

It is seen from Eq. (8.84a) that, in the region $0 < r < 1$, a always has a negative real part forcing the solution to always move in the $x - y$ plane towards the fixed-point $(x, y) = (0,0)$. Again, from Eq. (8.84b) it follows that z always moves towards $z = 0$. Consequently, in the linearized limit the solutions for (x, y, z) always converge to $(0,0,0)$ provided $0 < r < 1$.

A more comprehensive consideration [1] of the $(0,0,0)$ fixed-point shows that the general nonlinear Lorenz equation has the origin $(x, y, z) = (0,0,0)$ as a stable universal fixed-point in the region $0 < r < 1$. That is to say that in this region of r all initial points (x, y, z) eventual travel on trajectories sends them to the fixed-point at $(0,0,0)$. This behavior is similar to that found in the study of the logistic mapping for initial states in its lowest region $0 < \mu < 1$.

8.3.2 Fixed-Points for $1 < r < r_H = \dfrac{\sigma(\sigma + b + 3)}{\sigma - b - 1}$

Consider next looking for fixed-points of Eq. (8.80) in an expanded region of r. For these considerations of Eqs. (8.80) and (8.81) we will also search under the further condition that $x = y$. From Eq. (8.81) it then follows that [1]

$$0 = rx - y - xz = y(r - 1 - z), \tag{8.85a}$$

and

$$0 = y^2 - bz. \tag{8.85b}$$

Consequently, we find three possible fixed-points of Eq. (8.85) given by

$$(x, y, z) = (0,0,0), \tag{8.86a}$$

or

$$(x, y, z) = \left(\pm\sqrt{b(r-1)}, \pm\sqrt{b(r-1)}, r-1\right). \tag{8.86b}$$

The first in Eq. (8.86a) is at the origin and has already been treated earlier. The second two are non-zero fixed points whose stability properties must now be considered.

For the consideration of stability of the fixed-points, let us introduce some notations. In this regard, denote the fixed-point in Eq. (8.86a) by [1]

$$(x_1, y_1, z_1) = (0,0,0) \tag{8.87a}$$

and the fixed-points in Eq. (8.86b)

$$(x_\pm, y_\pm, z_\pm) = \left(\pm\sqrt{b(r-1)}, \pm\sqrt{b(r-1)}, r-1\right) \tag{8.87b}$$

Making a stability analysis as in Eq. (8.83) and (8.84), it is already found that the point in Eq. (8.87a) is a stable fixed-point only for $0 < r < 1$. We now show that the fixed-points in Eq. (8.87b) are stable only in the region $1 < r < r_H = \dfrac{\sigma(\sigma+b+3)}{\sigma-b-1}$ while the fixed-point in Eq. (8.87a) is an unstable fixed-point in the region $1 < r < r_H = \dfrac{\sigma(\sigma+b+3)}{\sigma-b-1}$. The transition with increasing r at $r = 1$ between these two types of stability is often termed a pitchfork bifurcation.

The stability of Eq. (8.87b) is determined by considering the nature of the solutions of Eq. (8.80) in their neighborhoods. Substituting the assumed solution forms $x = x_\pm + \delta x_\pm e^{at}$, $y = y_\pm + \delta y_\pm e^{at}$, and $z = z_\pm + \delta z_\pm e^{at}$ into Eq. (8.80) (Note: $x_\pm = y_\pm$ for this substitution) we find the eigenvalue equation for a. This is given by the matrix eigenvalue problem

$$\begin{vmatrix} a+\sigma & -\sigma & 0 \\ z_\pm - r & a+1 & x_\pm \\ x_\pm & x_\pm & -(b+a) \end{vmatrix} \begin{vmatrix} \delta x_\pm \\ \delta y_\pm \\ \delta z_\pm \end{vmatrix} = 0. \tag{8.88}$$

Taking the determinant of the matrix in Eq. (8.88) gives the solvability condition for the system of linear equations. In this way, it follows that [1]

$$a^3 + (\sigma+b+1)a^2 + (r+\sigma)ba + 2b\sigma(r-1) = 0 \tag{8.89}$$

Solving Eq. (8.89) for a, it is seen that the two solutions in Eq. (8.87b) are stable only in the region $1 < r < r_H = \dfrac{\sigma(\sigma+b+3)}{\sigma-b-1}$.

Notice that one of the three roots of Eq. (8.88) is pure imaginary and this enters into the consideration of the next transition with increasing r of the system at $r_H = \dfrac{\sigma(\sigma+b+3)}{\sigma-b-1}$. Specifically, it is found that at $r = r_H = \frac{\sigma(\sigma+b+3)}{\sigma-b-1}$ the system undergoes a so-called Hoff bifurcation to a chaotic state [1].

8.3.3 Solutions for $r_H = \frac{\sigma(\sigma+b+3)}{\sigma-b-1} < r$

The next region of increasing r is found to be the region $r_H = \frac{\sigma(\sigma+b+3)}{\sigma-b-1} < r$ containing chaotic solutions. In this region, an appropriate Lyapunov function can be defined which, again, acts as a measure of the divergence of the trajectories of the system in time. Specifically, it is found that trajectories having slightly different initial conditions quickly diverge with increasing time. For the details of this, the reader is referred to the literature [1–4].

8.3.4 Rossler Equations

Finally, as a point of interest regarding systems of differential equations exhibiting chaotic transitions, we note that an even simpler set of nonlinear differential equations than those of the Lorenz equations exists and displays many features similar to those of the logistic mapping. These are the so-called Rossler equations [14, 15] which were posed as a simplified version of the Lorenz system.

The Rossler equations are very similar to the Lorenz equations and are given by the set [1, 14, 15]

$$\dot{x} = -y - z, \tag{8.90a}$$

$$\dot{y} = x + ay, \tag{8.90b}$$

and

$$\dot{z} = b + xz - cz, \tag{8.90c}$$

where (x, y, z) are the space coordinates of the trajectories and a, b, c are constants [1]. Notice that the nonlinearity enters only in one place in Eq. (8.90) as the xz term in Eq. (8.90c). The Rossler equations can be parameterized to display a period doubling effect leading to the onset of chaos [1]. Following the initiation of chaos, the chaotic motion occurs on a chaotic attractor which exists within a bounded region of three-dimensional space.

REFERENCES

1. Strogatz, S. H. 1994. *Nonlinear Dynamics and Chaos* (Perseus Books, Reading).
2. Feldman, D. P. 2012. *Chaos and Fractals: An Elementary Introduction* (Oxford University Press, Oxford).
3. Rasband, S. N. 1990. *Chaotic Dynamics of Nonlinear Systems* (John Wiley & Sons, Inc, New York).
4. Nicolis, G. 1999. *Introduction to Nonlinear Science* (Cambridge University Press, Cambridge).
5. Arfken, G. B. and Weber, H. J. 2005. *Mathematical Methods for Physicists*, 6th Edition (Elsevier, Burlington).
6. Chen, S., Feng, S., Fu, W., and Zhang, Y. 2021. Logistic Map: Stability and Entrance to Chaos. *Journal of Physics Series* 2014, 012009. https://doi.org/10.1088/1742-6596/2014/1/012009.
7. Hirsch, M. W., Smale, S., and Devaney, R. L. 2013. *Differential Equations, Dynamical Systems, and an Introduction to Chaos* (Elsevier, Amsterdam).
8. Zhang, C. 2014. Cycles of the Logistic Map International. *Journal of Bifurcation and Chaos* 24, 1450005. https://doi.org/10.1142/S0218127414500059.
9. Feigenbaum, M. J. 1983. Universal Behavior in Nonlinear Systems. *Physica D* 7, 16–39.

10. Leonel, E. D. 2021. *Scaling Laws in Dynamical Systems* (Springer, Singapore).
11. Miranda, L. K. A., Kuwana, C. M., Huggler, Y. H., da Fonseca, A. K. P., Yoshida, M., de Olivera, J. A., and Leonel, E. D. 2021. A Short Review of Phase Transitions in a Chaotic System. *European Physical Journal: Special Topics*. https://doi.org/10.1140/epjs/s11734-021-00415-3.
12. Coppersmith, S. N. 1999. A Simpler Derivation of Feigenbaum's Renormalization Group Equation for the Period-Doubling Sequence. *American Journal of Physics* 67(1), 52–54.
13. Saha, P. and Strogatz, S. H. 1995. The Birth of Period Three. *Mathematics Magazine* 68(1), 42–47.
14. Rossler, O. E. 1976. An Equation for Continuous Chaos. *Physical Review Letters A* 57, 397–401.
15. Rossler, O. E. 1976. Different Types of Chaos in Two Simple Differential Equations. *Zeitschrift fur Naturforschung A* 31, 1661–1670.

9 Fractals

In the present chapter, we look at the characteristic dimensions of a variety of physical systems and discuss how this dimensionality affects their properties [1–11]. In this regard, we commonly classify physical systems as 1–dimensional (e.g., polymers), 2–dimensional (e.g., surfaces), or 3–dimensions (e.g., bulk materials) and are familiar with certain of their highly dimensional dependent properties. For example, the heat capacities and transport properties of materials are found to be strongly dependent on the dimensionality of the material, i.e., whether a material is a polymer, a thin film or surface, or a bulk material [8–11]. Similarly, the dispersion relations of the excitations in a material or the general types of particle dynamics observed in a system are known to be highly dependent on dimension [1, 8–11]. As we shall see, however, there are some physical systems which are not easily put into the classification of being 1–, 2–, or 3–dimensional [1, 3]. These provide exceptions to the general rules of the three integer dimensions. Nevertheless, in an expanded view of dimension, such exceptional materials can be regarded as being of a new type of fractional dimension. They are known as fractals, characterized by a fractal dimension [1].

Examples of such systems that can be characterized by non-integer fractal dimensions are a variety of disordered and growth systems and geometric features associated with nonlinear motion [1–11]. Polymer based materials and amorphous materials and random rough surfaces in particular display characteristics which often exhibit properties which are not easily classified as being 1–, 2–, or 3–dimensional [1]. In addition, systems developed from growth processes may exhibit properties that are best thought of as inter-dimensional in nature [1–11]. Thermodynamic properties of such fractal materials consequently display unusual temperature dependences not found in non-fractal materials [12–19].

In this regard, for some systems the characterization of the system by the dimension in which it is embedded is a gross simplification [1–19], and many materials exist which have features that are more properly understood as being less than 1–dimensional, being between 1– and 2–dimensional, or being between 2– and 3–dimensional [12–19]. The geometries of these types of materials can often be classified in terms of the ideas of fractal geometry and the related idea of fractal dimension [1–9]. Specifically, the fractal dimension may range over the set of real numbers rather than being limited to just the positive integers, i.e., 0,1,2,3.... In this sense a system with fractal dimension 2.1 may look like a 2–dimensional system but be a little more complex, or a system of fractal 2.9–dimension looks a little less complex than a fully 3-dimensional system [1]. As with 1–, 2–, and 3–dimensional systems we shall see, in addition, that fractals can be found as either ordered in structure or randomly disordered in structure [1–19].

There are a number of ways that the fractal dimension can be defined [1, 3, 8, 19–21]. Generally, however, each definition is made such that when applied to a conventional 1–, 2–, or 3–dimensional system it yields a fractal dimension of 1–, 2–, and 3–dimensions. To begin the discussions, we start with the definition of the so-called capacity fractal dimension [1, 3, 5–8], denoted as d_{cap}. This is defined in terms of a covering of the geometric object [1, 3]. Later, we shall define a number of other fractal dimensions [3, 8]. These arise in the context of the types of systems and the properties we wish to study. The capacity fractal dimension [3] we define now is one of the most commonly used definitions of fractal dimension.

Specifically, if we are given a geometric object or point set A in the space $\mathbb{R}^m$ (i.e., the set of m–tuples involving the real numbers $\mathbb{R}$) and if the distances between points in this space are defined as the standard Euclidean distance, one can find a minimum or smallest covering of A by m–dimensional cubes of side ϵ. If there are $N(\epsilon)$ cubes of side ϵ in the minimum covering, then the fractal dimension d_{cap} is defined in terms of the covering by the limiting process [3]

DOI: 10.1201/9781032674919-9

$$d_{cap} = \lim_{\epsilon \to 0} \frac{\ln N(\epsilon)}{\ln \frac{1}{\epsilon}}. \tag{9.1}$$

In the following, we shall first show that when Eq. (9.1) is applied to standard 0–,1–,2–,3–dimensional features of classical geometry, it correctly gives d_{cap} values, respectively, of 0,1,2,3 [1, 3]. This is then followed by the application of the capacity dimension to the study of a number of important examples of fractal systems.

9.1 CAPACITY FRACTAL DIMENSION

In Figure 9.1 schematics are shown of 0–,1–,2–,3–dimensional objects and their minimal coverings by cubes of sides ϵ. These are all found to have fractal dimensions equal to their standard geometric dimensions [1, 3]. To see this, first consider using Eq. (9.1) to compute the fractal dimension of the point in Figure 9.1a. We find for the single point contained within a box of side ϵ that $N(\epsilon) = 1$ and its fractal dimension is then [1, 3]

$$d_0 = \lim_{\epsilon \to 0} \frac{\ln N(\epsilon)}{\ln \frac{1}{\epsilon}} = \lim_{\epsilon \to 0} \frac{\ln 1}{\ln \frac{1}{\epsilon}} = 0, \tag{9.2a}$$

so that the fractal dimension agrees with the actual space dimension of the point.

Similarly, consider computing from Eq. (9.1) the fractal dimension of the line segment of length L in Figure 9.1b. It follows that the line segment is contained within $N(\epsilon) = \frac{L}{\epsilon}$ boxes of side ϵ so that its fractal dimension becomes [1, 3]

$$d_1 = \lim_{\epsilon \to 0} \frac{\ln N(\epsilon)}{\ln \frac{1}{\epsilon}} = \lim_{\epsilon \to 0} \frac{\ln \frac{L}{\epsilon}}{\ln \frac{1}{\epsilon}} = \lim_{\epsilon \to 0} \frac{\ln L + \ln \frac{1}{\epsilon}}{\ln \frac{1}{\epsilon}} = 1, \tag{9.2b}$$

which is also its spatial dimension.

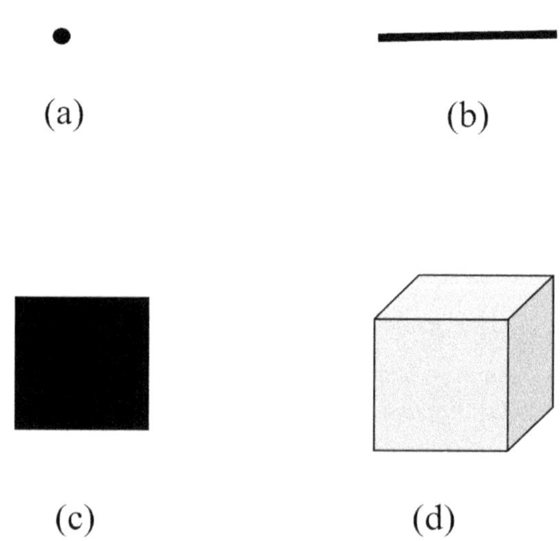

FIGURE 9.1 Schematics of: (a) a point, (b) a solid line segment, (c) a solid square area, and (d) a solid cube volume.

From Eq. (9.1) it also is seen that the fractal dimension of the square of area L^2 in Figure 9.1c is obtained by covering the square with $N(\epsilon)=\left(\frac{L}{\epsilon}\right)^2$ boxes of side ϵ. Consequently, the fractal dimension of the square is [1, 3]

$$d_2 = \lim_{\epsilon\to 0}\frac{\ln N(\epsilon)}{\ln\frac{1}{\epsilon}} = \lim_{\epsilon\to 0}\frac{\ln\frac{L^2}{\epsilon^2}}{\ln\frac{1}{\epsilon}} = \lim_{\epsilon\to 0} 2\frac{\ln L + \ln\frac{1}{\epsilon}}{\ln\frac{1}{\epsilon}} = 2, \tag{9.2c}$$

also being its spatial dimension.

Finally, for the three-dimensional cube in Figure 9.1d we find that $N(\epsilon)=\left(\frac{L}{\epsilon}\right)^3$ for a covering by cubes of side ϵ and the fractal dimension is [1, 3]

$$d_3 = \lim_{\epsilon\to 0}\frac{\ln N(\epsilon)}{\ln\frac{1}{\epsilon}} = \lim_{\epsilon\to 0}\frac{\ln\frac{L^3}{\epsilon^3}}{\ln\frac{1}{\epsilon}} = \lim_{\epsilon\to 0} 3\frac{\ln L + \ln\frac{1}{\epsilon}}{\ln\frac{1}{\epsilon}} = 3. \tag{9.2d}$$

It is found in all cases that the fractal dimension reduces to the standard dimensionality in the consideration of a point, a line, a square, and a cube.

We now turn to a consideration of examples of systems having exceptional fractal geometries as determined by Eq. (9.1). These include systems with both high and low fractal dimensions. This is followed by discussions of other definitions of fractal dimension that have been introduced into the study of fractals. Such definitions have been developed to facilitate or emphasize particular aspects of the physical properties associated with fractal dimensions. At the end of the chapter theoretical and experimental illustrations of the dependence of physical properties of some particular systems on fractal dimensions are discussed.

9.1.1 Fractal Dimension: Unusual Fractal Geometries

In this subsection the fractal dimension of a number of unusual geometric systems are addressed [1, 3, 8]. These act as illustrations of the application of Eq. (9.1) to classify specific objects having geometries seen to fall between those of traditional $0-$, $1-$, $2-$, and $3-$dimensional systems. In a first example, a discussion is given of Cantor dust [1, 3]. This is essentially a fractal dust formed from small particulates, and it is shown to have a fractal dimension between 0 and 1. Following this, examples are given of types of snowflake or condensation models developed in the $2-$dimensional plane [1, 3]. These provide illustrations of geometries associated with growth processes found in physical, chemical, and biological systems [1, 3, 14]. The models considered are shown to have a fractal dimension between 1 and 2 and exhibit patterns which are neither characterizable as $1-$ or $2-$dimensional in nature.

We begin with a discussion of the Canto dust [1, 3] which is created through a process of iteration. Consider the fractal system generated in Figure 9.2 by a sequence of transformations denoted $n = 0,1,2,\ldots,\infty$. Here we start in the initial, $n = 0$, generation step with a one-dimensional line segment of length L. For the next, $n = 1$, generation step we remove the center $1/3$ of the segment introduced in the $n = 0$ step. The result of this transformation is two separated line segments one-third of the length of the original segment. This is then followed by the $n = 2$ generation step in which we remove the center $1/3$'s of the two remaining segments of the $n = 1$ step. This leaves four separated line segments which are each one-ninth the length of the original segment at the start of the

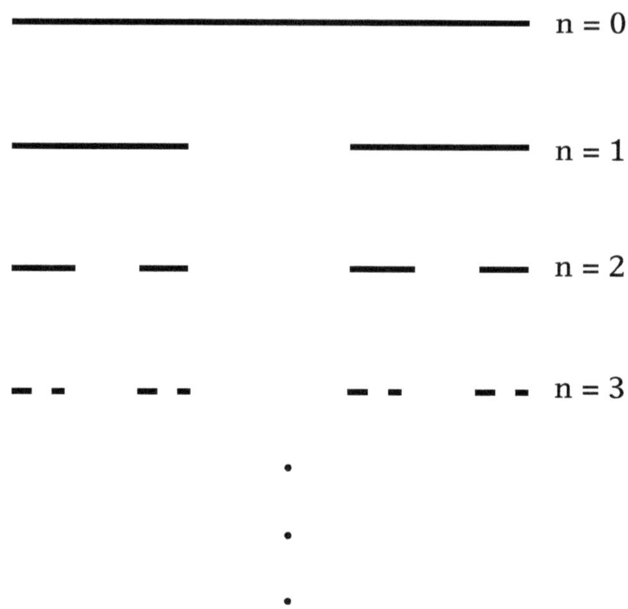

FIGURE 9.2 Schematic for the generation of Cantor dust. On the top line labeled $n = 0$ we start with a solid line segment of length L. In the next line labeled $n = 1$ a center segment of length $\frac{L}{3}$ is removed from the segment of length L. In the next line labeled $n = 2$ a center segment of length $\frac{L}{9}$ has been removed from each of the two segments of length $\frac{L}{3}$ in the $n = 1$ step. This is followed by the next line labeled $n = 3$ in which center segments of length $\frac{L}{27}$ are removed from each of the four segments of length $\frac{L}{9}$ in the $n = 2$ step. The process is continued indefinitely, resulting in the generation of a Cantor dust.

iterations. In this way, the process is iterated over and over so that in the limit as $n \to \infty$ an infinite dust of infinitesimal particles is generated. Very quickly we go from a line segment to a dust of small particles.

The remaining dust generated in this way is known as Cantor dust [1, 3]. It is found, following this process, that we begin with a system that is clearly $1-$dimensional and end with a dust that appears a little closer to a $0-$dimensional collection of points than a continuum $1-$dimensional line segment. In this regard, let us calculate the fractal dimension of the Cantor dust using the definition of d_{cap} in Eq. (9.1).

To determine the fractal dimension of the Cantor dust [1, 3], notice in going from $n = 0$ to $n = 1$ that for a covering of the generated system with size $\epsilon = \frac{1}{3}L$, the two covered segments give $N(\epsilon) = 2$, and consequently $\frac{1}{\epsilon} = \frac{3}{L}$. (Note that at this stage in our process the covering of the set is very crude. However, quickly with increasing n the covering of the generated dust becomes better and better in quantifying the nature of the dust.) Continuing the iteration process applied to the system in this way, it follows that at the mth iteration in the process of dust creation $N(\epsilon) = [2]^m$, $\epsilon = \left[\frac{1}{3}\right]^m L$, and $\frac{1}{\epsilon} = \frac{3^m}{L}$. Applying these in Eq. (9.1), it follows that the fractal dimension of the system is given for $\epsilon \to 0$ or $m \to \infty$ by [1, 3]

$$d_{cap} = \lim_{m \to \infty} \frac{\ln[2]^m}{\ln[3]^m - \ln L} = \lim_{m \to \infty} \frac{\ln[2]^m}{\ln[3]^m} = \frac{\ln 2}{\ln 3} = 0.6309. \tag{9.3}$$

In this regard, we see that the fractal dimension of the Cantor dust is, in fact, less than that of the $1-$dimensional line segment but more than that of the $0-$dimensional point. This difference in dimensionality should also manifest itself in the physical properties displayed by the system. This is discussed later.

Next, we consider a system which has a higher fractal dimension than that of the Cantor dust. In this regard, an interesting fractal is the so-called Koch curve [1, 3]. This is a system which has a fractal dimension a little greater than that of a $1-$dimensional line segment. The generation of the Koch curve is by an iterative process which is described in Figure 9.3.

To obtain the Koch fractal we begin in Figure 9.3a with our $n = 0$ starting point which is a straight-line segment of length L. This, of course, is a $1-$dimensional system. From the line segment we generate the structure in Figure 9.3b of the $n = 1$ iteration. This is composed of four line segments each of length $\frac{1}{3}L$. Consequently, for this transformation let us take the rough covering with the properties $\epsilon = \frac{1}{3}L$ and $N(\epsilon) = 4$. The next $n = 2$ transformation generates the structure in Figure 9.3c. Here the geometry is composed of $N(\epsilon) = 4^2 = 16$ segments and the length of each line segment is given by $\epsilon = \left(\frac{1}{3}\right)^2 L$. Proceeding in this way, we find that for the mth iteration the Koch curve is composed of $N(\epsilon) = 4^m$ segments and the length of each of the line segments is given by $\epsilon = \left(\frac{1}{3}\right)^m L$. Consequently, $\frac{1}{\epsilon} = 3^m \frac{1}{L}$.

We see, again, that as $m \to \infty$ the covering of the system better represents the curve. Finally, applying these results in Eq. (9.1), the fractal dimension is given by [1, 3]

$$d_{cap} = \lim_{m\to\infty} \frac{\ln\left[4\right]^m}{\ln\left[3\right]^m - \ln L} = \lim_{m\to\infty} \frac{\ln\left[4\right]^m}{\ln\left[3\right]^m} = \frac{\ln 4}{\ln 3} = 1.2619. \tag{9.4}$$

It is seen that, in fact, the Koch curve is a little more than a line segment but still much less than a $2-$dimensional object. This is not surprising as at each step of our generation process a little length has been introduced to the curve so that it has become more filling of the plane containing the curve. Next, we consider some higher dimensional systems that are of interest in growth and crystallization processes [1, 3, 14, 19–21].

A similar investigation can be given for a type of snowflake pattern generated in the $2-$dimensional plane [1, 3]. This provides a structure which is between $1-$ and $2-$dimensions so that the geometry is more complex than that of a $1-$dimensional system but less complex than that of a fully $2-$dimensional system. Unlike the Koch curve, however, it is much closer to the geometry of a $2-$dimensional object. It shall be seen later that these types of patterns are of interest in growth or condensation processes and also in studies of biological populations [1, 14].

Specifically, consider the system generated by the process shown schematically [1, 3] in Figure 9.4. To generate the pattern, we begin with the square in Figure 9.4a. This is our $n = 0$ step in the generation process, and the figure is composed of a single shaded square with a side of unit length. In the first or $n = 1$ step, shown in Figure 9.4b, we divide the shaded square in Figure 9.4a into nine sub-squares, and in the process one of the sub-squares is whitened. The length of the side of the sub-squares in the new pattern is then $\frac{1}{3}$ that of the square in Figure 9.4a, and the new pattern is composed of 8 shaded sub-squares and the one white sub-square. This results in the 9 sub-square figure shown in Figure 9.4b. In the next process we generate the $n = 2$ step shown in Figure 9.4c from that in Figure 9.4b. This is done by considering each of the 8 shaded sub-squares in Figure 9.4b and performing the operation of subdividing them into 9 smaller squares while making one of each of their subdivisions white. The length of the side of a shaded square in the new pattern is then

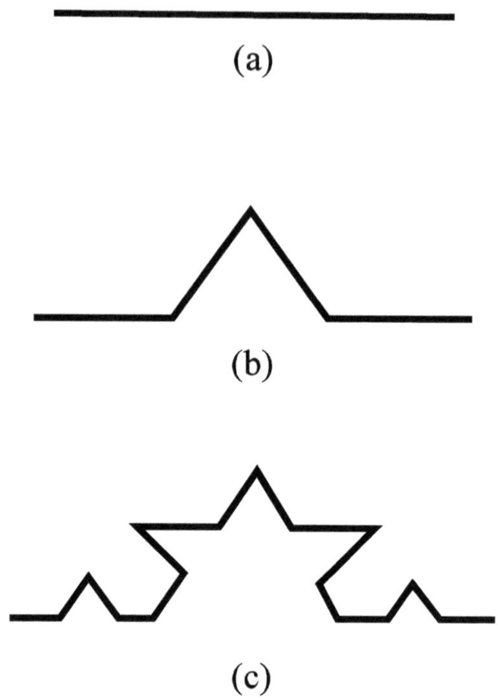

FIGURE 9.3 Schematic for the generation of the Koch curve: a) the $n = 0$ line segment of length L, b) the $n = 1$ curve of 4 line segments of length $\frac{L}{3}$, and c) the $n = 2$ curve of 16 line segments of length $\frac{L}{9}$. Continuing this process indefinitely generates the Koch curve.

$\left(\frac{1}{3}\right)^2$ and the pattern is composed of 8^2 shaded squares. Continuing in this way an infinitely facetted snowflake pattern is generated in the 2 - dimensional plane.

Considering the fractal generated from the process in Figure 9.4, we see that continuing the generation process to the mth iterative step results in a structure of $N(\epsilon) = 8^m$ shaded squares each of side $\epsilon = \left(\frac{1}{3}\right)^m$ so that $\frac{1}{\epsilon} = 3^m$. Consequently, from Eq. (9.1) it follows that the fractal dimension is given by [1, 3]

$$d_{cap} = \lim_{\epsilon \to 0} \frac{\ln N(\epsilon)}{\ln \frac{1}{\epsilon}} = \lim_{m \to 0} \frac{\ln 8^m}{\ln 3^m} = \frac{\ln 8}{\ln 3} = 1.8928. \tag{9.5}$$

This is found to be less than 2−dimensions but still considerably more than 1−dimension. The resulting patterning, nevertheless, fails to completely fill a 2−dimensional space so is of a lesser dimensionality.

Notice that all of the fractals considered so far were types of so-called self-similar geometries. Specifically, in these geometries a mapping exists from the system back into itself preserving the geometric pattern. Another way of looking at this is that they were all generated by well-defined

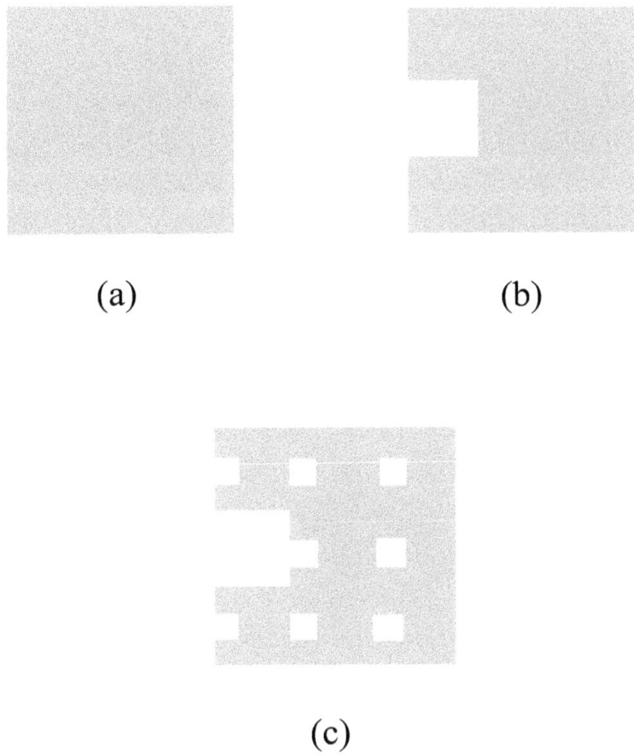

FIGURE 9.4 A fractal is generated from a solid square of area. In a) we start with a solid square area which is completely shaded. In b) the square in a) is broken into 9 equal sub-squares and one sub-square on the center left edge is made white. In c) the shaded sub-squares in b) are each broken into 9 equal sub-squares and one sub-square on their center left edge is made white. This process is repeated indefinitely to generate the fractal pattern.

unchanging iterative algorithms. This self-similarity, however, can be relaxed in the study of fractal systems by invoking processes involving the introduction of randomness into the system. In this regard, we can change from the consideration of self-similar systems to those which only give a certain general appearance of self-similarity but are, in fact, not self-similar. For these geometries the fractal dimension is computed in much the same way as in our earlier considerations.

For example, consider the generation of the system [1, 3] in Figure 9.5. This is a variation of that generated in Figure 9.4 in which the position of the whitened squares now appears randomly at different places in each iteration of the generation. The system in Figure 9.5, however, exhibits the same fractal dimension as that in Figure 9.4 even though the two patterns are not the same. The new pattern in Figure 9.5 exhibits a degree of randomness in the positioning of its white sub-squares. In this regard, the idea of fractal and fractal dimension are extended to random systems which, while not self-similar, give the appearance of an approximate self-similarity [1–8].

9.2 GROWTH PROCESSES

An example of a fractal system in nature that exhibits a degree of self-similarity is found in the growth of trees and bacterial cultures [1, 4], the development of cracks and fracture patterns [1, 4, 11], or the population expansion of cities [14]. In these systems, growth tends to start at some central point and extends outward in the development of various dendritic features. These types of expansions can be characterized in terms of fractals exhibiting a variety of fractal dimensions

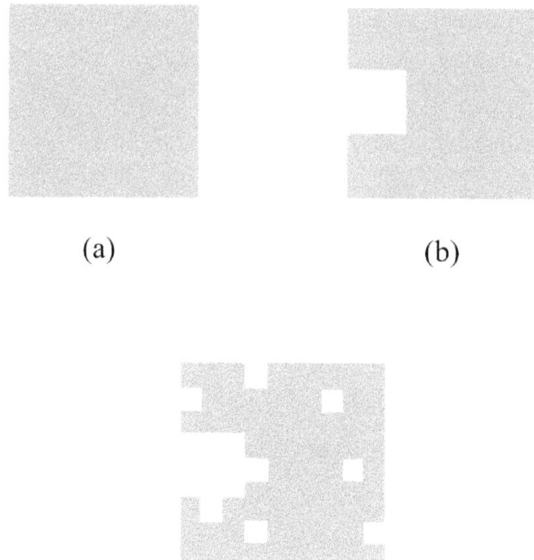

FIGURE 9.5 A fractal is generated from a solid square of area in a process which is similar to that in Figure 9.4. In a) we start with a solid square area which is completely shaded. In b) the square in a) is broken into 9 equal sub-squares and one sub-square on the center left edge is made white. In c), however, the shaded sub-squares in b) are each broken into 9 equal sub-squares and now arbitrarily one sub-square is made white. This process is repeated indefinitely to generate the fractal pattern.

dependent on the mechanisms of growth. They are examples of approximately self-similar systems which incorporate degrees of randomness in their generation processes [1].

A simple model of such a type of dendric pattern formed in a self-similar system [1] is shown in Figure 9.6. In the formation of this pattern, we begin in Figure 9.6a with a shaded square of unit side [1]. This is our $n = 0$ step in the process. In the first or $n = 1$ step, shown in Figure 9.6b, we divide the shaded square in Figure 9.6a into nine sub-squares and whiten four of them as shown in the figure. The length of the side of the sub-squares in the new pattern is then $\frac{1}{3}$ of that of the square in Figure 9.6a, and the new pattern is composed of 5 shaded sub-squares and 4 white sub-squares [1]. This results in the 9 sub-square figure shown in Figure 9.6b. In the next step, we generate the $n = 2$ figure shown in Figure 9.6c from that in Figure 9.6b. This is done by considering each of the 5 shaded sub-squares in Figure 9.4b and performing the operation of subdividing them into 9 smaller squares while making 4 of each of their subdivisions white. The length of the side of a shaded square in the new pattern is then $\left(\frac{1}{3}\right)^2$ and the pattern is composed of 5^2 shaded squares. Continuing in this manner, an infinitely facetted growth pattern is generated in $2-$dimensional space. The resulting pattern is shown to have fractal dimension

$$d_{cap} = \lim_{\epsilon\to 0}\frac{\ln N(\epsilon)}{\ln\frac{1}{\epsilon}} = \lim_{m\to 0}\frac{\ln 5^m}{\ln 3^m} = \frac{\ln 5}{\ln 3} = 1.465. \tag{9.6}$$

The pattern generated in Figure 9.6 provides a feature reminiscent of dendritic growth patterns found in cracks and snowflakes [1, 11]. These are generally open structures with approximately self-similar arms extending from a central location. In nature, this type of pattern is also observed

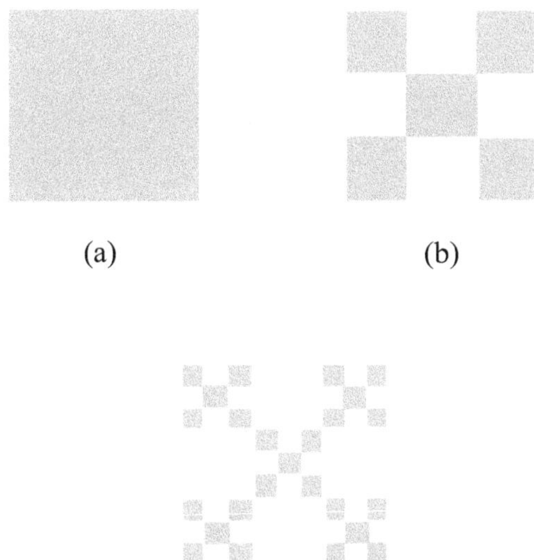

(a) (b)

(c)

FIGURE 9.6 A dendritic fractal is generated from a solid square of area. In a) we start with a solid square area which is completely shaded. In b) the square in a) is broken into 9 equal sub-squares and four sub-squares (one on the center edge of each of the square sides) are made white. In c) the shaded sub-squares in b) are each broken into 9 equal sub-squares and four sub-squares (one on the center edge of each of the square sides of the shaded squares in Figure 9.6b) are made white. This process is repeated indefinitely to generate the dendritic fractal pattern.

in the growth of vegetation [1] or in the population sprawl of cities [14]. In this regard, the patterning in Figure 9.7 showing a photo of a juniper displays a type of self-similarity in the development of its branches from their trunks to their endings. This growth is also typified by a fractal dimension. Later, we shall have more to say about the fractal nature of growth processes. For now, however, we will look at some alternative measures of fractal dimensions that have been developed in the study of physical and biological processes.

9.3 OTHER *MEANS* OF DETERMINING FRACTAL DIMENSION

In the earlier treatment we have considered the capacity fractal dimension of a geometric object [1–4]. This generates a single fractional dimension characterizing the general geometry of a self-similar or approximately self-similar structure [1]. In some situations, however, this characterization may be too general, and other methods have been devised to arrive at more useful numerical measures [3]. These often take into account local spatial variations in geometry in the generation of a single numerical dimension. Additionally, measures can also be formulated which allow for the variation of dimensions over a structure [1, 3, 4]. These allow the dimensions to be functions of position over the system. In this regard, a variety of methods for calculating the fractal dimension of a system have been offered, often heavily depending on the particular discipline involved in the study of its random and self-similar objects.

In the capacity dimension (also sometimes referred to as the Hausdorff [1, 3, 8, 9] dimension), we define the fractal dimension by the limiting form

$$d_{cap} = \lim_{\epsilon \to 0} \frac{\ln N(\epsilon)}{\ln \frac{1}{\epsilon}} \tag{9.7}$$

FIGURE 9.7 A photograph of the dendritic fractal pattern in the growth of a juniper. Each of the branchings of the juniper possesses a degree of self-similarity with the others in the juniper.

where $\mathrm{N}(\epsilon)$ is the smallest number of m – dimensional cubes of side ϵ which completely cover the object being studied. This measure depends on counting cubes containing points of the fractal, but one problem with the measure is that it does not account for the variation in the density of points between the counted cubes. Some measures have been formulated to better take this variation into consideration [3].

Specifically, the amount of the fractal system contained within each cube in the covering used to compute Eq. (9.7) can vary from cube to cube [3, 8]. To account for this, one can defined $\{C_i\}$ as the covering set used in calculating Eq. (9.7). Here C_i denotes the ith cube of the minimal covering in Eq. (9.7) of m – dimensional cubes of side ϵ. Next, determine the set $\{P_i\}$ where P_i is that fraction of the points in the fractal object that are contained within the particular cube C_i. For this system and covering, we then define the information dimension of the fractal as [3]

$$d_{inf} = \lim_{\epsilon\to 0}\frac{I(\epsilon)}{\ln\frac{1}{\epsilon}} \quad (9.8a)$$

where the information carried in the fractal is given by

$$I(\epsilon) = \sum_{i=1}^{N(\epsilon)} P_i \ln\frac{1}{P_i}. \quad (9.8b)$$

Notice, in this regard, that $P_i \ln\frac{1}{P_i}$ measures the information about the fractal contained within the C_i cube. We see then that Eq. (9.8) better accounts for the occupancy of the cubes in the minimal covering than does Eq. (9.7).

Note that it is found from Eq. (9.8) that in the limit [3] that $P_i = P = \frac{1}{N(\epsilon)}$ (i.e., for P_i is independent of i),

$$I(\epsilon) = \sum_{i=1}^{N(\epsilon)} P \ln\frac{1}{P} = \ln N(\epsilon). \quad (9.8c)$$

As a result, in this limit the information dimension reduces to the capacity dimension. In general, it is found also that [3]

$$d_{inf} < d_{cap}. \quad (9.9)$$

Consequently, a more refined estimate lowers the computed value of the fractal dimension [3].

Another way that the spatial variation over the fractal has been accounted for is by computing the so-called correlation fractal dimension [1, 3, 8]. In this method, we evaluate the correlator defined by

$$C(\epsilon) = \lim_{N\to\infty}\frac{1}{N^2}\sum_{i\neq j=1}^{N}\Theta\left(\epsilon - \left|\vec{r}_i - \vec{r}_j\right|\right) \quad (9.10)$$

where $\vec{r}_i$ and $\vec{r}_j$ are point positions in the fractal point set, N are the number of points in the fractal point set, ϵ is the radius of the correlator, and the Heaviside function is given by [3, 5]

$$\Theta(\alpha) = 0 \text{ if } \alpha < 0 \quad (9.11)$$

$$= 1, \text{ otherwise.}$$

In terms of $C(\epsilon)$ the correlator, fractal dimension is then defined as [3]

$$d_{corr} = \lim_{\epsilon\to 0}\frac{\ln C(\epsilon)}{\ln\epsilon}. \quad (9.12)$$

This provides another measure of fractal dimension which allows for the variation of the density of the fractal point set. In general, it is found between the three dimensions discussed here that [3]

$$d_{corr} \le d_{inf} \le d_{cap}. \quad (9.13)$$

The study of the relationship of these three definitions of fractal dimension can itself also be generalized to form a single characterizing dimension which includes each of the three.

A measure of fractal dimension that has also been employed is the Renyi dimension [3, 8]. This is written in terms of the probabilities defined in Eq. (9.8). In this definition, the fractal dimension is given by a variable form written as [3, 8]

$$d_\alpha = \lim_{\epsilon \to 0} \frac{1}{1-\alpha} \frac{\ln \sum_{i=1}^{N(\epsilon)} P_i^\alpha}{\ln \frac{1}{\epsilon}} \tag{9.14}$$

where $N(\epsilon)$ is the smallest number of m –dimensional cubes of side ϵ which completely cover the object being studied and α is varied depending on the interests of the application. It is seen that by varying α a number of different fractal dimensions are arrived at. As we shall see, by varying α, the Renyi dimension allows us to relate the dimension d_{cap}, d_{inf}, and d_{corr} to each other [3, 8].

For example, in the limit $\alpha \to 1$ the Renyi dimension can be shown to reduce to the information dimension [3]. This is seen by applying L'Hopital's rule to the limiting form as $\alpha \to 1$ of the ratio in Eq. (9.14). For this application, the derivative of the denominator $1-\alpha$ in the $\alpha \to 1$ limit is -1, and the derivative of the numerator for $\alpha \to 1$ is given from

$$\lim_{\alpha \to 1+} \frac{d}{d\alpha} \ln \sum_{i=1}^{N(\epsilon)} P_i^\alpha = \lim_{\alpha \to 1+} \frac{d}{d\alpha} \ln \sum_{i=1}^{N(\epsilon)} e^{\alpha \ln P_i} = \lim_{\alpha \to 1+} \frac{1}{\sum_{i=1}^{N(\epsilon)} P_i^\alpha} \sum_{i=1}^{N(\epsilon)} \ln P_i e^{\alpha \ln P_i} \tag{9.15a}$$

$$= \lim_{\alpha \to 1+} \frac{1}{\sum_{i=1}^{N(\epsilon)} P_i^\alpha} \sum_{i=1}^{N(\epsilon)} P_i^\alpha \ln P_i = \sum_{i=1}^{N(\epsilon)} P_i \ln P_i$$

It then follows from Eq. (9.14) and L'Hopital's rule that [3]

$$d_{\alpha \to 1} = \lim_{\epsilon \to 0} - \frac{\ln \sum_{i=1}^{N(\epsilon)} P_i \ln P_i}{\ln \frac{1}{\epsilon}} = \lim_{\epsilon \to 0} \frac{I(\epsilon)}{\ln \frac{1}{\epsilon}} = d_{inf}. \tag{9.15b}$$

In addition to this limiting form, for $\alpha = 0$ it follows from Eq. (9.14) that in the case that $P_i = P$ is independent of i the fractal dimension d_0 is given by [3]

$$d_{\alpha \to 0} = \lim_{\epsilon \to 0} \frac{\ln N(\epsilon)}{\ln \frac{1}{\epsilon}} = d_{cap}, \tag{9.16}$$

and for $\alpha = 2$ the dimension $d_{\alpha \to 2}$ can also be shown [3] to be related to the measure $d_{corr} = \lim_{\epsilon \to 0} \frac{\ln C(\epsilon)}{\ln \epsilon}$ defined in Eq. (9.12).

In this regard, the Renyi dimension is useful for interpolating between a variety of previously defined dimensions of fractal structures. Applying this interpolation, it can quite generally be shown that [3]

$$d_{\alpha_1} \geq d_{\alpha_2} \text{ for } \alpha_1 < \alpha_2. \tag{9.17}$$

This is consistent with the inequalities given earlier in Eq. (9.13).

Finally, another definition of fractal dimension is the so-called Lyapunov dimension [1, 3, 8]. The method of this definition arrives at a fractal dimension based on the Lyapunov exponents associated with the stability of the fractal system. In this regard, remember that for the logistic mapping the

Lyapunov exponents of the chaotic attractor were positive [1]. This indicated the absence of stability on the attractor. The Lyapunov exponents typified the trajectories of the system as being in a constant state of diverging aperiodic motion confined to the chaotic attractor. For higher dimensional chaotic attractors, the Lyapunov exponents are more complicated than those of the logistic mapping but nevertheless can be used in a formulation describing the fractal nature of the chaotic attractor and the motion of the system confined to it. In this way, the use of the Lyapunov exponents has been made to define a fractal dimension of general strange attractors. The reader is referred to the literature for a further discussion of the Lyapunov fractal dimension [8].

9.4 SOME APPLICATION OF FRACTALS

In this section, some applications of fractal dimensions in the study of fractal geometry found in physical, growth, and biological processes are summarized [1–4, 12–22]. We begin with a continuation of the studies of chaotic dynamics presented in Chapter 8. The systems focused on there are dissipative systems [1] having a dynamics described by nonlinear difference or differential equations which contract spatial volumes as functions of time. (Note that chaos also is found in some non-dissipative, Hamiltonian, systems [1]. This, however, is not a topic here, and the reader is referred to the literature for discussions of these [1–4, 12–22].) In the treatment of such types of equations of motion we encountered an unexpectedly complex dynamics in which, following an initial period of transient behavior, the system settled down to steady state motions characterized by fixed-points and confining regions known as chaotic attractors [1–4]. This convergence of the systems to limiting motions about points or bounded regions of attraction is a direct manifestation of the dissipative properties of the dynamical equations [1].

Specifically, we found that the longtime motions of nonlinear systems fell into two categories depending on the regions of parameter space being studied [1–4]. In the first case, the dynamics involved a region of periodic motion associated with a discrete set of fixed-points of the nonlinear mapping. A second case focused on a region of aperiodic motion of the system confined to a region of space known as a chaotic or strange attractor. Of interest here is the development of the understanding of the geometry of chaotic attractors and how they contain the trajectories of the motion in space. To begin, though, we will revisit the notion of dissipative motion, characterizing the settling of the dynamics onto the fixed-points and chaotic attractors. This is followed by a treatment of the nature of the fractal geometry of chaotic attractors.

9.4.1 Dissipative Nature of the Chaotic Systems

Our first consideration is on the condensation of the system to its attractors [1]. This is associated with the dissipative properties of the mapping which tends to send an initial finite volume in the system to a limiting geometric structure of zero volume. In this sense, the attractors of dissipative systems are bounded regions in the range of motion, while chaos in non-dissipative systems occurs in the absence of attractors [1]. In the case of contraction to a strange or chaotic attractor, the nature of the reduced dimension of the fractal geometry of the attractor is then an important consideration in understanding the system dynamics. It represents an essential aspect of its geometry.

To see an example of the volume contraction in a dissipative system, consider the Lorenz mapping [1] given in Eq. (8.80). The mapping is a set of coupled equations for the dynamic variables (x, y, z) as functions of time, and the equations of motion are given by

$$\dot{x} = \sigma(y - x), \tag{9.18a}$$

$$\dot{y} = rx - y - xz, \tag{9.18b}$$

and

$$\dot{z} = xy - bz, \tag{9.18c}$$

where $\sigma, r, b > 0$ are constant parameters used to set the dynamics of the system [1]. Defining $\vec{r} = x\hat{i} + y\hat{j} + z\hat{k}$ the set in Eq. (9.18) can be rewritten in a more convenient form as

$$\dot{\vec{r}} = \vec{f} = \sigma(y - x)\hat{i} + (rx - y - xz)\hat{j} + (xy - bz)\hat{k}. \tag{9.19}$$

The point of our specific interest now is in the determination of the effects of Eq. (9.19) on a general volume element in the space of the variables (x, y, z).

Consider [1] the initial volume element $V = \Delta x \Delta y \Delta z$ in the form of a rectangular cuboid, and its development in time under the action of Eq. (9.18). Denote the outer surface of V as S and the outward unit normal to the surface by $\hat{n}$. Here $\hat{n}$ is defined at each point on the surface. It follows that the outward normal component of $\dot{\vec{r}}$ at any point on the surface is then given by the form [1]

$$\dot{r}_n = \hat{n} \cdot \dot{\vec{r}} = \hat{n} \cdot \vec{f} \tag{9.20a}$$

so that in a time dt the change in the volume V is given by

$$\begin{aligned} V(dt) &= V + \oint_S \dot{r}_n dt dS = V + \oint_S \dot{r}_n dt dS \\ &= V + \oint_S \hat{n} \cdot \vec{f}\, dS dt = V + \int_V \nabla \cdot \vec{f} dV dt, \end{aligned} \tag{9.20b}$$

where we integrate over the entire surface enclosing the volume V. Here in the last equality in Eq. (9.20b) we have used the divergence theorem [1, 5] to convert the surface integral to a volume integral. From Eq. (9.20b) it then follows that

$$\frac{dV}{dt} = \frac{V(dt) - V}{dt} = \int_V \nabla \cdot \vec{f} dV \tag{9.21}$$

so that the rate of change of the volume element is directly related to the volume integral. Consequently, we see that the nature of the volume change with time is determined by the divergence of $\vec{f}$ which must be negative for a contraction of the volume [1].

From Eq. (9.19) it follows that

$$\begin{aligned} \nabla \cdot \vec{f} &= \nabla \cdot \left[\sigma(y - x)\hat{i} + (rx - y - xz)\hat{j} + (xy - bz)\hat{k}\right] \\ &= -\sigma - 1 - b < 0. \end{aligned} \tag{9.22}$$

By setting our range of interest to $\sigma, r, b > 0$ we have effectively limited the dynamics to be that of a dissipative system [1]. It, consequently, is found from Eq. (9.21) that

$$\frac{dV}{dt} < 0 \text{ with } V(t) = Ve^{-(\sigma+1+b)t} \tag{9.23}$$

and the volume contracts at each step of the iteration to approach a set of measure zero as $t \to \infty$.

A similar dissipation is found in the Rossler mapping [1] but only over a more limited range of parameter space. In the Rossler mapping, the equations of motion are given by [1]

$$\dot{x} = -y - z, \tag{9.24a}$$

$$\dot{y} = x + ay, \tag{9.24b}$$

and

$$\dot{z} = b + xz - cz, \tag{9.24c}$$

where (x, y, z) are the space coordinates of the trajectories and a, b, c are constants [1]. Here

$$\nabla \cdot \vec{f} = \nabla \cdot \left[-(y+z)\hat{i} + (x+ay)\hat{j} + (b+xz-cz)\hat{k} \right] \tag{9.25}$$

$$= 0 + a + x - c,$$

and it is found that for a negative divergence $a + x < c$. In this regard, the region of dissipative behavior is more limiting and does not include the complete region over which the system of equations is studied [1].

As an example of a 2-dimensional dynamical system exhibiting dissipation, we next consider the so-called Henon mapping [1, 3]. This mapping is defined in the 2-dimensional space of variables (x, y) and is given by the difference equations [1]

$$x_{n+1} = y_n + 1 - a x_n^2 \tag{9.26a}$$

$$y_{n+1} = b x_n \tag{9.26b}$$

where a and b are parameters setting the nature of the motion. It can be shown that under appropriate conditions the Henon mapping is dissipative, exhibiting a strange or chaotic attractor [1]. We now discuss some of the interesting properties of the dissipation.

To begin with, we show that the Henon mapping is dissipative. In this regard, consider the structure of the general mapping

$$x_{n+1} = f_1(x_n, y_n) \tag{9.27a}$$

$$y_{n+1} = f_2(x_n, y_n). \tag{9.27b}$$

For this form we determine the effects of applying a small variation $(\Delta x, \Delta y)$ to (x_n, y_n) by expanding Eq. (9.27) to first order in a Taylor series [1, 3]. In this expansion, it follows that Eq. (9.27) becomes

$$x_{n+1} + \Delta x' = f_1(x_n + \Delta x, y_n + \Delta y) \tag{9.28a}$$

$$= f_1(x_n, y_n) + \Delta x \frac{\partial f_1(x_n, y_n)}{\partial x_n} + \Delta y \frac{\partial f_1(x_n, y_n)}{\partial y_n}$$

and

$$y_{n+1} + \Delta y' = f_2(x_n + \Delta x, y_n + \Delta y) \tag{9.28b}$$

$$= f_2(x_n, y_n) + \Delta x \frac{\partial f_2(x_n, y_n)}{\partial x_n} + \Delta y \frac{\partial f_2(x_n, y_n)}{\partial y_n}$$

Using Eq. (9.28) we next relate the variation of the area, $A = \Delta x \Delta y$, about (x_n, y_n) to the area A' about (x_0', y_0') mapped onto.

To understand how $A = \Delta x \Delta y$ is mapped into A', we first consider how the variation $(\Delta x, 0)$ about (x_n, y_n) maps under Eq. (9.28). It follows from Eq. (9.28a) that under this mapping [1, 3]

$$\Delta x' = \Delta x \frac{\partial f_1(x_n, y_n)}{\partial x_n}. \tag{9.29a}$$

and similarly, from Eq. (9.28b) that

$$\Delta y' = \Delta x \frac{\partial f_2(x_n, y_n)}{\partial x_n}. \tag{9.29b}$$

Consequently, from a consideration of $\Delta y = 0$ it is found that

$$(\Delta x', \Delta y') = \left(\frac{\partial f_1(x_n, y_n)}{\partial x_n}, \frac{\partial f_2(x_n, y_n)}{\partial x_n} \right) \Delta x. \tag{9.30}$$

Likewise, it follows that the variation $(0, \Delta y)$ about (x_n, y_n) under Eq. (9.28) maps into

$$(\Delta x', \Delta y') = \left(\frac{\partial f_1(x_n, y_n)}{\partial y_n}, \frac{\partial f_2(x_n, y_n)}{\partial y_n} \right) \Delta y. \tag{9.31}$$

Consequently, the two vector $\Delta \vec{r}_1 = (\Delta x, 0)$ and $\Delta \vec{r}_2 = (0, \Delta y)$ are, respectively, found to map into the two vectors [1, 3]

$$\Delta \vec{r}_1^{\,'} = \left(\frac{\partial f_1(x_n, y_n)}{\partial x_n}, \frac{\partial f_2(x_n, y_n)}{\partial x_n} \right) \Delta x \tag{9.32a}$$

and

$$\Delta \vec{r}_2^{\,'} = \left(\frac{\partial f_1(x_n, y_n)}{\partial y_n}, \frac{\partial f_2(x_n, y_n)}{\partial y_n} \right) \Delta y. \tag{9.32b}$$

The resultant vectors in Eq. (9.32) then determine the area A' of the parallelogram about (x_0', y_0') generated under the mapping.

The area of the square defined by $\Delta \vec{r}_1 = (\Delta x, 0)$ and $\Delta \vec{r}_2 = (0, \Delta y)$ is obtained from the vector identity

$$A = \left| \Delta \vec{r}_1 \times \Delta \vec{r}_2 \right| = \Delta x \Delta y, \tag{9.33a}$$

while the area of the parallelogram mapped into by Eq. (9.27) is given from the same identity by

$$A' = \left| \Delta \vec{r}_1^{\,'} \times \Delta \vec{r}_2^{\,'} \right| = \left| \frac{\partial f_1(x_n, y_n)}{\partial x_n} \frac{\partial f_2(x_n, y_n)}{\partial y_n} - \frac{\partial f_1(x_n, y_n)}{\partial y_n} \frac{\partial f_2(x_n, y_n)}{\partial x_n} \right| \Delta x \Delta y \tag{9.33b}$$

$$= \left| \frac{\partial f_1(x_n, y_n)}{\partial x_n} \frac{\partial f_2(x_n, y_n)}{\partial y_n} - \frac{\partial f_1(x_n, y_n)}{\partial y_n} \frac{\partial f_2(x_n, y_n)}{\partial x_n} \right| A.$$

It is seen that the nature of the area increases or decreases in transforming A into A' is set by the factor [1]

$$\left| \frac{\partial f_1(x_n, y_n)}{\partial x_n} \frac{\partial f_2(x_n, y_n)}{\partial y_n} - \frac{\partial f_1(x_n, y_n)}{\partial y_n} \frac{\partial f_2(x_n, y_n)}{\partial x_n} \right|. \tag{9.34}$$

This factor is just the Jacobian matrix of the system and is commonly applied to change variables in integration transformations [1, 5].

Upon evaluating Eq. (9.34) for the mapping in Eq. (9.26), it is found that

$$\left| \frac{\partial f_1(x_n, y_n)}{\partial x_n} \frac{\partial f_2(x_n, y_n)}{\partial y_n} - \frac{\partial f_1(x_n, y_n)}{\partial y_n} \frac{\partial f_2(x_n, y_n)}{\partial x_n} \right| = |b|. \tag{9.35}$$

Consequently, it follows that for [1]

$$|b| < 1 \tag{9.36}$$

the Henon mapping is dissipative. In the region of $|b| < 1$, areas of the system are mapped into smaller areas. Depending on the value of the other mapping parameter a, eventually the trajectories of the mapping can develop chaotic motion focused about a chaotic attractor [1]. In this respect, the chaotic attractor is ultimately a fractal geometry limiting the motion of the system trajectories.

We next turn to a discussion of the fractal geometry of the chaotic attractors developed under the $1-, 2-,$and $3-$dimensional mappings studied earlier.

9.4.2 Fractal Dimensions of Chaotic Attractors

We next consider the nature of the fractal geometry of the chaotic attractors for the nonlinear mappings we have considered in this and the previous chapter. These have involved dynamical systems characterized by $1-, 2-,$ and $3-$dimensional mappings. In general, the chaotic attractors in each of these systems arise through dissipative processes which limit their dynamics to sets with quite different geometric properties than those embedding the attractors. This difference is determined by a study of the fractal dimensions of the attractors, and as we shall see, the resulting fractal dimensions are found to be always less than the dimensionality of the mappings generating the dynamics of the system. We begin with the logistic mapping, which is the simplest mapping displaying a chaotic attractor.

The first mapping of our considerations was the logistic mapping [1, 3] given in Eq. (8.1) and used to generate the dynamics shown in Figure 8.2. This is a $1-$ dimensional nonlinear mapping of the $x-$axis into itself considered in the space (μ, x) where μ is a parameter setting the motion. The mapping was shown to exhibit a period-doubling process as μ increased from 0 that eventually led to a transition to chaotic motion [1] at $\mu_\infty = 3.569946$. Following the transition of the system to chaos, it was found with increasing μ that the chaos was briefly interrupted intermittently [1] by small windows of periodic behavior up to $\mu = 4$. This marked the extreme limit of our treatment of the system dynamics of the mapping.

The fractal dimension of the chaotic attractor for the logistic mapping has been studied by all the methods introduced earlier for its determination [23]. These include the evaluation of d_{cap} in Eq. (9.1), d_{inf} in Eq. (9.8), and for d_{corr} in Eqs. (9.10) through (9.12); and the generation of all of these is based on the extensive determination of numerical trajectory data. This requires a significant amount of data in order to develop a reasonable certainty of the result. In addition, the ε of the cube sides for the evaluation of the limits in these definitions must be varied over a large range to arrive at an accurate value of the dimension. With these considerations at $\mu_\infty = 3.569946$ it is found [23] for the logistic mapping that $d_{cap} = 0.538$, $d_{inf} = 0.517$, and $d_{corr} = 0.5 \pm 0.005$. Consequently, the chaotic attractor of the logistic mapping is seen to be even more tenuous than that of the Cantor dust which was shown in Eq. (9.3) to have $d_{cap} = 0.6309$. In this regard, the logistic attractor is more like a set of isolated points scattered along the x-axis than the Cantor dust, and much less of a line segment.

The next more complex geometry we considered was that of the Henon mapping [1, 3, 23], characterized by the system of equations in Eq. (9.26). This is defined over the 2-dimensional space (x, y) with the dynamics controlled by a parameter set (a, b). In this regard, it was found that the system exhibits dissipative effects for $|b| < 1$. Due to the two parameters (a, b) setting the motion, the nature of the dynamics of the mapping is more complex to describe than that of the logistic mapping where the dynamics is set by the single parameter μ.

In the region of dissipative effects, the Henon mapping displays a range of the parameter a with fixed-points and with a period-2 effect [1]. These leads through a period-doubling process to chaotic motion with a chaotic attractor and intermittent windows in the chaos in which periodic motion is found [1]. In this regard, the Henon mapping resembles the logistic mapping. The fractal dimension of the chaotic attractor has been determined [23] for d_{cap} and d_{corr}. Specifically, for the chaotic attractor at $(a = 1.4, b = 0.3)$ it is numerically found for the Henon mapping [23] that $d_{cap} = 1.26$ and $d_{corr} = 1.21 \pm 0.01$. Consequently, the chaotic attractor is a little more than a line segment in space but much less than a solid area.

Continuing to increase the system complexity, the next more complex geometry considered was that of the Lorenz equations [1] defined in Eq. (9.18). These display motions which are distinctly different from the logistic and Henon mappings. The Lorenz equations are a set of differential equations over the 3-dimensional space (x, y, z) characterized by a parameter set $\sigma, r, b > 0$. The system is found to exhibit dissipative effects in the region defined by $-\sigma - 1 - b < 0$.

The Lorenz system, as we have discussed, has a fixed-point region, a period-2 region, and then goes over directly to a transition to chaos [1]. This leads for $r_H = \dfrac{\sigma(\sigma + b + 3)}{\sigma - b - 1} < r$ to a chaotic motion with a chaotic attractor. Specifically, the chaotic attractor has been studied numerically, and it is found for the Lorenz equations [23] that $d_{cap} = 2.06 \pm 0.01$ and $d_{corr} = 2.05 \pm 0.01$. Consequently, the attractor is a little more than a solid area in space but much less than a solid volume.

Related to the Lorenz equations is a simplification of them (given in Eq. (9.24)) known as the Rossler system of differential equations [1, 3]. These are defined in 3-dimensions (x, y, z) and are characterized by a parameter set a, b, and c. The Rossler system has a complex dynamics because of the three parameter characterization and the increased number of dimensions. It exhibits some of the interesting period-doubling to chaos effects found in the logistic mapping [1]. These lead to chaotic motion with a chaotic attractor and intermittently to small windows of periodic dynamics within the region of chaos [1]. Specifically, for the parameters $a = 0.1$, $b = 0.1$, and $c = 14$ the numerically determined fractal dimension [23] is $d_{cap} = 2.015 \pm 0.005$. As with the Lorenz equations, this is a little more than the 2-dimensional nature found in a solid area.

In general, we see that the mappings and differential equations displaying chaotic solutions we have studied have been defined in 1-, 2-, and 3-dimensions. The chaotic behaviors they display are found to be confined about bounded regions of space known as the chaotic attractors of the system. These chaotic attractors possess fractal geometry with a fractal dimension $d_{fractal}$ which is fractional in value and such that in a d-dimensional dynamic system it is found that $d - 1 < d_{fractal} < d$. This is then a statement of the essential mathematics of the systems treated.

We next turn to a discussion of some of the physical, chemical, and biological applications that have been made of the ideas of chaotic dynamics and the development of their understanding in terms of fractal structures. These are focused on the view that the trajectories of chaotic motion unfold geometrically in bounded regions of space characterized by fractal attractors. It was the picture developed mathematically in the models introduced in Chapter 8 and earlier in this chapter. For each model, a brief discussion is now given of the physical processes providing the basis upon which the models studied were originally formulated.

9.5 APPLICATIONS OF CHAOS AND ITS RELATIONSHIP TO FRACTAL ATTRACTORS

One of the earliest treated chaotic systems was that of the logistic mapping [1–4]. This mapping originally evolved from a consideration of the study of population dynamics. In this regard, the simplest model of population was proposed by Malthus and considered the population dynamics as a development based solely on a rate of either population increase or decrease. This view determined that the population changed at a rate proportional to the existing level of population, increasing indefinitely or otherwise decreasing to zero [4].

This, of course, turns out to be an oversimplified view, and the first correction to the oversimplification presently will bring us to the study of the logistic equation. As we have seen, the dynamics of the logistic equation can be extremely complex so that we seem to proceed from an oversimplified view to what appears to be an overly complex process. Let us then turn to a discussion of the simple considerations which led us to the logistic equation as a description of population dynamics.

The Malthus-based treatment of population considered the population at a series of equal time steps so that the population at the $(n+1)^{th}$ time step P_{n+1} was related to that at the nth time step P_n recursively by [1, 4]

$$P_{n+1} = \mu P_n. \tag{9.37}$$

Note, in this regard, that for our considerations the variable P_n is formulated as a type of measure of population appropriate to the system being treated [1, 4]. In addition, the constant $\mu > 0$ is a parameter setting the rate of growth or decay in the system and relating the generated population at any time step to that at the previous step.

The solutions of Eq. (9.37) for $n > 0$ is then given by the form

$$P_n = P_0 \mu^n. \tag{9.38}$$

It follows from the solution in Eq. (9.38) that for the case $0 < \mu < 1$ the population decreases to zero with increasing n, and in the case that $\mu > 1$ the population increases exponentially with increasing n. As we have noted, this is a highly simplistic description because populations exhibit many types of fluctuations other than just simply increasing or decreasing. The reason for this is that it is unreasonable to regard the parameter μ as a constant during each time step. Rather, it must be dependent on a number of factors so that $\mu(P,\ldots\ldots)$ is really a complicated function of the population and the environment around the population. (Note, here in $\mu(P,\ldots\ldots)$ the variable P represents the population at an unspecified time.)

Rather than making a detailed study of the factors affecting population, let us assume a form for $\mu(P,\ldots\ldots)$ which limits the unlimited growth of population in a simple way. The easiest consideration in this regard is to treat $\mu(P,\ldots\ldots) \approx \mu(P)$ and to Taylor expand $\mu(P)$ in P so that in a truncated series

$$\mu(P) = \alpha - \beta \mathrm{P}. \tag{9.39}$$

Here $\alpha, \beta > 0$ and the sign of the second term is chosen to limit the population growth. We shall take the simple limiting form in Eq. (9.39) as a general statement for the development of the population in our corrected model. This, of course, is an approximation of the population growth.

For our correction to Eq. (9.37) we next replace μ in Eq. (9.37) with the form $\mu(P) = \alpha - \beta P$ in Eq. (9.39). Our resulting new model of population dynamics then takes the form

$$P_{n+1} = (\alpha - \beta P_n) P_n = \alpha \left(1 - \frac{\beta}{\alpha} P_n\right) P_n. \tag{9.40}$$

If we now set $P_n = \gamma x_n$, it follows in the new set of population variables $\{x_n\}$ that Eq. (9.40) becomes

$$x_{n+1} = \alpha \left(1 - \frac{\beta\gamma}{\alpha} x_n\right) x_n, \tag{9.41}$$

and choosing $\gamma = \frac{\alpha}{\beta}$ then gives

$$x_{n+1} = \alpha (1 - x_n) x_n. \tag{9.42}$$

This is just the logistic equation in Eq. (8.1) with μ re-denoted as α.

We see then that the logistic equation arises from a very simple and realistic restriction on the form of the growth rate. In this regard, it is well known that populations cannot exhibit unlimited growth so that the basic form of the logistic equation is a most reasonable population model. In a still simple model, it displays ranges of periodic behaviors, and these are found to naturally arise solely from the restrictions on the rate of growth. Such periodic behaviors are found to be completely absent in the simple Malthus treatment of population. Despite its simplicity, the logistic mapping also generates highly unusual chaotic behaviors exhibited by its solutions. These are initially unexpected.

In regard to the application of the ideas of fractals in the study of populations, it is interesting to also note that the ideas of fractal geometry not only are applied to population studies characterizing the time trajectories of population density on a chaotic attractor, e.g., in the dynamics of the logistic mapping. They have, in addition, found application in the characterization of, for example, the spatial spreading of a population in a city or on a petri culture [1, 4, 14]. This is part of, for example, a morphological study of population and acts as a point of comparison in the classification of the growth characteristics of different populations or physical systems.

These types of morphological treatments can further be developed to characterize cracks and crack propagations as well as other types of general growth processes [1–4, 11, 14]. In addition, they are important in the study of phase transitions [12, 19]. At the critical point of a second-order phase transition, the fluctuations in the thermodynamic system are found to exhibit fractal geometry. Here, the correlation length, which is commonly measured in scattering from the system, diverges at the critical point of the phase transition. The divergence leaves the system without a natural length scale (i.e., the scale set by the correlation length is absent) and provides the fluctuations of the system with a scale-invariant self-similar geometric structure. This absence of a natural scale is a characteristic of its fractal geometry.

Population studies and growth processes are not the only systems in which chaotic motion and fractal geometry are important aspects of the system dynamics. Other types of physical and chemical systems can include a variety of processes displaying variable rates of change arising in the context of a nonlinear spatial or temporal interaction. These give rise in some cases to behaviors reminiscent of those observed in the logistic equation. In this regard, many types of physical, chemical, or electrical dynamics incorporate nonlinearities in their interactions which then cause them to display regions of chaos [1–4].

Early types of systems exhibiting chaos also focused on fluid mechanics, which is known to have important nonlinear aspects to its dynamics [1–4, 7, 10]. The combination of physical flow with a variety of chemical equilibrium processes has led in some instances to systems exhibiting flow patterns of different chemical natures [4]. (Note: An example of such a chemical-fluid system is observed experimentally in the Belousov-Zhabotinsky reaction and theoretically in the Brusselator Model [1–4]. Both of these systems are based on a chemical system which can develop various spatial and temporal patterns arising from the nonlinear chemical and fluid dynamics of its constituent chemicals. We shall say more about these model systems at the end of this chapter.) Transitions between various patterns in such systems can be made based on the physical parameters of the materials forming them. As well, there are many systems found in basic classical and quantum mechanics which, because of their nonlinear natures, develop chaotic behaviors.

We shall next turn to a consideration of a few additional examples of such other types of systems which have been important in the development of the study of chaos. These involve developments in the context of the more complex models of chaos that we have studied. Here only the highlights are given, and the reader is referred to the literature for a more detailed listing [1–4].

9.5.1 Turbulence in a Slab of Mercury

An early interest in the process of period-doubling to chaos involved the study of the development of turbulence in fluids [1, 4]. Fluid mechanics involves systems generally exhibiting a variety of types of nonlinearities so that they provide natural systems in which to study chaotic transitions. A common example of this, which can involve only a single chemical species, is Rayleigh-Benard convection found in planar layers of fluid [1, 4]. In this process convective cells based on the upward and downward motion of fluid driven by a temperature gradient are observed in the fluid layer. In this regard, some of the earliest studies of chaos have focused on nonlinear fluid mechanics as well as processes involving chemical reactions in fluids [4]. We shall begin by considering a fluid involving a single chemical species.

In one early experiment illustrating chaotic processes, the convective properties of a slab of mercury were studied [1, 4]. For these studies, a horizontal layer of mercury was heated from below, creating a temperature gradient between the top and the bottom of the slab of mercury. The state of the temperature gradient was characterized by a fluid mechanics parameter known as the Rayleigh number, r. As the temperature gradient and its respective Rayleigh number r were increased in the system, a transition was made to a state of convective motion involving a one-dimensional pattern of horizontally alternating upward and downward regions of convective motion of mercury along the slab [1].

In the initial transition from a uniform slab of mercury to a patterning of convective upward and downward moving cells, the temperature at any point in the convective slab initially remained constant. However, with increasing r another transition was made to a state in which at a point in the convective layer there was a periodic change in temperature with time. As r was increased further, transitions to temperature states of higher periodicity in time were made. These period-doubling transitions occurred until a state of chaos (i.e., a turbulent motion) was reached [1]. This process is highly reminiscent of the features found in the logistic mapping, and the mercury slab, in fact, provided one of the early examples of the transition to chaos.

One of the important applications of fluid dynamics is in the study of weather and atmospheric turbulence. This was a focus which led to some of the earliest ideas of chaos.

9.5.2 Weather and the Study of the Lorenz Equations

In another study of chaos, the Lorenz equations were originally developed as a simplified model of convection rolls in the atmosphere [1–4]. The Lorenz system was a seminal study of chaotic systems and was introduced early on in the treatment of chaotic dynamics. In this regard, it was fashioned as a type of minimalist representation of what is essentially going on in fluid systems which are ultimately subject to turbulence. It has a long history with many applications to real and contrived systems meant to model its features. More recently, it has been used to illustrate features of laser dynamics, the dynamic properties of dynamos, and in certain contrived models of waterwheels and similar mechanical devices [1].

In another category of models, some models of chaos have been introduced solely to provide a simplification in the study of the general ideas of chaos. These tend to be formulated with a focus on determining the specifics of what goes on in the motion along trajectories contained within the chaotic attractors of dynamical systems.

9.5.3 Models Developed to Aid in the Understanding of Chaotic Dynamics

The other models we have treated (i.e., the Henon mapping and the Rossler equations) were introduced as simplifications of the dynamics of the Lorenz model or as simplified models to help develop an understanding of the dynamics and geometry of chaotic attractors [1–4, 6–8, 10]. They have had less application as direct models of realistic chaotic physical systems but provide an understanding of the behaviors of more complex models of realistic systems.

A problem of importance in the approach offered by these models was a determination of the dynamics on the chaotic attractor. A focus here regards the divergence of trajectories with slightly different initial conditions and how the subsequent motion of these divergent trajectories could be confined to a bounded chaotic attractor. By studying simplified models, the dynamics on the attractor were shown to exhibit a variety of stretching, contractings, and remappings of the system dynamics over and over again into the bounded attractor [1–4, 6–8, 10]. These processes aid in confining trajectories which nevertheless exhibited exponential divergences from close initial states of the system. For the details of these types of discussions the reader is referred to the literature [1 –4, 6–8, 10].

9.5.4 The Belousov-Zhabotinsky Reaction and the Brusselator Model

As a final system of interest, chemical reactions in fluid systems are next mentioned in the context of chaos [1–4, 13]. These represent the combination of two types of dynamical processes, combining both spatial and chemical changes. They are interesting as they provide a variety of physical changes offering colorful displays of chaotic properties and generating various spatial patterns with interesting time evolutions.

The Belousov-Zhabotinsky reaction [1–4, 13] is from a class of non-equilibrium chemical reactions involving malonic acid, bromate ions, and a catalyst of cerous or ferrous ions. Originally, the system was found to exhibit periodic color changes in time over a fixed time scale of minutes. From this seminal study, a variety of systems have been experimented on, which are formed by injecting chemicals into reaction vessels or formed as a layer in a petri dish. From these studies, an interesting set of pattern formations which evolve in time are generated. These involve periodic behaviors in time and space, and ultimately a chaotic state has been observed in them. The reactions provide an important example of chaos generated from both nonlinear fluid mechanics and nonlinear chemical equilibrium.

The Brusselator Model, on the other hand, is an early theoretical model developed to understand nonlinear chemical equilibrium [1–4, 13]. It, consequently, goes into the understanding of some of the processes in the Belousov-Zhabotinsky reaction-based systems [1–4, 13]. Specifically, the model looks at a system in which the dynamics are set by a series of chemical reactions. These have the general form

$$A \xrightarrow{k_1} X, \tag{9.43a}$$

$$B + X \xrightarrow{k_2} Y + C, \tag{9.43b}$$

$$2X + Y \xrightarrow{k_3} 3X, \tag{9.43c}$$

and

$$X \xrightarrow{k_4} D. \tag{9.43d}$$

Here A, B, C, D, X, Y are chemical species and k_1, k_2, k_3, k_4 their reaction rate constants. The focus in this set of ideal reactions is on the chemical species X, Y which are the interesting dynamical variables of the system. The other species A, B, C, D are usually kept at fixed concentrations in the study of Eq. (9.43).

In the configuration of the reaction in Eq. (9.43), the rate equations for the dynamical variables X, Y are then given by [1–4, 13]

$$\frac{d[X]}{dt} = k_1[A] - k_2[B][X] + k_3[X]^2[Y] - k_4[X] \tag{9.44a}$$

and

$$\frac{d[Y]}{dt} = k_2[B][X] - k_3[X]^2[Y]. \tag{9.44b}$$

These follow based on the laws of chemical equilibrium. Here $[A]$, $[B]$, $[C]$, $[D]$ are the constant concentrations of A, B, C, D and $[X]$,$[Y]$ are the time varying concentrations of X, Y. Consistent with our earlier treatment of the nonlinear systems of differential equations, the study of Eq. (9.44) begins with a determination of the fixed points of Eq. (9.44).

In a first study [1–4, 13] of Eq. (9.44) we see that by setting $\frac{d[X]}{dt} = \frac{d[Y]}{dt} = 0$ a fixed-point of the rate equation, it is obtained. The fixed points then are obtained as

$$[X]_0 = \frac{k_1}{k_4}[A] \tag{9.45a}$$

and

$$[Y]_0 = \frac{k_2 k_4}{k_1 k_3}[A]. \tag{9.45b}$$

This study can be continued as in our earlier treatment to a determination of the stability of these fixed points and to a generation of a further more complex geometric and time dependence. For this, the reader is referred to the literature for a more detailed consideration of the Brusselator system [1–4, 13].

9.6 FRACTONS

As a last consideration of fractals, we look at some of the properties of excitations in systems that have natural fractal geometries rather than the fractal geometries of chaotic attractors arising from nonlinear dynamics [15–22]. Unlike the earlier discussed dynamic systems which incorporated the ideas of fractal geometry in understanding the chaotic properties of their nonlinear dynamics, systems with natural fractal geometry can also support physical excitations which are affected by the fractal nature of the system. Some examples of such systems which can exhibit a natural fractal geometry are those involving polymers, geometric disorder, or which have origins in growth processes. The excitations developed in these types of natural fractal systems are known as fractons, and the fracton dynamics on naturally occurring fractals may be either linear or nonlinear in nature [1–4, 13].

Next, we briefly summarize some of the basic studies associated with the fracton excitations, focusing on their thermodynamic properties. In this regard, fracton excitations involving phonons, magnons, and other types of topological excitations have been investigated [22]. Here, however, we shall limit ourselves to address the vibrational properties of some studies made on silica aerogels [16], though the basic theory was generated for polymers, glasses, and other general materials with fractal geometry [15]. In these studies, only the low-temperature properties are treated.

In 1–, 2–, and 3–dimensional crystalline materials, it is known that in the Debye approximation the low frequency vibrational density of states has the form

$$D_d(\omega) \propto \omega^{d-1} \tag{9.46a}$$

where d is the dimension of the crystalline material. This form characterizes many of the very low-temperature limits of statistical properties arising from phonons in the system, e.g., such common properties as the specific heat and thermal conductivity. In this regard, it is found in the Debye approximation that the specific heat has the form

$$C_d(\omega) \propto T^d \tag{9.46b}$$

where again d is the dimension of the crystalline material.

In fractal systems [15, 16] we might expect that the fractal dimension $\bar{d}$ computed from the distribution of mass in space would replace the space dimension d in Eq. (9.46). (Note, here that the mass fractal dimension $\bar{d}$ is defined by the relationship [15, 16]

$$M(r) \propto r^{\bar{d}} \tag{9.47}$$

where $M(r)$ is the mass of the system within a radius r. This is seen to be an extension of the mass-radius relationship for homogeneous materials for the case of a 1–, 2–, or 3–dimensional material. Such a replacement, however, would account for the fractal structure only partially filling the space in which it is embedded. Consequently, the specific heat would exhibit a fractional power law with a fractional power related to the mass fractal dimension.

This, however, is an oversimplification of the problem [15, 16]. The oversimplification is not surprising as the excitations in a system are determined to a great extent not only by the Hamiltonian of the system but also by the boundary conditions on the modal solutions. These can be quite involved for an excitation on a fractal network. Consequently, the focus of our considerations should be on obtaining a correct expression for the density of states in Eq. (9.46b).

In this regard, however, it is found that a so-called spectral fractal dimension $\tilde{d}$ can be measured in the system [15, 16]. This is obtained from the density of states using the form

$$D_{\tilde{d}}(\omega) \propto \omega^{\tilde{d}-1}. \tag{9.48a}$$

In terms of the spectral dimension, it then follows that the specific heat is given by

$$C_{\tilde{d}}(\omega) \propto T^{\tilde{d}}. \tag{9.48b}$$

Here $\tilde{d}$ is again a fractional dimension, but it is generally different from $\bar{d}$.

Similarly, relations are also found for the thermal conductivities of fractal systems which at low temperatures have a power law dependence on the temperature [15, 16]. These are again determined from the spectral fractal dimension obtained from the density of states. Generally, the relationship between $\tilde{d}$ and the mass fractal dimension $\bar{d}$ can be complex. However, in some materials, studies have been made which provide a relationship [16] between the two dimensions $\tilde{d}$ and $\bar{d}$.

As an example of this general approach, a variety of experimental studies have been made on silica aerogels [16]. These gels are examples of fractal materials in which both the spectral and mass fractal dimensions have been determined. In these systems the specific heat was measured [16] to have a $T^{1.4}$ temperature dependence below $1^{\circ}K$ and the thermal conductivity was measured [16] to have a $T^{1.2}$ temperature dependence between $1^{\circ}K$ and $2^{\circ}K$. Both these results indicate that the fractal geometry characterizing the material is a little more than $1-$dimensional in its nature.

REFERENCES

1. Strogatz, S. H. 1994. *Nonlinear Dynamics and Chaos* (Perseus Books, Reading).
2. Feldman, D. P. 2012. *Chaos and Fractals: An Elementary Introduction* (Oxford University Press, Oxford).
3. Rasband, S. N. 1990. *Chaotic Dynamics of Nonlinear Systems* (John Wiley & Sons, Inc, New York).
4. Nicolis, G. 1999. *Introduction to Nonlinear Science* (Cambridge University Press, Cambridge).
5. Arfken, G. B. and Weber, H. J. 2005. *Mathematical Methods for Physicists*, 6th Edition (Elsevier, Burlington).
6. Chen, S., Feng, S., Fu, W., and Zhang, Y. 2021. Logistic Map: Stability and Entrance to Chaos. *Journal of Physics Series* 2014, 012009. https://doi.org/10.1088/1742-6596/2014/1/012009.
7. Hirsch, M. W., Smale, S., and Devaney, R. L. 2013. *Differential Equations, Dynamical Systems, and an Introduction to Chaos* (Elsevier, Amsterdam).
8. Theile, J. 1990. Estimating Fractal Dimension. *Journal of the Optical Society of America A* 7, 1056–1073.
9. Edgar, G. 2008. *Measure, Topology, and Fractal Geometry* (Springer, New York).
10. Semyon, M., Bakhanova, Y., Kazakov, A., Pusuluri, K., and Shilnikov, A. 2020. Homoclinic Chaos in the Rossler Model. *Chaos* 30, 113126.
11. Levey-Vehel, J. and Lutton, E. 2005. *Fractals in Engineering: New Trends in Theory and Applications* (Springer, London).
12. Suzuki, M. 1983. Phase Transitions and Fractals. *Progress in Theoretical Physics* 69, 65–76.
13. Gyorgyi, L. and Field, R. 1992. A Three-Variable Model of Deterministic Chaos in Belousov-Zhabotinsky Reaction. *Nature* 355, 808–810.
14. D'Acci, L. 2019. *The Mathematics of Urban Morphology* (Birkhauser, Springer Nature, Switzerland).
15. Alexander, S., Laemans, C., Orbach, R., and Rosenberg, H. M. 1983. Fracton Interpretation of Vibrational Properties of Cross-Linked Polymers, Glasses, and Irradiated Quartz. *Physical Review B* 28, 4615–4619.
16. De Goer, A. M., Calemczuk, R., Salac, B., Bon, J., and Bonjour, E. 1989. Low-Temperature Energy Excitations and Thermal Properties of Silica Aerogels. *Physical Review B* 40, 8327–8335.
17. Razdan, A. 2007. On Non-Extensive Nature of Thermal Conductivity. *Pramana Journal of Phsyics* 68, 61–65.
18. Tel, T. 1988. Fractals. Multifractals, and Thermodynamics: An Introduction. *Zeitschrift fur Naturforschung A* 43, 1154–1174.
19. Gaspard, P. *Chaos, Scattering, and Statistical Mechanics* (Cambridge University Press, Cambridge).
20. Kinsner, W. A 2007. Unified Approach to Fractal Dimensions. *International Journal of Cognitive Information and Natural Intelligence* 1(4), 26–46.
21. Genzor, J., Gendiar, A., and Nishino, T. 2016. Phase Transition of the Ising Model on Fractal Lattice. *Physical Review E* 93, 012141.
22. Nandkishore, R. M. and Hermele, M. 2019. Fractons. *Annual Review of Condensed Matter Physics* 10, 295–313.
23. Grassberger, P. and Procaccia, I. 1983. Characterization of Strange Attractors. *Physical Review Letters* 50, 346–349.

10 Quantum Monte Carlo Simulation

In this chapter we develop some interesting techniques that have been applied in Monte Carlo based computer simulation treatments of the thermodynamics of quantum mechanical systems [1–3], i.e., the focus is on the recent development of methodologies for performing so-called quantum Monte Carlo simulation studies. In general quantum mechanical systems are difficult to treat by standard Monte Carlo methods because quantum systems are composed of non-commuting operators rather than variables characterized as real or complex numbers [1–3]. The noncommutativity of the quantum operators is found to significantly increase the complexity of the system to be numerically studied over those of classical mechanical systems. These difficulties have been found to require the development of a number of methods to facilitate the formulation of rapid and accurate algorithms for the generation of precise numerical results on the thermodynamics of quantum systems.

In this regard, quantum Monte Carlo simulation has been a topic of intense study over many years so that the present chapter is meant to only offer an introduction to the most basic ideas of the simulation techniques and to illustrate their applications on a number of interesting basic quantum many-body problems [1]. Consequently, the present discussions are not intended to provide a detailed review of all the techniques available to develop algorithms for the simulation of many-body problems but only to offer a flavor of what is available. The applications, for example, of quantum Monte Carlo to systems of identical fermions as well as their development in the study of magnetic materials are not a consideration here as these systems require special considerations peculiar to their specialized features. For these and other recent topics of quantum Monte Carlo application the reader is referred to the literature (e.g., see for example the collection of papers in the book of Ref. [3]).

In the considerations presented here the focus is on the application of path-integral techniques in the development of algorithms for the study of the thermodynamics of quantum systems [1]. In this approach the quantum mechanical problem of a many-body system written in terms of a Hamiltonian operator in a d-dimensional space is rewritten in terms of a path integral in $d+1$-dimensional space expressed solely in terms of classical mechanical variables [1, 2]. Consequently, a 3-dimensional problems is rewritten as a 4-dimensional path integral, a 2-dimensional problems is rewritten as a 3-dimensional path integral, etc. Since the path integrals are written in terms of complex (i.e., classical mechanical) variables, the complexity of the non-commuting operators in the quantum problem is exchanged for the relatively less complex problem of an essentially classical mechanical system with an additional dimension. While the evaluation of the path integral may be complex it nevertheless is treatable by standard Monte Carlo methods.

In the course of the treatment, we shall find that the path integral formulation for the thermodynamic properties of a quantum system (i.e., its partition function) is closely related to the path integral formulation for the quantum mechanical dynamics of the system (i.e., its time propagator) [4]. The two problems are essentially related to one another by a process of analytical continuation, and the development of each technique is to a large extent paralleled in the development of the other. Specifically, in the path integral formulation of the quantum mechanical dynamics of a system, the path integral propagator is written in terms of the time evolution operator of the system [4]. This develops the dynamics of the system as a sequence of infinitesimal time steps [4]. Similarly, in the path integral formulation of the thermodynamics of quantum mechanical systems, the partition function of the system is expressed in terms of a path integral written as a sequence of incremental infinitesimal temperature changes made in the development of the temperature in the system [4].

DOI: 10.1201/9781032674919-10

This essentially regards the temperature as a type of imaginary time variable. Between these two limits for the analytical continuation of the path integral, we find the development of the time and temperature dependent correlation functions of the system. In these the time or temperature is analytically continued to a complex number so that the system is subject to both time and temperature increments in its development.

Following the development of the path integral methodology for determining the partition function of quantum many-body systems, its applications to some basic many-body problems are discussed. The focus is on the illustration of early made considerations of simple systems and to important successes the methods have provided in the numerical understanding of fundamental problems in many-body quantum systems. In this regard, as a first problem we consider the thermodynamics of a vibrational chain of atoms [2, 3]. Specifically, this is a generalization of the standard textbook problem of a linear chain of atoms connected by springs as extended to the treatment of a linear chain of quantum mechanical atoms interacting by nonlinear interatomic potentials. It represents one of the most basic problems that can be treated by quantum Monte Carlo methodologies.

For a second illustration, the more complex problem involved in the treatment of the physical properties of liquid helium is considered [1, 5–9]. This requires a discussion of the important topic of superfluidity [5–9] and the physics associated with macroscopic quantum states of matter. To begin, a review is made of some of the basic properties of the bosonic system liquid helium ^4He with a focus on the lambda transition and the development of the superfluid state [5–9]. Here the bosonic nature of the fluid arises from the nuclear total spin zero of the ^4He isotopes. Due to the identical nature of the ^4He atoms and the symmetry requirements of the system under the labeling of the particles, a variety of interesting physical properties are observed in the liquid system [1]. This introduces a new feature into the many-body considerations as, unlike the chain of atoms, the indistinguishable nature of the particles becomes an important determinator of the liquid properties [2, 3].

The discussion of liquid helium begins with a consideration of the early treatment of the properties of liquid ^4He provided by the two fluid model [5–7] and the Ginsberg theory of the second order transition to the superfluid state [5–7]. This leads to an introduction to the basic ideas of the nature of the excitations, the excitation spectra in the superfluid state, and how these contribute to the superfluid properties of helium exhibited below the lambda transition [5, 7]. Ultimately, a consideration is made of how the path integral formulation must be modified to treat Bose-Einstein systems in its application to the study of the excitations in liquid helium [1]. In this context, a brief review of recent work on the quantum Monte Carlo applied to the determination of some of the important properties of the liquid helium superfluid is given [1].

To conclude, a related problem of interest is considered. This is the problem of the superfluid state exhibited by the fermionic system of liquid ^3He [7, 10, 11]. In this liquid the ^3He nuclei are spin $\frac{1}{2}$ fermions, and their resultant superfluidity properties come from mechanisms similar to those accounting for the properties of electronic superconductors [10, 11]. (Notice that, unlike the electron systems, the ^3He liquid does not carry a charge so that its superfluidity involves its viscous properties.) However, while in low-temperature superconductors the superconductivity is from s-wave pairing interactions, the pairing interactions in superfluid ^3He arise from p-wave pairing interactions. Both of these mechanisms for a transition to a macroscopic quantum state of the systems are quite different from those at play in the superfluidity exhibited by ^4He. For a comparison between the two systems, a brief review is given of ^3He with an introduction to its theory and properties. These are contrasted with the theory and properties found in the study of liquid ^4He [7].

10.1 DEVELOPMENT OF THE PATH INTEGRAL FORMULATION FOR THE QUANTUM PARTITION FUNCTION

We begin the study of the thermodynamics of quantum many-body systems with a discussion of how the quantum partition function, expressed in terms of quantum mechanical operators, can be rewritten as a path integral involving only classical mechanical variables [1–3]. The resulting path integral is expressed in terms of real and complex numbers so that it and the thermodynamic averages obtained from it are treatable using standard Monte Carlo methods based on random number generation. On the other hand, the original quantum partition function involves non-commuting quantum operators so that it and the averages obtained from it are not directly treatable using random number techniques of Monte Carlo simulation.

In the path integral approach [1–3] the complexity of dealing with a system characterized by non-commuting operators is exchanged for a problem involving a consideration of a higher dimensional classical system. We start with a discussion entailing the rewrite of the quantum partition function into the form of a path integral. This is followed by an application of the path integral approach to a simple system.

Specifically, consider a many-body quantum mechanical Hamiltonian H and its associated quantum mechanical eigenvalue problem given by [1–3]

$$H|\psi_i\rangle = \epsilon_i |\psi_i\rangle \tag{10.1}$$

where $|\psi_i\rangle$ is the wavefunction of the ith eigenmode of the system, $\in_i$ is the energy of the ith eigenmode, and i is an index running over the eigenmodes. In terms of the eigenvectors and eigenvalues defined in Eq. (10.1), the partition function of the system is given by [1–6]

$$Z = Tr\, e^{-\beta H} \tag{10.2}$$

where Tr is a trace over a complete set of states of the system, $\beta = \frac{1}{k_B T}$, T is the temperature, and k_B is the Boltzmann constant. The Hamiltonian H is typically composed of a variety of non-commuting operators which make the evaluation of the trace in Eq. (10.2) a complicated process. This process is greatly simplified by rewriting Eq. (10.2) as a path integral, expressed in classical mechanical variables [1–3].

In our considerations of the many-body partition function in Eq. (10.2) we must also focus on some important types of systems, exhibiting invariance of H under particle permutation symmetries [1, 5–7]. These are quantum systems formed of identical particles. A consequence of these symmetries is found in the restrictions they introduce to the wavefunctions of the Hamiltonian. As a result of the restrictions placed on the system by the permutation symmetries, three types of systems are commonly encountered in condensed matter physics.

In the first type, we consider photonic-like systems which are characterized by a distribution of particles in which the absolute number of particles in the system is not conserved. This is typified by particle statistics in the form of a Planck distribution and is applied in photonic and phononic systems [5, 6]. In a second type, we find systems composed of identical bosonic particles in which, as an important component in the determination of the partition function, the wavefunction is symmetric under particle permutation symmetry and in which the total number of particles in the system is conserved [5, 6]. A third type of system is formed of identical fermions which have an antisymmetric permutation symmetry of the wavefunction as an important factor in determining the partition function [5–7]. For such fermion systems the number of particles is always conserved. The fermion systems of condensed matter physics, however, are found to require a special treatment for the determination of the partition function and its properties by simulation methods. These will not be considered here so that for this study the reader is referred to the literature [1, 6, 7, 10, 11].

We begin our discussion with a formulation for Eq. (10.2) appropriate for the treatment of atomic and molecular vibrational systems in which the identical nature of the atoms and molecules is not important [1–3]. This provides initially a helpful simplification in the problem of developing the partition function of the system in terms of a path integral formulation. Following this, a discussion of the wavefunction restrictions needed to develop the partition function for Bose-Einstein systems is discussed. This leads to a study of the low-temperature properties of the bosonic system of superfluid liquid helium.

10.1.1 The Path Integral

A transformation of the partition function of the many-body system in Eq. (10.2) can be made in which we initially do not treat the system as being composed of identical particles. This facilitates its placement into the form of a path integral [1–3]. The procedure is initiated by first rewriting Eq. (10.2) into the form of a product of exponentials with infinitesimally small arguments. In this way, it follows that [1–4]

$$Z = Tr\, e^{-\beta H} = Tr\left[e^{-\frac{\beta}{M}H} \right]^M = Tr\, e^{-\frac{\beta}{M}H} e^{-\frac{\beta}{M}H} e^{-\frac{\beta}{M}H} \,\ldots\ldots\ldots e^{-\frac{\beta}{M}H} \quad (10.3)$$

where M is an integer considered in the limit in which $M \to \infty$.

In the representation in Eq. (10.3) the partition function is envisioned as arising from an infinite sequence of infinitesimal steps of temperature increments $\frac{\beta}{M}$. Given this form, the partition function in Eq. (10.3) is next expressed in a basis of position space states of the particles in the system. Specifically, for a system of $N \to \infty$ particles contained within a volume V these basis states are of the form

$$|\vec{R}\rangle = |\vec{r}_1, \vec{r}_2, \vec{r}_3 \ldots, \vec{r}_N\rangle \quad (10.4a)$$

where $\{\vec{r}_1, \vec{r}_2, \vec{r}_3 \ldots, \vec{r}_N\}$ are the position vectors of the N particles. The resulting basis set $\{|\vec{R}\rangle\}$ of states is a complete orthonormal basis for which the identity operator is defined as [1–3]

$$\int |\vec{R}\rangle\, d\vec{R} \langle \vec{R}| = 1. \quad (10.4b)$$

where the integral in $d\vec{R}$ is over the set of variables $\{\vec{r}_1, \vec{r}_2, \vec{r}_3 \ldots, \vec{r}_N\}$ ranging throughout the entire volume V. It then follows from Eq. (10.3) that [1–3]

$$Z = Tr\, e^{-\beta H} = \int d\vec{R} \langle \vec{R} \,|\, e^{-\beta H} \,|\, \vec{R}\rangle \quad (10.4c)$$

represents the partition function of the many-body system.

Next, we express the exponentials $e^{-\frac{\beta}{M}H}$ of the product in Eq. (10.3) in terms of the position space variables defined in Eq. (10.4). In this way the matrix elements of $e^{-\frac{\beta}{M}H}$ represented in the position space basis are obtained as [1]

$$\langle \vec{R}_i \,|\, e^{-\frac{\beta}{M}H} \,|\, \vec{R}_j\rangle = \sum_l \langle \vec{R}_i \,|\, \psi_l\rangle e^{-\frac{\beta}{M}\epsilon_l} \langle \psi_l \,|\, \vec{R}_j\rangle \quad (10.5)$$

where l runs over the eigenstates defined in Eq. (10.1), $\left|\vec{R}_i\right\rangle = \left|\vec{r}_{1,i}, \vec{r}_{2,i}, \vec{r}_{3,i} \ldots, \vec{r}_{N,i}\right\rangle$ and $\left|\vec{R}_j\right\rangle = \left|\vec{r}_{1,j}, \vec{r}_{2,j}, \vec{r}_{3,j} \ldots, \vec{r}_{N,j}\right\rangle$ are two position space configurations in the N particle position state basis set $\left\{\left|\vec{R}\right\rangle\right\}$, and the indices i and j label the particular basis states. Applying this notation in the total partition function in Eq. (10.3), it then follows that [1]

$$Z = Tre^{-\beta H} = Tr\, e^{-\frac{\beta}{M}H} e^{-\frac{\beta}{M}H} e^{-\frac{\beta}{M}H} \ldots\ldots\ldots e^{-\frac{\beta}{M}H} \tag{10.6}$$

$$= \int \prod_{i=1}^{M} d\vec{R}_i \langle \vec{R}_i \mid e^{-\frac{\beta}{M}H} \mid \vec{R}_{i+1} \rangle$$

$$= \int \left(\prod_{i=1}^{M} d\vec{R}_i \right) \langle \vec{R}_1 \mid e^{-\frac{\beta}{M}H} \mid \vec{R}_2 \rangle \langle \vec{R}_2 \mid e^{-\frac{\beta}{M}H} \mid \vec{R}_3 \rangle \ldots\ldots \langle \vec{R}_M \mid e^{-\frac{\beta}{M}H} \mid \vec{R}_1 \rangle$$

where $\left|\vec{R}_{M+1}\right\rangle = \left|\vec{R}_1\right\rangle$ and the integrals in $d\vec{R}_i$ run over all the configurations of the N particles.

For the evaluation of Eq. (10.6) we now focused on the determination of the matrix elements of the form $\langle \vec{R}_i | e^{-\frac{\beta}{M}H} | \vec{R}_{i+1} \rangle$ from which it is composed. These can be written as [1]

$$\langle \vec{R}_i \mid e^{-\frac{\beta}{M}H} \mid \vec{R}_{i+1} \rangle = \sum_l \langle \vec{R}_i \mid \psi_l \rangle \langle \psi_l \mid e^{-\frac{\beta}{M}H} \mid \psi_l \rangle \langle \psi_l \mid \vec{R}_{i+1} \rangle \tag{10.7}$$

$$= \sum_l \langle \vec{R}_i \mid \psi_l \rangle \langle \psi_l \mid e^{-\frac{\beta}{M}(K+U)} \mid \psi_l \rangle \langle \psi_l \mid \vec{R}_{i+1} \rangle$$

$$= \sum_l \left\langle \vec{R}_i \mid \psi_l \right\rangle e^{-\frac{\beta}{M}\epsilon_l} \left\langle \psi_l \mid \vec{R}_{i+1} \right\rangle$$

where we have used Eq. (10.1), and the non-commuting operators K and U are, respectively, the kinetic and potential energy operators forming the Hamiltonian, i.e., $H = K + U$. In order to treat the matrix elements of the exponential operator $e^{-\frac{\beta}{M}(K+U)}$ we next apply a standard result from quantum mechanics.

Specifically, the Baker-Campbell-Hausdorff formula states that for the operators A, B, C, it follows that [1–4]

$$e^A e^B = e^C \tag{10.8a}$$

where

$$C = A + B + \frac{1}{2}[A, B] + \frac{1}{12}\left[A, [A, B]\right] + \frac{1}{12}\left[B, [A, B]\right] + \ldots. \tag{10.8b}$$

Consequently, from Eqs. (10.7) and (10.8) we find that

$$e^{-\frac{\beta}{M}K} e^{-\frac{\beta}{M}U} = e^{-\frac{\beta}{M}(K+U) + \frac{1}{2}\left(\frac{\beta}{M}\right)^2 [K,U] + O\left(\left(\frac{\beta}{M}\right)^3\right)} \tag{10.9a}$$

which for $\frac{\beta}{M} \ll 1$ gives

$$e^{-\frac{\beta}{M}(K+U)} \approx e^{-\frac{\beta}{M}K} e^{-\frac{\beta}{M}U} \tag{10.9b}$$

to leading order in $\frac{\beta}{M}$.

In this regard, from the Baker-Campbell-Hausdorff formula it follows that C in Eq. (10.8b) for our problem in Eq. (10.7) is a rapidly convergent series as $\frac{\beta}{M} \to 0$. Consequently, from Eqs. (10.7) and (10.9b) it is then found that [1]

$$\left\langle \vec{R}_i \middle| e^{-\frac{\beta}{M}H} \middle| \vec{R}_{i+1} \right\rangle \approx \left\langle \vec{R}_i \middle| e^{-\frac{\beta}{M}K} e^{-\frac{\beta}{M}U} \middle| \vec{R}_{i+1} \right\rangle = \left\langle \vec{R}_i \middle| e^{-\frac{\beta}{M}K} \middle| \vec{R}_{i+1} \right\rangle e^{-\frac{\beta}{M}U(\vec{R}_{i+1})}. \tag{10.10}$$

For the evaluation of the remaining matrix element $\left\langle \vec{R}_i \mid e^{-\frac{\beta}{M}K} \mid \vec{R}_{i+1} \right\rangle$ we start by defining k –space basis vectors corresponding to the particle position space vectors basis $|\vec{R}\rangle = |\vec{r}_1, \vec{r}_2, \vec{r}_3 \ldots, \vec{r}_N\rangle$ defined in Eq. (10.4a). In this regard, the k –space basis vectors are defined as

$$|\vec{K}\rangle = |\vec{k}_1, \vec{k}_2, \vec{k}_3 \ldots, \vec{k}_N\rangle \tag{10.11}$$

and represent wavevector states of the particles labeled by $i = 1, 2, 3, \ldots, N$.

If we further define the vector notations

$$\vec{R} = (\vec{r}_1, \vec{r}_2, \vec{r}_3 \ldots, \vec{r}_N) \tag{10.12a}$$

and

$$\vec{K} = (\vec{k}_1, \vec{k}_2, \vec{k}_3 \ldots, \vec{k}_N) \tag{10.12b}$$

so that

$$\vec{K} \cdot \vec{R} = \vec{k}_1 \cdot \vec{r}_1 + \vec{k}_2 \cdot \vec{r}_2 + \ldots + \vec{k}_N \cdot \vec{r}_N, \tag{10.12c}$$

it then follows within this notation that [1]

$$\langle \vec{R} \mid \vec{K} \rangle = \frac{1}{V^{N/2}} e^{-i\vec{K}\cdot\vec{R}} = \frac{1}{V^{N/2}} e^{-i(\vec{k}_1 \cdot \vec{r}_1 + \vec{k}_2 \cdot \vec{r}_2 + \ldots + \vec{k}_N \cdot \vec{r}_N)}. \tag{10.12d}$$

From this notation we next evaluate Eq. (10.10). In order to do this, we consider the system to exist in a cubic box of side L so that the possible k –space basis vectors are of the form

$$\vec{k} = \frac{2\pi}{L}(l\hat{i} + m\hat{j} + n\hat{k}) \tag{10.12e}$$

where l, m, n are integers.

In terms of these definitions, it then follows from Eq. (10.10) that [1]

$$\langle \vec{R}_i | e^{-\frac{\beta}{M}K} | \vec{R}_{i+1}\rangle = \sum\nolimits_{n,m} \langle \vec{R}_i | \vec{K}_n\rangle e^{-\frac{\beta}{M}K} \langle \vec{K}_m | \vec{R}_{i+1}\rangle \tag{10.12f}$$

$$= \sum\nolimits_{\{\vec{k}_{1,n},\vec{k}_{2,n},..\}} \langle \vec{R}_i | \vec{K}_n\rangle e^{-\frac{\beta}{M}\frac{\hbar^2}{2m}\left[k_{1,n}^2+k_{2,n}^2+k_{3,n}^2+...+k_{N,n}^2\right]} \langle \vec{K}_n | \vec{R}_{i+1}\rangle$$

$$= \frac{1}{V^N}\sum\nolimits_{\{\vec{k}_1,\vec{k}_2,...\}} e^{-\frac{\beta\hbar^2}{M2m}\left[k_1^2+k_2^2+k_3^2+\cdots k_N^2\right]} *$$

$$* e^{-i\left[\vec{k}_1\cdot(\vec{r}_{1,i}-\vec{r}_{1,i+1})+\vec{k}_2\cdot(\vec{r}_{2,i}-\vec{r}_{2,i+1})+...+\vec{k}_N\cdot(\vec{r}_{N,i}-\vec{r}_{N,i+1})\right]}$$

$$= \prod_{j=1}^{N}\left\{\frac{1}{V}\sum\nolimits_{\vec{k}_j} e^{-\frac{\beta}{M}\frac{\hbar^2}{2m}k_j^2 - i\vec{k}_j\cdot(\vec{r}_{j,i}-\vec{r}_{j,i+1})}\right\}.$$

Note that here we have used the fact that the planewave basis states are eigenstates of the kinetic energy operator, which has been taken to be of the free particle form $K = \frac{\hbar^2 k^2}{2m}$ with $\vec{k}$ the single particle wavevector and m the single particle effective mass. Also, notice that after the third equality we have dropped the index n.

Consequently, we now need to evaluate

$$\frac{1}{V}\sum\nolimits_{\vec{k}_j} e^{-\frac{\beta}{M}\frac{\hbar^2}{2m}k_j^2 - i\vec{k}_j\cdot(\vec{r}_{j,i}-\vec{r}_{j,i+1})} \approx \int \frac{d^3k}{(2\pi)^3}\exp\left[-\frac{\beta}{M}\frac{\hbar^2}{2m}k^2 - i\vec{k}\cdot\left(\vec{r}_{j,i}-\vec{r}_{j,i+1}\right)\right] \tag{10.13}$$

where the sum on $\vec{k}_j$ is approximated by an integral which can be treated using standard identities of the Gaussian integrals. From this, it follows that [1]

$$\langle \vec{R}_i | e^{-\frac{\beta}{M}K} | \vec{R}_{i+1}\rangle = \left[\frac{1}{\sqrt{4\pi\lambda\tau}}\right]^{3N} \prod_{j=1}^{N}\exp\left[-\frac{1}{4\lambda\tau}\left(\vec{r}_{j,i}-\vec{r}_{j,i+1}\right)^2\right] \tag{10.14}$$

$$= \left[\frac{1}{\sqrt{4\pi\lambda\tau}}\right]^{3N} \exp\left[-\frac{1}{4\lambda\tau}\left(\vec{R}_i-\vec{R}_{i+1}\right)^2\right]$$

where $\lambda = \frac{\hbar^2}{2m}$ and $\tau = \frac{\beta}{M}$.

Applying Eq. (10.14) in Eq. (10.10), we then have

$$\langle \vec{R}_i | e^{-\frac{\beta}{M}H} | \vec{R}_{i+1}\rangle = \langle \vec{R}_i | e^{-\frac{\beta}{M}K} | \vec{R}_{i+1}\rangle e^{-\frac{\beta}{M}U(\vec{R}_{i+1})} \tag{10.15}$$

$$= \left[\frac{1}{\sqrt{4\pi\lambda\tau}}\right]^{3N} \exp\left[-\frac{1}{4\lambda\tau}\left(\vec{R}_i-\vec{R}_{i+1}\right)^2 - \tau U\left(\vec{R}_{i+1}\right)\right].$$

Consequently, from Eq. (10.6) it follows that [1]

$$Z = \int\prod_{i=1}^{M} d\vec{R}_i \langle \vec{R}_i | e^{-\frac{\beta}{M}H} | \vec{R}_{i+1}\rangle \tag{10.16}$$

$$= \int \prod_{i=1}^{M} d\vec{R}_i \left\{ \left[\frac{1}{\sqrt{4\pi\lambda\tau}} \right]^{3N} \exp\left[-\frac{1}{4\lambda\tau} \left(\vec{R}_i - \vec{R}_{i+1} \right)^2 - \tau U\left(\vec{R}_{i+1} \right) \right] \right\}$$

$$= \left[\frac{1}{\sqrt{4\pi\lambda\tau}} \right]^{3MN} \int \prod_{i=1}^{M} d\vec{R}_i \exp\left[-\frac{1}{4\lambda\tau} \left(\vec{R}_i - \vec{R}_{i+1} \right)^2 - \tau U\left(\vec{R}_{i+1} \right) \right]$$

where $\left|\vec{R}_{M+1}\right\rangle = \left|\vec{R}_1\right\rangle$. This expresses the partition function of the quantum system characterized by the operator $H = K + U$ in terms of a path integral expressed in the classical mechanics (i.e., real commuting) variables $\left\{\vec{R}_i\right\}$.

Notice that in the case that $U = 0$, the partition function can be directly evaluated using the result in Eq. (10.14). It follows from Eq. (10.14) for the case in which $M = 1$ that [1]

$$\langle \vec{R}_i \mid e^{-\beta K} \mid \vec{R}_{i+1} \rangle = \left[\frac{1}{\sqrt{4\pi\lambda\beta}} \right]^{3N} \exp\left[-\frac{1}{4\lambda\beta} \left(\vec{R}_i - \vec{R}_{i+1} \right)^2 \right]. \tag{10.17a}$$

Applying Eqs. (10.3), (10.4c), and (10.6), we then have from Eq. (10.17a) evaluated for $\left|\vec{R}_i\right\rangle = \left|\vec{R}_{i+1}\right\rangle = \left|\vec{R}_1\right\rangle$ that [1]

$$Z = \int d\vec{R}_1 \left\langle \vec{R}_1 \left| e^{-\beta K} \right| \vec{R}_1 \right\rangle \tag{10.17b}$$

$$= V^{3N} \left[\frac{1}{\sqrt{4\pi\lambda\beta}} \right]^{3N} = V^{3N} \left[\sqrt{\frac{mk_B T}{2\pi\hbar^2}} \right]^{3N}.$$

This is the standard Maxwell-Boltzmann partition function of a gas of classical particles. In this regard, we again note that the partition function is not corrected for identical particles. In the case of identical particles, Z in Eq. (10.17b) is divided by a factor of $N!$.

Before continuing further with the applications of the path integral formulation to more complex dynamical systems, it is interesting at this point to note the relationship between the quantum partition function in the path integral formulation and the dynamical evolution of the quantum system in time. Specifically, these two problems are closely related to each other by an analytic continuation of the time and temperature variables.

10.1.2 The Path Integral Formulation of Quantum Mechanics

Consider next the analytic continuation of Eq. (10.16) to the imaginary time axis. This turns the partition function into a Green's function for the propagation in time of the set of N particles. Consequently, the analytic continuation of each of these problems to the other (i.e., the partition function to the propagator or the propagator to the partition function) establishes an intimate connection between the problems of quantum dynamics and quantum statistical mechanics.

Specifically, we change the variable $\tau = \frac{\beta}{M}$, replacing it by $i\frac{t}{\hbar M}$. Under this substitution, it follows from Eq. (10.16) that [1]

$$\int \prod_{i=1}^{M} d\vec{R}_i \langle \vec{R}_i \mid e^{-i\frac{tH}{\hbar M}} \mid \vec{R}_{i+1} \rangle \tag{10.18a}$$

$$= \int \prod_{i=1}^{M} d\vec{R}_i \left\{ \left[\sqrt{\frac{-i\hbar M}{4\pi\lambda t}} \right]^{3N} \exp\left[i\frac{\hbar M}{4\lambda t}\left(\vec{R}_i - \vec{R}_{i+1}\right)^2 - i\frac{t}{\hbar M} U\left(\vec{R}_{i+1}\right) \right] \right\}$$

$$= \left[\sqrt{\frac{mM}{2\pi i\hbar t}} \right]^{3MN} \int \prod_{i=1}^{M} d\vec{R}_i \exp\left[\frac{i}{\hbar} M \frac{m\left(\vec{R}_i - \vec{R}_{i+1}\right)^2}{2t} - \frac{i}{\hbar}\frac{t}{M} U\left(\vec{R}_{i+1}\right) \right]$$

where again $|\vec{R}_{M+1}\rangle = |\vec{R}_1\rangle$. This represents the probability or transition amplitude [1–3]

$$\int d\vec{R}_1 \langle \vec{R}_1 | e^{-i\frac{tH}{\hbar}} | \vec{R}_1 \rangle \tag{10.18b}$$

of a particle propagating from the initial state $|\vec{R}_1\rangle$ to return back to $|\vec{R}_1\rangle$ in the time interval t. In this context, the path integral formulation of quantum statistical mechanics and the quantum mechanical time evolution operator are related to each other.

10.2 CHAIN OF ATOMS

As an example of another application of the path integral method in quantum statistical mechanics, we consider the basic problem of a chain of N atoms in which the atoms of the chain only interact with their nearest neighbors [2, 3]. This motion is made in such a manner that the atoms are constrained to move along the axis of the chain, and they are not allowed to pass through one another. Consequently, the atoms do not change their sequence along the chain.

The interaction between the neighboring atoms is through a Lennard-Jones 6–12 pair potential given by the form

$$\Phi(r) = 4\,\epsilon\left[\left(\frac{\sigma}{r}\right)^{12} - \left(\frac{\sigma}{r}\right)^{6}\right] \tag{10.19}$$

where r is the separation between a pair of nearest neighbor atoms along the chain which is of total length L. Periodic boundary conditions are applied between the ends of L so that topologically the chain is of the form of a ring. Notice that this problem is related to that of the harmonic chain of atoms commonly studied in basic solid state physics texts only now the atoms are coupled by anharmonic potentials so that the motion is a nonlinear dynamic.

The system is studied using a quantum Monte Carlo simulation in the path integral formulation, applying the methodology discussed earlier. From these methods the pressure, energy, and specific heat of the chain are computed as functions of temperature by applying standard thermodynamic identities to the partition function [2, 3]. In Table 10.1 we list the values of pressure and energy as functions of temperature computed in this way for a chain of $N = 10$ atoms. The values in the table were obtained from Ref. 2 by making an average of the data listed in that paper for simulation runs on $N = 10$ atom chains. As a comparision, in Table 10.1 results are also presented for the pressure and energy of the corresponding system obtained using standard Monte Carlo and studied in the context of the classical mechanical realization of the system [2]. The two results display a qualitative similarity to one another.

The results in Ref. 2 present an interesting early application of the methodology of quantum Monte Carlo to study a simple model. We shall next look at a much more ambitious treatment involving the use of quantum Monte Carlo to understand the properties of liquid helium superfluids [1]. First, however, a review of some of the basic properties of liquid helium is given and a discussion made of some of the earlier theories devised to understand these properties.

TABLE 10.1
Data of Pressure and Energy as a Function of Temperature for a Chain of $N = 10$ Atoms [2]

$\beta\epsilon$	$\frac{p\sigma}{\epsilon}$ (q.MC)	$\frac{p\sigma}{\epsilon}$ (Cl.MC)	$\frac{E}{N\epsilon}$ (q.MC)	$\frac{E}{N\epsilon}$ (Cl.MC)
10	1.46	0.92	–0.827	–0.901
5	1.93	1.78	–0.768	–0.806
3.333	2.54	2.58	–0.692	–0.715
2.5	3.16		–0.613	
2	3.75	4.02	–0.533	–0.538
1.429	4.80		–0.381	
1	6.79	7.10	–0.124	–0.120
0.667	9.23	9.80	0.261	0.281
0.333	14.89		1.321	
0.167	27.78		3.573	

Note: Results are presented for the quantum Monte Carlo (q.MC) applied on the quantum model and the standard Monte Carlo (Cl.MC) applied on its classical mechanical counterpart.

10.3 SUPERFLUIDS

In this section we look at studies focused on liquid helium (i.e., ^4He) and particularly at the theory of superfluids. Liquid helium ^4He is a system composed of nuclear spin zero particles which obey Bose-Einstein statistics, and these statistics account for many of the interesting phenomena exhibited by the liquid [5–7]. At low temperatures and at pressures below 25 atmospheres helium transforms to a liquid state which on a temperature-pressure phase diagram is found to separate into a region of Helium I and a region of Helium II fluid states. (See Figure 10.1a for a schematic of the Helium I and Helium II phase diagram.) While Helium I does not exhibit extraordinary properties, the lower temperature Helium II phase exhibits a variety of interesting properties arising from the quantum dynamics of the system [5–7]. Specifically, a property of great interest is the presence of a component of the fluid which exhibits an absence of viscosity leading to its characterization as a superfluid. In the following, we shall treat the origins of some of the interesting properties of the Helium II superfluid and of the transition between the Helium I and Helium II phases.

Superfluidity involves the apparent change in the viscous state of liquid helium that occurs at the lambda transition (i.e., below $T_\lambda = 2.17^\circ K$ at atmospheric pressure) at which Helium I changes to Helium II [5–7]. Specifically, at temperatures above the lambda transition, liquid helium is in the Helium I normal phase, which exhibits no superfluid properties, while at temperatures below the lambda transition, the Helium II phase exhibits a superfluid component. In this regard, the properties of Helium II can phenomenologically be viewed as behaving as if the system is composed of a mixture of two different types of helium fluids of different viscosities, while above the lambda transition there is only one Helium I normal fluid phase [5–7]. In the mixed phase of Helium II one fluid consists of normal (viscous) fluid, which has a nonzero viscosity, and the other fluid is a superfluid exhibiting zero viscosity.

Starting at the lambda temperature, then, the density of superfluid in the system is zero and as the temperature of the system decreases the superfluid fraction of the total fluid gradually increases until at zero temperature the system is completely composed of superfluid [1–7]. In this regard, early in the study of liquid helium a two-fluid model of the system was developed to explain Helium II properties below the lambda transition. This will be the first of the phenomenological models we shall consider. In the two-fluid model [5–7], the helium is treated as composed of a fraction x of

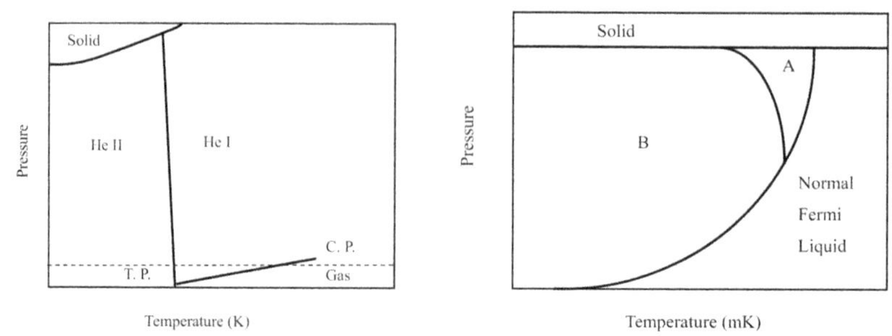

FIGURE 10.1 Schematic plot of: a) The phase diagram of Helium I and Helium II including the line of lambda transitions between the two states of the fluid. At the top of the phase diagram is a solid phase which occurs at pressures greater than 25 atmospheres, and at the bottom of the diagram is a gas phase. The dashed line in the figure indicates 1 atmosphere of pressure. At atmospheric pressure helium liquefies at $4.2^{\circ} K$. b) The phase diagram of liquid ^{3}He including lines of transitions between the normal fluid, the superfluid A-phase, and the superfluid B-phase. At the top of the phase diagram is a solid phase which occurs at pressures greater than 25 atmospheres. At atmospheric pressure helium liquifies at $4.2^{\circ} K$.

superfluid fluid and a fraction $(1-x)$ of normal fluid so that the total properties exhibited by Helium II are dependent on x.

This is followed by an introduction to the Ginzburg theory [5, 6, 8] of second order phase transitions. The Ginzburg theory is found to offer an explanation of the Helium I to Helium II transition in terms of a proposed form for the free energy of the system at its phase transition. It offers a qualitative explanation of a variety of the general features associated with the superfluid transition. In addition, the closely related Gross-Pitaevskii theory [6, 9] of Helium II is also presented as a development from studies of the weak interaction limit of the helium problem.

A justification of these phenomenological considerations is eventually given in terms of a discussion of quantum Monte Carlo studies of the thermodynamics from the Hamiltonian of the helium system [1]. This treats the full quantum nature of the system, including its nature as a strongly interacting Bose-Einstein fluid [6]. It explains many of the aspects of the system which were known experimentally and entered phenomenologically in earlier theoretical constructions.

In this regard, we shall eventually explain that since Helium II is composed of bosons, a new feature arises in the application of the quantum Monte Carlo as the path integral formulation developed in Section 10.1 needs to be modified to account for the symmetry properties of the many-body wavefunctions under the identical particle permutation symmetry [1]. Notice that, while permutation symmetry is easily considered for the system of particles with boson statistics, another system we shall be interested in near the end of this chapter is ^{3}He. Unlike ^{4}He, ^{3}He is a fermion system and as a result much more difficult to treat by simulation [1, 6]. Specifically, systems of particles with fermi statistics such as conduction electrons and liquid ^{3}He are much more difficult to treat in the path integral method [1, 3]. In light of this difficulty, for the path integral formulation treatment of identical fermions the reader is referred to the literature [1, 3], and we shall restrict our considerations of ^{3}He to non-simulation based studies [6, 7, 10, 11].

Our discussions start by outlining a review of some of the unusual properties of the liquid Helium II superfluid which distinguish it from the Helium I phase and by introducing some of the early so-called phenomenological treatments proposed to offer an explanation of these properties [5–7]. Specifically, a number of basic phenomenological theories for the Helium II thermodynamics and the treatment of the transition between the Helium II and Helium I phases are developed. After this, a consideration of quantum Monte Carlo techniques applied to the path integral formulation of the liquid helium partition function is given [1]. These place many of the phenomenological considerations on a firm theoretical basis involving first principle considerations of many-body physics. At the end of the chapter, considerations are given to the properties of liquid ^3He.

10.3.1 Two-Fluid Model

The behavior of the free energy of the two-fluid model of liquid helium in the vicinity of the lambda transition temperature is characterized in the context of a Ginzburg model for the free energy of the system at the Helium I to Helium II phase transition [5–9]. This formulation is developed from some general thermodynamic considerations of the observed properties of liquid helium, and the resulting theory [5, 6] then allows us to solve for a number of the physically interesting behaviors of Helium II. (In essence, the resulting Ginzburg model can be shown to yield a mean field treatment for the thermodynamics of the system.) This is all accomplished in the context of a so-called phenomenological theory which is developed in macroscopic variables characterizing the system, offering a qualitative treatment for the properties of the helium fluid. To begin the discussion of the two-fluid model, we then start by treating the theory for the system in which the normal component of the fluid is stationary, and the superfluid is involved in a zero-viscosity motion. The superfluidity below the lambda transition temperature is then postulated as coming from the zero-viscosity component of Helium II while the normal fluid component of Helium II is not part of the fluid flow.

Before we treat the Helium I and II phases, let us first go back to discuss some aspects of a general ideal gas of weakly interacting atoms. For the consideration of these types of systems we focus on the energy of atomic motions in the collection of gas atoms. In a weakly interacting gas such as water or carbon dioxide vapor, the motion of a particle in the bulk fluid is characterized by a single particle wavefunction and the total wavefunction of the gas is viewed as composed of these single particle states. (For example, in the extreme limit of this model we have the ideal gas of noninteracting particles.) Considering such a gas of weakly interacting particles, the wavefunction of the gas is then approximated in terms of single particle states $\psi_s(\vec{r})$ of the form

$$\psi_s(\vec{r}) = v_{\vec{k}} e^{i\vec{k}\cdot\vec{r}} \tag{10.20}$$

where $v_{\vec{k}}$ is the wavefunction amplitude, $\vec{k}$ is the wavevector of the particle, and $\vec{r}$ is a position vector in the volume of the fluid. The kinetic energy contribution of the single particle in Eq. (10.20) to the kinetic energy of the total gas then has the form [6, 2, 9]

$$-\frac{\hbar^2}{2m}\int d\vec{r}\psi_s^*(\vec{r})\nabla^2\psi_s(\vec{r}) = \frac{\hbar^2}{2m}\int d\vec{r}\nabla\psi_s^*(\vec{r})\cdot\nabla\psi_s(\vec{r}) = \frac{\hbar^2}{2m}\int d\vec{r}\left|\nabla\psi_s(\vec{r})\right|^2 \tag{10.21}$$

where m is the single particle mass. By analogy with the kinetic energy term in Eq. (10.21) we next consider how the motion of the superfluid component of Helium II could be accounted for in a treatment of the free energy of the Helium II system and its superfluid and normal fluid components.

For the case of the zero-viscosity superfluid component of Helium II, we generalize the form in Eq. (10.20) to develop a characterization of the contribution to the free energy of the system from the motion of the total superfluid component of Helium II. This is done by treating the superfluid component in the two-fluid model as being described by a macroscopic wavefunction $\psi(\vec{r})$ of the form [6–9]

$$\psi(\vec{r}) = \sqrt{n_0}\, e^{i\varphi(\vec{r})} \tag{10.22}$$

where n_0 is the particle density of the superfluid in Helium II and $\varphi(\vec{r})$ characterizes the spatially dependent phase of the superfluid. In this regard, we introduce the phase $\varphi(\vec{r})$ to describe the flow of the fluid much as $\vec{k}\cdot\vec{r}$ describes the motion of the single particle in Eq. (10.20). We see that, whereas $\psi_s(\vec{r})$ represents the motion of a single particle of the gas, $\psi(\vec{r})$ represents the motion of a macroscopic set of particles composing the so-called superfluid.

We then postulate that, by analogy with Eq. (10.21), the contribution to the free energy from the motion of the superfluid is given by the form

$$\frac{\hbar^2}{2m}\int d\vec{r}\left|\nabla\psi(\vec{r})\right|^2 \tag{10.23}$$

where m is the mass of an atom in the superfluid, e.g., the mass of ^4He in the Helium II phase. In this respect, while Eq. (10.23) determines the form of the kinetic energy contribution of the superfluid, it is important to note that it does not determine n_0 which is a temperature-dependent term that is only nonzero below the lambda temperature. In addition, we remind the reader that the normal fluid is stationary, so the normal fluid component does not contribute to the kinetic energy of Helium II due to flow. To determine n_0 then requires a further development of the free energy characterizing Helium II which shall presently be addressed.

As an additional important point regarding the motion of the systems, however, notice from Eq. (10.20) that in the dilute or ideal gas each particle in the gas has a motion characterized by a different wavevector $\vec{k}$. Consequently, the particles of the gas, considered in general, propagate by a phase incoherent motion. On the other hand, in the superfluid condensate characterized by Eq. (10.22) we see that all the many particles in the superfluid share the common phase $\varphi(\vec{r})$. This phase coherence of the superfluid component of Helium II has important consequences for the theoretical properties of Helium II, and these properties have been exhibited in experiments done on Helium II. Specifically, in this regard, the phase coherence in the superfluid is found to be similar to the photonic phase coherence observed in lasers.

Note, in addition, that an important example of the phase coherence in the superfluid phase is found in low-temperature Bose condensates studied in atomic physics [6, 7]. Specifically, one can prepare low-temperature superfluid collections of atoms. In a recent experiment [7] on such systems, two spatially separate collections of atoms were prepared and then allowed to pass through each other. As the two systems traveled through one another, a set of interference fringes was detected in the combined system. These fringes arose from the phase difference of the two original atomic collections [6, 7]. This offers a further justification for the form of the superfluid wavefunction in Eq. (10.22).

We shall next see how the free energy of the Helium II system can be phenomenologically developed in terms of the wavefunction in Eqs. (10.22) and (10.23). This will ultimately lead to an understanding of the transition between Helium I and Helium II in the neighborhood of the lambda temperature.

10.3.2 The Ginzburg Formulation for the Free Energy

For a consideration of Helium II in the neighborhood of the lambda transition, we next develop a theory for the superfluid particle concentration n_0 given as a function of temperature [6–9]. This will allow for the determination of the amplitude of the superfluid wavefunction given in Eq. (10.22).

To do this we assume that at temperatures near the lambda transition the free energy of the system can be Taylor expanded in terms of the small parameter $n_0 = \left|\psi(\vec{r})\right|^2$. Specifically, for $\nabla\psi(\vec{r}) = 0$ an acceptable form for the free energy F of the system is then given by [6–9]

$$F = F_0 + \int d\vec{r}\left[a(T)\left|\psi(\vec{r})\right|^2 + \frac{b}{2}\left|\psi(\vec{r})\right|^4\right] \tag{10.24a}$$

where F_0 is the constant free energy of the normal component of Helium I and II, and the second term on the right is the contribution from the superfluid. In Eq. (10.24a)

$$a(T) = a_0 (T - T_\lambda) \tag{10.24b}$$

where T_λ is the pressure-dependent transition temperature of the lambda transition, a_0 is a positive constant, and

$$b > 0 \tag{10.24c}$$

is also a constant.

Notice that in our study of Helium II near T_λ, the superfluid phase is only a small component of Helium II so that treating the total free energy as the sum of the superfluid free energy and the normal fluid free energy F_0 is a reasonable approximation in Eq. (10.24a). From a consideration of Eq. (10.24) we next determine n_0 as a function of temperature [8]. This is done at fixed $T \approx T_\lambda$ by minimizing the free energy form considered as a function of $\psi(\vec{r})$.

Minimizing Eq. (10.24a) with respect to $\psi(\vec{r})$ it then follows that [8]

$$n_0 = |\psi(\vec{r})|^2 = 0 \text{ for } T > T_\lambda \tag{10.25a}$$

and

$$n_0 = |\psi(\vec{r})|^2 = \frac{a_0 |T - T_\lambda|}{b} \text{ for } T < T_\lambda. \tag{10.25b}$$

Consequently, with the choice of Eq. (10.24) we see that a reasonable behavior of $n_0 = |\psi(\vec{r})|^2$ around the transition temperature is observed. Specifically, above T_λ there is no superfluid in the Helium I phase, and below T_λ Helium II contains a component of the superfluid phase which increases as the temperature decreases.

From Eqs. (10.24) and (10.25) the difference between the constant pressure specific heat of the superfluid and normal fluid phases is determined for a uniform bulk fluid by the thermodynamic identity [8]

$$\Delta C_P = -\frac{T}{V}\frac{\partial^2 (F - F_0)}{\partial T^2} = \frac{a_0^2}{b} T \approx \frac{a_0^2}{b} T_\lambda \text{ for } T < T_\lambda \tag{10.26a}$$

$$= 0 \text{ for } T > T_\lambda.$$

where V is the volume of the fluid and $T \approx T_\lambda$. From Eqs. (10.26a) and (10.25b) it then follows that

$$a_0 = \frac{\Delta C_P}{T_\lambda} \frac{1}{\dfrac{|\psi|^2}{T_\lambda - T}} \approx \frac{\Delta C_P}{T_\lambda} \frac{1}{\left|\dfrac{\partial n_0}{\partial T}\right|_{T_\lambda}} \tag{10.26b}$$

where $\lim_{T \to T_\lambda} \dfrac{|\psi|^2}{T_\lambda - T} \to \left|\dfrac{\partial n_0}{\partial T}\right|_{T_\lambda}$ and, similarly, from Eq. (10.25b) that

$$b = a_0 \frac{1}{\dfrac{|\psi|^2}{T_\lambda - T}} \approx a_0 \frac{1}{\left|\dfrac{\partial n_0}{\partial T}\right|_{T_\lambda}}. \tag{10.26c}$$

In the original paper of Ginzburg and Pitaevski [8] Eqs. (10.26b) and (10.26c) were used to obtain the estimated values $a_0 \approx 4.5\times10^{-17}$ erg/degree and $b \approx 4\times10^{-40}$ erg-cm^2.

Later work, however, has shown that the specific heat at the lambda transition is singular and Eq. (10.24) only provides a rough approximation for the behavior of liquid helium. Notice, however, that the validity of the Ginzburg results for the thermodynamics of the liquid helium systems in Eqs. (10.24) through (10.26) differs from that obtained from an application of the Ginzburg-Landau formulation [5–8] applied to the study of low-temperature superconductivity [5, 7]. In the Ginzburg-Landau formulation of the free energy of superconductors [5–7], the free energy is assumed to have the form in Eq. (10.24) and, in addition, is modified to include electrodynamic effects. The resulting formulation is found to offer a good representation of the superconducting transition in low-temperature superconductors. In the superconductors, as opposed to the liquid helium system, a mean field theory treatment is a better approximation than in the liquid helium systems. This is because the thermodynamic fluctuations in the low-temperature superconductors are smaller than those found in the Helium I and II systems [5, 7].

Combining Eqs. (10.23) and (10.24a), we then postulate a form for the free energy F of the superfluid helium state which includes spatial variations. This is given by [5–7]

$$F - F_0 = \int d\vec{r}\left[\frac{\hbar^2}{2m}\left|\nabla\psi(\vec{r})\right|^2 + a(T)\left|\psi(\vec{r})\right|^2 + \frac{b}{2}\left|\psi(\vec{r})\right|^4\right] \tag{10.27}$$

and includes both the motional contributions to the free energy as well as the contribution from the bulk condensate contribution n_0. A complete phenomenology of both the phase transition and the spatial variation of the liquid helium system is accessed in the free energy form so that a particular realization of the system described by Eq. (10.27) is then obtained by minimizing the free energy with respect to $\psi(\vec{r})$ and $\psi^*(\vec{r})$ subject to boundary conditions. (Note in this regard that $\psi(\vec{r})$ and $\psi^*(\vec{r})$ are two independent variables because they each separate into a real and an imaginary function, i.e., $\psi(\vec{r}) = f(\vec{r}) + ig(\vec{r})$).

Consequently, requiring that

$$\frac{\delta F}{\delta\psi} = 0 = \frac{\delta F}{\delta\psi^*} \tag{10.28}$$

we find [5–8]

$$-\frac{\hbar^2}{2m}\nabla^2\psi(\vec{r}) + a(T)\psi(\vec{r}) + b\left|\psi(\vec{r})\right|^2\psi(\vec{r}) = 0 \tag{10.29}$$

subject to boundary conditions on $\psi(\vec{r})$. The $\psi(\vec{r})$ determined from Eq. (10.29) is then substituted into Eq. (10.27) to obtain the free energy of the helium system.

Continuing the analogy between the particle states of the weakly interacting gas in Eq. (10.20) with the macroscopic state of the condensate in Eq. (10.22), we can make a consideration of the particle and condensate currents. Specifically, the single particle state in Eq. (10.20) for the dilute gas has a particle current density given in terms of the single particle wavefunction by the quantum mechanical expression [5–7]

$$\vec{j}_s = -i\frac{\hbar}{2m}\left[\psi_s^*\nabla\psi_s - \psi_s\nabla\psi_s^*\right] = \frac{\hbar}{m}\vec{k} = \frac{\vec{p}}{m}. \tag{10.30a}$$

By analogy with this formulation, let us define the particle current density of the condensate $\psi(\vec{r}) = \sqrt{n_0}e^{i\varphi(\vec{r})}$ in Eq. (10.22). In this way we express the superfluid particle current density in terms of its wavefunction as

$$\vec{j} = -i\frac{\hbar}{2m}\left[\psi^*\nabla\psi - \psi\nabla\psi^*\right] = \frac{\hbar}{m}n_0\nabla\varphi. \tag{10.30b}$$

Consequently, from this the velocity of a single particle in the condensate is given by [6, 8]

$$\vec{v} = \frac{\vec{j}}{n_0} = \frac{\hbar}{m}\nabla\varphi. \tag{10.30c}$$

As in the case of the single particle wavefunction, it is seen that the velocity of the particles in the condensate is related to the condensate phase $\varphi(\vec{r})$. In addition, note that by our assumption in Eq. (10.22) all the particles in the superfluid condensate have the same phase, i.e., they are in a coherent quantum state. This is not the case with the particles in the weakly interacting gas which are represented by Eq. (10.20) because $\vec{k}$ for each of the single particles can differ from particle to particle forming the gas.

10.3.3 Solutions of the Free Energy Form: Vortex States

As an example of the phase coherence of the condensate, we next consider the possibility of forming vortex states in the Helium II superfluid phase [6–8]. Vortices are solutions of Eq. (10.29) in which the superfluid exhibits a circulating flow about a linear core of normal fluid, and they may occur in the Helium II model singly or in multiple groups. In the following, we consider the solution of a single vortex with a core of normal fluid positioned along the z−axis of a bulk Helium II fluid. The vortex is a distortion of the uniform bulk state of the system so that it represents an excited state in the Helium II.

In the vortex solution the superfluid is found to circulate around the core of normal fluid in the plane perpendicular to the axis of the vortex. These types of vortex solutions are also found to occur in the Ginzburg-Landau theory of superconductivity [5] where they again originate in a macroscopic condensate phase composed of Cooper pairs of electrons. This is the superfluid state of the electronic system. In the superconductor version of the theory, vortices provide the basis for the distinction between type I and II superconductors [5, 7].

To understand the vortex solution in Helium II, consider Eqs. (10.29), (10.30b), and (10.30c) written in cylindrical coordinates (r,φ,z) where the z−axis is set to correspond to the normal fluid core of the vortex. From Eq. (10.29) it then follows for this geometry that [8]

$$-\frac{\hbar^2}{2m}\left[\frac{1}{r}\frac{\partial}{\partial r}\left(r\frac{\partial\psi}{\partial r}\right) + \frac{1}{r^2}\frac{\partial^2\psi}{\partial\varphi^2} + \frac{\partial^2\psi}{\partial z^2}\right] + a(T)\psi + b|\psi|^2\psi = 0 \tag{10.31}$$

is the Ginzburg equation defined about the vortex axis. A vortex solution is obtained from Eq. (10.31) by assuming a cylindrically symmetric form for $\psi(\vec{r})$ given by [8]

$$\psi(\vec{r}) = \sqrt{n_0(\vec{r})}e^{i\varphi(\vec{r})} = \psi_r(r)e^{in\varphi} \tag{10.32}$$

where n is an integer. We see from Eq. (10.32) that the radial distribution of superfluid about the core is characterized by the rotationally invariant form $|\psi(\vec{r})| = \psi_r(r)$. In this regard, note that $\varphi(\vec{r}) = n\varphi$ so that the wavefunction along a closed path enclosing the z−axis is a continuous periodic function in φ.

In the consideration of the superfluid phase, it is required that $a(T) < 0$ in order for the superfluid phase to exist. Under this condition, it then follows upon substituting the proposed vortex form in Eq. (10.32) into Eq. (10.31) that [8]

$$-\frac{\hbar^2}{2m}\left[\frac{\partial^2\psi_r}{\partial r^2}+\frac{1}{r}\frac{\partial\psi_r}{\partial r}-\frac{n^2\psi_r}{r^2}\right]+\left|a(T)\right|\psi_r+b\left|\psi_r\right|^2\psi_r=0. \tag{10.33}$$

To provide a simplification in the treatment of Eq. (10.33), we next make a change of variables by defining new variables $l=\frac{\hbar}{\sqrt{2m\left|a(T)\right|}}$, $\xi=\frac{r}{l}$, and $\psi_0=\sqrt{\frac{b}{\left|a(T)\right|}}\psi_r$. In terms of these new variables and with the further assumption that ψ_r is real, we obtain the vortex equation [7, 8]

$$\xi^2\frac{\partial^2\psi_0}{\partial\xi^2}+\xi\frac{\partial\psi_0}{\partial\xi}+\left(\xi^2-n^2\right)\psi_0-\xi^2\psi_0^{\,3}=0. \tag{10.34}$$

(Notice here that in the absence of the nonlinear term, Eq. (10.34) is just the cylindrical Bessel equation.)

The solution of the nonlinear equation Eq. (10.34) can be obtained numerically [8] and is found to smoothly interpolate between the limiting forms of the solutions given by

$$\lim_{\xi\to 0}\psi_0\to\left|\xi\right|^{|n|}\to 0 \tag{10.35a}$$

and

$$\lim_{\xi\to\infty}\psi_0\to 1-\frac{n^2}{\xi^2}\to 1. \tag{10.35b}$$

Note in this regard that an unphysical singular solution to Eq. (10.34) also exists and is discarded in order to meet the boundary conditions that the wavefunction is nonsingular.

It is important to note that the vortex solution obtained from Eq. (10.34) is not the lowest energy state of the system but describes an excitation of the system in which a filament of normal fluid is introduced into the vortex core. In addition, the solution only exists for the Helium II system because in the macroscopic condensate of Helium II all the particles carry the same coherent phase. This differs from the case of the normal fluid and that of the dilute or ideal gas, which are collections of particles, each of which carries their own phase.

As a final point about the vortex state, it is interesting to consider the velocity of the circulating superfluid about the core of the vortex. From Eq. (10.30c) and the expression for the gradient in cylindrical coordinates, we find for $\varphi(\vec{r})=n\varphi$ in Eq. (10.32) that the velocity of the superfluid as it circulates about the vortex core is given by [6, 8]

$$\vec{v}=\frac{\vec{j}}{n_0}=\frac{\hbar}{m}\nabla\varphi=n\frac{\hbar}{m}\frac{1}{r}\hat{\varphi}. \tag{10.36}$$

Note that the superfluid travels along circular trajectories centered on the core of normal material. In addition, for $r\to 0$, as the core of the vortex is approached, the tangential velocity diverges. This divergence, however, is offset by the limiting form of the radial function, i.e., $\psi_r(r)\to 0$. In addition, it follows from Eq. (10.36) that [6, 8]

$$\lim_{r\to\infty}\vec{v}\to 0, \tag{10.37}$$

i.e., the circulation velocity decays to zero at great distance from the vortex core.

10.4 ORIGINS OF SUPERFLUIDITY

The novel property characterizing the superfluid component of liquid helium is its absence of viscosity [5, 7]. Early in the study of superfluidity, a discussion of the origins of the zero viscosity was made by Landau [5] in terms of the excitation spectrum of Helium II. Specifically, Landau proposed that the viscous behavior of the superfluid would arise in a system in which the modes of excitation from the ground state had a particular form of the energy versus momentum dispersion relation [5]. The form of the excitation spectrum chosen was such that, below a certain threshold velocity relative to a rough surface, excitations could not be generated by the surface roughness interacting with the ground state of the Helium II. Specifically, this feature has a fundamental effect in contributing to the zero-viscosity displayed by the liquid below a critical threshold velocity.

To understand Landau's ideas, consider a fluid that is flowing down a tube. The viscosity of the fluid is manifest in the scattering of the fluid by the roughness of the tube walls. In a collision of the fluid with the walls, a certain amount of momentum is transferred to modal excitations of the liquid, and this then shows up as a change in the energy and velocity of the flow, i.e., a viscosity. Ultimately, in this way, a series of successive collisions can result in a friction which reduces the speed of the liquid moving down the tube. To understand these scattering processes, let us consider a collision event in which a single excitation is created in the fluid and determine its consequence to the fluid flow.

We treat a system in which the fluid moves with a speed v down a tube. First consider the single excitation scattering event in the rest frame of the fluid of mass $M \to \infty$ where the energy of the fluid is initially $E_0 = \frac{1}{2M}0^2 = 0$. After the interaction with the wall and the resulting creation of an excitation of momentum $\vec{p}$, the momentum of the fluid becomes [5–7]

$$\vec{P}_{fluid} = \vec{p} \tag{10.38a}$$

and the energy of the fluid becomes

$$E_{fluid} = E_0 + \epsilon(p) + \frac{1}{2M}p^2 \approx E_0 + \epsilon(p) = \epsilon(p). \tag{10.38b}$$

Here E_{fluid} is the energy in the fluid rest frame after the introduction of the excitation, E_0 is the ground state energy of the fluid before the excitation is created, and $\epsilon(p)$ is the energy of the mode excited in the fluid. Notice that in the last steps on the right of Eq. (10.38b) we have used the fact the M is very large and $E_0 = 0$.

Consider now this same event in the rest frame of the tube. The fluid is moving in the tube with a velocity $\vec{v}$. Consequently, in the rest frame of the tube, for a fluid of mass $M \to \infty$, the momentum of the fluid in the presence of the excitation $\vec{p}$ [5–7] is given by

$$\vec{P}_{tube} = \vec{P}_{fluid} + M\vec{v} = \vec{p} + M\vec{v} \tag{10.39a}$$

and the energy of the fluid in the presence of the excitation is

$$E_{tube} = \epsilon(p) + \frac{1}{2M}\left|\vec{p} + M\vec{v}\right|^2. \tag{10.39b}$$

Here $\vec{P}_{fluid}$ is from (10.38a) and the second term on the right of Eq. (10.39b) accounts for the additional kinetic energy from the relative motions of the two systems following the $\vec{p}$ momentum transfer from the tube. Note that Eq. (10.39b) can be further simplified as [5–7]

$$E_{tube} = \epsilon(p) + \frac{1}{2M}p^2 + \frac{1}{2M}\left[2M\vec{p}\cdot\vec{v} + M^2v^2\right] \tag{10.39c}$$

$$\approx \epsilon(p) + \vec{p}\cdot\vec{v} + \frac{1}{2}Mv^2$$

where we have used the large mass of the fluid (i.e., $M \to \infty$) in the last step in Eq. (10.39c).

Consequently, we see from Eq. (10.39c) that in the absence of the excitation of momentum $\vec{p}$ the energy of the system in the tube frame is

$$E_{tube}(absence) = \frac{1}{2}Mv^2 \tag{10.40a}$$

while in the presence of the excitation of momentum $\vec{p}$ the energy of the system in the tube frame is

$$E_{tube}(presence) = \epsilon(p) + \vec{p}\cdot\vec{v} + \frac{1}{2}Mv^2. \tag{10.40b}$$

It follows from Eqs. (10.40a) and (10.40b) that the system will support the presence of an excitation of momentum $\vec{p}$ provided that [5–7]

$$E_{tube}(presence) - E_{tube}(absence) = \epsilon(p) + \vec{p}\cdot\vec{v} < 0, \tag{10.41a}$$

i.e., if the creation of the excitation will lower the energy of the system. The introduction to the system of an excitation of momentum $\vec{p}$ will also be seen to result in a viscous slowing of the fluid.

Choosing $\vec{p}$ and $\vec{v}$ as oppositely directed vectors gives $\vec{p}\cdot\vec{v} = -|\vec{p}||\vec{v}|$ which optimally contributes to satisfying the inequality in Eq. (10.41a). For this condition, it then follows from Eq. (10.41a) that to create an excitation requires [5–7]

$$|\vec{v}| > \frac{\epsilon(p)}{|\vec{p}|}. \tag{10.41b}$$

We find that an excitation with a momentum $\vec{p}$ opposite to the velocity $\vec{v}$ will occur in the system provided that the relative velocity of the tube and fluid has a speed which is greater than $\frac{\epsilon(p)}{|\vec{p}|}$. Since the contributed momentum $\vec{p}$ is opposite to the velocity $\vec{v}$ the effect of the excitation is to slow the fluid.

As an example of Eq. (10.41b), notice that for a particle dispersion relation in the fluid of the form $\epsilon(p) = \frac{p^2}{2m}$, we find that [5–7]

$$|\vec{v}| > \frac{\epsilon(p)}{|\vec{p}|} = \frac{|\vec{p}|}{2m}. \tag{10.42}$$

Since we can choose $|\vec{p}|$ to be arbitrarily small, for some $|\vec{p}|$ this inequality is always satisfied for a free particle with a quadratic dispersion relation. As a result, a fluid with such a particle dispersion relation would always exhibit a nonzero viscosity. Consequently, the excitations of Helium II must differ from those of a gas of free particles.

In the original treatment of superfluidity in Helium II, Landau proposed a model dispersion relation [5] for the excitations from the ground state which is shown schematically in Figure 10.2. At low momenta the dispersion is a linear phonon type of dispersion which arises from the strongly scattered helium atom interactions. This linear part is represented by the low momentum form [8]

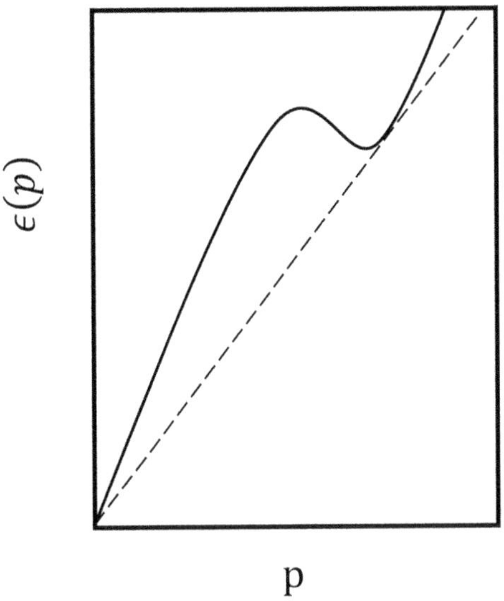

FIGURE 10.2 Schematic of the dispersion relation of energy versus momentum for the excitations in Helium II as proposed by Landau. The solid line is the proposed form of the dispersion relation. The dashed line is the lower bounding line $\epsilon\left(p\right)=v_c p$ given in Eq. (10.43c).

$$\epsilon\left(p\right)=cp \tag{10.43a}$$

where $c=2.4\times10^4\ cm/sec$. As the momentum is increased, a region of a local minimum of the dispersion relation is proposed which is found to be characterized by the form [8]

$$\epsilon\left(p\right)=\Delta+\frac{\left(p-p_0\right)^2}{2m^*} \tag{10.43b}$$

where $\frac{\Delta}{k_B}=8.7^\circ K$, $p_0=1.9\,Ang^{-1}$, $m^*=0.16m$, and m is the mass of the helium atom.

As proposed, the spectrum is bounded from below by the dashed line in Figure 10.2 that has the form

$$\epsilon\left(p\right)=v_c p. \tag{10.43c}$$

From Eq. (10.41b) we see that v_c is the critical velocity such that if the fluid moves through the tube with $\left|\vec{v}\right|>v_c$ it experiences a nonzero viscosity. At velocities less than v_c, however, there is a zero-viscosity flow.

The original proposal for the form of the dispersion in Eq. (10.43) was made considering the macroscopic thermodynamic properties of the Helium II phase, including its superfluidity. An early analytic theory offering a justification for the form of the dispersion relation, however, can also be offered. This is made on the basis of the so-called Gross-Pitaevskii equation [6–9] in a consideration of the microscopic Hamiltonian of a superfluid model.

10.5 GROSS-PITAEVSKII EQUATION

In the Gross-Pitaevskii equation [6, 7, 9] the ground state of a gas of atoms interacting with one another through delta-function atomic pair potentials is studied. The formulation is made in the context of the Hartree approximation in which the ground state wavefunction of the atomic gas is represented as a product of atomic wavefunctions which are of the same form for each of the atoms in the gas [6, 7]. This is then consistent with the atoms being bosons and for the limits of low temperatures and low densities. In this representation, each atom is part of the ground state of the system, and the ground state energy of the gas is computed using the variational method [7]. From the resulting Gross-Pitaevskii equation [6, 7, 9] a theory for the single excitation dispersion is subsequently developed.

To formulate the Gross-Pitaevskii equation we begin with the many-body Hamiltonian of a dilute fluid of bosons at zero temperature [6, 7, 9]. In the absence of an interaction with an outside field this has the general form

$$H = \sum_i \frac{p_i^2}{2m} + \frac{1}{2}\sum_{i\neq j} v\left(\left|\vec{r}_i - \vec{r}_j\right|\right) \tag{10.44}$$

where $i, j = 1,2,3,\ldots,N$ for $N \to \infty$ label the atoms of the gas and $v\left(\left|\vec{r}_i - \vec{r}_j\right|\right)$ is a pair potential describing the scattering between the atoms of the gas. To study the ground state of the system we assume a Hartree form for the ground state wavefunction given by the product [6, 7, 9]

$$\psi\left(\vec{r}_1,\vec{r}_2,\vec{r}_3,\ldots,\vec{r}_N\right) = \prod_{i=1}^{N}\varphi\left(\vec{r}_i\right) = \langle \vec{r}_1,\vec{r}_2,\vec{r}_3,\ldots,\vec{r}_N \mid \psi\rangle = \prod_{i=1}^{N}\langle \vec{r}_i \mid \varphi\rangle \tag{10.45}$$

where for each i the single particle wavefunctions $\varphi\left(\vec{r}_i\right)$ have the same functional form. (Notice that on the far right side of Eq. (10.45) we have adopted Dirac notation.) The functional form of the $\varphi\left(\vec{r}_i\right)$ is considered to be determined in the course of the treatment of H, during which the wavefunctions are normalized such that $\langle \psi \mid \psi\rangle = 1$ and $\langle \varphi \mid \varphi\rangle = 1$.

In terms of Eqs. (10.44) and (10.45) the ground state energy of the system is then given by [6, 7, 9]

$$\langle \psi \mid H \mid \psi\rangle = \langle \psi \mid \sum_i \frac{p_i^2}{2m} \mid \psi\rangle + \langle \psi \mid \frac{1}{2}\sum_{i\neq j} v\left(\left|\vec{r}_i - \vec{r}_j\right|\right) \mid \psi\rangle \tag{10.46a}$$

$$= \sum_i \frac{\hbar^2}{2m}\int \nabla_i\varphi^*\left(\vec{r}_i\right)\cdot\nabla_i\varphi\left(\vec{r}_i\right) + \frac{1}{2}\sum_{i\neq j}\int\int \varphi^*\left(\vec{r}_i\right)\varphi^*\left(\vec{r}_j\right)v\left(\left|\vec{r}_i - \vec{r}_j\right|\right)\varphi\left(\vec{r}_i\right)\varphi\left(\vec{r}_j\right)d\vec{r}_i d\vec{r}_j$$

$$= -N\frac{\hbar^2}{2m}\int \varphi^*\left(\vec{r}\right)\nabla^2\varphi\left(\vec{r}\right) + \frac{N(N-1)}{2}\int\int \varphi^*\left(\vec{r}\right)\varphi^*\left(\vec{r}'\right)v\left(\left|\vec{r} - \vec{r}'\right|\right)\varphi\left(\vec{r}\right)\varphi\left(\vec{r}'\right)d\vec{r}d\vec{r}'$$

where in the variational treatment of the ground state energy [9],

$$E\left(\psi\right) = \frac{\langle \psi \mid H \mid \psi\rangle}{\langle \psi \mid \psi\rangle} \tag{10.46b}$$

is minimized in terms of $|\psi\rangle$ with respect to the form chosen for the single particle wavefunctions $\varphi\left(\vec{r}\right)$. Considering the thermodynamics of the system, its free energy is then also given by extremizing the free energy form [9]

$$F\left(\psi\right) = \langle \psi \mid H \mid \psi\rangle - N\mu\langle \psi \mid \psi\rangle \tag{10.47}$$

considered as a function of ψ and where μ is the chemical potential.

From a consideration of Eq. (10.47) we find that [6, 7, 9]

$$\frac{\delta F}{\delta \psi^*} = 0 \tag{10.48}$$

$$= N \int d\vec{r} \left\{ -\frac{\hbar^2}{2m} \nabla^2 \varphi(\vec{r}) - \mu\varphi(\vec{r}) + (N-1)\int d\vec{r}' \left|\varphi(\vec{r})\right|^2 v\left(\left|\vec{r}-\vec{r}'\right|\right)\varphi(\vec{r}) \right\} \delta\varphi^*(\vec{r})$$

where we use the fact that $\varphi(\vec{r})$ being complex allows us to treat $\varphi(\vec{r})$ and $\varphi^*(\vec{r})$ as independent variables. In this regard, a consideration of $\frac{\delta F}{\delta \psi} = 0$ just yields the complex conjugate of Eq. (10.48).

A hard-core potential used to describe the scattering between the atoms of the gas is often taken to have the form [9]

$$\begin{aligned} v\left(\left|\vec{r}-\vec{r}'\right|\right) &= \frac{4\pi\hbar^2}{m} a_s \delta\left(\left|\vec{r}-\vec{r}'\right|\right) \\ &= g\delta\left(\left|\vec{r}-\vec{r}'\right|\right) \end{aligned} \tag{10.49}$$

where a_s is the so-called s-wave scattering length characterizing the strength of the atomic scattering and $g = \frac{4\pi\hbar^2}{m} a_s$. Applying Eq. (10.49) in Eq. (10.48), we then find that [9]

$$-\frac{\hbar^2}{2m}\nabla^2\varphi(\vec{r}) + Ng\left|\varphi(\vec{r})\right|^2 \varphi(\vec{r}) = \mu\varphi(\vec{r}) \tag{10.50}$$

which is a common form of the Gross-Pitaevskii equation [6, 7, 9]. Notice the interesting comparison of Eq. (10.50), obtained from the free energy form in Eq. (10.47), with Eq. (10.29), obtained from the Ginzburg form for the free energy.

10.5.1 An Example

As an example, consider the special case of an atomic gas composed of planewave states of the form $\varphi(\vec{r}) = \frac{1}{\sqrt{V}} e^{i\vec{k}\cdot\vec{r}}$ where V is the volume of the gas. Substituting this into Eq. (10.50) we find that [6, 7, 9]

$$\frac{\hbar^2 k^2}{2m} + \frac{N}{V} g = \mu(\vec{k}) \tag{10.51}$$

where $\mu(\vec{k})$ is the chemical potential of the ground state in which each atom translates with wavevector $\vec{k}$. The chemical potential is, consequently, found to be shifted from that of the system at rest (i.e., $\mu(\vec{k}) = \frac{N}{V} g$) by the particle kinetic energy $\frac{\hbar^2 k^2}{2m}$.

10.5.2 Time-Dependence

By similar considerations to those given in obtaining the free energy form in Eq. (10.50), an extension to consider the time-dependent free energy can be made. From these, it is shown that an appropriate partial differential equation for the time-dependent system is given by [6, 7, 9]

$$-\frac{\hbar^2}{2m}\nabla^2\varphi(\vec{r},t)+Ng\left|\varphi(\vec{r},t)\right|^2\varphi(\vec{r},t)=i\hbar\frac{\partial}{\partial t}\varphi(\vec{r},t). \tag{10.52}$$

This is an important partial differential equation of mathematical physics known as the nonlinear Schrodinger equation [7]. Due to its nonlinearity, it has a variety of interesting properties including the presence of soliton solutions and a variety of solutions exhibiting the characteristics of chaotic dynamics. We shall next look at an interesting form of perturbational dynamical solutions of Eq. (10.52) taken about the stationary ground state. These are obtained in the so-called Bogoliubov approximation [7, 9].

We begin by looking at solutions which are obtained as perturbations of the planewave forms considered in Eq. (10.51) for the stationary case of the ground state in which $\vec{k}=0$. Specifically, we consider [6, 7, 9]

$$\varphi(\vec{r},t)=\varphi_0(\vec{r},t)+\delta\varphi(\vec{r},t) \tag{10.53a}$$

where $\varphi_0(\vec{r},t)$ is a ground state orbital of the $\vec{k}=0$ system given by

$$\varphi_0(\vec{r},t)=\frac{1}{\sqrt{V}}e^{-i\frac{\mu}{\hbar}t} \tag{10.53b}$$

and $\delta\varphi(\vec{r},t)$ is a perturbation of the system.

Upon then substituting Eq. (10.53) into Eq. (10.52), it follows that [6, 7, 9]

$$i\hbar\frac{\partial}{\partial t}\delta\varphi=-\frac{\hbar^2}{2m}\nabla^2\delta\varphi+Ng\left(2\left|\varphi_0\right|^2\delta\varphi+\varphi_0^2\delta\varphi^*\right) \tag{10.54a}$$

$$=-\frac{\hbar^2}{2m}\nabla^2\delta\varphi+\frac{N}{V}g\left(2\delta\varphi+e^{-2i\mu t}\delta\varphi^*\right)$$

and similarly [6, 7, 9]

$$-i\hbar\frac{\partial}{\partial t}\delta\varphi^*=-\frac{\hbar^2}{2m}\nabla^2\delta\varphi^*+Ng\left(2\left|\varphi_0\right|^2\delta\varphi^*+\left(\varphi_0^*\right)^2\delta\varphi\right) \tag{10.54b}$$

$$=-\frac{\hbar^2}{2m}\nabla^2\delta\varphi^*+\frac{N}{V}g\left(2\delta\varphi^*+e^{2i\mu t}\delta\varphi\right).$$

These represent two equations to be solved for the variables $\delta\varphi$ and $\delta\varphi^*$, and the solvability condition gives the dispersion relation of the low energy excitations of the system.

In a matrix format, Eq. (10.54) is next written as the expression [6, 7, 9]

$$\begin{vmatrix} i\hbar\frac{\partial}{\partial t}+\frac{\hbar^2}{2m}\nabla^2-2\frac{N}{V}g & -\frac{N}{V}ge^{-2i\frac{\mu}{\hbar}t} \\ -\frac{N}{V}ge^{2i\frac{\mu}{\hbar}t} & -i\hbar\frac{\partial}{\partial t}+\frac{\hbar^2}{2m}\nabla^2-2\frac{N}{V}g \end{vmatrix}\begin{vmatrix}\delta\varphi \\ \delta\varphi^*\end{vmatrix}=0. \tag{10.55a}$$

To solve Eq. (10.55a) let us assume that the variations in $\delta\varphi(\vec{r},t)$ and $\delta\varphi^*(\vec{r},t)$ are given by the forms

$$\delta\varphi(\vec{r},t) = e^{-i\frac{\mu}{\hbar}t}\left[ae^{i\vec{q}\cdot\vec{r}}e^{-i\omega t} - b^{*}e^{-i\vec{q}\cdot\vec{r}}e^{i\omega t}\right] \tag{10.55b}$$

and

$$\delta\varphi^{*}(\vec{r},t) = e^{i\frac{\mu}{\hbar}t}\left[a^{*}e^{-i\vec{q}\cdot\vec{r}}e^{i\omega t} - be^{i\vec{q}\cdot\vec{r}}e^{-i\omega t}\right] \tag{10.55c}$$

where a and b are new unknowns used to characterize $\delta\varphi(\vec{r},t)$ and $\delta\varphi^{*}(\vec{r},t)$. Upon substituting Eqs. (10.55b) and (10.55c) into Eq. (10.55a) we then find that

$$\begin{vmatrix} \frac{\hbar^2}{2m}q^2 + \frac{N}{V}g - \hbar\omega & -\frac{N}{V}g \\ -\frac{N}{V}g & \frac{\hbar^2}{2m}q^2 + \frac{N}{V}g + \hbar\omega \end{vmatrix} \begin{vmatrix} a \\ b \end{vmatrix} = 0 \tag{10.55d}$$

expresses the problem in terms of the new unknowns a and b. Note that in arriving at Eq. (10.55d) we have also used the result from Eq. (10.51) that in the stationary ground state $\mu(\vec{k}=0) = \frac{N}{V}g$.

Applying the solvability condition to Eq. (10.55d), it then follows that for a solution to exist [6, 7, 9]

$$\left[\frac{\hbar^2}{2m}q^2 + \frac{N}{V}g - \hbar\omega\right]\left[\frac{\hbar^2}{2m}q^2 + \frac{N}{V}g + \hbar\omega\right] - \left(\frac{N}{V}g\right)^2 = 0. \tag{10.56}$$

From this we arrive at the dispersion relation given by the form

$$\hbar\omega = \pm\sqrt{\frac{\hbar^2}{2m}q^2\left(\frac{\hbar^2}{2m}q^2 + 2g\frac{N}{V}\right)}. \tag{10.57}$$

The result in Eq. (10.57) is, consequently, seen to give the frequencies of the modes of excitations from the ground state of the system in its rest configuration.

Notice that the dispersion relation in Eq. (10.57) is consistent with the Landau postulate [5] for the form required of the excitation spectrum from the ground state in order that the fluid exhibit superfluidity [6, 7, 9]. In this regard, for $q \to 0$

$$\hbar\omega \to \sqrt{\frac{\hbar^2}{m}g\frac{N}{V}}|q| \tag{10.58a}$$

which is linear in $|q|$, and for $q \to \infty$

$$\hbar\omega \to \frac{\hbar^2}{2m}q^2 \tag{10.58b}$$

which approaches a quadratic in $|q|$. This allows for it to mimic the dispersion relation proposed in Eq. (10.43).

10.6 PATH-INTEGRAL QUANTUM MONTE CARLO STUDY OF HELIUM II

We next return to a discussion of the quantum Monte Carlo as a means of computing the quantum statistical mechanics of a system of interacting particles [1]. Our earlier considerations regarded the particles of the system as distinguishable particles distributed according to the Maxwell-Boltzmann distribution. While this is appropriate for atoms bound in a lattice in which the tunneling between different lattice sites is small, it becomes a poor approximation for identical particles in the liquid state and particularly in a highly correlated quantum fluid characterized as a macroscopic condensate.

For example, in the superfluid state of Helium II it is necessary to regard the identical particles (i.e., the atoms) in the system as obeying Bose statistics. In such a system the identical nature of the particles requires a treatment going beyond the simple classical statistical mechanical remedies applied to resolve the Maxwell paradox of identical classical mechanical particles [5, 6]. It involves an account of the full symmetry properties of the Hamiltonian and its corresponding wavefunctions under the permutations of the indices labeling the particles in the system. This accounting shows up as a reformulation of the path-integral methodology developed for a quantum Monte Carlo simulation of the system.

In applying the path integral method to the study of Helium I and II, it needs to be modified to account for the identical nature of the helium atoms and their resultant Bose statistics. To understand this, consider that we are given the atomic position basis [1]

$$\left|\vec{R}\right\rangle = \left|\vec{r}_1, \vec{r}_2, \vec{r}_3 \ldots, \vec{r}_N\right\rangle \tag{10.59a}$$

in which the helium atoms are labeled by subscript indices $1,2,3,\ldots,N$ and located in space by the position vector

$$\vec{R} = \left(\vec{r}_1, \vec{r}_2, \vec{r}_3 \ldots, \vec{r}_N\right). \tag{10.59b}$$

One can permute the atomic position labels with a permutation operator P such that

$$P\left|\vec{R}\right\rangle = \left|\vec{r}_{P1}, \vec{r}_{P2}, \vec{r}_{P3} \ldots, \vec{r}_{PN}\right\rangle \tag{10.59c}$$

and

$$P\vec{R} = \left(\vec{r}_{P1}, \vec{r}_{P2}, \vec{r}_{P3} \ldots, \vec{r}_{PN}\right). \tag{10.59d}$$

Since the Hamiltonian of the identical particles is invariant under this relabeling or permutation symmetry, it follows that the wavefunctions of a Bose system of particles must also be invariant under the permutation. This requires that the wavefunction of the bosonic system be of the form [1]

$$\psi_B\left(\vec{r}_1, \vec{r}_2, \vec{r}_3 \ldots, \vec{r}_N\right) = \frac{1}{\sqrt{N!}} \sum_P P\psi_{MB}\left(\vec{r}_1, \vec{r}_2, \vec{r}_3 \ldots, \vec{r}_N\right) \tag{10.60}$$

$$= \frac{1}{\sqrt{N!}} \sum_P \psi_{MB}\left(\vec{r}_{P1}, \vec{r}_{P2}, \vec{r}_{P3} \ldots, \vec{r}_{PN}\right)$$

where $\psi_B\left(\vec{r}_1, \vec{r}_2, \vec{r}_3 \ldots, \vec{r}_N\right) = \langle \vec{r}_1, \vec{r}_2, \vec{r}_3 \ldots, \vec{r}_N | \psi_B \rangle$ is the particle wavefunction treated as distributed by Bose statistics, $\psi_{MB}\left(\vec{r}_1, \vec{r}_2, \vec{r}_3 \ldots, \vec{r}_N\right) = \langle \vec{r}_1, \vec{r}_2, \vec{r}_3 \ldots, \vec{r}_N | \psi_{MB} \rangle$ is the wavefunction of the particles treated as distinguishable Maxwell-Boltzmann particles, and the sum is over the set of all the possible permutations $\{P\}$ of the particle labeling. For this process, notice from Eq. (10.60) that $\langle \psi_B | \psi_B \rangle = \frac{1}{N!} \sum_P \langle \psi_{MB} P | P\psi_{MB} \rangle = 1$ accounts for the normalization of the new symmetrized boson wavefunction.

The introduction of Bose statistics into the partition function defined in Eq. (10.6) is then given from a consideration of the form [1]

$$Z = Tre^{-\beta H} = Tr\, e^{-\frac{\beta}{M}H} e^{-\frac{\beta}{M}H} e^{-\frac{\beta}{M}H} \ldots\ldots\ldots e^{-\frac{\beta}{M}H} \tag{10.61}$$

where now the trace and matrix products are expressed in the complete basis of boson wavefunctions $\{\psi_B(\vec{r}_1,\vec{r}_2,\vec{r}_3\ldots,\vec{r}_N) = \langle \vec{r}_1,\vec{r}_2,\vec{r}_3\ldots,\vec{r}_N | \psi_B\}$. These involve the symmetrized forms in Eq. (10.60) rather than the unsymmetrized Maxwell-Boltzmann wavefunctions $\{\psi_{MB}(\vec{r}_1,\vec{r}_2,\vec{r}_3\ldots,\vec{r}_N) = \langle \vec{r}_1,\vec{r}_2,\vec{r}_3\ldots,\vec{r}_N | \psi_{MB}\rangle\}$. Since the symmetrized boson wavefunctions involve a sum over a set of permuted Maxwell-Boltzmann wavefunctions, the difficulty of handling the identical particle problem is markedly increased over that of the case of distinguishable particles. Nevertheless, by a consideration of the factors $e^{-\frac{\beta}{M}H}$ a path integral representation can be made, and this has been studied for Helium II.

Developed in this symmetrized format, a quantum Monte Carlo treatment has recently been given [1] for the Helium II system based on Eq. (10.61). This involved a study of the thermodynamic properties of the superfluid system and has provided a verification of many of the earlier phenomenological considerations. In these treatments, the facilitation of the numerical evaluation of the path integral required the development of several additional numerical techniques for processing and generating the simulated data. In the following we shall only summarize a few features of the simulation studies which are of interest in regard to our earlier discussion of the properties of the Helium II superfluid. For the details of these studies the reader is referred to the review given by Ceperley [1].

A first interesting point from Ref. 1 is that by a study of the scattering structure factor computed in the quantum Monte Carlo, the single particle excitation dispersion relation can be generated. Results for the dispersion relation obtained in this manner are shown in Figure 10.3. The plot shows simulation data points presented against a curve of the dispersion measured from experimental data. The simulation results are found to agree well with experiment and are a vindication of the Landau conjecture [5] made to explain zero viscosity.

As an additional point we show in Figure 10.4 a plot of the specific heat and energy in the vicinity of the lambda transition [12]. Again, the agreement between the simulation and the experimental data is found to be good. The simulation and experimental results are consistent with a singular specific heat at the phase transition. This contradicts the finite discontinuity of the specific heat in the prediction of the mean field theory (Ginzburg) treatment of the free energy [1, 8, 12].

Next, we turn to a brief consideration of the superfluidity exhibited by liquid ^3He. This is a system of atoms which exhibits superfluidity, but, due to its total nuclear spin $\frac{1}{2}$, obeys Fermi-Dirac statistics. Consequently, the origins of the Fermi-Dirac superfluid state in ^3He are from quite a different set of mechanisms than those resulting in the Bose-Einstein superfluid state of ^4He. In this regard, the application of the quantum Monte Carlo to identical particles with Fermi-Dirac statistics has many complications in the development of path integral methods and will not be discussed here [1–3].

10.7 SUPERFLUIDITY IN LIQUID ^3He

The superfluidity exhibited by ^3He originates through quite a different set of physical processes than those found in the superfluid ^4He system [6, 7, 10, 11]. This is because the particles in ^3He are fermions with spin $\frac{1}{2}$ nuclei whereas the particles of ^4He are bosons with spin zero nuclei. In this regard, ^3He has a superfluid state which initiates at temperatures of order of 10^{-3} times lower than

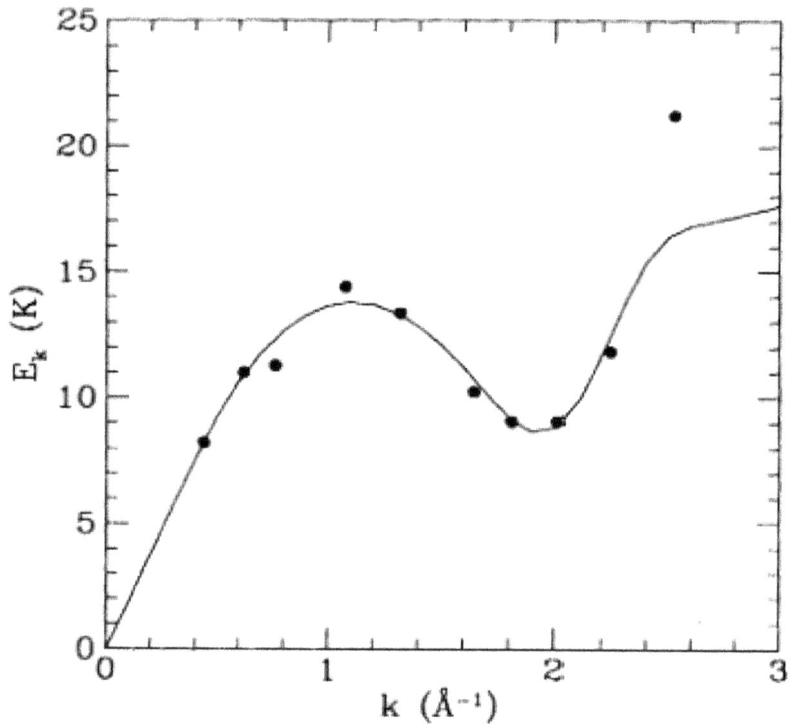

FIGURE 10.3 Plot of the energy versus momentum of the single particle excitation spectrum in Helium II. The line is from experimental data and the dots are from quantum Monte Carlo studies [1]. (Reprinted figure with permission from [1]. Copyright 1995, American Physical Society.)

those found in the boson system, with an origin related to that of the superfluidity observed in low-temperature superconductors [7].

A schematic of the low-temperature pressure-temperature phase diagram of ^3He is shown in Figure 10.1b where it is seen that there are actually two different superfluid phases, i.e., a region of an A and a region of a B superfluid phase [7]. These regions arise from interactions between pairs of atoms in the fluid which are different from the s-wave pairing interactions observed in a majority of low-temperature electronic superconducting materials. Ultimately it is found that the pairing in ^3He arises from a mechanism of p-wave pairing interactions. To understand the differences between the electronic superconductors, the Helium II phase and ^3He superfluids, we begin with a brief review of electronic superconductors based on s-wave pairing. This is followed by a treatment of the new pairing mechanisms found in the ^3He phases, with a focus on their differences from the superfluidity processes in Helium II.

10.7.1 S-Wave Pairing in Superconductors

Consider a system of noninteracting fermions at zero temperature. In wavevector space the Hamiltonian of the system has the form [7]

$$H = \sum_{\vec{k},\sigma} \left[\epsilon(k) - \mu\right] c^{\dagger}_{\vec{k}\sigma} c_{\vec{k}\sigma} \tag{10.62}$$

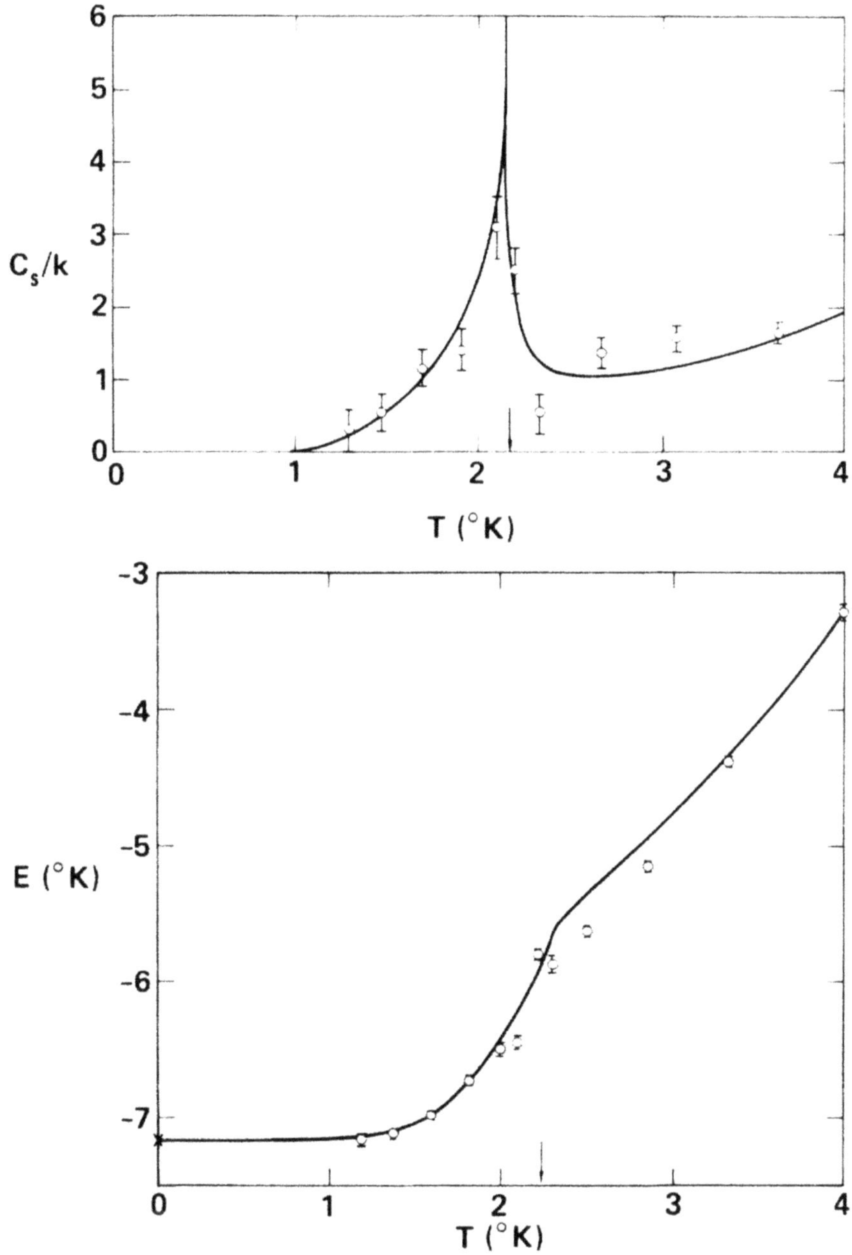

FIGURE 10.4 Plot of: (a) the specific heat versus temperature at the Helium I–Helium II lambda transition and (b) the energy versus temperature at the Helium I–Helium II lambda transition. The points are the quantum Monte Carlo simulation data [12]. (Reprinted figure with permission from [12]. Copyright 1986, American Physical Society.)

where $\epsilon(k) = \dfrac{\hbar^2 k^2}{2m}$ is the free particle energy dispersion, m is the effective mass of each of the particles, $\sigma = \uparrow, \downarrow$ is the particle spin (i.e., typically the electron spin in electronic superconductors), μ is the chemical potential of the system of fermions, and $c^{\dagger}_{\vec{k}\sigma}$ and $c_{\vec{k}\sigma}$ are the fermion creation and destruction operators obeying the fermion anticommutation relations

$$\left\{c_{\vec{k}\sigma}, c^{\dagger}_{\vec{k}'\sigma'}\right\} = \delta_{\vec{k}\sigma,\vec{k}'\sigma'} \tag{10.63a}$$

$$\left\{c^{\dagger}_{\vec{k}\sigma}, c^{\dagger}_{\vec{k}'\sigma'}\right\} = 0 \tag{10.63b}$$

and

$$\left\{c_{\vec{k}\sigma}, c_{\vec{k}'\sigma'}\right\} = 0. \tag{10.63c}$$

For the system described by Eq. (10.62) we shall further consider that at zero temperature the states with $\epsilon(k) < \mu$ are occupied by particles while those with $\epsilon(k) > \mu$ have no occupying particles.

By applying the Heisenberg equations of motion, we find that

$$i\hbar\frac{d}{dt}c^{\dagger}_{\vec{k}\sigma} = \left[c^{\dagger}_{\vec{k}\sigma}, H\right] = -\left[\epsilon(k) - \mu\right]c^{\dagger}_{\vec{k}\sigma} \tag{10.64a}$$

accounts for the free particle dynamics. Similarly, it also follows that

$$i\hbar\frac{d}{dt}c_{\vec{k}\sigma} = \left[c_{\vec{k}\sigma}, H\right] = \left[\epsilon(k) - \mu\right]c_{\vec{k}\sigma}. \tag{10.64b}$$

Consequently, the time-dependence of the particle creation and destruction operators is of the form [7]

$$c^{\dagger}_{\vec{k}\sigma}(t) = c^{\dagger}_{\vec{k}\sigma}(0)e^{i\frac{\epsilon_{\vec{k},\sigma}}{\hbar}t} \tag{10.64c}$$

and

$$c_{\vec{k}\sigma}(t) = c_{\vec{k}\sigma}(0)e^{-i\frac{\epsilon_{\vec{k},\sigma}}{\hbar}t} \tag{10.64d}$$

where $\epsilon_{\vec{k},\sigma} = \left[\epsilon(k) - \mu\right]$.

Adopting a matrix notation, Eq. (10.62) can then be rewritten as [7]

$$H = \sum_{\vec{k}} \begin{vmatrix} c^{\dagger}_{\vec{k}\uparrow} & c_{-\vec{k}\downarrow} \end{vmatrix} \begin{vmatrix} H_{\vec{k}\uparrow,\vec{k}\uparrow} & H_{\vec{k}\uparrow,-\vec{k}\downarrow} \\ H_{-\vec{k}\downarrow,\vec{k}\uparrow} & H_{-\vec{k}\downarrow,-\vec{k}\downarrow} \end{vmatrix} \begin{vmatrix} c_{\vec{k}\uparrow} \\ c^{\dagger}_{-\vec{k}\downarrow} \end{vmatrix} \tag{10.65a}$$

$$= \sum_{\vec{k}} \begin{vmatrix} c^{\dagger}_{\vec{k}\uparrow} & c_{-\vec{k}\downarrow} \end{vmatrix} \begin{vmatrix} \epsilon(\vec{k}) - \mu & 0 \\ 0 & -\left[\epsilon(\vec{k}) - \mu\right] \end{vmatrix} \begin{vmatrix} c_{\vec{k}\uparrow} \\ c^{\dagger}_{-\vec{k}\downarrow} \end{vmatrix}$$

where

$$H_{\vec{k}'\sigma,\vec{k}''\sigma'} = \langle \vec{k}', \sigma \mid H \mid \vec{k}'', \sigma' \rangle \tag{10.65b}$$

is the matrix element between states of wavevectors $\vec{k}'$ and $\vec{k}''$ with respective spin states $\sigma = \uparrow, \downarrow$ and $\sigma' = \uparrow, \downarrow$. (Note that in the last equality in Eq. (10.65a) we have omitted a constant term. This does not affect the determination of the particle dynamics.) In this format it then follows that for the free particle state [7]

$$H = \sum_{\vec{k}} \begin{vmatrix} c_{\vec{k}\uparrow}^{\dagger} & c_{-\vec{k}\downarrow} \end{vmatrix} \begin{vmatrix} \epsilon(\vec{k}) - \mu & 0 \\ 0 & -\left[\epsilon(\vec{k}) - \mu\right] \end{vmatrix} \begin{vmatrix} c_{\vec{k}\uparrow} \\ c_{-\vec{k}\downarrow}^{\dagger} \end{vmatrix} \tag{10.65c}$$

where we note the absence of off-diagonal terms in the matrix. Next, we shall see that by a small modification of the Hamiltonian in Eq. (10.65c) a superconducting state arises.

The superfluid or electronic superconducting state of the system is associated with a bonding interaction between pairs of electrons in the system characterized by the quantum numbers $(\vec{k},\uparrow)$ and $(-\vec{k},\downarrow)$. These types of interactions lead to a bonding between singlet pairs of electrons which in their bound state are known as Cooper pairs [7]. The formation of such Cooper pairs requires the introduction into our 2×2 matrix formulation in Eq. (10.65c) of additional off-diagonal interactions. These additional terms create a spin-dependent interaction accounting for s-wave scattering between the opposite spin electrons forming the Cooper pairs. In addition, it also allows for the Cooper pairs generated in the system to interact with one another, stabilizing the electrons in the system into the form of a macroscopic quantum wavefunction [6, 7]. As we shall see, these pairing interactions can be generalized to other (non-electron systems) of identical fermions resulting in further types of superfluid behaviors.

With the introduction of a spin-dependent pairing interaction, the description of the fermion systems changes to that of a mixture of spin states. This is characterized by a matrix equation for the Hamiltonian given by the form [6, 7]

$$H_s = \sum_{\vec{k}} \begin{vmatrix} c_{\vec{k}\uparrow}^{\dagger} & c_{-\vec{k}\downarrow} \end{vmatrix} \begin{vmatrix} \epsilon(\vec{k}) - \mu & \Delta(\vec{k}) \\ \Delta^*(\vec{k}) & -\left[\epsilon(\vec{k}) - \mu\right] \end{vmatrix} \begin{vmatrix} c_{\vec{k}\uparrow} \\ c_{-\vec{k}\downarrow}^{\dagger} \end{vmatrix} \tag{10.66}$$

where $\Delta(\vec{k})$ and $\Delta^*(\vec{k})$ introduce a weak s-wave scattering in the superfluid. The solution for the dynamics of Eq. (10.66) is then obtained by making a linear transformation to a new set of fermion operators written in terms of the electron fermion operators. The transformation results in a Hamiltonian which is of a diagonal form in the new set of fermion operators and such that the thermodynamics of the superconductor is easily expressed in terms of the new basis of fermions.

In order to formulate the linear transformation which simplifies the Hamiltonian in Eq. (10.66), first let us determine the eigenvalues E of the matrix in Eq. (10.66). These are determined from the condition that the determinant of the matrix [7]

$$\begin{vmatrix} \epsilon(\vec{k}) - \mu - E & \Delta(\vec{k}) \\ \Delta^*(\vec{k}) & -\left[\epsilon(\vec{k}) - \mu\right] - E \end{vmatrix} \tag{10.67a}$$

is zero. In this way, it follows that there are two eigenvalues of the system given by

$$E = \pm\, \epsilon_{\vec{k}}^{s} \tag{10.67b}$$

where $\epsilon_{\vec{k}}^{s} = \sqrt{\left(\epsilon(\vec{k}) - \mu\right)^2 + |\Delta|^2}$.

Next, we define a new set of fermion operators of the form [7]

$$\begin{vmatrix} b_{\vec{k}\uparrow} \\ b^{\dagger}_{-\vec{k}\downarrow} \end{vmatrix} = U^{\dagger} \begin{vmatrix} c_{\vec{k}\uparrow} \\ c^{\dagger}_{-\vec{k}\downarrow} \end{vmatrix} = \begin{vmatrix} u^{*}_{\vec{k}} & v^{*}_{\vec{k}} \\ -v_{\vec{k}} & u_{\vec{k}} \end{vmatrix} \begin{vmatrix} c_{\vec{k}\uparrow} \\ c^{\dagger}_{-\vec{k}\downarrow} \end{vmatrix} = \begin{vmatrix} u^{*}_{\vec{k}} c_{\vec{k}\uparrow} + v^{*}_{\vec{k}} c^{\dagger}_{-\vec{k}\downarrow} \\ -v_{\vec{k}} c_{\vec{k}\uparrow} + u_{\vec{k}} c^{\dagger}_{-\vec{k}\downarrow} \end{vmatrix} \tag{10.68a}$$

and [7]

$$\begin{vmatrix} b^{\dagger}_{\vec{k}\uparrow} & b_{-\vec{k}\downarrow} \end{vmatrix} = \begin{vmatrix} c^{\dagger}_{\vec{k}\uparrow} & c_{-\vec{k}\downarrow} \end{vmatrix} U = \begin{vmatrix} u_{\vec{k}} c^{\dagger}_{\vec{k}\uparrow} + v_{\vec{k}} c_{-\vec{k}\downarrow} & -v^{*}_{\vec{k}} c^{\dagger}_{\vec{k}\uparrow} + u^{*}_{\vec{k}} c_{-\vec{k}\downarrow} \end{vmatrix} \tag{10.68b}$$

where

$$U = \begin{vmatrix} u_{\vec{k}} & -v^{*}_{\vec{k}} \\ v_{\vec{k}} & u^{*}_{\vec{k}} \end{vmatrix} \tag{10.68c}$$

and

$$U^{\dagger} = \begin{vmatrix} u^{*}_{\vec{k}} & v^{*}_{\vec{k}} \\ -v_{\vec{k}} & u_{\vec{k}} \end{vmatrix} \tag{10.68d}$$

are a set of unitary matrices, i.e., $UU^{\dagger} = U^{\dagger}U = 1$. Note that the unitary condition of the transformation requires that

$$\left|u_{\vec{k}}\right|^2 + \left|v_{\vec{k}}\right|^2 = 1, \tag{10.69}$$

and for the transformation to diagonalize the Hamiltonian in Eq. (10.66) we must further set [7]

$$u_{\vec{k}} = u^{*}_{\vec{k}} = \sqrt{\frac{1}{2}\left(1 + \frac{\epsilon\left(\vec{k}\right) - \mu}{\epsilon^{s}_{\vec{k}}}\right)} \tag{10.70a}$$

and

$$v_{\vec{k}} = \frac{\Delta^{*}}{|\Delta|}\sqrt{\frac{1}{2}\left(1 - \frac{\epsilon\left(\vec{k}\right) - \mu}{\epsilon^{s}_{\vec{k}}}\right)}. \tag{10.70b}$$

The resulting new fermion operators generated by Eq. (10.68), in addition, are seen to obey the standard fermion anticommutation relations given by [7]

$$\left\{b_{\vec{k}\sigma}, b_{\vec{k}'\sigma'}\right\} = \delta_{\vec{k}\sigma,\vec{k}'\sigma'} \tag{10.71a}$$

$$\left\{b^{\dagger}_{\vec{k}\sigma}, b^{\dagger}_{\vec{k}'\sigma'}\right\} = 0 \tag{10.71b}$$

and

$$\left\{b_{\vec{k}\sigma}, b_{\vec{k}'\sigma'}\right\} = 0. \tag{10.71c}$$

Consequently, the new operators describe a new basis of fermion particles formed in the system by the introduction of the off-diagonal interactions.

Applying U and $U^\dagger$ in Eq. (10.66) we can rewrite the Hamiltonian in terms of the new fermion operators. In this way it then follows that [7]

$$H_s = \sum_{\vec{k}} \begin{vmatrix} c^\dagger_{\vec{k}\uparrow} & c_{-\vec{k}\downarrow} \end{vmatrix} \begin{vmatrix} \epsilon(\vec{k}) - \mu & \Delta(\vec{k}) \\ \Delta^*(\vec{k}) & -[\epsilon(\vec{k}) - \mu] \end{vmatrix} \begin{vmatrix} c_{\vec{k}\uparrow} \\ c^\dagger_{-\vec{k}\downarrow} \end{vmatrix} \tag{10.72}$$

$$= \sum_{\vec{k}} \begin{vmatrix} c^\dagger_{\vec{k}\uparrow} & c_{-\vec{k}\downarrow} \end{vmatrix} UU^\dagger \begin{vmatrix} \epsilon(\vec{k}) - \mu & \Delta(\vec{k}) \\ \Delta^*(\vec{k}) & -[\epsilon(\vec{k}) - \mu] \end{vmatrix} UU^\dagger \begin{vmatrix} c_{\vec{k}\uparrow} \\ c^\dagger_{-\vec{k}\downarrow} \end{vmatrix}$$

$$= \sum_{\vec{k}} \begin{vmatrix} b^\dagger_{\vec{k}\uparrow} & b_{-\vec{k}\downarrow} \end{vmatrix} \begin{vmatrix} \epsilon^s_{\vec{k}} & 0 \\ 0 & -\epsilon^s_{\vec{k}} \end{vmatrix} \begin{vmatrix} b_{\vec{k}\uparrow} \\ b^\dagger_{-\vec{k}\downarrow} \end{vmatrix},$$

and the original Hamiltonian has become diagonalized. The Hamiltonian in the new fermion modes can then be written as [7]

$$H_s = \sum_{\vec{k}} \begin{vmatrix} b^\dagger_{\vec{k}\uparrow} & b_{-\vec{k}\downarrow} \end{vmatrix} \begin{vmatrix} \epsilon^s_{\vec{k}} & 0 \\ 0 & -\epsilon^s_{\vec{k}} \end{vmatrix} \begin{vmatrix} b_{\vec{k}\uparrow} \\ b^\dagger_{-\vec{k}\downarrow} \end{vmatrix} \tag{10.73}$$

$$= \sum_{\vec{k}} \left[\epsilon^s_{\vec{k}} b^\dagger_{\vec{k}\uparrow} b_{\vec{k}\uparrow} - \epsilon^s_{\vec{k}} b_{-\vec{k}\downarrow} b^\dagger_{-\vec{k}\downarrow} \right]$$

$$= \sum_{\vec{k}} \left[\epsilon^s_{\vec{k}} b^\dagger_{\vec{k}\uparrow} b_{\vec{k}\uparrow} + \epsilon^s_{\vec{k}} b^\dagger_{-\vec{k}\downarrow} b_{-\vec{k}\downarrow} \right]$$

where in the last equality we again ignore a constant term which has no effect on the particle dynamics. From the far right of Eq. (10.73) it follows by analogy with Eq. (10.64) that [7]

$$b^\dagger_{\vec{k}\sigma}(t) = b^\dagger_{\vec{k}\sigma}(0) e^{i\frac{\epsilon^s_{\vec{k}}}{\hbar}t} \tag{10.74a}$$

and

$$b_{\vec{k}\sigma}(t) = b_{\vec{k}\sigma}(0) e^{-i\frac{\epsilon^s_{\vec{k}}}{\hbar}t} \tag{10.74b}$$

are the time-dependent forms of the new particle creation and destruction operators.

Notice that operators of the form $b^\dagger_{\vec{k}\sigma}$ create fermion excitations from the superconductor ground state while operators of the form $b_{\vec{k}\sigma}$ destroy fermion excitations generated from the superconductor ground state. Consequently, since the ground state $|G.S.\rangle$ of the superconductor contains no excitations, it must satisfy the defining condition that [7]

$$b_{\vec{k}\sigma} |G.S.\rangle = 0 \tag{10.75}$$

for all wavevectors and for $\sigma = \uparrow$ or $\downarrow$. These conditions are satisfied by the form

$$|G.S.\rangle = \prod_{\vec{k}} \left(u_{\vec{k}} - v^*_{\vec{k}} c_{\vec{k}\uparrow} c_{-\vec{k}\downarrow} \right) |0\rangle \tag{10.76a}$$

where the product is over all $\vec{k}$, and $|0\rangle$ is the vacuum of the electron creation and destruction operators defined by

$$c_{\vec{k}\sigma}|0\rangle = 0 \tag{10.76b}$$

for all $\vec{k}$ and $\sigma = \uparrow$ or $\downarrow$. The ground state, consequently, represents the lowest energy configuration of the superconductor in which none of the new fermion excitations are present, and the chemical potential sets the number of electrons present in the system forming the ground state. In addition, notice that because of the form of the energy dispersion relation in Eqs. (10.67b) and (10.73), there is an energy gap $|\Delta|$ between the ground state and the excitation spectrum of the new fermions in the superconductor [7].

Given the total number of electrons in the system, the ground state solution and the zero-temperature chemical potential of the superconductor can be computed. For a spatially isotropic superconductor, these are determined in terms of [7]

$$n_{\vec{k}\uparrow} = \langle G.S.| c^{\dagger}_{\vec{k}\uparrow} c_{\vec{k}\uparrow} | G.S.\rangle = \tag{10.77a}$$

$$= \langle G.S.| \left(u_{\vec{k}} b^{\dagger}_{\vec{k}\uparrow} - v_{\vec{k}} b_{-\vec{k}\downarrow}\right)\left(u_{\vec{k}} b_{\vec{k}\uparrow} - v^{*}_{\vec{k}} b^{\dagger}_{-\vec{k}\downarrow}\right) | G.S.\rangle = \left|v_{\vec{k}}\right|^2$$

where $n_{\vec{k}\uparrow}$ is the number of electrons in the state of wavevector $\vec{k}$ and spin $\uparrow$. It then follows that the total number of electrons N in the system is given by [7]

$$N = \sum_{\vec{k}} \left(n_{\vec{k}\uparrow} + n_{\vec{k}\downarrow}\right) \tag{10.77b}$$

where $n_{\vec{k}\uparrow} = n_{\vec{k}\downarrow}$ and the sum is over all possible wavevectors. Applying the result in Eq. (10.70b) for $v_{\vec{k}}$ then allows us to evaluate the zero-temperature chemical potential of the system when it is in its ground state configuration.

As a final point, we note that in the model Hamiltonian of Eq. (10.66) the off-diagonal terms Δ and Δ^{*} for the Cooper pair interactions can be determined self-consistently. Since these interactions between the Cooper pairs in the system are scattering interactions in which one Cooper pair scatters from its fellows, we expect that scattering interactions in Eq. (10.66) should be approximately of the form [7]

$$\Delta = -A \sum_{\vec{k}} \langle G.S.| c_{-\vec{k}\downarrow} c_{\vec{k}\uparrow} | G.S.\rangle, \tag{10.78a}$$

i.e., it should be proportional to the intensities of the presence of Cooper pairs in the system. Here A is a constant of proportionality, $\langle G.S.| c_{-\vec{k}\downarrow} c_{\vec{k}\uparrow} | G.S.\rangle$ is a matrix amplitude giving the intensity of the Cooper pair forming operator $c_{-\vec{k}\downarrow} c_{\vec{k}\uparrow}$ in the system, and we approximate the many-body interaction between the Cooper pairs in the system by the sum over $\vec{k}$. In evaluating Eq. (10.78a) we see that [7]

$$\langle G.S.| c_{-\vec{k}\downarrow} c_{\vec{k}\uparrow} | G.S.\rangle = \langle G.S.| \left(v^{*}_{\vec{k}} b^{\dagger}_{\vec{k}\uparrow} + v_{\vec{k}} b_{-\vec{k}\downarrow}\right)\left(u_{\vec{k}} b_{\vec{k}\uparrow} - v^{*}_{\vec{k}} b^{\dagger}_{-\vec{k}\downarrow}\right) | G.S.\rangle \tag{10.78b}$$

$$= -u_{\vec{k}} v^{*}_{\vec{k}}$$

so that Eq. (10.78a) reduces to

$$\Delta = A\sum_{\vec{k}} u_{\vec{k}} v_{\vec{k}}^{*} = A\sum_{\vec{k}} \frac{\Delta}{2\epsilon_{\vec{k}}^{s}} = A\sum_{\vec{k}} \frac{\Delta}{2\sqrt{\left(\epsilon\left(\vec{k}\right)-\mu\right)^{2}+|\Delta|^{2}}}. \tag{10.78c}$$

This determines the energy gap at zero temperature and is easily generalized to systems at nonzero temperatures. Notice, that $\Delta = 0$ is a solution of Eq. (10.78c) which yields a normal electron gas solution. However, there can also be nonzero Δ solutions, and these yield superconducting states which are in general lower states of free energy than the normal metal solutions.

10.7.2 Ginzburg-Landau Free Energy of Electronic Superconductors

As in the discussions of the helium superfluid near the lambda transition based on Eqs. (10.24) and (10.27), a similar treatment can be made for the free energy of the transition between the normal metal and superconducting state. Specifically, in the bulk superconductor an acceptable form for the free energy F of the system is given by [5–7, 10, 11]

$$F = F_0 + \int d\vec{r}\left[a(T)\left|\psi(\vec{r})\right|^{2} + \frac{b}{2}\left|\psi(\vec{r})\right|^{4}\right] \tag{10.79a}$$

where F_0 is the free energy of the normal non-superconducting metal, $\psi(\vec{r})$ is a macroscopic wavefunction defined such that $n(\vec{r}) = \left|\psi(\vec{r})\right|^{2}$ is the density of Cooper pairs in the superconductor, and the second term on the right is the free energy contribution from the superfluid. Again in Eq. (10.79a) for $a_0 > 0$

$$a(T) = a_0\left(T - T_c\right) \tag{10.79b}$$

where T_c now is the critical temperature of the superconducting transition, and

$$b > 0 \tag{10.79c}$$

is a constant. Consequently, in the case that for $T > T_c$ the Cooper pair density is

$$\left|\psi(\vec{r})\right|^{2} = 0, \tag{10.80a}$$

while for $T < T_c$ the nonzero pair density is

$$\left|\psi(\vec{r})\right|^{2} = \frac{a_0\left|T - T_c\right|}{b}. \tag{10.80b}$$

Notice in Eq. (10.79) that, similar to the liquid helium formulation, we have chosen the macroscopic wave function of the superconductor to be complex, so that due to its complex nature the wave function has of the form

$$\psi(\vec{r}) = n(\vec{r})e^{i\varphi(\vec{r})} \tag{10.80c}$$

where the phase again is related to the flow of supercurrent [5–7, 10, 11].

The formulation of the superconducting free energy is seen to give a similar description of the properties of a superconductor in the vicinity of the critical temperature as that of the Ginzburg formulation for the helium system. However, in one point, it is quite different from the Helium

I–Helium II formulation. At low temperatures, due to the diminished role of thermal fluctuations in the low-temperature superconducting materials, Eq. (10.79) gives a more accurate description of the superconducting system. In this regard, the superconducting specific heat at the critical temperature is more of the form of a discontinuity rather than of the lambda singularity displayed by the liquid helium system. Note that, on the other hand, high-temperature superconductors are more like the liquid helium system as, due to the greater thermal fluctuations in these systems, the Ginzburg-Landau theory gives a poorer representation of their properties.

In addition, one can generalize the helium free energy form in Eq. (10.27) to a similar form for the superconductor free energy associated with a space-dependent macroscopic wavefunction. Based on Eq. (10.79), this yields a free energy given by [5–7, 10, 11]

$$F - F_0 = \int d\vec{r}\left[\frac{\hbar^2}{2m}\left|\nabla\psi(\vec{r})\right|^2 + a(T)\left|\psi(\vec{r})\right|^2 + \frac{b}{2}\left|\psi(\vec{r})\right|^4\right]. \tag{10.81}$$

In the superconducting system $\frac{\hbar^2}{2m}$ is now treated as a parameter characterizing the spatial flexibility of the macroscopic wavefunction.

The generalization in Eq. (10.81) allows for the description of superconducting surfaces, Type I and Type II superconductors [6, 7], and the flow of supercurrents. In this regard, the current density of the superconductor, generalized from the helium form in Eq. (10.30b), is given by [6, 7]

$$\vec{j} = -i\frac{\hbar}{2m}\left[\psi^*\nabla\psi - \psi\nabla\psi^*\right] \tag{10.82a}$$

For the case that $\psi(\vec{r}) = n_0 e^{i\varphi(\vec{r})}$, it then follows from Eq. (10.82a) that

$$\vec{j} = \frac{\hbar}{m}n_0\nabla\varphi \tag{10.82b}$$

Again, in Eq. (10.82) the ideas of vortex solutions and surface currents can be shown to apply to the superconducting system. This justifies the choice in Eq. (10.80c) of the complex form of the wavefunction or order parameter of the macroscopic state of the electron or Cooper pairs.

10.7.3 P-Wave Pairing in ^3He

Unlike ^4He which has a nucleus of total spin zero, the nucleus of ^3He is in a state of total spin $\frac{1}{2}$. Consequently, liquid ^3He is a fluid obeying Fermi-Dirac statistics, with properties similar to those found in superconducting electron systems, rather than those found to arise in the Bose-Einstein statistics of liquid ^4He [7, 10, 11]. Notice, in this regard, that as in electronic superconductors the interactions in ^3He involve pairs of atoms, and these in turn interact with all the other pairs of atoms formed in the system. The ^3He interactions, consequently, give rise to a macroscopic quantum state of the atoms in the system and a superfluidity involving the flow of the neutral uncharged helium atoms. Now, however, the pairing is not between singlet spin states arising from s-wave scattering but is found to be between triplet spin states arising from p-wave scattering.

This difference in pairing is due to the dissimilarity of the origins of the pairing interactions between the two types of superconductors. In low-temperature electronic superconductors, pairing is mediated by electron-phonon scattering, while in ^3He the pairing arises from residual attractive interactions coming from the otherwise hard-core scattering between the atoms in the fluid. In the following, a brief sketch is made of the theory of ^3He, emphasizing its similarity to the earlier treatment of s-wave superconductivity. For the details of the treatment, the reader is referred to the literature [10, 11].

Let us begin our study of superfluid ^3He, with a consideration of the model of a gas of free noninteracting fermionic particles. In the absence of interactions between the atoms of the gas, the system is characterized by a single particle Hamiltonian of the form [7, 10, 11]

$$H = \tag{10.83}$$

$$= \sum_{\vec{k}} \begin{vmatrix} c^{\dagger}_{\vec{k}\uparrow} & c^{\dagger}_{\vec{k}\downarrow} & c_{-\vec{k}\uparrow} & c_{-\vec{k}\downarrow} \end{vmatrix} \begin{vmatrix} \epsilon(\vec{k})-\mu & 0 & 0 & 0 \\ 0 & \epsilon(\vec{k})-\mu & 0 & 0 \\ 0 & 0 & -[\epsilon(\vec{k})-\mu] & 0 \\ 0 & 0 & 0 & -[\epsilon(\vec{k})-\mu] \end{vmatrix} \begin{vmatrix} c_{\vec{k}\uparrow} \\ c_{\vec{k}\downarrow} \\ c^{\dagger}_{-\vec{k}\uparrow} \\ c^{\dagger}_{-\vec{k}\downarrow} \end{vmatrix}.$$

This form is a generalization of Eq. (10.65) which allows for the inclusion into our considerations of a greater number of possible spin states of the system. Consequently, when off-diagonal pairing interactions are introduced into the model, the scattering between a wider variety of particle spin states is possible. This will allow us an increase in the possible off-diagonal spin interactions, which is essential for the development in the system of the theory of triplet as opposed to singlet pairing.

Specifically, following the treatment of Eq. (10.66), we now consider the introduction of off-diagonal terms into Eq. (10.83). These off-diagonal terms will allow for triplet pairing interactions between the atoms in the fluid. In this regard, it is found that for p-wave triplet pairing, the Hamiltonian has the general form [7, 10, 11]

$$H_p = \tag{10.84}$$

$$= \sum_{\vec{k}} \begin{vmatrix} c^{\dagger}_{\vec{k}\uparrow} & c^{\dagger}_{\vec{k}\downarrow} & c_{-\vec{k}\uparrow} & c_{-\vec{k}\downarrow} \end{vmatrix} \begin{vmatrix} \epsilon(\vec{k})-\mu & 0 & \Delta_{\uparrow\uparrow}(\vec{k}) & \Delta_{\uparrow\downarrow}(\vec{k}) \\ 0 & \epsilon(\vec{k})-\mu & \Delta_{\downarrow\uparrow}(\vec{k}) & \Delta_{\downarrow\downarrow}(\vec{k}) \\ \Delta^{*}_{\uparrow\uparrow}(\vec{k}) & \Delta^{*}_{\downarrow\uparrow}(\vec{k}) & -[\epsilon(\vec{k})-\mu] & 0 \\ \Delta^{*}_{\uparrow\downarrow}(\vec{k}) & \Delta^{*}_{\downarrow\downarrow}(\vec{k}) & 0 & -[\epsilon(\vec{k})-\mu] \end{vmatrix} \begin{vmatrix} c_{\vec{k}\uparrow} \\ c_{\vec{k}\downarrow} \\ c^{\dagger}_{-\vec{k}\uparrow} \\ c^{\dagger}_{-\vec{k}\downarrow} \end{vmatrix}$$

where the off-diagonal spin interactions

$$\begin{vmatrix} \Delta_{\uparrow\uparrow}(\vec{k}) & \Delta_{\uparrow\downarrow}(\vec{k}) \\ \Delta_{\downarrow\uparrow}(\vec{k}) & \Delta_{\downarrow\downarrow}(\vec{k}) \end{vmatrix} \qquad 0(10.85a)$$

considers a greater variety of scattering interactions between the different spin states of the atomic pairs. From the generalized form in Eq. (10.85a) the possibility of both singlet and triplet spin scattering states now exists in the system.

Next, we extend the discussions leading to Eq. (10.78a) to the present treatment of Eq. (10.84). In this way, we relate the off-diagonal pairing interaction in the Hamiltonian to the presence in the fluid of a large number of paired particle states. From this consideration, notice that at zero temperatures [7, 10, 11]

$$\Delta_{\alpha\beta}\left(\vec{k}\right) \propto \langle G.S. | c_{-\vec{k}\alpha} c_{\vec{k}\beta} | G.S. \rangle \tag{10.85b}$$

so that the presence of bound pairs in the system is self-consistently determined from the solutions of the Hamiltonian in Eq. (10.84). In doing this, it follows that some important properties are related to the pairing. Specifically, from the properties of the anticommutations of the atomic creation and destruction operators, it follows that [7, 10, 11]

$$\Delta_{\alpha\beta}\left(\vec{k}\right) = -\Delta_{\beta\alpha}\left(-\vec{k}\right) \tag{10.85c}$$

for the elements of Eq. (10.85a). This symmetry enters into distinguishing between singlet and triplet atomic pairing in the fluid.

It is helpful at this point to write Eq. (10.85) in the complete matrix basis formed from the 2×2 Pauli matrices and the identity matrix. In this format, it follows that a complete set of 2×2 matrices is $\{1,\sigma_x,\sigma_y,\sigma_z\} = \{1,\vec{\sigma}\}$ where 1 is the 2×2 unit matrix and σ_i for $i = x, y, z$ are the Pauli spin matrix components. Taking into consideration the symmetry properties in Eq. (10.85c), in terms of this matrix basis, the general form of Eq. (10.85a) is then represented by [7, 10, 11]

$$\begin{vmatrix} \Delta_{\uparrow\uparrow}\left(\vec{k}\right) & \Delta_{\uparrow\downarrow}\left(\vec{k}\right) \\ \Delta_{\downarrow\uparrow}\left(\vec{k}\right) & \Delta_{\downarrow\downarrow}\left(\vec{k}\right) \end{vmatrix} = i\left[\Delta\left(\vec{k}\right) + \vec{d}\left(\vec{k}\right)\cdot\vec{\sigma}\right]\sigma_y = \begin{vmatrix} -d_x\left(\vec{k}\right) + id_y\left(\vec{k}\right) & \Delta\left(\vec{k}\right) + d_z\left(\vec{k}\right) \\ -\Delta\left(\vec{k}\right) + d_z\left(\vec{k}\right) & d_x\left(\vec{k}\right) + id_y\left(\vec{k}\right) \end{vmatrix} \tag{10.86a}$$

where

$$\Delta\left(\vec{k}\right) = \Delta\left(-\vec{k}\right), \tag{10.86b}$$

and

$$\vec{d}\left(\vec{k}\right) = -\vec{d}\left(-\vec{k}\right). \tag{10.86c}$$

From Eqs. (10.84) and (10.86), it is found that $\Delta\left(\vec{k}\right)$ is associated with singlet spin and even parity space states while $\vec{d}\left(\vec{k}\right)$ is associated with triplet spin and odd parity space states. Consequently, for a treatment of triplet pairing, we must focus on the properties of the system associated with $\vec{d}\left(\vec{k}\right)$, ignoring the $\Delta\left(\vec{k}\right)$ terms.

A transformation to a new set of fermion operators is next made by choosing a unitary operator U which diagonalizes the matrix in Eq. (10.84). The diagonalization leads to four energy eigenvalues given by the form $\pm\epsilon_n^p$ where $n = 1,2$. In terms of U and $U^\dagger$ it then follows that [7, 10, 11]

$$b_{\vec{k}n} = \sum_\sigma \left(u^*_{\vec{k}\sigma n} c_{\vec{k}\sigma} - v^*_{\vec{k}\sigma n} c^\dagger_{-\vec{k}n}\right) \tag{10.87}$$

where the set of new creation and destruction operators $\{b^\dagger_{\vec{k}n}\}$ and $\{b_{\vec{k}n}\}$ obey fermion anticommutation rules, and $u^*_{\vec{k}\sigma n}$, $v^*_{\vec{k}\sigma n}$ are elements of U and $U^\dagger$. The operators $\{b^\dagger_{\vec{k}n}\}$ create excited states from the ground state of the system, whereas the operators $\{b_{\vec{k}n}\}$ destroy excitations that are present in the system. The ground state $|G.S.$ is once again defined by the condition that for all $\vec{k}$ and n

$$b_{\vec{k}n}|G.S.\rangle = 0, \tag{10.88}$$

i.e., the ground state represents the state of the system in the absence of any excitations arising from the set of $\left\{b^{\dagger}_{\vec{k}n}\right\}$ operators [7, 10, 11].

In the context of the model in Eqs. (10.84) through (10.86), two forms of the model have been put forth to explain the superconductivity of ^{3}He. In both models

$$\Delta\left(\vec{k}\right)=0, \tag{10.89}$$

so that no component of singlet pairing is present in them.

In this regard, one model is the Balian-Werthamer model [7, 10, 11]. For the Balian-Werthamer model, the vector $\vec{d}\left(\vec{k}\right)$ is assumed to have the form

$$\vec{d}\left(\vec{k}\right)\propto\left(\sin\theta_{\vec{k}}\cos\varphi_{\vec{k}},\sin\theta_{\vec{k}}\sin\varphi_{\vec{k}},\cos\theta_{\vec{k}}\right). \tag{10.90}$$

where $\left(\sin\theta_{\vec{k}}\cos\varphi_{\vec{k}},\sin\theta_{\vec{k}}\sin\varphi_{\vec{k}},\cos\theta_{\vec{k}}\right)$ are the angular polar coordinates of $\vec{k}$. This is thought to model the B-phase of liquid ^{3}He.

A second model is the Anderson-Brinkman-Morrel model [7, 10, 11]. In the Anderson-Brinkman-Morrel model, the vector $\vec{d}\left(\vec{k}\right)$ is assumed to have the form

$$\vec{d}\left(\vec{k}\right)\propto\left(\sin\theta_{\vec{k}}\left(\cos\varphi_{\vec{k}}+i\sin\varphi_{\vec{k}}\right),0,0\right). \tag{10.91}$$

This provides a representation of the A-phase of liquid ^{3}He.

By pursuing the self-consistent solution of the p-wave Hamiltonian defined in Eqs. (10.84) through (10.86) for the two forms in Eqs. (10.90) and (10.91), a complete theory of the thermodynamics of the A and B states of liquid ^{3}He is developed. This is done for the most part by following the methods used in our earlier discussions of the thermodynamics of the s-wave singlet system. For further details of this, the reader is referred to the literature [7, 10, 11].

REFERENCES

1. Ceperley, D. M. 1995. Path Integrals in the Theory of Condensed Helium. *Reviews of Modern Physics* 67, 279–353.
2. McGurn, A. R., Ryan, P., Maradudin, A. A., and Wallis, R. F. 1989. Quantum Monte Carlo Calculation of the Thermodynamic Functions of a Lennard-Jones Chain of Atoms. *Physical Review B* 40, 2407–2413.
3. McGurn, A. R. 1993. *Path-Integral Quantum Monte Carlo Studies of the Static and Time-Dependent Thermodynamics of the Vibrational Properties of Crystals in Quantum Monte Carlo Methods in Condensed Matter Physics*, ed. M. Suzuki (World Scientific, Singapore).
4. Feynman, R. P. and Hibbs, A. R. 1965. *Quantum Mechanics and Path Integrals* (McGraw-Hill Book Company, New York).
5. Landau, L. D. and Lifshitz, E. M. 1958. *Statistical Physics* (Pergamon Press LTD, London).
6. Huang, K. 2010. *Introduction to Statistical Physics*, 2nd Edition (CRC Press, Boca Raton).
7. Annett, J. F. 2004. *Superconductivity, Superfluids and Condensates* (Oxford University Press, New York).
8. Ginzburg, V. L. and Piteavskii, I. P. 1958. On the Theory of Superfluidity. *Soviet Physics JETP* 34(7), 858–861.
9. Rogel-Salazar, J. 2013. The Gross-Pitaevskii Equation and Bose-Einstein Condensates. *European Journal of Physics* 34, 247–250.
10. Vollhardt, D. and Wolfle, D. 1990. *The Superfluid Phases of Helium 3* (Taylor & Francis, London).
11. Lee, M. D. 1997. The Extraordinary Phases of Liquid ^{3}He. *Reviews of Modern Physics* 69, 645–665.
12. Ceperley, D. M. and Pollock, E. L. 1986. Path-Integral Computation of the Low-Temperature Properties of Liquid ^{4}He. *Physical Review Letters* 56, 351–354.

11 Scaling Theory

In this chapter we develop some interesting ideas of scaling theory and their applications in the study of disorder [1–3]. This is done by considering a type of geometric phase transition known as a percolation transition. Percolation theory studies geometric systems which exhibit geometric transformations similar in aspect to many of the transformations found in thermodynamic systems undergoing second order phase transitions [1–3]. For example, ferromagnetic materials are thermodynamic systems which are observed as functions of temperature to change continuously from an ordered to a disordered magnetic state as they pass through a second order phase transition. Unlike many thermodynamic systems, however, the systems of percolation theory only involve a study of geometry and the connectivity of lattices subject to various connective disorders [2]. In this regard, percolation systems are much easier to treat using theoretical techniques and computer simulation methods than are thermodynamic models [1–4]. The models studied in percolation theory are also helpful in understanding the properties of alloys, biological systems, and the properties of porous media [1–4].

In percolation theory we consider a lattice which is periodic in space with bonds connecting the nearest neighbor sites of the lattice. For this system there are generally two types of disorder studied [1, 2]. These include bond and site random disorder. In bond percolation theory a certain fractional concentration p of bonds are randomly present in the system and a certain fractional concentration $1-p$ of bonds are randomly absent from the system. A consideration is then made of the geometric properties of the clusters of sites connected to one another through a sequence of paths of connecting bonds. Of particular interest to us is the nature and statistical properties of infinite clusters of connected sites as a function of the fractional bond concentration.

Note that when $p = 1$ all the sites are part of one infinite cluster, whereas when $p = 0$ each site is isolated so that none of the sites are part of an infinite cluster. It then follows that there must exist some fractional concentration $0 < p_c < 1$ known as the critical concentration at which a single infinite cluster first appears in the system [1, 2]. Above p_c there is always an infinite cluster in the system whereas below p_c no infinite clusters are present in the system. In this regard, the system exhibits a phase transition from a system with an infinite cluster to one without infinite clusters. The phase transition is a type of second order transition as the probability of a site being in an infinite cluster continuously goes to zero at p_c. (See Figure 11.1 for a schematic representation of a typical plot of the probability of a site being in an infinite cluster as a function of p [1–5].) In this geometric transition, the fractional bond concentration then plays a role similar to that of the temperature in magnetic phase transitions between an ordered and a disordered magnetic phase.

In site percolation theory a certain fractional concentration p of sites are randomly present in the system and a certain fractional concentration $1-p$ of sites, known as vacancies, is absent from the system. In this formulation, bonds are only present between nearest-neighbor occupied lattice sites and not between an occupied site and a vacancy or between nearest-neighbor vacancies. A consideration is then made of the geometric properties of clusters of occupied sites connected to one another by paths of connecting bonds. Again, it then follows that there exists a critical fractional concentration of sites $0 < p_c < 1$ at which a single infinite cluster first appears in the system [1, 2]. Above p_c there is always an infinite cluster in the system whereas below p_c no infinite clusters are present in the system. Consequently, a geometric transition exists between the region of infinite clusters $p > p_c$ and the region of only finite clusters $p < p_c$. The phase transition is again a type of second order transition as the probability of a site being in an infinite cluster continuously goes to zero at p_c. (See again Figure 11.1 for a schematic representation of a typical plot of the probability of a site being in an infinite cluster as a function of p.)

DOI: 10.1201/9781032674919-11

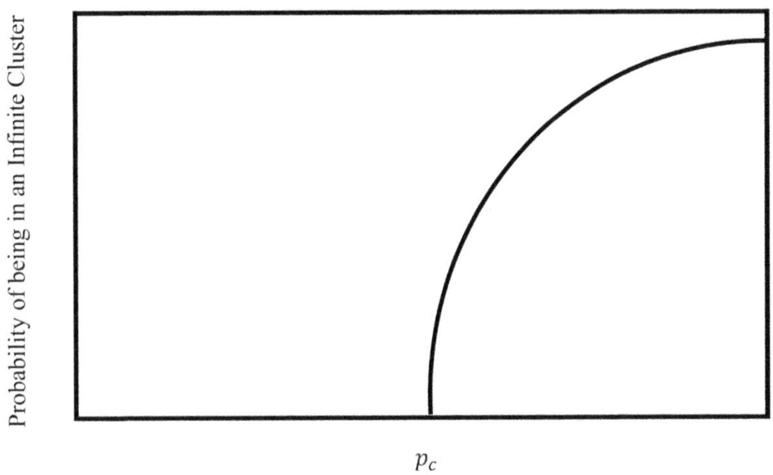

FIGURE 11.1 A schematic plot of the probability of an occupied site being in an infinite cluster as a function of the bond (site) fractional concentration for a typical bond (site) percolation problem. Note that the probability of a site being in an infinite cluster ranges on the vertical axis between 0 to 1 and the bond (site) fractional percolation concentration ranges on the horizontal axis from 0 (left) to 1 (right).

To begin the discussions of the scaling treatment of percolation theory, we will consider a crude theory for the determination of the critical percolation concentration p_c for a square lattice subject to bond percolation. This will provide the gist of some of the basic ideas at play in the scaling theory study of disordered materials [1–3].

11.1 EXAMPLE BOND PERCOLATION ON THE SQUARE LATTICE

As an example of a crude estimate of the critical concentration of bonds on the square lattice, consider the following discussion: assume in the context of random bond percolation theory that we are given a square lattice of sites with nearest neighbor bonds between the lattice sites. In Figure 11.2a we represent the square lattice system in which a fractional concentration p of the nearest neighbor bonds is randomly present and a fractional concentration of bonds $1 - p$ are randomly removed from the system. For this system, let us divide the square lattices in Figure 11.2a into two sublattices as shown in Figure 11.2b where one sublattice is denoted by open circles and the other is denoted by closed circles. In the following, we shall show that by studying the connectivity properties of the sites on the closed circle sublattice, an estimate of the critical percolation concentration of the bond percolation p_c problem can be made [1, 2, 4]. This involves an application of scaling theory [1–3].

Consider the connectivity of the sublattice of closed circles in Figure 11.2b which are also formed in a square lattice [1]. Specifically of interest are the connecting paths of bonds existing between two nearest-neighbor closed circles on the closed circle sublattice. To further simplify the considerations, however, let us focus on the unit square of four lattice sites shown in Figure 11.3a. This is a small subregion of the lattice in Figure 11.2b. However, it will allow us to develop ideas of connectivity as they are limited to the four sites of the unit square in the lattice. The generalization of these to larger regions of the lattice is straightforward.

In the cluster in Figure 11.3a the sites along the edges of the square are nearest neighbor lattice sites, and our considerations are restricted to connecting paths between the two closed circle sites within the cluster of four sites. In this regard, in Figure 11.3a the two nearest-neighbor sites on the

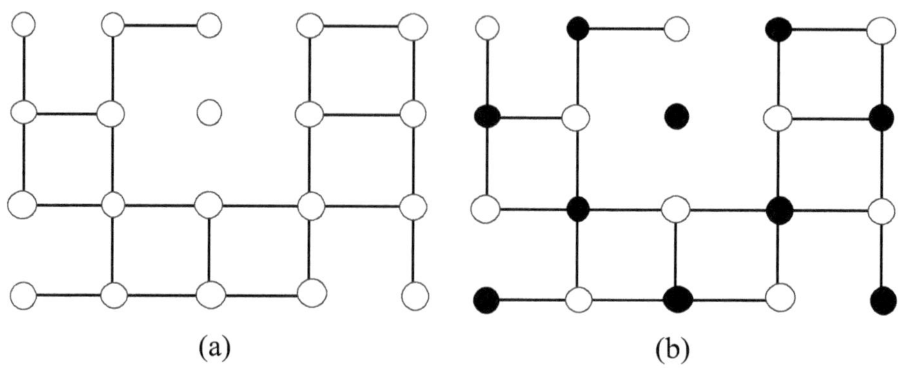

FIGURE 11.2 Schematic of the bond percolation problem on the square lattice: a) a representation of the square lattice in which a fraction of bonds p are present so that a fraction $1-p$ are absent, b) the same system as in a) but with a separation into a sublattice of open circle sites and a sublattice of closed circle sites.

closed circle sublattice are considered to have a sublattice bond or connection between them only if they both connect to at least one of the two open circle sublattice sites of the cluster [1].

We next consider the system connectivity in the closed circle sublattice in Figure 11.2a, under these restrictions. For example, Figure 11.3a shows the case when all four bonds are present in the cluster. This occurs in terms of the lattice bond concentration p with probability p^4. In Figure 11.3b we see the case in which three lattice bonds are present. This occurs with probability $p^3(1-p)$ and offers a connection between the closed circles which can be made in 4 different possible ways. In Figure 11.3c we see the case in which only two lattice bonds are present. This occurs with probability $p^2(1-p)^2$ and offers a connection which can be made in 2 different ways. All other possible configurations fail to provide a path between the two nearest neighbor closed sites on the lattice [1].

From these considerations, in the treatment of the cluster of four lattice sites we find that the probability p' of a bond being present between the closed circles on the sublattice is given by [1]

$$p' = p^4 + 4p^3(1-p) + 2p^2(1-p)^2 = -p^4 + 2p^2. \tag{11.1}$$

This then relates the fractional concentrations of the nearest neighbor bonds between the lattices in Figures 11.2a and 11.2b and between the closed circle sublattice. In a sense we have approximately mapped the square lattice problem in Figures 11.2a and 11.2b, involving the bond disorder between the open and closed circled sites, into the new bond disordered problem in Figure 11.4, involving bond disorder on the square lattice formed by the closed circle sites. Whereas the problem in Figures 11.2a and 11.2b had a bond percolation probability p the new bond percolation probability of the system in Figure 11.4 has a percolation probability p'.

Notice, however, that the bond percolation problem on the original lattice of nearest-neighbor site separation d with bond concentration p is approximately mapped into a new bond percolation problem on a new lattice [1] of nearest-neighbor site separation $\sqrt{2}d$ with a bond concentration p'. This is an important point for us as we see that the mapping includes a length scale change of the system. Note that the mapping between the two lattices is only approximate as it only considers including connective paths involving the presence of two bonds. Paths of higher connective lengths, running out and back to the unit square, have been excluded from the considerations. These paths involve higher order powers in p and are assumed small.

An interesting consideration in these discussions is the fixed-points of the mapping in Eq. (11.1), i.e., the point at which $p = p'$. These represent concentrations in the system in which the properties of the system appear to be independent of the change of length scale between the two systems [1]. At the fixed points, then, the system becomes independent of the scale upon which it is being considered.

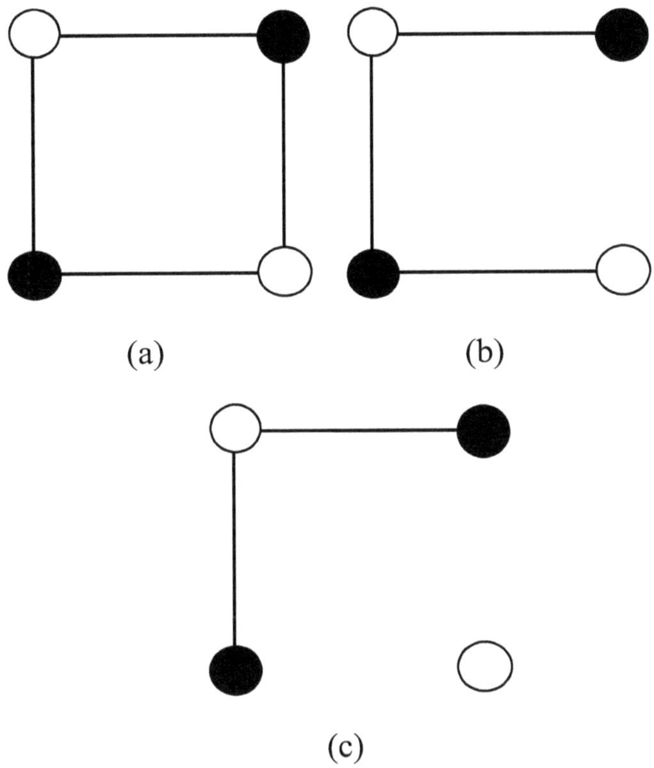

FIGURE 11.3 Schematic of a unit square of the bond percolation square lattice in Figure 11.2b. In a) all four bonds are present between the nearest-neighbor sites. In b) three bonds are present between the nearest-neighbor sites. In c) two bonds connect the two filled circles.

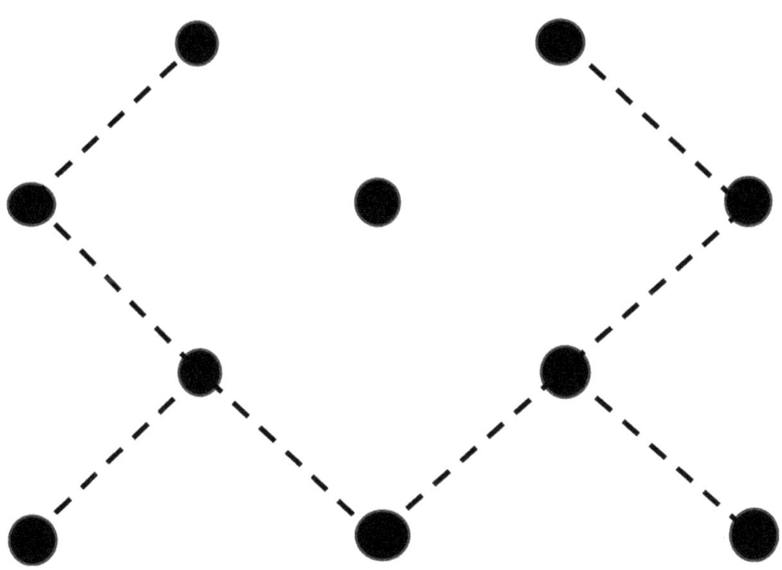

FIGURE 11.4 The connectivity of the sublattice of filled circle sites. The dashed lines represent the new nearest-neighbor bonds obtained from consideration of Figure 11.2b.

The fixed-points of Eq. (11.1) are found to be located at [1]

$$p(p-1)(p^2+p-1)=0 \tag{11.2a}$$

so that

$$p_c = 0, 1, \frac{-1 \pm \sqrt{5}}{2} \approx 0.62, \tag{11.2b}$$

and may be understood as follows: the fixed-point at $p_c = 0$ is the case in which no bonds are present in the lattice. Consequently, since no bonds are present in the original lattice, no bonds can be present in the sublattice model. In this regard, the length scale of the lattice is not significant to the connectivity because there is no connectivity on any length scale. Similarly, for the fixed point $p_c = 1$ all the bonds are present in the original system so that all the bonds must be present in the sublattice. Again, the length scale of the system is really irrelevant to the consideration as everything is always connected.

The fixed-point at $p_c = \frac{-1+\sqrt{5}}{2} \approx 0.62$, however, is of greater significance [1]. This fixed-point is an approximation of the bond concentration at which the bond percolation on the original lattice of nearest-neighbor site separation d and bond concentration p_c is mapped into a bond percolation problem on a new lattice of nearest-neighbor site separation $\sqrt{2}d$ and bond concentration p_c. At this point, the system loses a fixed dependence on a particular length scale. In this regard, the fixed points should relate to the onset of an infinite fractal cluster of connected points in the system [5]. Specifical the infinite cluster should exhibit the length scale independence of a fractal geometry. (As a final point, we note that the fixed-point at $\frac{-1-\sqrt{5}}{2} < 0$ is unphysical and represents an unimportant solution for our considerations.)

Notice, as we shall presently see, that the critical percolation concentration for the formation of infinite clusters on the square lattice can be calculated exactly [1] and is found to be $p_c = \frac{1}{2}$. In light of the many approximations involved in our estimate, the difference between our result and the exact result is not surprising.

In this regard, by allowing for the introduction of more parameters into the percolation models and for the applications of further iterative mappings, the theory can be greatly improved. This develops into a full renormalization group study of the problem. For these considerations, the reader is referred to the literature [1–3].

We next turn to an important aspect of the percolation model in the neighborhood of the critical concentration. This involves the study of the so-called critical exponents of the transition. However, before this treatment, a brief consideration of the exact value of the critical percolation concentration will be made.

11.1.1 Exact Critical Bond Percolation Concentration on the Square Lattice

The exact critical bond concentration for the bond random problem on the square lattice can be determined by applying a plausibility argument based on geometric considerations [2, 3].

Specifically, consider the random bond problem illustrated in Figure 11.2a. In this system, a fraction p of bonds are present in the lattice and a fraction $q = 1 - p$ of bonds are absent from the lattice. We shall refer to this bond percolation problem as the old-lattice problem. Next, consider a bond percolation problem on another square lattice generated from the old-lattice problem in Figure 11.2a.

The new square lattice problem is generated from the old-lattice problem in Figure 11.2a as follows: first place a site of the new-lattice at the center of each of the unit squares of the old-lattice in Figure 11.2a. Next, define nearest-neighbor bonds between the sites on the new-lattice [2]. This is done so that if two nearest-neighbor sites of the new-lattice are separated by a bond of the old-lattice, we do not connect the two new-lattice sites by a new-lattice nearest-neighbor bond. Otherwise, if the two nearest-neighbor new-lattice sites are not separated by a bond on the old-lattice, we connect the two new-lattice sites with a new-lattice nearest-neighbor bond [2]. The resulting new-lattice and the bonds defined upon it form a bond random percolation problem on the new square lattice. In the new-lattice bond percolation problem, the bonds are present in the new-lattice with probability q and absent in the new-lattice with probability $p = 1 - q$. This is opposite to the conditions for the bond problem on the old-lattice and follows directly from how we have constructed the new-lattice problem from the old-lattice problem [2].

We see in the old-lattice problem that if $p > p_c$, then the bonds on the old-lattice form an infinite connected cluster from one side of the infinite system to the other. This infinite cluster would block the formation of a similar infinite cluster on the new-lattice [2]. Therefore, on the new-lattice we must have $q < q_c$ where q_c is the critical percolation concentration on the new lattice. Similarly, if $q > q_c$ in the new-lattice, an infinite cluster must exist on the new-lattice and this would block infinite cluster formation in the old-lattice problem [2]. Consequently, no infinite cluster would exist in the old-lattice problem, and we must have $p < p_c$. In these discussions, we find that an infinite cluster on the new-lattice would block the formation of infinite clusters on the old-lattice and vice versa.

When the old-lattice is at $p = p_c$ and transitioning to a system of only finite clusters, the new-lattice must be transitioning to a system with an infinite cluster [2]. This is due to the blockage effects in the presence of an infinite cluster. Consequently, at the critical point, due to these opposite transitions, we find both $p = p_c$ and $q = q_c$. Since both the old and the new systems are square lattices subject to bond percolation, we expect that $p_c = q_c$ for the two bond problems [2]. But because $q = 1 - p$ in both lattice systems, it follows that $p_c = q_c = \frac{1}{2}$.

Next, we look at some of the properties of the percolation problems in the neighborhood of the critical percolation concentration. These have been of interest particularly in terms of their relationship to the study of critical exponents in thermodynamic materials.

11.2 CRITICAL EXPONENTS

Early on in the study of percolation theory, a close similarity was realized between the geometric transitions of percolating systems, considered as functions of the bond or site fractional concentrations, and the thermodynamics of a second order phase transition in magnetic systems studied as a function of temperature [2, 3]. This similarity has been validated in both types of systems through a variety of interesting analytic and computer simulation studies [4, 5]. In this section, we will present a summary of some of the similarities between the two systems in the neighborhood of the critical percolation concentration and in the neighborhood of the Curie temperature of a ferromagnetic material [2, 3].

In the second order phase transition of a ferromagnetic material, as the temperature is lowered the magnetization (i.e., the magnetic ordered parameter) is found to go continuously from zero in the paramagnetic state to a nonzero value indicating the degree of ordering in the ferromagnetic phase [2]. Similarly, in percolating systems a continuous transition of the probability of a site being found in an infinite cluster (i.e., the order parameter of the percolation problem) is studied as a function of the fractional bond or site concentration. In this regard, it is found for fractional bond or site concentrations below a critical concentration that the system has no infinite cluster, i.e., there is zero probability of a site being in an infinite cluster. Above the critical concentration [2], however, there is an infinite cluster, i.e., there is a nonzero probability of a site being in an infinite cluster. As we shall see there are a number of other similarities between the two systems. First, however, let us consider a summary of some of the interesting characterizations of the ferromagnetic transition and how these can be extended to the percolation system. Specifically, we shall look at the description of the system through its critical exponents. These characterize the properties of the system at the phase transition.

In the neighborhood of the transition or Curie temperature T_c of a ferromagnetic system the system goes from a paramagnetic state to a ferromagnetic state [2, 3]. Specifically, a type of power law behavior of the magnetization $M(T)$ is found which is given by the general form [2, 3]

$$M(T) = 0 \text{ for } T > T_c \tag{11.3}$$

$$= M_0 (T_c - T)^{\beta} \text{ for } T < T_c$$

where M_0 is a constant and $\beta > 0$ is a critical exponent indicating the nature of the power law dependence near T_c. Closely related to this power law dependence is the power law dependence of the magnetic susceptibility $\chi = \frac{\partial M}{\partial H}$ where H is an external magnetic field. In the vicinity of the Curie temperature, the magnetic susceptibility then has the form [2, 3]

$$\chi(T) = \chi_0' (T - T_c)^{-\gamma} \text{ for } T > T_c \tag{11.4}$$

$$= \chi_0 (T_c - T)^{-\gamma} \text{ for } T < T_c$$

where χ_0 and χ_0' are constants and $\gamma > 0$ is a critical exponent.

In addition, the ferromagnetic free energy in the absence of an external field has a temperature dependent form [2]

$$G(T). \tag{11.5a}$$

In the neighborhood of the Curie temperature, it is found to be given by [2]

$$G(T) = G_0 |T - T_c|^{2-\alpha} \tag{11.5b}$$

where $\alpha > 0$ is the critical exponent.

A final set of critical exponents that will be of interest in our discussion of the ferromagnetic system is those for the spin correlation function and its correlation length for the spins carrying the atomic magnetic moments of the material [2, 3]. These exponents enter heavily into the study of scattering processes from the ferromagnet. In this regard, the correlation lengths are defined, considering only the z-component of spin, by the correlation function

$$\langle S_z(0) S_z(\vec{r}) \rangle - \langle S_z(0) \rangle \langle S_z(\vec{r}) \rangle = \langle S_z(0) S_z(\vec{r}) \rangle - \langle S_z(0) \rangle \langle S_z(0) \rangle \tag{11.6}$$

where $S_z(\vec{r})$ is the z-component of spin at position $\vec{r}$, and $\langle \ \rangle$ is a thermodynamic average of the system. As $r = |\vec{r}| \to \infty$ it is found that the correlation function assumes the general limiting form [2]

$$\langle S_z(0) S_z(\vec{r}) \rangle - \langle S_z(0) \rangle \langle S_z(0) \rangle \to \frac{1}{r^{d-2+\eta}} \exp\left(-\frac{r}{\xi} \right) \tag{11.7a}$$

where η is a critical exponent and ξ is the spin correlation length. Here we see that ξ acts to define a particular length scale in the description of the system.

In the vicinity of the Curie temperature, it is found that the correlation length develops a power law dependence on the temperature, having the form [2, 3]

$$\xi = \xi_0 \left|T - T_c\right|^{-\nu} \tag{11.7b}$$

where ξ_0 is a constant and $\nu > 0$ is a critical exponent. Consequently, from Eq. (11.7b) we find that the correlation length diverges at the Curie temperature, and the correlator in Eq. (11.7a) takes on a power law form. In this sense it becomes independent of the length scale defined by the correlation length, and as a result exhibits a pure power law form. The geometry of the system at the Curie temperature then is found to be of the form of a fractal [2–5], i.e., it becomes scale invariant.

The critical properties and their critical exponents exhibited by ferromagnetic materials are then defined in Eqs. (11.3) through (11.7). As a particular example of the critical exponents for an exactly solvable magnetic model, in the case of the 2−dimensional Ising model the following values [2, 3] are found: $\alpha = 0, \beta = \frac{1}{8}, \gamma = \frac{7}{4}, \eta = \frac{1}{4}$, and $\nu = 1$. We shall next turn to a consideration of the analogues of these exponents in the critical properties of percolation theory systems.

11.2.1 Critical Exponents for the Percolation Problem

In our discussions, we begin with a consideration of the site percolation problem for a system in which a fractional concentration of sites p of the system is occupied or present in the system [1–3]. From this model, we then develop an analogy of the properties of the percolating system with the properties of the ferromagnetic system. We start by developing an analogy for the magnetization and its form considered in Eq. (11.3).

Specifically, we define $P(p)$ as the fraction of occupied sites that belong to an infinite connected cluster at a fractional site concentration p. (In this regard, note that it can be shown that only one infinite cluster can exist in the system [1–3].) It is found in the neighborhood of the critical concentration p_c that [2]

$$\begin{aligned} P(p) &= 0 \text{ for } p < p_c \\ &= P_0 (p - p_c)^{\beta} \text{ for } p > p_c \end{aligned} \tag{11.8}$$

where P_0 is a constant and $\beta > 0$ is the critical exponent of the site percolation problem.

Another quantity of interest for site percolation is $n_s(p)$ defined as [2]

$$n_s(p) = \frac{\textit{the number of connected site} - \textit{clusters of s sites}}{N} \tag{11.9}$$

where $N \to \infty$ is the number of lattice sites. In terms of this it follows that

$$Nn_s(p)s = \textit{the number of sites in connected site} - \textit{clusters of s sites} \tag{11.10}$$

so that

$$n_s(p)s = \frac{\textit{the number of sites in connected site} - \textit{clusters of s sites}}{N}. \tag{11.11}$$

From Eqs. (11.9) through (11.11) and the definition of $P(p)$ it then follows that [2]

$$\sum_s n_s(p)s + pP(p) = p \tag{11.12}$$

where the sum over s ranges over the finite clusters, and $pP(p)$ is the probability that a general lattice site (not necessarily an occupied site) is part of an infinite cluster. Consequently, Eq. (11.12) is a statement of the probability that a lattice site of the system is either a member of a finite or an infinite cluster of sites.

Continuing our considerations of the geometry of the clusters in the site percolation system, from Eq. (11.11) we see that the mean cluster size of the finite clusters $S(p)$ is given by [2]

$$S(p) = \frac{\sum_s s^2 n_s(p)}{\sum_s s n_s(p)} \tag{11.13}$$

where again the sums are over the finite clusters. This form is the analogy of Eq. (11.4) in the magnetic system. Specifically, in the vicinity of p_c it is found that [2]

$$S(p) = S_0'(p - p_c)^{-\gamma} \text{ for } p > p_c \tag{11.14}$$

$$= S_0(p_c - p)^{-\gamma} \text{ for } p < p_c$$

where S_0 and S_0' are constants and $\gamma > 0$ is a critical exponent.

In addition, the sum [2]

$$G(p) = \sum_s n_s(p) \tag{11.15a}$$

where s sums over the clusters in the system are the analogy of the free energy of the ferromagnetic system. In the neighborhood of the critical site concentration, it is found to be of the form [2]

$$G(p) = G_0|p - p_c|^{2-\alpha} \tag{11.15b}$$

where $\alpha > 0$ is the critical exponent.

In addition to these critical exponent forms, one can also define a spatially dependent function known as the pair connectedness function $C(p,r)$ between the connected sites of a cluster. This function is also found to exhibit critical behaviors at the critical percolation concentration that are of power law forms. The connectedness function is the probability that two occupied sites, which are separated by a distance r of belong to the same cluster. It is found to exhibit the general functional behavior [2]

$$C(p,r) \to \frac{1}{r^{d-2+\eta}} \exp\left(-\frac{r}{\xi}\right) \tag{11.16a}$$

for $r \to \infty$ where the correlation function is again characterized by a correlation length of the form

$$\xi = \xi_0|p - p_c|^{-\nu}, \tag{11.16b}$$

and where ξ_0 is a constant. In these expressions η and $\nu > 0$ are two new critical exponents which are analogous to those in Eq. (11.7) for the magnetic system. We again see that at $p = p_c$ the site connections arrange themselves to form an infinite fractal cluster with their correlations characterized by power law behaviors [3, 5].

For a particular example of the critical exponents of a percolation model, in the case of 2−dimensional site percolation on the square lattice the following values are found [2]:

$\alpha = -\frac{2}{3}, \beta = \frac{5}{36}, \gamma = \frac{43}{18}, \eta = \frac{5}{24}$, and $\nu = \frac{4}{3}$. As an interesting point, in both the magnetic and the percolation theory models the critical exponents are found to satisfy certain identities. These so-called scaling relationships include [2] two important ones which state that:

$$\alpha + 2\beta + \gamma = 2 \tag{11.17a}$$

and

$$d\nu = 2 - \alpha \tag{11.17b}$$

where d is the dimension of the system. Both the 2-dimensional Ising models and the percolation systems we have studied are seen to obey these identities.

In the case of the magnetic system, the relationships in Eq. (11.17) arise from the fact that the specific heat, magnetization, magnetic susceptibility, and spin correlation functions are all ultimately derived from the free energy and partition function of the system. As we shall see later, similar underlying relationships occur in the geometric properties of the percolation problem.

We next present an example of scaling for a site percolation problem. This allows us to develop a scaling treatment for the determination of the critical fraction concentration and the critical exponents of the percolation system.

11.3 EXAMPLE SITE PERCOLATION ON THE TRIANGLE LATTICE

An example of site percolation theory and the application of the ideas of scaling theory in its study is provided by a consideration of the triangle lattice [2]. (As a reference, in Figure 11.5a we show a schematic of the triangle lattice with all its sites present.) For site percolation on this lattice, we consider a system in which only a fractional concentration p of sites are randomly present throughout the lattice so that a fractional concentration $1-p$ of sites are randomly removed from the lattice. (An example of a configuration of such a percolation system is shown schematically in Figure 11.5b.) The connectivity properties of the resulting system are then determined as functions of p. Specifically, our objective in the following discussions will be to develop a scaling mapping from this disordered system to another disordered system formed on a greater length scale, having a fractional concentration p' of sites present. By the consideration of the fixed points of this transformation, a discussion of the critical concentration as well as some of the critical exponents of the

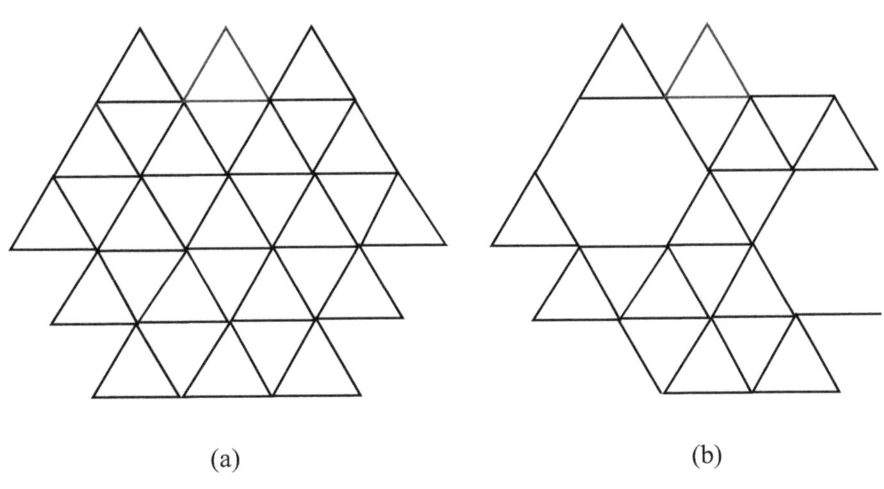

FIGURE 11.5 A schematic of the triangle lattice: a) with all the sites present, b) with a fraction p of sites randomly present and a fraction $1-p$ of sites randomly removed.

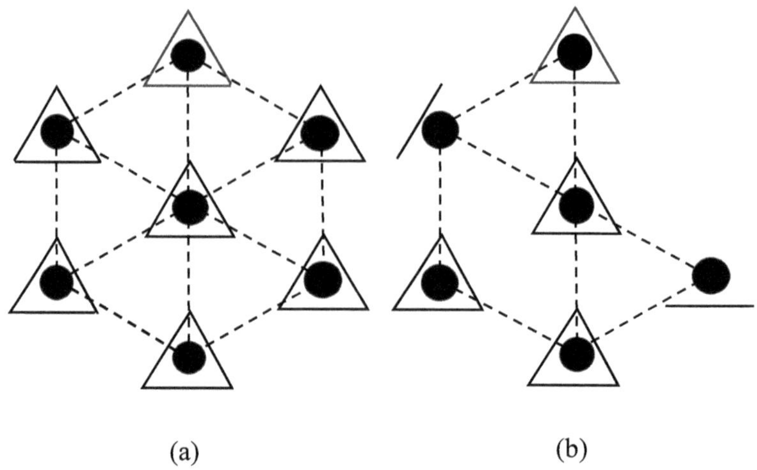

FIGURE 11.6 Schematic of the triangle lattice broken up into sub-triangles: a) the triangle lattice without disorder and b) the triangle lattice with site disorder. Notice that black dots have been added to the center of some of the triangles (triangles shown by solid lines). These form the sites of the new-lattice which is a triangle lattice drawn on a larger scale and they are connected by dashed nearest-neighbor bonds on the new-lattice. The old-lattice sites are at the vertices of the triangles (triangles shown by solid lines). These are from the original triangle lattice in Figure 11.5.

site percolation problem will be developed. The discussion somewhat parallels the treatment in Eqs. (11.1) and (11.2) for the bond percolation problem.

In Figure 11.6a we have redrawn the system in Figure 11.5a adding an emphasis to certain triangles by placing dots at the center of the emphasized triangles [2]. Notice that when continued throughout the lattice in Figure 11.6a the dots form a triangle lattice of their own. In this regard, in Figure 11.6a we show the lattice of dots connected by nearest-neighbor bonds (dashed lines). We next make a mapping from the site random problem on the lattice in Figure 11.5b to a site random problem formulated on the triangle lattice of dots shown in Figure 11.6b. In this regard, the site disordered problem in Figure 11.5b will be mapped onto a new site disordered problem generated in the triangular lattice in Figure 11.6b.

To define the mapping of the old-lattice to the new-lattice, we consider that a dot site on the lattice in Figure 11.6b is occupied or unoccupied according to the following rules [2]. If the three sites on the triangle containing a dot of the new-lattice have 2 or 3 occupied sites, then the dot on the new triangle lattice will be considered to be an occupied site of Figure 11.6b. Otherwise, the new site will be treated as an unoccupied site of the lattice in Figure 11.6b.

These rules are illustrated in Figure 11.7a for the case of a single triangle containing a dot in which all three vertices of the triangle are occupied. Consequently, the dot at the center of the triangle must be occupied, and the probability of this occurring is p^3. In Figure 11.7b only two of the triangle vertices are occupied so that the dot at the center of the triangle is again occupied. The probability of this occurring is $p^2(1-p)$, and the configuration in Figure 11.7b can occur in 3 different ways. These are the only configurations of the triangle which have an occupied central dot.

It then follows that the probability $p'$of the central dot in Figure 11.6 or 11.7 being present on the new lattice is given in terms of p by [2, 6, 7]

$$p' = p^3 + 3p^2(1-p). \tag{11.18}$$

Consequently, Eq. (11.18) maps the old-site problem in which the site concentration is p onto the new-site problem of site concentration p' on the new triangle lattice, and for this mapping between the two lattices it follows that the set of fixed-points is given by

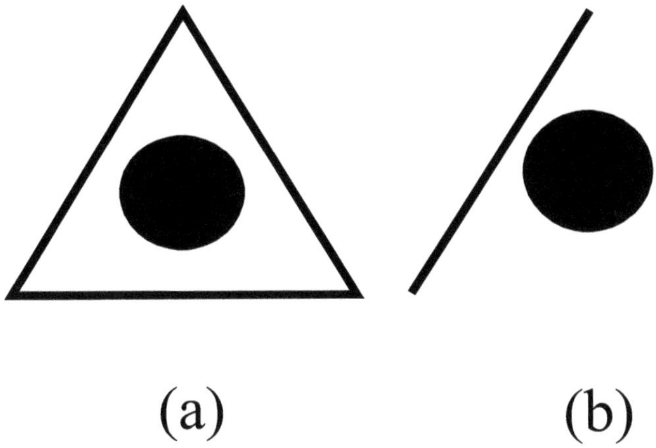

FIGURE 11.7 Schematic of the occupancy properties of the sites on the new-lattice in Figure 11.6. In a) three vertices of the old-lattice triangle are occupied so the black dot is present in the new-lattice and b) two vertices of the old-lattice triangle are present, so the black dot is present in the new-lattice. Otherwise, for other occupancies of the vertices the black dot is absent, i.e., it is an unoccupied site.

$$p' = p = 0, \frac{1}{2}, 1. \tag{11.19}$$

In addition, the scales of the two systems are found to be different. If the p site percolation problem is composed of triangles of side d, then the p' site percolation problem is composed of triangles of side $\sqrt{3}d$.

Considering the nature of the three fixed-points we find: for the case in which $p' = p = 0$ there are no sites on either the p or the p' lattice. On the other hand, in the case in which $p' = p = 1$ all the sites are present on both the p and the p' lattice. In the case that $p' = p = \frac{1}{2}$ there is a single infinite cluster characterized by a scale-independent fractal geometry [5–7].

Unlike the bond percolation problem, the exact solution of the site percolation problem on the triangle lattice gives the critical fractional concentration of sites on the triangle lattice $p_c = \frac{1}{2}$, and this agrees with the result from our crude mapping. In the site problem, then our crude result happens to agree with the exact treatment. Next, we turn to a consideration of some of the critical exponents of the system and see how these can be approximated in the context of scaling theory [2, 3, 6, 7].

11.3.1 Critical Exponents of the Correlation Function

In a consideration of the correlation lengths between the p and p' systems, we recognize that the p' lattice is $\sqrt{3}$ times larger than the lattice in the p system. Consequently, it follows that expressing the correlation lengths in units of the lattice constant between each of the two lattices [2, 7]

$$\sqrt{3}\xi'(p') = \xi(p). \tag{11.20}$$

If we adopt a power law form in the neighborhood of the critical percolation concentration, we find that

$$\xi(p) = \xi_0 |p - p_c|^{-\nu}. \tag{11.21}$$

Applying this to both sides of Eq. (11.20) it then follows that [2, 7]

$$p' - p_c = \left(\sqrt{3}\right)^{\frac{1}{\nu}} \left(p - p_c\right). \tag{11.22}$$

This in turn can be inverted for ν.

Solving Eq. (11.22) for ν we find [2, 7]

$$\nu = \frac{\ln\sqrt{3}}{\ln\dfrac{p' - p_c}{p - p_c}} = \frac{\ln\sqrt{3}}{\ln\left.\dfrac{dp'}{dp}\right|_{p_c}} \tag{11.23}$$

where $\left.\dfrac{dp'}{dp}\right|_{p_c} = \dfrac{3}{2}$ is obtained from Eq. (11.18). In this way from Eq. (11.23) we obtain $\nu = 1.35$ whereas the best accepted value is $\nu = \dfrac{4}{3}$. This provides an example of a scaling theory treatment of at least one of the critical exponents.

We shall now consider the relationship between bond percolation on a lattice and the thermodynamics of a particular magnetic model known as the q-state Potts model [8, 9]. This provides a detailed mapping between the two statistical models which allows us to treat and relate the theory of critical exponents at the phase transitions in both of these models.

11.4 RELATIONSHIP OF PERCOLATION THEORY TO THE Q-STATE POTTS MODEL

In this final section we look at the relationship of the percolation theory problems to a class of thermodynamic models of magnetic ordering [2, 8, 9]. In particular, we will show that the bond percolation problem can be mapped onto the problem of the statistical mechanics of the q-state Potts model in the $q \to 1$ limit. This allows us to develop a connection between the various thermodynamic properties of the Potts model and the geometric properties of a percolation system. These relationships have been described earlier in Section 11.2 in the context of a brief discussion of critical exponents. Now for the case of bond percolation we will develop a formal linkage between the geometric models and the statistical mechanical models. Other relationships have also been found between various types of geometric and thermodynamic models, and we will refer the reader to the literature for a treatment of these further discussions [8, 9].

We begin with a brief description of the q-state Potts model.

11.4.1 Q-State Potts Model

The q-state Potts model is a generalization of the Ising model to a consideration involving spins greater than spin-$\dfrac{1}{2}$. In this regard, the Hamiltonian of the Potts model is given by the form [8]

$$H = -J\sum_{\langle i,j\rangle}\left[q\delta_{s_i,s_j} - 1\right] = \sum_{\langle i,j\rangle} H_{ij} \tag{11.24a}$$

where the nearest-neighbor bonds of the model are

$$H_{ij} = -J\left[q\delta_{s_i,s_j} - 1\right], \tag{11.24b}$$

$\langle i,j\rangle$ denotes a sum over pairs of nearest-neighbor lattice sites, the spins are of the form $s_i, s_j = 1, 2, 3, \ldots, q$ for an integer q, and δ_{s_i,s_j} is the Kronecker delta. It then follows that [2, 8]

$$H_{ij} = -J\left[q-1\right] \text{ for } s_i = s_j \tag{11.25}$$

$$= J \text{ for } s_i \neq s_j$$

so that when $J > 0$ the system is ferromagnetic and when $J < 0$ the system is antiferromagnetic. Also, it is interesting to note that for the case in which $q = 2$, Eqs. (11.24) and (11.25) are seen to reduce to the spin $-\frac{1}{2}$ Ising model.

The Boltzmann factor associated with each of the individual nearest-neighbor bonds in Eq. (11.24) is of the form [8]

$$e^{-\beta \times H_{ij}} = A + B\delta_{s_i, s_j} \tag{11.26}$$

where use is made here of the identity $\left(\delta_{s_i, s_j}\right)^2 = \delta_{s_i, s_j}$. In the case that $s_i = s_j$ it then is found that

$$e^{\beta J[q-1]} = A + B, \tag{11.27a}$$

and in the case that $s_i \neq s_j$ we have

$$A = e^{-\beta J}. \tag{11.27b}$$

Consequently, from Eq. (11.27) it follows that

$$B = e^{\beta J[q-1]} - e^{-\beta J} \tag{11.27c}$$

so that with this notation [8]

$$e^{-\beta \times H_{ij}} = e^{\beta J[q-1]}\left\{e^{-\beta Jq} + \left[1 - e^{-\beta Jq}\right]\delta_{s_i, s_j}\right\} \tag{11.28}$$

provides a simple form for the Boltzmann factors associated with the nearest-neighbor bonds in the model.

In this notation, the partition function of the Potts model is then given by [8]

$$Z_q = Tr\, e^{-\beta \sum_{\langle i,j\rangle} \mathcal{H}_{ij}} = Tr \prod_{\langle i,j\rangle} e^{-\beta \times H_{ij}} \tag{11.29}$$

$$= e^{\beta J[q-1]N_b}\, Tr \prod_{\langle i,j\rangle} \left\{e^{-\beta Jq} + \left[1 - e^{-\beta Jq}\right]\delta_{s_i, s_j}\right\}$$

where N_b is the number of nearest-neighbor pairs of sites in the model and the trace is over $\{s_l\}$. It is also interesting to note that the product in Eq. (11.29) is essentially a set of geometric factors associated with the temperature dependent coefficients $e^{-\beta Jq}$ and $1 - e^{-\beta Jq}$. This will be useful in establishing the connection between the geometric properties of the percolation models and the thermodynamic properties of the Potts model.

If for $J > 0$ we define [8, 9]

$$p = 1 - e^{-\beta Jq} \tag{11.30a}$$

it follows that $0 \le p \le 1$, and consequently we also have

$$1 - p = e^{-\beta J q} \tag{11.30b}$$

where again $0 \le 1 - p \le 1$. In terms of the new variables in Eq. (11.30), it follows from Eq. (11.29) that the Potts model partition function becomes

$$Z_q = e^{\beta J[q-1]N_b}\, Tr \prod_{\langle i,j \rangle} \left(1 - p + p\delta_{s_i,s_j}\right) \tag{11.31}$$

where the trace is over the spins $s_i, s_j = 1, 2, 3, \ldots, q$ for an integer q. As shall be seen later, p will ultimately be associated with the fractional nearest-neighbor bond concentration on an associated percolation problem obtained from the Potts model in Eq. (11.31). Similarly, $1 - p$ will be related to nearest-neighbor bond vacancies in the system. At this point, however, it is interesting to note that [8, 9]

$$Tr \prod_{\langle i,j \rangle} \left(1 - p + p\delta_{s_i,s_j}\right) \tag{11.32}$$

appears as a purely geometric factor involving weighting p and $1 - p$.

We next discuss the evaluation of the geometric factors in Eq. (11.32), providing them with a geometric interpretation in terms of cluster connectivity. This will lead to the development of lattice percolation theory in terms of the thermodynamics of the Potts model on the lattice and finally to an association of the critical exponents of the lattice problem with those of the thermodynamic problem [8, 9].

11.4.2 Lattice Clusters and the q-State Potts Model: Critical Exponents

We begin by considering the evaluation [8] of the product in Eq. (11.32). It involves the form $\prod_{\langle i,j \rangle} \left(1 - p + p\delta_{s_i,s_j}\right)$ where $\langle i, j \rangle$ runs over all the possible nearest-neighbor pairings of sites over the entire lattice [8], and we will denote the number of these pairings over the infinite lattice by $N_b \to \infty$.

First consider the single factor $\left(1 - p + p\delta_{s_i,s_j}\right)$ for the pairing of the i and j nearest-neighbor sites. We will treat this as a sum of two terms [8, 9]. The first term is $0 \le 1 - p \le 1$ which we shall interpret as the probability that the bond connecting the two nearest-neighbor sites is absent [8]. The second term is $p\delta_{s_i,s_j}$ which we interpret as the probability $0 \le p \le 1$ that a bond is present times δ_{s_i,s_j} which in turn determines the bond presence in terms of the spin values on the two sites it connects. In this interpretation the product $\prod_{\langle i,j \rangle} \left(1 - p + p\delta_{s_i,s_j}\right)$ is then a sum of 2^{N_b} products involving the $1 - p$ and $p\delta_{s_i,s_j}$. In the products composing the sum, the factors of $1 - p$ come from the bond vacancies [8]. On the other hand, the factors of $p\delta_{s_i,s_j}$ represent the possible presence of particular bonds in the lattice [8]. In regard to $p\delta_{s_i,s_j}$, we find that if $\delta_{s_i,s_j} = 1$ the bond between the ith and j^{th} sites is present with probability p. But if $\delta_{s_i,s_j} = 0$ the product term is zero and does not contribute in the sum of 2^{N_b} terms. In this way the product $\prod_{\langle i,j \rangle} \left(1 - p + p\delta_{s_i,s_j}\right)$ generates a sum over the variety of weighted configurations of the possible bond arrangements in the system. The trace in Eq. (11.32) then generates all the possible configurations of the percolation lattice problem weighted by factors involving the probability of the configuration's occurrence in the system and ultimately accounting for the multiplicity of the q-state Potts model spin states [8, 9].

Considering the interpretation of the $\prod_{\langle i,j\rangle}\left(1-p+p\delta_{s_i,s_j}\right)$ product in this light, we see that for a particular configuration of spins $\{s_l\}$ it can be written as a sum of 2^{N_b} individual product terms each of which has the form

$$\left(1-p\right)^{n_{vac}}\, p^{n_{occ}}\, D_{n_{occ},n_{vac}}\left(\left\{\delta_{s_i,s_j}\right\}\right). \tag{11.33}$$

Here n_{occ} is the number of bonds present between nearest neighbor sites, n_{vac} is the number of bonds absent between nearest neighbor sites, and $D_{n_{occ},n_{vac}}\left(\left\{\delta_{s_i,s_j}\right\}\right)$ is a product function of the nonzero $\left\{\delta_{s_i,s_j}\right\}$ present in the particular term under consideration. The factor in Eq. (11.33) is then one of the products entering into the sum for $\prod_{\langle i,j\rangle}\left(1-p+p\delta_{s_i,s_j}\right)$that was discussed in the previous paragraph [8]. Notice, in addition, that in the context of our various definitions

$$N_b = n_{occ} + n_{vac} \tag{11.34}$$

for our present considerations.

If we sum over all the terms of the form in Eq. (11.33) for fixed n_{occ} and n_{vac}, we find

$$\left(1-p\right)^{n_{vac}}\, p^{n_{occ}} \sum_{\langle i,j\rangle} D_{n_{occ},n_{vac}}\left(\left\{\delta_{s_i,s_j}\right\}\right) \tag{11.35}$$

where $\sum_{\langle i,j\rangle} D_{n_{occ},n_{vac}}\left(\left\{\delta_{s_i,s_j}\right\}\right)$ sums over all the distinct nonzero configurations weighted by $\left(1-p\right)^{n_{vac}}\, p^{n_{occ}}$. Notice that these configurations can be denoted as $C\left(n_{occ},n_{vac}\right)=\sum_{\langle i,j\rangle} D_{n_{occ},n_{vac}}\left(\left\{\delta_{s_i,s_j}\right\}\right)$ and their occurrence is weighted with the probability factor

$$P\left(C\left(n_{occ},n_{vac}\right)\right)=\left(1-p\right)^{n_{vac}}\, p^{n_{occ}}. \tag{11.36}$$

With this notation, Eq. (11.31) then becomes [8, 9]

$$Z_q \propto e^{\beta J[q-1]N_b}\, Tr \sum_{C\left(n_{occ},n_{vac}\right)} P\left(C\left(n_{occ},n_{vac}\right)\right) \prod_{Occupied} \delta_{s_i,s_j} \tag{11.37a}$$

$$= e^{\beta J[q-1]N_b} \sum_{C\left(n_{occ},n_{vac}\right)} P\left(C\left(n_{occ},n_{vac}\right)\right) Tr \prod_{Occupied} \delta_{s_i,s_j}$$

where the product $\prod_{Occupied} \delta_{s_i,s_j}$ is over the occupied bonds subject to the constraint of fixed n_{occ} and n_{vac}. Notice that the trace is made over all the possible configurations of $\{s_i\}$.

If we recognize that the spins on each isolated connected cluster must all have the same value (i.e., $\delta_{s_i,s_j}=1$ over the entire connected cluster), we can rewrite Eq. (11.37a) as [8]

$$Z_q \propto e^{\beta J[q-1]N_b} \sum_{C\left(n_{occ},n_{vac}\right)} P\left(C\left(n_{occ},n_{vac}\right)\right) \sum_{\alpha_1}\sum_{\alpha_2}\cdots\sum_{\alpha_{n_c}} 1 \tag{11.37b}$$

where $\alpha_i = 1,2,3,..,q$ is the value of the spin on the ith connected cluster of the $C\left(n_{occ},n_{vac}\right)$ configuration, and the system has $n_c = n_c\left(C\left(n_{occ},n_{vac}\right)\right)$ distinct connected clusters. Each cluster in the system has q possible spin realizations, so the sums over the α_i can be evaluated to give

$$Z_q \propto e^{\beta J[q-1]N_b} \sum_{C\left(n_{occ},n_{vac}\right)} P\left(C\left(n_{occ},n_{vac}\right)\right) q^{n_c} \tag{11.38a}$$

and taking the $q \to 1$ limit with $q = 1 + (q-1)$ we find [8]

$$Z_q = e^{\beta J[q-1]N_b} \sum_{C(n_{occ},n_{vac})} P(C(n_{occ},n_{vac}))[1+(q-1)]^{n_c} \tag{11.38b}$$

$$\to e^{\beta J[q-1]N_b} \sum_{C(n_{occ},n_{vac})} P(C(n_{occ},n_{vac}))[1+n_c(q-1)]$$

$$\propto e^{\beta J[q-1]N_b}\left[1+(q-1)\frac{\sum_{C(n_{occ},n_{vac})} P(C(n_{occ},n_{vac}))n_c}{\sum_{C(n_{occ},n_{vac})} P(C(n_{occ},n_{vac}))}\right]$$

where we retained first order powers of $|q-1| \ll 1$.

From the partition function, the free energy then follows from the relationship

$$F_q = -k_B T \ln Z_q. \tag{11.39}$$

In this regard, we find to leading order in $(q-1)$ that up to an ignored constant [8]

$$\frac{\ln Z_q}{N_s} = \beta J(q-1)\frac{N_b}{N_s} + (q-1)\frac{\sum_{C(n_{occ},n_{vac})} P(C(n_{occ},n_{vac}))n_c}{N_s \sum_{C(n_{occ},n_{vac})} P(C(n_{occ},n_{vac}))} \tag{11.40a}$$

where N_s is the number of sites on the lattice and

$$n_{clusters} = \frac{N_{clusters}}{N_s} = \frac{\sum_{C(n_{occ},n_{vac})} P(C(n_{occ},n_{vac}))n_c}{N_s \sum_{C(n_{occ},n_{vac})} P(C(n_{occ},n_{vac}))} \tag{11.40b}$$

is the average number of clusters in the system per lattice site. Here we have applied the relationship $N_{clusters} = \frac{\sum_{C(n_{occ},n_{vac})} P(C(n_{occ},n_{vac}))n_c}{\sum_{C(n_{occ},n_{vac})} P(C(n_{occ},n_{vac}))}$ giving the average number of clusters in the system. The free energy per site is then obtained as

$$f = \frac{F_q}{N_s} = -k_b T\frac{\ln Z_q}{N_s} = -k_b T(q-1)\left[\beta J\frac{N_b}{N_s} + n_{clusters}\right] \tag{11.41a}$$

and can now be solved for $n_{clusters}$ in terms of f.

Solving for $n_{clusters}$ in Eq. (11.41a) we find [8]

$$n_{clusters} = -\beta\left[\frac{f}{(q-1)} + \frac{N_b}{N_s}J\right]. \tag{11.41b}$$

It is seen from Eq. (11.41b) that $n_{clusters}$, which is a fundamental quantity of percolation theory, is related to the $q-1 \to 0$ limit for the q-states Potts model through the Potts model free energy [2, 8, 9]. In this regard, from renormalization group and computer simulation studies of its thermodynamics, it is known that the Potts model has a power law behavior at its Curie temperature that is of the form [8]

$$f \to f_0 |T - T_c|^{2-\alpha_P} \tag{11.42}$$

where $2-\alpha_P$ is then the Potts model critical exponent associated with the free energy and $\alpha_P > 0$ is the critical exponent for the specific heat of the Potts model [2, 8].

To further establish the behavior of the percolation system in the neighborhood of the critical percolation concentration, from Eq. (11.30a) we have the relationship $p = 1 - e^{-\beta Jq}$ between the bond concentration on the lattice in the percolation problem and the corresponding temperature of the Potts model. Consequently, solving for p in terms of T it follows that

$$k_B T = -\frac{Jq}{\ln(1-p)} \tag{11.43}$$

so that

$$T - T_c = -\frac{Jq}{k_B}\left[\frac{1}{\ln(1-p)} - \frac{1}{\ln(1-p_c)}\right] = -\frac{Jq}{k_B}\left[\frac{1}{\ln(1-p_c-(p-p_c))} - \frac{1}{\ln(1-p_c)}\right]. \tag{11.44}$$

Expanding in $p - p_c \to 0$, it then follows from Eq. (11.44) that to leading order in $(p - p_c)$ we have

$$T - T_c = -\frac{Jq}{k_B}\frac{1}{\left[\ln(1-p_c)\right]^2(1-p_c)}(p-p_c) \propto p_c - p. \tag{11.45}$$

From Eqs. (11.41) through (11.45) we then find that in the neighborhood of the divergence at the critical percolation concentration (i.e., at the Curie temperature of the Potts model)

$$n_{clusters} \propto |p - p_c|^{2-\alpha_P}. \tag{11.46}$$

This is the critical exponent form introduced earlier in Eq. (11.15) of Section 11.2.

We conclude with an outline of the treatments of the probability of a site being in an infinite cluster and of the site cluster correlation function. These are again obtained in the context of the mapping between percolation theory and the q-state Potts model. They require the reformulation of the Potts model to include the effects of an external magnetic field [8].

11.4.3 Probability of a Site Being in an Infinite Cluster and the Cluster Correlation Function

In order to treat some of the more interesting properties of the percolation system, we need to return to the q-state Potts model formulation and revise it to include the effects of an applied magnetic field. Specifically, we modify the model Hamiltonian in Eq. (11.24) to be of the form [8]

$$H = -J\sum_{<i,j>}\left[q\delta_{s_i,s_j} - 1\right] - \sum_i h\left[q\delta_{s_i,1} - 1\right] = \sum_{<i,j>} H_{ij} - \sum_i h\left[q\delta_{s_i,1} - 1\right] \tag{11.47}$$

where h is the applied magnetic field and notice in the magnetic field term that $\delta_{s_i,1}$ is only nonzero when $s_i = 1$. The magnetic field interaction, consequently, involves a sum over spin sites so that it can be straightforwardly introduced into the theory. Notice that for $h = 0$ the system returns to the original form in Eq. (11.24).

It then follows that with the introduction of the field terms the partition function in Eq. (11.37) becomes [8]

$$Z_q \propto e^{\beta J[q-1]N_b} e^{\beta h[q-1]N_s} \times \tag{11.48}$$

$$\times \sum_{C(n_{occ},n_{vac})} P(C(n_{occ},n_{vac}))\, Tr \prod_{Occupied} \delta_{s_i,s_j} \prod_l e^{\beta hq\left[\delta_{s_l,1}-1\right]}$$

where we have applied the identity to the field terms the identity [8]

$$\prod_i \left(e^{\beta hq\delta_{s_i,1}} e^{-\beta h}\right) = \prod_i \left(e^{\beta hq\delta_{s_i,1}} e^{-\beta hq} e^{\beta hq} e^{-\beta h}\right) \tag{11.49}$$

$$= e^{\beta h[q-1]N_s} \prod_i e^{\beta hq\left[\delta_{s_i,1}-1\right]}.$$

From Eq. (11.48) we find that Eq. (11.38) takes the form

$$Z_q \propto e^{\beta J[q-1]N_b} e^{\beta h[q-1]N_s} \times \tag{11.50}$$

$$\times \sum_{C(n_{occ},n_{vac})} P\left(C\left(n_{occ},n_{vac}\right)\right) \sum_{\alpha_1} \sum_{\alpha_2} \cdots \sum_{\alpha_{n_c}} \prod_{k=1}^{n_c} \left\{ \prod_{i\in\alpha_k} e^{\beta hq\left[\delta_{s_i,1}-1\right]} \right\}$$

$$= e^{\beta J[q-1]N_b} e^{\beta h[q-1]N_s} \times$$

$$\times \sum_{C(n_{occ},n_{vac})} P\left(C\left(n_{occ},n_{vac}\right)\right) \sum_{\alpha_1} \sum_{\alpha_2} \cdots \sum_{\alpha_{n_c}} \prod_{k=1}^{n_c} e^{\beta hqn(\alpha_k)\left[\delta_{\alpha_k,1}-1\right]}$$

where $n(\alpha_k)$ is the number of sites in the α_k connected cluster. Notice that we find that both equations in Eqs. (11.49) and (11.50) reduce to their original forms when $h = 0$.

Considering the terms in Eq. (11.50) given by [8, 9]

$$\sum_{\alpha_1} \sum_{\alpha_2} \cdots \sum_{\alpha_{n_c}} \prod_{k=1}^{n_c} e^{\beta hqn(\alpha_k)\left[\delta_{\alpha_k,1}-1\right]} \tag{11.51}$$

we see that when $\alpha_i = 1$

$$e^{\beta hqn(\alpha_i)\left[\delta_{\alpha_i,1}-1\right]} = 1 \tag{11.52a}$$

and when $\alpha_i = 2,3,\ldots,q$

$$e^{\beta hqn(\alpha_i)\left[\delta_{\alpha_i,1}-1\right]} = e^{-\beta hqn(\alpha_i)}. \tag{11.52b}$$

Consequently, Eq. (11.51) can be written in the form

$$\sum_{\alpha_1} \sum_{\alpha_2} \cdots \sum_{\alpha_{n_c}} \prod_{k=1}^{n_c} e^{\beta hqn(\alpha_k)\left[\delta_{\alpha_k,1}-1\right]} = \prod_{k=1}^{n_c} \left[1+(q-1)e^{-\beta hqn(\alpha_k)}\right]. \tag{11.53}$$

Applying this in Eq. (11.50) and considering the limit $(q-1) \to 0$, it follows that [8]

$$Z_q = e^{\beta J[q-1]N_b} e^{\beta h[q-1]N_s} \sum_{C(n_{occ},n_{vac})} P\left(C\left(n_{occ},n_{vac}\right)\right) \prod_{k=1}^{n_c} \left[1+(q-1)e^{-\beta hqn(\alpha_k)}\right] \tag{11.54}$$

$$\approx e^{\beta J[q-1]N_b} e^{\beta h[q-1]N_s} \sum_{C(n_{occ},n_{vac})} P\left(C\left(n_{occ},n_{vac}\right)\right)\left[1+\left(q-1\right)\sum_{k=1}^{n_c} e^{-\beta hqn(\alpha_k)}\right]$$

$$\propto e^{\beta J[q-1]N_b} e^{\beta h[q-1]N_s} \left\{1+\left(q-1\right)\frac{\sum_{C(n_{occ},n_{vac})}\left\{P\left(C\left(n_{occ},n_{vac}\right)\right)\sum_{k=1}^{n_c} e^{-\beta hqn(\alpha_k)}\right\}}{\sum_{C(n_{occ},n_{vac})} P\left(C\left(n_{occ},n_{vac}\right)\right)}\right\}$$

From this it then follows that as $(q-1)\rightarrow 0$ we have

$$\frac{\ln Z_q}{N_s} = \beta J\left(q-1\right)\frac{N_b}{N_s} + \beta h\left(q-1\right) + \left(q-1\right)\frac{\sum_{C(n_{occ},n_{vac})}\left\{P\left(C\left(n_{occ},n_{vac}\right)\right)\sum_{k=1}^{n_c} e^{-\beta hqn(\alpha_k)}\right\}}{N_s\sum_{C(n_{occ},n_{vac})} P\left(C\left(n_{occ},n_{vac}\right)\right)} \quad (11.55)$$

and

$$f = \frac{F_q}{N_s} = -k_b T\frac{\ln Z_q}{N_s} = \quad (11.56)$$

$$= -k_b T\left(q-1\right)\left[\beta J\frac{N_b}{N_s} + \beta h + \frac{\sum_{C(n_{occ},n_{vac})}\left\{P\left(C\left(n_{occ},n_{vac}\right)\right)\sum_{k=1}^{n_c} e^{-\beta hqn(\alpha_k)}\right\}}{N_s\sum_{C(n_{occ},n_{vac})} P\left(C\left(n_{occ},n_{vac}\right)\right)}\right].$$

Again, in the limit that $h = 0$ Eq. (11.56) reduces back to Eq. (11.41a).

The magnetization of the Potts model is then obtained from the free energy through the standard thermodynamics relationship [8] in which the magnetization in the limit that $h \rightarrow 0$ is given by

$$m = -\frac{df}{dh} = \quad (11.57)$$

$$= \left(q-1\right)\left[1+\frac{\sum_{C(n_{occ},n_{vac})}\left\{P\left(C\left(n_{occ},n_{vac}\right)\right)\sum_{k=1}^{n_c}\left[-qn\left(\alpha_k\right)\right] e^{-\beta hqn(\alpha_k)}\right\}}{N_s\sum_{C(n_{occ},n_{vac})} P\left(C\left(n_{occ},n_{vac}\right)\right)}\right]$$

and as $(q-1)\rightarrow 0$ and $h \rightarrow 0$ it follows that [8]

$$\frac{m}{q-1} = 1+\frac{\sum_{C(n_{occ},n_{vac})}\left\{P\left(C\left(n_{occ},n_{vac}\right)\right)\sum_{k=1}^{n_c}\left[-n\left(\alpha_k\right)\right] e^{-\beta hn(\alpha_k)}\right\}}{N_s\sum_{C(n_{occ},n_{vac})} P\left(C\left(n_{occ},n_{vac}\right)\right)} \quad (11.58)$$

$$= \lim_{h\rightarrow 0}\left\{\frac{\sum_{C(n_{occ},n_{vac})}\left\{P\left(C\left(n_{occ},n_{vac}\right)\right)\sum_{k=1}^{n_c}\left[-n\left(\alpha_k\right)\right] e^{-\beta hn(\alpha_k)}\right\}}{N_s\sum_{C(n_{occ},n_{vac})} P\left(C\left(n_{occ},n_{vac}\right)\right)}\right\}$$

$$= 1+\frac{1}{N_s}\left[-N_f\right] = \frac{N_s - N_f}{N_s}$$

where N_s is the number of lattice sites and N_f is the number of sites in finite clusters. Consequently, the probability of a site being in an infinite cluster of the system is

$$P_\infty = \frac{N_s - N_f}{N_s} = \frac{m}{q-1} \tag{11.59}$$

for the limit that $(q-1) \to 0$ and $h \to 0$.

Since [2, 8]

$$m \propto (T_c - T)^\beta \tag{11.60}$$

and from Eq. (11.45) $T_c - T \propto p - p_c$, it follows from Eqs. (11.58) through (11.60) that [8]

$$P_\infty \propto (p - p_c)^\beta. \tag{11.61}$$

This relates the power law dependence above the critical percolation concentration of the probability of a spin being located in an infinite cluster to the q-state Potts model critical exponent β in Eq. (11.60).

As a final point, we note that the probability of two spins being part of the same cluster can be computed from the Potts model [8] by introducing a site dependent external field h_i into the Hamiltonian in Eq. (11.47) [7, 8]. By generalizing Eq. (11.57) in this way, we find the second derivative [8]

$$\frac{\partial^2 f}{\partial h_i \partial h_j} = g(i,j) \tag{11.62}$$

where the function $g(i,j)$ is between the lattice site i and j. Through the study of $g(i,j)$ the relationships between the correlation functions we discussed earlier in Eqs. (11.6), (11.7), and (11.16) are obtained. These link the critical exponents η and ν of the q-state Potts model to those of the corresponding lattice percolation problem [7].

REFERENCES

1. McComb, W. D. 2004. *Renormalization Methods: A Guide for Beginners* (Oxford University Press, Oxford).
2. Plischke, M. and Bergersen, B. 1989. *Equilibrium Statistical Physics* (Prentice Hall, New Jersey).
3. McGurn, A. R. 2023. *An Introduction to Condensed Matter Physics for the Nanosciences* (CRC Press, Oxon).
4. Malthe-Sorenssen, A. 2024. *Percolation Theory Using Python* (Springer, Cham, Switzerland).
5. Strogatz, S. H. 1994. *Nonlinear Dynamics and Chaos* (Perseus Books, Reading).
6. Stauffer, D. 1979. Scaling Theory of Percolation Clusters. *Physics Reports* 54, 1–74.
7. Klien, W., Stanley, E. H., Reynold, P. J., and Coniglio, A. 1978. Renormalization Group Approach to the Percolation Properties of the Triangular Ising Model. *Physical Review Letters* 17, 1145–1148.
8. Berlinsky, A. J. and Harris, A. B. 2019. *Statistical Mechanics* (Springer Nature, Switzerland, AG).
9. Fortuin, C. M. and Kasteleyn, P. W. 1972. On the Random-Cluster Model: I. Introduction and Relation to Other Models. *Physica* 57, 536–564.

12 Suggested Simple Problems and Computer Projects

1. Develop a program for the generation of an ordered sequence of random numbers x_n where $n = 1,2,3,\ldots..$ in the interval $x_n \in [0,1]$. Statistically test the randomness of the numbers generated.
2. Use the sequence generated in I. to evaluate the integral $\int_0^1 \cos(x)dx$.
3. Use the sequence generated in 1. To evaluate the integral $\int_0^1\int dxdy\cos(x)\cos(y)$.
4. Write a Monte Carlo algorithm to determine the average energy and specific heat as functions of temperature of an Ising spin in an external field with Hamiltonian $H = \sigma h$. Here $\sigma = \pm 1$ and h is an externally applied field.
5. Repeat 4. for two interacting Ising spins in an external magnetic field. The Hamiltonian is now $H = -J\sigma_1\sigma_2 + h(\sigma_1 + \sigma_2)$ where $J > 0$.
6. Repeat 4. For four interacting Ising spins in an external magnetic field. The Hamiltonian is now $H = -J\sigma_1\sigma_2 - J\sigma_2\sigma_3 - J\sigma_3\sigma_4 - J\sigma_1\sigma_4 + h(\sigma_1 + \sigma_2 + \sigma_3 + \sigma_4)$ where $J > 0$.
7. Write a Monte Carlo algorithm for the energy and specific heat as a function of temperature of the Ising Hamiltonian:

$$H = -J\sum_{\langle i,j\rangle}\sigma_i\sigma_j + h\sum_i \sigma_i$$

   where $J, h > 0$, $\langle i,j\rangle$ indicates a sum over nearest neighbor pairs on a 5 by 5 square array of the Ising spins, $\sigma_i = \pm 1$. Compute the average value of the center spin as a function of temperature. Compute the average energy of the system as a function of temperature.
8. Redo 7. taking the interaction of the center spin with its nearest neighbors to be $J' > 0$. Compute the average value of the center spin as a function of temperature. Compute the average energy of the system as a function of temperature.
9. a) Consider the five events $\{x_1, x_2, x_3, x_4, x_5\}$ which occur with respective probabilities $\{p_1 = 0.2, p_2 = 0.2, p_3 = 0.2, p_4 = 0.2, p_5 = 0.2\}$ and compute the information entropy of the distribution.
   b) Consider the five events $\{x_1, x_2, x_3, x_4, x_5\}$ which occur with respective probabilities $\{p_1 = 0.1, p_2 = 0.3, p_3 = 0.2, p_4 = 0.3, p_5 = 0.1\}$ and compute the information entropy of the distribution.
   c) Consider the five events $\{x_1, x_2, x_3, x_4, x_5\}$ which occur with respective probabilities $\{p_1 = 0.1, p_2 = 0.1, p_3 = 0.6, p_4 = 0.4, p_5 = 0.1\}$ and compute the information entropy of the distribution.
   d) Consider the five events $\{x_1, x_2, x_3, x_4, x_5\}$ which occur with respective probabilities $\{p_1 = 0.1, p_2 = 0.1, p_3 = 0.1, p_4 = 0.1, p_5 = 0.6\}$ and compute the information entropy of the distribution.
   e) Consider the five events $\{x_1, x_2, x_3, x_4, x_5\}$ which occur with respective probabilities $\{p_1 = 0.0, p_2 = 0.0, p_3 = 1.0, p_4 = 0.0, p_5 = 0.0\}$ and compute the information entropy of the distribution.

DOI: 10.1201/9781032674919-12

10. Compute the critical exponent of the correlation length for bond percolation on the square lattice based on our discussions in Chapter 11.
11. Write a computer simulation to compute the critical exponent of the correlation length for bond percolation on the square lattice considered in Chapter 11.
12. Consider the $q = 4$ Potts model defined on a 10×10 square lattice with nearest neighbor exchanges $J > 0$, write a Monte Carlo simulation for the energy versus temperature.
13. Compute the specific heat for the system in Problem 12.
14. Determine the information entropy of: (a) a fair coin with the probability of heads being 0.5 and the probability of tails being 0.5, (b) a biased coin with the probability of heads being 0.6 and the probability of tails being 0.4.
15. Consider a bond dilute two-dimensional ferromagnetic Ising model on the triangle lattice in which a fraction p of nearest neighbor bonds are $J > 0$, the remain $1 - p$ fraction of nearest neighbor bonds is zero, and there are no further neighbor bonds. Solve for the energy as a function of temperature of this system in the annealed model limit.
16. Consider a bond disordered two-dimensional ferromagnetic Ising model on the triangle lattice in which a fraction p of nearest neighbor bonds are $J > 0$, the remain $1 - p$ fraction of nearest-neighbor bonds is $J' > 0$, and there are no further neighbor bonds. Solve for the energy as a function of temperature of this system in the annealed model limit.
17. Write a computer simulation for energy as a function of temperature for the quenched limit solution of the system in Problem 15.
18. Write a computer simulation for energy as a function of temperature for the quenched limit solution of the system in Problem 16.
19. Consider an array of seven infinite slabs of dielectric constant $\varepsilon > 0$ and thickness d which are periodically arrayed with the slab faces perpendicular to the x –axis and the separation between neighboring slabs is d. Outside of the slabs is vacuum. Compute the transmission through the array for an electromagnetic plane wave incident from infinity on the array and traveling along the x –axis.
20. Repeat Problem 19 for the case in which the dielectric constant of the center slab is changed from ε to 4ε.
21. Consider the mapping over the interval $x \in [0,1]$ for $\mu \in [0,2]$ given by

    $$x_{n+1} = \mu x_n \text{ for } x_n < \frac{1}{2}$$

    $$= \mu(1 - x_n) \text{ for } \frac{1}{2} \leq x_n,.$$

    Determine its map of trajectories plotted in x versus μ as was done in the text for the logistic mapping, indicating its fixed points and chaotic attractors.
22. Solve for the thermodynamics of the one-dimensional Ising model given by $H = -J \sum_{i=1}^{\infty} \sigma_i \sigma_{i+1}$ for $\sigma_j = \pm 1$ and $J > 0$. Determine the energy and specific heat as functions of the temperature.
23. Solve for the thermodynamics of the three-dimensional electron gas. Determine the energy and specific heat as functions of the temperature.
24. Determine the fractal dimension of the Sierpinski triangle shown in Figure 12.1. In this figure, we start with a triangle, which is then broken into three triangles, which in turn are broken into three triangles. This process is continued indefinitely.
25. Consider a dust formed by starting with a line segment, breaking the line segment up into seven equal parts and keeping only the outer two parts. This process is then applied to the two remaining parts. The process is repeated indefinitely. Compute the fractal dimension of the dust resulting from this process.

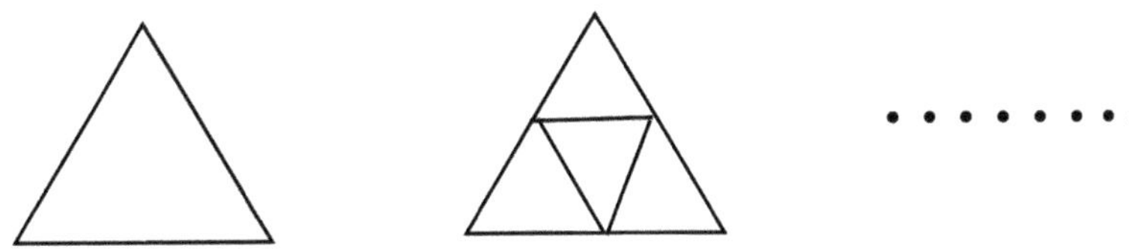

FIGURE 12.1 The first two iterations of the Sierpinski triangle.

26. Consider the electron gas of spin up and spin down electrons at zero temperature. Compute the paramagnetic susceptibility of the gas in an applied external magnetic field.
27. Consider the two-dimensional Ising model defined with nearest neighbor couplings on the triangle lattice. Using mean field theory compute the average energy and specific heat as a function of temperature.
28. Consider the Ginzburg-Landau free energy for superconductors. This has the form

$$f_s = f_n + \alpha(T)\left|\psi(\vec{x})\right|^2 + \frac{1}{2}\beta\left|\psi(\vec{x})\right|^4 + \frac{1}{m}\left|-i\hbar\nabla - \frac{q}{c}\vec{A}\right|^2$$

where f_n is the constant free energy per volume of the electrons, $\alpha(T) = a_0(T - T_c)$, $a_0 > 0$ is constant, T_c is the critical temperature, $\beta > 0$ is a constant, m is the effective electron mass, q is the effective charge, and $\vec{A}$ is the magnetic vector potential of an applied field. For a complex order parameter $\psi(\vec{x}) = \left|\psi(\vec{x})\right|e^{i\theta(\vec{x})}$, find an equation for $\psi(\vec{x})$ and for the current density $\vec{j}(\vec{x})$.
29. For the system in Problem 28, compute the current density at the planar interface between a superconductor and a vacuum.
30. Maximize the entropy of a probability distribution $p(x)$ defined as nonzero only over the intervals $x \in [a,b]$ and $x \in [c,d]$ where $a < b < c < d$ and there are no other constraints on the distribution.

Index

I

J

K

L

M

N

O

P

Q

R

S

T

V

W

For Product Safety Concerns and Information please contact our EU
representative GPSR@taylorandfrancis.com
Taylor & Francis Verlag GmbH, Kaufingerstraße 24, 80331 München, Germany

www.ingramcontent.com/pod-product-compliance
Ingram Content Group UK Ltd.
Pitfield, Milton Keynes, MK11 3LW, UK
UKHW052218190926
475702UK00006B/1075

* 9 7 8 1 0 3 2 6 6 9 8 1 6 *